WITHDRAWN

Ancient Marine Reptiles

Ancient Marine Reptiles

edited by

Jack M. Callaway
Department of Biology
Texas A&M International University
Laredo, Texas

Elizabeth L. Nicholls
Royal Tyrrell Museum of Palaeontology
Drumheller
Alberta, Canada

Academic Press

San Diego   London   Boston   New York   Sydney   Tokyo   Toronto

Cover illustration: "Mosasaur and loons" © 1989 William Stout.

This book is printed on acid-free paper. ∞

Copyright © 1997 by ACADEMIC PRESS

All Rights Reserved.
No part of this publication may be reproduced or transmitted in any form or by any means, electronic or mechanical, including photocopy, recording, or any information storage and retrieval system, without permission in writing from the publisher.

Academic Press
525 B Street, Suite 1900, San Diego, California 92101-4495, USA
http://www.apnet.com

Academic Press Limited
24-28 Oval Road, London NW1 7DX, UK
http://www.hbuk.co.uk/ap/

Library of Congress Cataloging-in-Publication Data

Ancient marine reptiles / edited by Jack M. Callaway, Elizabeth L. Nicholls.
      p.   cm.
  Includes index.
  ISBN 0-12-155210-1 (alk. paper)
    1. Marine reptiles, Fossil. I. Callaway, Jack M. II. Nicholls, Elizabeth L., date.
QE861.A53   1997
567.9--dc20   96-41835
                                                CIP

PRINTED IN THE UNITED STATES OF AMERICA
97 98 99 00 01 EB 9 8 7 6 5 4 3 2 1

Contents

Contributors	*xiii*
Preface	*xvii*
Foreword	*xix*

PART I
Ichthyosauria

Introduction

Text	3
References	11

1. The Paleobiogeography of *Shastasaurus*

Introduction	17
Shastasaurus neubigi N. Sp. from the German Muschelkalk	20
Systematic Paleontology	22
Description	23
Discussion	35
Paleobiogeography of *Shastasaurus*	39

v

Conclusions	39
Summary	41
References	42

2. A New Look at *Mixosaurus*

Introduction	45
Species and Distribution	46
The Cranium	48
Appendicular Skeleton	52
Axial Skeleton	53
Conclusions	54
Summary	55
References	56

3. A Transitional Ichthyosaur Fauna

Introduction	61
Materials and Methods	63
Williston Lake	63
Material Collected	65
Discussion	77
Summary	78
References	78

4. Temporal and Spatial Distribution of Tooth Implantation in Ichthyosaurs

Introduction	81
Materials and Methods	82
Terminology	82
Tooth Implantation and Replacement in Ichthyosaurs	87
Discussion	95
Summary	100
References	101

PART II
Sauropterygia

Introduction

Text	107
References	115

5. Paleobiogeography of Middle Triassic Sauropterygia in Central and Western Europe

Introduction	121
Sauropterygia from the Lower Anisian of the Vicentinian Alps	123
Comparison with Faunal Elements from the Lower Muschelkalk, Including *"Proneusticosaurus" silesiacus* Volz	125
Taxonomic Conclusions	129
A Comparison of Sauropterygia in the Germanic and Alpine Triassic	132
Summary and Conclusions	139
References	140

6. Morphological and Taxonomic Clarification of the Genus *Plesiosaurus*

Introduction	145
Systematic Paleontology	149
Description	151
Discussion	178
Summary	181
References	182

7. Comparative Cranial Anatomy of Two North American Cretaceous Plesiosaurs

Introduction	191
Comparison of *Libonectes* and *Dolichorhynchops*	194
Significance of the Similarities between *Libonectes* and *Dolichorhynchops*	207
Conclusions	208
Summary	210
References	211

PART III
Testudines

Introduction

Text	219
References	221

8. Distribution and Diversity of Cretaceous Chelonioids

Introduction	225
Classification of the Cretaceous Chelonioids	225
Discussion	234
Conclusion	238
Summary	239
References	240

9. *Desmatochelys lowi,* a Marine Turtle from the Upper Cretaceous

Introduction	243
Age of the Material	245
Previous Work	245
Systematic Paleontology	246
Description	246
Discussion	253
Age and Distribution	255
Summary	256
References	257

10. The Paleogeography of Marine and Coastal Turtles of the North Atlantic and Trans-Saharan Regions

Introduction	259
Marine Turtles of the Desmatochelyidae, Osteopygidae, Cheloniidae, and Dermochelyidae	260
Marine Turtles of the Pelomedusidae	271
Conclusions	272
Summary	275
References	275

PART IV
Mosasauridae

Introduction

Text	281
References	288

11. A Phylogenetic Revision of North American and Adriatic Mosasauroidea

Introduction	293
Polarity Decisions	294
Phylogenetic Analysis	296
Consensus Trees and Preferred Hypothesis of Relationships	297
Character Analysis	297
Conclusions	321
Revised Phylogeny	321
Summary	323
References	325

12. Ecological Implications of Mosasaur Bone Microstructure

Introduction	333
Terrestrial versus Aquatic Bone Microstructure	334
Methods	336
Bone Microstructure of Adult *Clidastes*	337
Bone Microstructure of Adult *Platecarpus*	338
Bone Microstructure of Adult *Tylosaurus*	344
Conclusion	347
Summary	350
References	351

PART V
Crocodylia

Introduction

What Is a Marine Reptile?	357
The Physiology of Marine Crocodilians	358
The Teleosauridae	359
The Metriorhynchidae	364
The Pholidosauridae	366
The Dyrosauridae	367
Conclusions	369
References	370

13. The Marine Crocodilian *Hyposaurus* in North America

Introduction	375
Materials and Methods	376
Systematic Paleontology	377
Referred Specimens	378
Tentatively Referred Specimens	390
Paleobiology of *Hyposaurus*	390
Biostratigraphic Considerations	394
Summary	395
References	395

PART VI
Faunas, Behavior, and Evolution

Introduction

Swimming Behavior	401
Tooth Form and Prey Preference	403
Faunas Through Time	405
References	418

14. Marine Reptiles and Mesozoic Biochronology

Introduction	423
Mesozoic Marine Reptiles and Biochronology	423
Taxonomic Problems	426
Distributional Problems	427
Neglect	428
Examples	428
Summary	431
References	432

15. Tithonian Marine Reptiles of the Eastern Pacific

Introduction	435
Geographic and Geologic Occurrence	436
Systematics	439
Tithonian Marine Reptile Assemblage in the Neuquén Basin	442
Paleobiogeographic Distribution	444
Summary	446
References	447

16. Morphological Constraints on Tetrapod Feeding Mechanisms: Why Were There No Suspension-Feeding Marine Reptiles?

Introduction	451
Marine Tetrapod Predatory Guilds	452
Plankton Availability and Upwelling Zones	454
Prey Capture in Suspension Feeders	455
Recognizing a Fossil Suspension-Feeding Tetrapod	456
Neomorphies of the Mammalian Pharynx	458
Suspension Feeding in Marine Tetrapods	460
Conclusion	463
Summary	464
References	465

17. Mesozoic Marine Reptiles as Models of Long-Term, Large-Scale Evolutionary Phenomena

Introduction	467
Completeness of the Fossil Record	470

Better Known Transitions	473
Development	477
Trends	481
Observations	485
Summary	486
References	486

Index   491

Contributors

Numbers in parentheses indicate the pages on which the authors' contributions begin.

Gorden L. Bell, Jr. (281, 293)
Museum of Geology
South Dakota School of Mines
Rapid City, South Dakota 57701

Eric Buffetaut (357)
CNRS
Laboratoire de Paléontologie des Vertébrés
Université Paris 6
Paris 75252 Cedex 05, France

Jack M. Callaway (3, 45)
Department of Biology
Texas A&M International University
Laredo, Texas 78041
and Rochester Institute of Vertebrate Paleontology
Penfield, New York 14526

Kenneth Carpenter (191)
Department of Earth Sciences
Denver Museum of Natural History
City Park
Denver, Colorado 80205

Robert M. Carroll (467)
Redpath Museum
McGill University
Montréal, Quebec
H3A 2K6, Canada

Rachel Collin (451)
Department of Zoology
University of Washington
Seattle, Washington 98195

Robert K. Denton, Jr. (375)
Johnson and Johnson Consumer Worldwide Research
Applied Science Department
Skillman, New Jersey 08558

James L. Dobie (375)
Department of Zoology and Wildlife Science
Auburn University
Auburn, Alabama 36849

David K. Elliott (243)
Department of Geology
Northern Arizona University
Flagstaff, Arizona 86001

Contributors

Marta Fernandez (435)
Departamento Paleontología Vertebrados
Museo de Ciencias Naturales
Universidad Nacional de La Plata
1900 La Plata, Argentina

Zulma Gasparini (435)
Departamento Paleontología Vertebrados
Museo de Ciencias Naturales
Universidad Nacional de La Plata
1900 La Plata, Argentina

Hans Hagdorn (121)
Muschelkalkmuseum
D-74653 Ingelfingen
Germany

Ren Hirayama (225)
Faculty of Information
Teikyo Heisei University
Chiba 290-01, Japan

Stéphane Hua (357)
CNRS
Laboratoire de Paléontologie des Vertébrés
Université Paris 6
Paris 75252 Cedex 05, France

J. Howard Hutchinson (243)
Museum of Paleontology
University of California
Berkeley, California 94720

Grace V. Irby (243)
Department of Geology
Northern Arizona University
Flagstaff, Arizona 86001

Christine M. Janis (451)
Department of Geophysical Sciences
The University of Chicago
Chicago, Illinois 60637

Spencer G. Lucas (423)
New Mexico Museum of Natural History
Albuquerque, New Mexico 87104

Chris McGowan (61)
Department of Vertebrate Paleontology
Royal Ontario Museum
Toronto Ontario
M5S 2C6 Canada;
and Department of Zoology
University of Toronto
Ontario M5S 1A1, Canada

Judy A. Massare (401)
Department of Earth Sciences
College at Brockport
State University of New York
Brockport, New York 14420;
and Rochester Institute of
  Vertebrate Paleontology
Penfield, New York 14526

Richard T. J. Moody (259)
School of Geological Sciences
Kingston University
Kingston-upon-Thames
Surrey KT1 2EE
United Kingdom

Ryosuke Motani (81)
Department of Vertebrate Paleontology
Royal Ontario Museum
Toronto, Ontario
M5S 2C6, Canada;
and Department of Zoology
University of Toronto
Ontario M5S 1A1, Canada

Elizabeth L. Nicholls (219)
Royal Tyrrell Museum of Palaeontology
Drumheller, Alberta, Canada T0J 0Y0

David C. Parris (375)
Bureau of Natural History
New Jersey State Museum
Trenton, New Jersey 08625

Olivier C. Rieppel (107, 121)
Department of Geology
The Field Museum of Natural History
Chicago, Illinois 60605

Contributors

P. Martin Sander (17)
Institut für Paläontologie
Universität Bonn
Bonn 1, D-53115, Germany

Amy Sheldon (333)
Rochester Institute of Vertebrate
  Paleontology
Penfield, New York 14526

Glenn W. Storrs (145)
Geier Collections and Research Center
Cincinnati Museum of Natural History
Cincinnati, Ohio 45202;
and Department of Geology
University of Bristol
Bristol BS8 1RJ, United Kingdom

Preface

The last book devoted solely to aquatic reptiles was Samuel Wendell Williston's *Water Reptiles of the Past and Present,* published in 1914. Although Williston's book has often been cited by professional paleontologists and biologists, much of its content was directed toward a general audience. Williston's survey was broad in scope, covering both freshwater and marine reptiles as well as extant and extinct groups. However, the vast amount of new information on aquatic reptiles that has accumulated over the past eighty years has essentially precluded a simple revision of Williston's volume by a single author; we make no pretensions that this work is an attempt to do so. Instead, we have opted to limit the current volume to marine fossil reptiles. Even within this limited range, it has been impossible to cover all groups of marine reptiles in equal detail; some groups are not discussed at all, simply because not enough is known. Nonetheless, we provide a contemporaneous cross section of the types of research being conducted on extinct marine reptiles.

The introductory chapter provides a historical overview of marine reptile studies. Such studies played a critical role in the development of nineteenth-century concepts of pre-Darwinian evolution, extinction, and homology.

The origin and phylogeny of ichthyosaurs remain obscure, as discussed in the introduction to Part I, but new, quality fossil material is shedding light on these enigmatic reptiles. Studies of ichthyosaurs of Triassic age in particular are undergoing a much-needed and welcomed renaissance; thus, three of the four ichthyosaur chapters are devoted to those of Triassic age.

Part II updates our understanding of the phylogenetic relationships of sauropterygians within the Diapsida. Many of the problems faced in plesiosaur phylogeny result from insufficient data on cranial comparative anatomy, and some of these problems are addressed.

Turtles and crocodiles are the only groups whose fossil records extend beyond the Mesozoic. New fossil material is described for the first time in Parts III and V.

Turtle systematics have been extensively revised over the past two decades, although placement of some marine taxa remains controversial.

Part IV summarizes current views of the phylogenetic history of mosasaurs, including a new, more detailed cladistic hypothesis of mosasaur phylogeny. A new technique using mosasaur bone microstructure to assist in paleoecological interpretations is described; there are exciting implications for its future use.

And finally, Part VI looks at new ways to study marine reptiles. Biostratigraphic techniques and interpretations, a physiological analysis of feeding strategies used by marine reptiles, and a much-needed look at some important marine reptile faunas from the southern hemisphere are all discussed. The final chapter addresses how Mesozoic marine reptiles can be used as models of long-term, large-scale evolutionary phenomena.

Throughout the book we have tried to encourage a sense that paleobiologists face controversies in the classification of marine reptiles. The anatomical modifications that accompanied the transition from a terrestrial environment to a marine environment have obscured many of the key characteristics used in the phylogenetic analysis of reptiles. This has been particularly true in the case of limbs.

We hope that this volume will give the reader an appreciation of the research that is under way and some of the many problems that have yet to be resolved. More importantly, we hope that the information in this work will provide some inspiration for future research.

We are grateful to our authors for willingly sharing their research efforts and for contributing their time and effort to put together this volume.

Elizabeth L. Nicholls and Jack M. Callaway

Foreword

BEFORE THE DINOSAUR: THE HISTORICAL SIGNIFICANCE OF THE FOSSIL MARINE REPTILES

MICHAEL A. TAYLOR

INTRODUCTION

This first-ever modern volume on fossil marine reptiles is a key event marking today's revival---indeed, a new Golden Age---of research on these extraordinary animals. Now is a good time to reassess their importance in the history of our science. Marine reptiles have long suffered, in popular and specialist accounts, by being overshadowed by the dinosaurs. This seems unfair. In the first Golden Age of their discovery, during the late eighteenth century and the early nineteenth century, the marine reptiles made a major contribution to the development of vertebrate paleontology. This science barely existed when fossil crocodiles were discovered in the upper Lias of Yorkshire in 1758 (Benton and Taylor, 1984) and when the famous skull of *Mosasaurus* was discovered at Maastricht, now in the Netherlands, in 1780 (Lingham-Soliar, 1995; Buffetaut, 1987). Yet, a few decades later, after the ichthyosaur and plesiosaur had been described, it was already a mature and productive science when Richard Owen finally described the order Dinosauria in 1842 (Torrens, 1992).

This foreward aims to illustrate the development of vertebrate paleontology by an examination of early work on the Mesozoic marine reptiles, especially that done in England c. 1810-1840, and to assess the marine reptiles' contribution to the later development of the science from scientific and popular viewpoints. I do not have space to discuss the important role of French and German workers on marine reptiles (for which see Buffetaut, 1987). Nor am I trying to search narrowly for "firsts," an always dangerous and often sterile attempt. Plainly, the development of vertebrate paleontology relied not only on the marine reptiles, but also on other important groups, such as the Paleozoic fishes, Tertiary mammals of the Paris Basin, and Pleistocene cave faunas (Buffetaut, 1987). Finally, in this text *marine reptiles* should be taken as comprising Mesozoic and Cenozoic forms and not living groups.

Repository Abbreviations

The abbreviations used for the institutions referred to in the text are as follows: BGS, British Geological Survey, Keyworth, Nottinghamshire, England; NHM, Department of Palaeontology, Natural History Museum, Cromwell Road, London; NMW, Department of Geology, National Museum of Wales, Cathays Park, Cardiff; and OXFUM, Hope Library, University Museum, Parks Road, Oxford, England.

CONYBEARE, DE LA BECHE, AND THE MARINE REPTILES

My particular case study is the research carried out during the early 1820s by the Anglican cleric Rev. William Daniel Conybeare (1787-1857), then of Bristol, at first in collaboration with Henry De la Beche (1796-1855), and later on his own. It was one of the very first pieces of research in British vertebrate paleontology, alongside the Pleistocene cave studies of the Rev. William Buckland (1784-1856). Its subjects were the ichthyosaurs and plesiosaurs from the lower Liassic rocks (uppermost Triassic, and Hettangian and Sinemurian stages of the Jurassic) of southwestern England, especially those from Lyme Regis in Dorset, where De la Beche lived for a while (De la Beche and Conybeare, 1821; Conybeare, 1822, 1824).

De la Beche and Conybeare (1821) reexamined available ichthyosaur material, improving on previous work. At the same time, they identified miscellaneous unidentified bones and partial skeletons as belonging to a new form, which they named *Plesiosaurus*. Among these bones was a hitherto mysterious fossil skeleton from Nottinghamshire described by the antiquarian William Stukeley (1687-1765) (Stukeley, 1719; his drawing is the oldest published illustration of a marine reptile skeleton). Conybeare's 1822 paper carried on the study of both ichthyosaurs and plesiosaurs, including an isolated plesiosaurian head (Figure 1) found by Thomas Clark, Jr. (1792-1864), in the lower Lias of Street, Somerset. This was the first of many important specimens of ichthyosaurs and plesiosaurs from that area, collected by the eccentric Thomas Hawkins (1810-1889) and by other workers (Storrs and Taylor, 1996; Taylor, 1989a; Howe et al., 1981; Bowen, 1854:84; Hawkins, 1834, 1840).

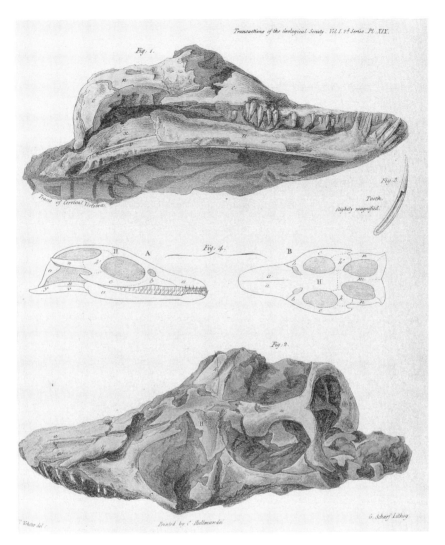

Figure 1. Plate 19 from Conybeare (1822), the first known plesiosaur head (now BGS GSM 26035). This specimen was prepared by the sculptor Chantrey and lithographed for publication, exemplifying the advanced infrastructure supporting English vertebrate paleontology in the 1820s. Photograph courtesy of the Bristol City Museums and Art Gallery.

Conybeare's reconstruction of the plesiosaur from more or less disarticulated material had been criticized, so he was pleased when the Anning family, fossil collectors of Lyme Regis (Torrens, 1995; Taylor and Torrens, 1987), discovered the first complete skeleton in 1823 (Figure 2). It "confirmed the justice of my former conclusions in every essential point connected with the organisation of the skeleton"

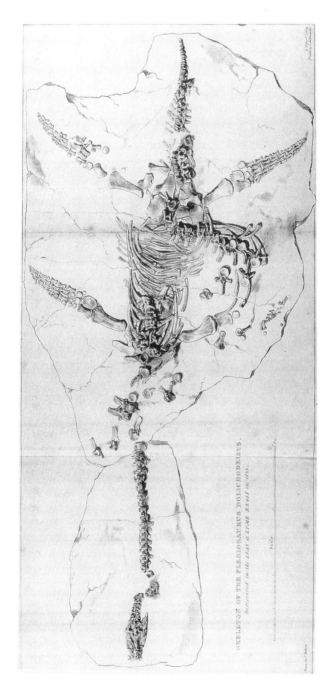

Figure 2. Plate 48 from Conybeare (1824), based on a drawing by T. Webster of the first complete plesiosaur skeleton, found in 1823 by the Anning family. It is now NHM 22656, the type specimen of *Plesiosaurus dolichodeirus* Conybeare, 1824. Photograph courtesy of the Bristol City Museums and Art Gallery.

(Conybeare, 1824:381). He wrote to De la Beche on March 4, 1824. After reporting the new discovery to the Bristol Institution's associated Philosophical and Literary Society (Taylor, 1994), he had gone down to London to await the arrival of the specimen, by then sold to the Duke of Buckingham (Torrens, 1995; Taylor and Torrens, 1987):

> The following Friday was the anniversary meeting of the Geol[ogical] Soc[iety] [of London] at which [William Buckland] was to be elected its President, & the specimen of *Plesiosaurus* being placed by the Duke [of Buckingham] in his hands for scientific investigation he had shipped it off to be deposited for a time in the Geol[ogical] Soc[iety]'s house, where he charged me to meet it on pain of its falling into the hands of Sir Ev[erar]d H[ome]. This summons of course I did not neglect... When I came to Town I found the specimen delayed in the Channel [the later reference to "Aeolus & Neptune" indicates that contrary winds were hindering the sailing ship carrying the plesiosaur, presumably from Lyme to the Port of London via the English Channel and Thames Estuary] nor did it arrive till 10 days afterwards, but I shewed the drawing wh[ich] was quite sufficiently clear for every essential purpose at the R[oyal] S[ociety] Club on the Thursday, having gone up on that day to be there present. I lectured Sir H[umphrey] D[avy (1778-1829), then President of the Royal Society] on the satisfaction it afforded me to have my discoveries *wh[ich] had been questioned* thus confirmed, Sir E. H. being placed next to him. Wollaston [William Hyde Wollaston, doctor and chemist (1766-1828)] was so interested that he reduced the drawing with his Camera lucida & I made my Beast roar as loud as Buckland's Hyenas. Friday's dinner at the Geol[ogical] [Society] was one of the pleasantest public meetings I have ever attended. Buckland as the new Pres[iden]t was put to his oratory, & some dozen of us talked in our turn, but in place of the usual trash on such occasions every one of us had some interesting facts connected with the management of the Society or the progress of the science to communicate. We adjourned to the Society's rooms at ½ past eight, and there I lectured on my M[on]st[e]r & Buckland on the Stonesfield bones [of the dinosaur *Megalosaurus*, this paper later published as Buckland (1824)]... The long interval which I waited at the pleasure of Aeolus & Neptune for the arrival of the anxiously expected case I spent pleasantly enough.... At last the important package arrived, & after wasting a day in vainly attempting to move it up stairs to the room of meeting at the G[eological] S[ociety] by the aid of ten men, we were constrained to unpack it in the entrance passage. [There follows a detailed discussion of anatomy.] ...On Friday 19

the G[eological] S[ociety] was as full as the room c[oul]d hold, to see & hear all about it & as our talk was dressed out thro' Mr. Brookes [Joshua Brookes (1761-1833), owner of the Brookesian Museum, a major comparative anatomical collection (Desmond, 1989)] aid with quantities of skeletons of recent lacertae we cut a formidable figure. Webster [Thomas Webster (1772-1844), Geological Society curator and engraver] is drawing, Chantry [sic] [Sir Francis Legatt Chantrey (1781-1841), fashionable sculptor] modelling and we are getting up the thing in high style for the Vol[ume of the *Transactions of the Geological Society*, i.e., Conybeare, 1824] now forthcoming.... (NMW 84.20G.D302; emended from versions in Lang, 1939; North, 1935, 1957)

This letter is highly illuminating: as I will discuss, it shows the incipient infrastructure of vertebrate paleontology, and how firmly Conybeare was rooted in the "gentlemanly geology" of the time.

THE INFRASTRUCTURE OF THE MARINE REPTILES

For vertebrate paleontology to mature as a science, its proponents needed to develop an infrastructure. This comprised the organizational and technical solutions to the problems of collecting, preparing, preserving, replicating, illustrating, and publishing specimens. All these areas were being actively developed in the 1820s, at a time of transition from the isolated, amateur naturalist to the quasi-professional geologist (Rudwick, 1985; Allen, 1976, 1985).

Field Collecting

In Britain, despite the lack of "badland" terrain, direct field collecting was much easier during the early and middle nineteenth century than ever before or since. There were many available coastal exposures and quarries. Unlike today, there were few coastal defense works to obscure outcrops that were actively eroding and so especially productive. Small local quarries were far more widely distributed than today's much more highly concentrated industry. Indeed, it may not be an exaggeration to say that every parish had as wide a range of lime, marl, sand, and stone pits, and every canal its puddling pits, as the local geology allowed. Moreover, the fact that quarrying was done mostly by hand labor meant that quarrymen could, given an incentive such as cash payment, spot and set aside (or point out for excavation by the collector) fossils before they went for burning or breaking into road metal. In contrast, today's machinery is much more economical but destroys many fossils before they are ever seen.

On top of all this, the economic growth associated with the industrial and agricultural revolutions stimulated the demand for stone, lime, and aggregate of all kinds, and the enormous expansion of new roads, canals, and railways over Britain meant many fresh cuttings and borrow pits, from which came fossils such as the ichthyosaur from Isambard Kingdom Brunel's Great Western Railway (Taylor, 1994). However, in the longer run, these improvements in transportion all too often meant the invasion of alien building materials and lime from outside the parish or even the county, with the resulting closure of local quarries.

These quarries and outcrops were, nevertheless, valueless without the regular presence of collectors to exploit these opportunities. Thus, the rise and fall of interest in geology (Allen, 1976) had an indirect effect on the range and quality of fossils being recovered, via the number and activity of collectors. These factors go a long way toward explaining the British heyday of marine reptile collecting in the nineteenth century versus its almost complete collapse in the middle twentieth century.

Collectors could find and dig out their own specimens, or get them indirectly from quarrymen (who might have to be suitably briefed and primed with beer money). Many were "amateurs" in the modern sense, although this is, to some extent, an anachronistic distinction (Taylor, 1994; Cleevely and Chapman, 1992; Allen, 1985) and in any case many, like Thomas Hawkins, eventually sold their collections to museums. However, some of the most important collectors were undoubtedly professionals making a living from their finds, the most famous of these being the Annings of Lyme Regis, who made some of the key discoveries of British paleontology. This family's best-known member is Mary Anning, Jr. (1799-1847), but her mother, Mary, Sr., "Molly," and her brother Joseph, were also active participants in the business up to 1825 (Torrens, 1995; Taylor and Torrens, 1987, 1995). (For these and other early collectors of English marine reptiles, see also Howe et al., 1981; Delair, 1969; Cumberland, 1829.) However, even the Annings found it an insecure trade and their business became possible at all only in the first two decades of the nineteenth century. The development of Lyme as a holiday resort in the late eighteenth century provided the basis for a fossil industry to spring up as soon as geology became a fashionable Romantic interest (Taylor and Torrens, 1987; Rupke, 1983; Fowles, 1982; Lang, 1939). Other commercial collectors were equally subject to local and national fashion.

Museums, Collections, and Libraries

A major constraint was the availability of collections. Conybeare and De la Beche may have been hindered in their use of private collections by problems of access, availability, and research facilities (Taylor, 1994). Collections in private hands were also at special risk of being eventually lost to science by means of dispersal through sale or destruction when the collector lost interest, died, or ran into financial trouble. The British Museum, the sole national museum, and the museums at the Universities of Oxford and Cambridge did secure several important collections, notably those made by Thomas Hawkins, as well as many Anning specimens (Torrens, 1995; Taylor, 1989a; Cleevely, 1983). However, during the early nineteenth century, the purchase of major acquisitions by the British Museum required considerable lobbying. This was partly because of the need to secure Parliamentary funding and partly because of its staff's attitude (Cleevely and Chapman, 1992). Certainly it was not nearly as enthusiastic an institutional collector as the new provincial museums and institutions then being established, such as the Bristol Institution for the Advancement of Science, Literature and the Arts which opened in 1823, Conybeare and De la Beche being among the founder members. These institutions were a major advance. Not only did they, at least theoretically, ensure the long-term safety of the collections which they housed, but they allowed less wealthy members to club together to buy fossils, books, and equipment which they could not individually afford. Indeed, a major attraction was the otherwise impossibly expensive libraries of current books and journals, as crucial a resource as the collections themselves, and required before almost any serious work could be done (Taylor and Torrens, 1987).

Preparation and Replication

Before study, the specimens had to be freed of the rock, always a major constraint in vertebrate paleontology, as seen by the more recent impact of new techniques like acid preparation (Whybrow, 1985). Amateurs often did their own preparation, sometimes very well, although Thomas Hawkins and perhaps others were too willing to go too far and mix and match specimens to create composites (Taylor, 1989a; McGowan, 1989, 1990). In fact, the Bristol surgeon, comparative anatomist, and paleontologist Henry Riley (1797-1848) is reported as commenting on the 1823 plesiosaur skeleton that the

> Duke of Buckingham... wishing, unfortunately, to obtain a view of the other side of the animal, which was on its back, caused a considerable portion of the lias in which it was embedded to be chiselled away, during which the specimen was considerably injured. (Anonymous, 1832:[4])

The original plesiosaur head from Street was, interestingly, consigned to the sculptor Chantrey "to be relieved of some of its incrustations" (Bowen, 1854:84). Of course, the actual work may not necessarily have been done by Chantrey himself, but by one of the assistants or stoneworkers in his workshop who were skilled in replicating sculpture in marble and other stones (Potts, 1980). However, Chantrey's official Geological Society obituary remarked (Murchison, 1842:638):

> We have benefited from his sound advice... on the best means of preserving organic remains, which presented difficulties from their size, their condition, or the nature of the rock in which they were imbedded; and upon several occasions he has assisted us by superintending the moulding of osteological specimens which have been brought to this country, and of which it was important to obtain casts. Indeed at all times was his assistance freely given when it could be useful, and his chisel even has been employed in dissecting from their matrix the bones of fossil reptiles.

Conybeare's letter also recorded how Chantrey was "modelling" the new plesiosaur skeleton. Although this, strictly speaking, implies that Chantrey was sculpting a model or replica, perhaps of clay, I suggest that Conybeare really meant that Chantrey (or rather his assistant) was casting the specimen in plaster. The production of three-dimensional casts of actual specimens was, and still is, a vital means of visual communication in paleontology (incidentally omitted by Rudwick [1976b] in his important study of two-dimensional methods). High skills were, and are, needed to make good replicas with rigid plaster molds in multiple pieces, compared to today's soft rubber or silicone molds. However, such skills were routinely practiced in a studio such as Chantrey's where plaster casts were made from clay originals of such things as portrait busts (Penny, 1992:213-214). A Parliamentary inquiry of 1836 into the British Museum elicited a testimonial as to the excellence of Chantrey's fossil casts (this time of the Tertiary mammal *Megatherium*; Anonymous, 1836:218-219). Alternatively, the work could be contracted to outside professional specialists, at least later in the century (see Potts, 1980:7-11, and Read, 1982:49-78, for this and other practical details of sculptors' workshops).

All this suggests a transfer of preparation, conservation, and casting technology from Chantrey's sculpting and casting workshop to the geological community. This link presumably came about through Chantrey's involvement in the scientific world, including membership of the Royal and Geological societies, and is further exemplified by his numerous sculptural commissions for scientific societies, as well as his personal geological collection---indeed, he is presumably the "Mr Chantrey" who donated casts of fossil saurians to the Yorkshire Philosophical Society Museum (now Yorkshire Museum) in 1823 and 1831 (Cleevely, 1983:80; Potts, 1980).

Further research is needed to confirm this suggested technology transfer, and to elucidate the wider development of such skills within the geological community in England, as well as further afield in places such as Cuvier's laboratory in Paris. For example, one alternative route of technology transfer, especially appropriate to a multidisciplinary provincial museum such as Bristol, might be via the casting of antiquities such as statuary. Yet the history of fossil preparation and replication, so crucial to the development of the science, has received little attention, with the notable exception of reviews by Whybrow (1985) and Howie (1986). This seems a case of the tendency of historians to neglect the concrete object or "manufact" in favor of the written document and publication, which has, for example, led to the downgrading of Mary Anning, Jr., because she produced specimens rather than scientific papers (Torrens, 1995).

Publication and Illustration

Dissemination of research findings required illustrated journals. Again, Conybeare and De la Beche exploited the most advanced techniques available. The Geological Society of London, founded in 1807, started publishing its well-illustrated *Transactions* in 1811. At first, it used expensive engraving (e.g., De la Beche and Conybeare, 1821), but soon it shifted to the newer and cheaper technique of lithography (Figures 1 and 2; Rudwick, 1976b, 1993) in time for Conybeare's 1824 paper for which Webster drew the master figure, as we know from Conybeare's letter. Furthermore, Wollaston was trying out his new camera lucida (Rudwick, 1976b:194fn.), an early attempt to increase the accuracy of scientific drawings.

Perhaps most interestingly of all, Webster's drawings for Conybeare (1824) included side-view reconstructions of "complete" ichthyosaur and plesiosaur skeletons, not only as they were preserved but, in a conceptual and heuristic leap, reconstructed as they had existed in the living animals. These are of enormous interest, being almost the first examples of this typical convention of vertebrate paleontological communication, apparently preceded only by Cuvier's drawings of Tertiary mammals from the Paris Basin. Indeed, they have never been superseded by modern drawings in the case of *Plesiosaurus dolichodeirus*, with the exception of the recent reconstructions of the head by Storrs (Chapter 6).

Networking

To use an admittedly anachronistic term, Conybeare and colleagues were past masters at networking within the elite science community of the time. As a gentleman geologist active in the Geological Society of London (Rudwick, 1976a, 1985), Conybeare had links with other important centers, such as the Royal Society, as is clear from his letter, and seemingly the Royal College of Surgeons, for he acknowledged an unnamed "friend" for anatomical information, for example, on the jaw adductor musculature (Conybeare, 1822:110fn., 114fn.). Hugh Torrens (personal communication, 1992) suggests that this was William Clift (1775-1849), the conservator (i.e., curator) of the College's Hunterian Museum (Desmond, 1989; Dobson, 1954). Clift was a capable researcher and illustrator. His son-in-law Richard Owen (1804-1892) commented that from 1814 to shortly before 1849 most work on the "higher classes of animals... [was] more or less indebted to Mr. Clift, either for his determination of the fossils described... or for his accurate and beautiful figures..." (Dobson, 1954:130).

Georges Cuvier (1769-1832), in Paris, was another major contact, both directly and via his assistant, the Irish naturalist Joseph B. Pentland (1797-1873), with the exchange of information and specimens. Pentland must indeed have been the "friend resident in Paris" who, requesting anonymity (De la Beche and Conybeare, 1821:593), drew attention to Cuvier's unpublished observations on an ichthyosaurian mandible (Sarjeant and Delair, 1980; Delair and Sarjeant, 1978).

Effective networking would also have been taking place within the wider provincial community of amateur geologists, undoubtedly including the marine reptile enthusiasts among them (for an example of a Salisbury collector of Chalk fossils and flints around 1810, see Torrens, 1990). However, such networking had a strong vertical dimension within the nineteenth-century English geological community: the amateurs who either collected fossils themselves or purchased them from professional collectors, whether individually or collectively in their museums, were "expected" deferentially to place the specimens at the disposal of the metropolitan elite such as Conybeare or, later, Richard Owen, who would pronounce authoritatively upon them (for the geological community, see Rudwick, 1985; for the vertebrate paleontologist Gideon Mantell, see Cleevely and Chapman, 1992). This scientific-social stratification, and the resulting stresses, have not been specifically explored for workers on marine reptiles, although they must undoubtedly have existed.

GEOLOGICAL THEORY AND THE MARINE REPTILES

Theoretical infrastructure is vital to the interpretation of any fossil. In the 1820s, the necessary work had largely been done. William Smith (1769-1839) had developed the basic use of fossils in stratigraphy and relative dating, and started the elucidation, at least in England, of the Mesozoic rocks which were the sources of the marine reptiles. Georges Cuvier and other palaeontologists had developed other necessary further concepts: extinction, and the change of faunas with time; the functional analysis of fossils as living animals; and the comparison of their anatomy with a range of living animals (Buffetaut, 1987; Rudwick, 1976a). Conybeare and De la Beche could therefore take all this, more or less as it stood, and place the ichthyosaur and plesiosaur into the context of these new concepts.

Although by no means perfect---for example, in their idea that the marine reptiles blew out spouts of water, like whales, and in not resolving the problem of the ichthyosaurian caudal fin---Conybeare and De la Beche reached a level of detailed understanding and completeness of description perhaps surpassed in British vertebrate paleontology only by Richard Owen a decade or more later. Their work was also important in being among the first detailed research on fossil tetrapods radically different from living ones. Until then, with the notable exception of Cuvier's work on pterosaurs (Wellnhofer, 1991), most of the only *competent* work on large fossil tetrapods had been on animals not grossly different from their living relatives, such as Tertiary and Quaternary mammals, mosasaurs---large marine lizards---from Belgium, and crocodiles from the Jurassic of Yorkshire or Normandy (Buffetaut, 1983, 1987).

The basic theme of Conybeare and De la Beche's work is the documentation of the surprising transformation of a basically terrestrial reptilian Bauplan---to use the modern term---to two divergent and extreme adaptations to marine life. Emphasizing form and function, they use what, at first, seem surprisingly modern concepts (though some would have been familiar to, for example, naval architects or doctors, and thus to other scientifically educated contemporaries): function of jaw adductor musculature, crossed-ply fibrous sheets in overlapped sutures, directional stability during aquatic locomotion, and optimization of strength-to-mass ratios. They went on to relate all this to the animals' lifestyles, as Conybeare remarked about the plesiosaur (1824:388-389):

> It may perhaps have lurked in shoal water along the coast, concealed along the sea-weed, and raising its nostrils to a level with the surface from a considerable depth, may have found a secure retreat from the assaults of dangerous enemies; while the length and flexibility of its neck may have compensated for the want of strength in its jaws and its incapacity for swift motion through the water, by the suddenness and agility of the attack which they enabled it to make on every animal fitted for its prey, which came within its extensive sweep.

Conybeare and De la Beche's work was thus in part a Cuvierian functional analysis of fossils as living organisms (Porter, 1977:169-170). Yet it was also un-Cuvierian in being a classic example of "natural theology," using biological adaptation as evidence for a divine Creator in those days before Darwinian natural selection. Far from being monstrous, the ichthyosaur's and plesiosaur's deviations from the "normal" anatomy of reptiles were specifically designed by the Creator for the animals' aquatic life. In his popular sixth Bridgewater treatise, explicitly commissioned to show how the "power, wisdom and goodness of God [were] manifested in His creation" (Rupke, 1983; Buckland, 1836:185-186, 202-214), Buckland commented:

> The introduction of these animals [the ichthyosaur and the almost equally aquatic platypus], of such aberrations from the type of their respective orders, to accommodate deviations from the usual habits of these orders, exhibits an union of compensative contrivances, so similar in their relations, so identical in their objects, and so perfect in the adaptation of each subordinate part, to the harmony and perfection of the whole; that we cannot but recognise throughout them all, the workings of one and the same eternal principle of Wisdom and Intelligence, presiding from first to last over the total fabric of Creation. (Buckland, 1836:186)

Indeed, for Church of England clerics like Conybeare or Buckland, "the value of geology was judged by its congruence with the existing tradition of learning and by its relevance to a clerical education"---in other words, the only education "acceptable" to the Oxford- and Cambridge-influenced English elite (Rupke, 1983:21). Membership of an Oxford or Cambridge college, even as an undergraduate, let alone a Fellow, presupposed membership of the Church of England. Moreover, England then lacked nationally funded posts comparable to those in France. English researchers without private means might find that clerical appointments were the only way in which they could maintain at least a part-time level of activity without the gross loss of social status (and the often poor salaries!) suffered by paid curators (Taylor, 1994). Thus, Buckland was given a place in Christ Church Cathedral, Oxford, in 1825 to prevent his loss to science in a country living (Edmonds, 1978).

The descriptions of the ichthyosaur and plesiosaur made real intellectual impacts (Buffetaut, 1987; Rupke, 1983; Rudwick, 1976a). They reinforced the evidence for the importance of extinction as a factor in the history of life. It was becoming increasingly apparent that the major differences between extinct and living faunas were real, as advocated by Cuvier, who proposed periodic catastrophic mass extinctions, the latest being the biblical Deluge. These new fossil finds were plainly not merely species or genera closely related to known living forms, which might have suffered local extinctions, but rather radically different from known living forms. Thus, it was significantly harder to argue that such large and supposedly "extinct" groups were still living in unexplored corners of the globe. Clearly, therefore, the "secondary period [i.e., Mesozoic] represented a long succession of epochs each with its characteristic plants and animals" (Wilson, 1972:84).

The marine reptiles' supposedly cold-blooded, reptilian physiology was added to the evidence, such as fossil plants, then being used to infer a warm Mesozoic world heated by internal subterranean warmth, and which had not yet lost enough heat to yield the colder climate held to suit the warm-blooded mammals of the Tertiary. They contributed to Buckland's idea of Divine providence, allowing for a progressive change of climate toward a world ultimately suited for habitation by people, and the later evolutionary version clearly popularized by Robert Chambers (1844) in *Vestiges of the Natural History of Creation*.

In contrast, the controversial principle of uniformitarianism advocated by the

Figure 3. "Awful Changes," Henry De la Beche's cartoon of 1830 mocking Lyellian uniformitarianism. The text reads: "Awful Changes. Man found only in a fossil state---Reappearance of Ichthyosauri. 'A change came o'er the spirit of my dream.'" Byron. A Lecture.---"You will at once perceive," continued Professor Ichthyosaurus, "that the skull before us belonged to some of the lower order of animals[:] the teeth are very insignificant [,] the power of the jaws trifling, and altogether it seems wonderful how the creature could have procured food." NMW 84.20G.367; photograph courtesy of the National Museum of Wales.

Scots lawyer Charles Lyell (1797-1875) led him to suggest that the time of the marine reptiles could return:

> Then might those genera of animals return, of which the memorials are preserved in the ancient rocks of the continents. The huge iguanodon might reappear in the woods, and the ichthyosaur in the sea.... (Lyell, 1830:23, Vol. 1)

This inspired De la Beche's famous cartoon "Awful Changes" (Figure 3), of Professor *Ichthyosaurus*---presumably Lyell himself---lecturing on the unimpressive dentition of a fossil *Homo sapiens* (Rupke, 1983; McCartney, 1977; Rudwick, 1975).

Finally, the marine reptiles were the dominant predators in the Lower Jurassic marine ecosystem preserved at Lyme Regis and Street. With the exception of Buckland's Quaternary hyena dens, this was the first ecosystem to be reconstructed

Figure 4. An early draft of "Duria Antiquior," Henry De la Beche's cartoon of c. 1830, later printed to raise money for Mary Anning (McCartney, 1977; Taylor and Torrens, 1987; Torrens, 1995). The first such reconstruction of an extinct ecosystem, complete with feeding and locomotor adaptations, taphonomic phenomena, and nutrient recycling in the form of droppings ready to become Bucklandian coprolites (these last omitted from the published print!). NMW 84.20G.368; photograph courtesy of the National Museum of Wales.

on paleobiological evidence. It lent itself well to visual representation, as in De la Beche's reconstruction painting of about 1830, "Duria Antiquior," which must be the first graphic representation of any past ecosystem, complete with trophic relationships (Figure 4; Rudwick, 1989:241; Rupke, 1983; McCartney, 1977). Such illustrations were nevertheless not then regarded as acceptable for formal academic publications, but they were circulated semiprivately, and Buckland greatly valued them as aids for his teaching at Oxford (McCartney, 1977; e.g., the enlarged copy of "Awful Changes" in OXFUM).

EVOLUTION AND THE MARINE REPTILES

What role did the marine reptiles have in the early debates about evolution? The ichthyosaur and plesiosaur were originally interpreted in a nonevolutionary manner. Their names were directly derived from the pre-evolutionary concept of the static Great Chain of Being, an explanation of ordered diversity within Divine Creation (Taylor, 1994; Taylor and Torrens, 1987). *Ichthyosaur* comes from the Greek for "fish reptile" reflecting its position in the Chain between fish and reptiles, and *plesiosaur* comes from the Greek for "nearer [to] reptile," fitting between ichthyosaurs and reptiles such as lizards and crocodiles. However, the Great Chain of Being, as applied to the animal kingdom, was already an obsolescent concept because of internal inadequacies (Appel, 1987; Rolfe, 1985).

An alternative synthesis was emerging in the form of Lamarckian transmutation, which De la Beche and Conybeare (1821:560-561fn.) sharply attacked as

> ...an idea so monstrous, and so completely at variance with the structure of the peculiar organs considered in the detail... and no less so with the evident permanency of all animal forms, that nothing less than the credulity of a material philosophy could have been brought for a single moment to entertain it---nothing less than its bigotry to defend it.

Already highlighted by Rupke (1983) and Rudwick (1976a:154-155), this attack seems, remarkably enough, to be the first British discussion in print of the evolutionary ideas of Jean-Baptiste de Monet, Chevalier de Lamarck (1744-1829), at least in connection with fossil vertebrates. At first, such a critique seems surprising, given its apparent irrelevance to marine reptiles: Lamarck never attempted an in-depth discussion of paleontological discoveries in his evolutionist (rather than purely taxonomic) writings (Corsi, 1988:160). However, we should rather regard it as a defense of the Great Chain as such, and by implication of the English status quo of Church and State, in the face of the linked scientific, social-radical, and atheistic implications of Lamarckism, all repugnant to Conybeare and his gentlemanly geological friends (Taylor, 1994; Desmond, 1987, 1989).

The marine reptiles are otherwise almost completely absent from the early evolutionary debates before and immediately after the publication of *The Origin of Species* in 1859. Perhaps we should not be surprised. Thanks to the vagaries of the fossil record, most of the major groups of marine reptiles are extinct, lacking modern descendants, with their evolutionary affinities often unclear and doubtful even today, unlike those of a classic "missing link" such as *Archaeopteryx*'s fitting well between birds and reptiles. However, crocodilians do have living representatives familiar to nineteenth-century zoologists, which may explain why they are the major exception, if at first only through the rather fanciful speculations of Étienne Geoffroy Saint-Hilaire (1772-1844).

Geoffroy suggested that the fossil crocodilians had been transformed into modern forms, but later went even further and suggested that *Teleosaurus* was a kind of intermediate link between crocodilians and mammals (Corsi, 1988; Buffetaut, 1987; Appel, 1987; Geoffroy Saint-Hilaire, 1825). He went on to sketch a possible evolutionary sequence, with the transformations driven by environmentally influenced teratological change rather than purely Lamarckian functionally adaptive change:

> ...il suffira, bien que très-imparfaitement sans doute, de rappeler une série progressive, comme la suivante, par exemple: *Icthyosaurus* [sic], *Plesiosaurus, Pterodactylus, Mososaurus* [sic], *Teleosaurus, Megalonix* [sic], *Megatherium, Anoplotherium, Paleotherium* [sic], etc. (Geoffroy Saint-Hilaire, 1828:215)

Was Geoffroy being entirely serious here, one wonders? Possibly he deliberately chose his (to us) rather odd lineup of genera to provoke Cuvier (cf. Corsi, 1988:290), as, except for Geoffroy's own *Teleosaurus*, most were named by or otherwise intimately associated with Cuvier or his English colleague Conybeare. At any rate, Geoffroy was later to change his mind about *Teleosaurus* at least, this time seeing it as an intermediate between ichthyosaurs and modern crocodiles (Buffetaut, 1987).

Richard Owen's "discovery" of the dinosaurs in 1842 had roots in prior work on the marine reptiles. He grouped together the few dinosaurs then known, mostly from scrappy material, and reassessed them as advanced quadrupedal animals (Owen, 1842). He then used the apparent degeneration of such magnificently sophisticated quadrupeds into modern reptiles as ammunition against progressionist interpretations of the history of life. Owen's analysis of dinosaurs, with its ideological and nationalist undertones, was attacked by Thomas H. Huxley in the 1850s and finally refuted by the more complete bipedal skeletons of the 1870s (Haste, 1993; Torrens, 1992, 1993; Desmond, 1982, 1989). However, I wonder whether Owen would have gained widespread acceptance for his ideas, without his personal prestige and without Conybeare and De la Beche's clear prior demonstration of the presence of large, now extinct fossil reptiles unlike anything living today. Nor must it be forgotten that most of the 1842 paper was actually the *second* part of a "Report" on British fossil reptiles (Owen, 1840, 1842). In the first part, Owen had submitted the first major review of the marine reptiles since Conybeare and De la Beche's work. By then Owen had modified Geoffroyan transcendental anatomy to a more static theory of homology and of the vertebrate archetype which was more acceptable to conservative opinion. Thus, in the "Report," he attacked Geoffroy's transformist speculations about marine reptiles and, by implication, also Lamarckian progressionism. One reason Owen adduced was that these large ectothermic reptiles were now obsolete thanks to changes in climate (Desmond, 1989:324; Rupke, 1983, 1994).

THE PUBLIC AND THE MARINE REPTILES

The marine reptiles' major role in early and middle nineteenth-century vertebrate paleontology is reflected in the science's popular image. These large, extinct, and splendidly gruesome saurians seized the newly Romantic public imagination (Rupke, 1983; Porter, 1978). The resulting works seem to have written not of *dinosaurs*---a word only coined in 1842---but rather of *saurians* of all kinds, in which the marine reptiles featured prominently. Indeed, the current overemphasis on dinosaurs seems to have distorted our understanding of nineteenth-century perceptions, and a more correct appreciation gives more weight to the marine reptiles.

Early visual representations of the "age of monsters," such as "Duria Antiquior," often used the marine reptiles if they pretended to any accuracy---even if the most famous portrayal, John Martin's frontispiece of battling sea-saurians for Thomas Hawkins' bizarre *The Book of the Great Sea Dragons*, hardly aspires to technical accuracy (Hawkins, 1840; see also illustrations in Rudwick, 1992). This book incidentally gave the marine reptiles one of their more enduring popular names, the Great Sea Dragons, repeated, for example, in Bristol City Museums and Art Gallery's eponymous exhibition in 1989 (Taylor, 1989b). Battling sea-dragons surfaced again, notably as the "Dragons of the prime,/That tare each other in their slime" in Tennyson's 1850 poem *In Memoriam* (Gould, 1992; Rupke, 1983), and yet again in Jules Verne's 1864 novel *Voyage au centre de la terre* (Buffetaut, 1987).

Another interesting indicator is: which animals were used in caricature? Such beasts have to be instantly recognizable, with accepted "characters"; thus, the British political cartoonist Steve Bell uses distinctive animals such as penguins and pandas. When he caricatured the 1987 British General Election, he chose large dinosaurs versus small Mesozoic mammals to symbolize party political conflict (Bell, 1987). In contrast, when De la Beche's cartoons required prehistoric animals, he usually chose marine reptiles, as in "Awful Changes" (Figure 3; McCartney, 1977; Rudwick, 1975). This wasn't just personal predilection: the ichthyosaur and plesiosaur were among the few well-attested extinct animals available, and they were, of course, visually more distinctive than the pre-1842 reconstructions of dinosaurs as overgrown lizards.

Even after Owen "reconstructed" the dinosaurs with an elephantine stance and thereby conferred on them a newly distinct visual identity, the dinosaurs seem usually to have been lumped in with the other extinct saurians, albeit as a rather larger terrestrial variety. Indeed, to the extent that dinosaurs were not generally perceived as a separate category, at least in the early and middle Victorian era, it seems anachronistic to talk of public attitudes toward dinosaurs, and more correct to refer to attitudes to extinct "saurians." Thus, in his enormously popular protoevolutionary work *Vestiges of the Natural History of Creation,* the highly competent journalist Robert Chambers (1844:97-99) does not distinguish between the dinosaurs and other "saurians" nearly as sharply as a modern book would. Again, in his popular book *Zoological Recreations*, W. J. Broderip (1849:326-380) hardly uses the word *dinosaur* in an even-handed survey of the Mesozoic "dragons" of land, sea, and air.

Perhaps the most startling example of our misleading concentration, with hindsight, on dinosaurs *sensu stricto* is the famous group of life-size replicas at the Crystal Palace Park in south London. This was a kind of prototype theme park of the mid-1850s, simultaneously entertaining and educational in the best Victorian tradition. These replicas, created by Waterhouse Hawkins with Richard Owen's advice in the mid-1850s, are not, contrary to the general impression given by popular dinosaur books over the years, a group of dinosaurs. Rather, the dinosaurs *are* there, but only as members of a whole (and never finished) range of almost all the then known extinct large amphibians, reptiles, and mammals, including various ichthyosaurs, plesiosaurs, crocodilians, and a mosasaur wallowing in the surrounding lake, with appropriately reconstructed geological strata as a backdrop (Figure 5; for the whole story, see Doyle and Robinson, 1993, 1995, and McCarthy and Gilbert, 1994).

It seems that this early experiment in paleontological education needed more labeling. The public, at least in the 1890s---well after the American finds of dinosaurs!---still allegedly lacked the basic background knowledge to identify these replicas as extinct saurians, let alone whether they were dinosaurs or not (R. S. Owen, 1895:398, 398fn., 399fn., Vol. 1):

> ...the popular mind was divided as to whether these images were inferior imitations, on a large scale, of certain animals at the Zoological Gardens ... creations of some eccentric person's imagination... [or] placed there with the pious purpose of setting clearly before the eyes of the public, as a

Figure 5. A possibly diagrammatic sketch of "The Secondary [i.e., Mesozoic] Island" of c. 1855 at the Crystal Palace, south London; far from being confined to dinosaurs, these reconstructions included a much fuller range of fossil saurians, including several marine reptiles. Original in Anonymous (1877:27); photograph courtesy of the Trustees of the National Museums of Scotland.

terrible warning, the fantastic visions sometimes seen by such as are in the habit of indulging too freely in spirituous liquors.

Today the public often persists, despite the best efforts of fussy academics, in its willful insistence that ichthyosaurs, plesiosaurs, mammal-like reptiles, and sundry other extinct beasts (sometimes, in my experience, even including mammoths and other pachyderms) are just as much "dinosaurs" as *Megalosaurus* and *Diplodocus*. Plainly, *dinosaur*, in popular usage, has become equivalent to the Victorian *saurian*. It seems still too early for specialists to abandon the word *dinosaur*, like *Brontosaurus*, to popular usage and replace it in technical discussion with a more specific clade name. Yet most popular "dinosaur" books focus much more sharply on the Dinosauria *sensu stricto* than, I suspect, the average publisher or reader might well wish, presumably because their authors take the academically much stricter definition of *dinosaur*. This may be one reason why, with a few honorable exceptions (e.g., McGowan, 1991; Benton, 1990; Halstead and Halstead, 1984), popular books on marine reptiles are so regrettably rare.

THE MARINE REPTILES TODAY

Research on marine reptiles, from the middle nineteenth century onward to the 1960s, settled down to what was mostly a worthy, rather than exciting, series of discoveries and taxonomic descriptions. Nevertheless, among (but not exhausting) the major highlights are: Thomas Henry Huxley's middle nineteenth-century work on the evolution of fossil crocodilians (Desmond, 1982); the discovery of ichthyosaurs, in the Lower Lias of England and the Upper Lias of Germany, with embryos and with soft parts such as fin outlines (Buffetaut, 1987; Pearce, 1846); the reidentification of placodonts, not as fishes but as a distinct group of reptiles (Nosotti and Pinna, 1989); Alfred Leeds' huge collection from the Oxford Clay of England (Leeds, 1956); and the great expansion of our knowledge of Cretaceous marine reptiles from Belgium, North America, and eventually South America and Australia, such as the giant pliosaur *Kronosaurus* (Romer and Lewis, 1959). This led to one of history's more ironic moments, when Edward Drinker Cope (1840-1897) and Othniel Charles Marsh (1831-1899) had their famous argument about which end of an elasmosaur was the front end. This may have played an early part in their animosities, and certainly exacerbated their later feuding and competition to discover those very dinosaurs of the American West that soon came to overshadow the marine reptiles (Storrs, 1984).

Conybeare's earlier appreciation notwithstanding, detailed functional analyses and paleobiological appreciations of any fossil reptile were rare, until the American S. W. Williston published his thoughtful synthesis *Water Reptiles of the Past and*

Present (Williston, 1914). Still the only book-length review of marine reptiles, it was full of good sense, such as a skeptical look at the supposedly aquatic lifestyle of the dicynodont *Lystrosaurus* (now confirmed by King, 1991). Soon after this, the British worker D. M. S. Watson published his classic study of the locomotion of plesiosaurs (Watson, 1924), using osteological evidence to attempt an unprecedentedly detailed reconstruction of the musculature and limb action of a fossil reptile, and interpreting it in hydromechanical terms. Watson (1951) went on to discuss the form, function, and evolution of marine reptiles in his classic *Paleontology and Modern Biology*, which linked evidence from fossil vertebrates to then current thinking on adaptation and evolution. By then, work on marine reptiles had almost died out completely, but researchers such as S. P. Welles, followed in the next two decades by L. B. Halstead, C. McGowan, E. Buffetaut, J.-M. Mazin, and J. A. Robinson---to name just a few---kept the field alive until the start of today's new Golden Age.

It is not only a Golden Age of research. New finds of marine reptiles have been acquired and old ones renovated in museums from Bristol through London to Brussels, reflecting the overwhelming importance of marine reptiles rather than dinosaurs in the fossil reptile faunas of many areas; plesiosaurs are famous again, supposedly the most likely candidates for the fabulous Monster of Loch Ness (Gibson and Heppell, 1988; Scott and Rines, 1975); ichthyosaurs are displayed *in situ* in the United States; Lyme Regis is once again the home of a small community of professional fossil collectors, selling their ichthyosaurs to visitors in the tradition of Mary Anning; and amateur collectors are increasingly following in their Victorian and Edwardian forebears' footsteps in recovering fossils from quarries and coastal exposures. Perhaps Lyell was right in his cyclical view of history.

SUMMARY

Contemporary emphasis on dinosaurs should not be allowed to obscure the historical significance of the fossil marine reptiles in stimulating the early development of vertebrate paleontology. The early work on ichthyosaurs and plesiosaurs in Britain by W. D. Conybeare and colleagues exemplifies the development of vertebrate paleontology as a maturing science with a significant practical and theoretical infrastructure. The marine reptiles were especially important in "natural theology" antievolutionary interpretations of the fossil record and, conversely, in the case of the crocodiles, in early transmutationist interpretations by Geoffroy Saint-Hilaire. Early popular interpretations of vertebrate paleontology, notably the Crystal Palace reconstructions of c. 1855, gave full weight to the marine reptiles among the extinct "saurians."

ACKNOWLEDGMENTS

I am very grateful to Dr. J. M. Callaway and Dr. E. L. Nicholls for inviting me to contribute, and to the Trustees of the National Museums of Scotland for support. I began this project on study leave at Oxford University from Leicestershire Museums, Arts and Records Service, when holding a Leverhulme Trust Research Fellowship; I thank Dr. T. S. Kemp (University Museum) and Dr. E. A. Newsholme (Merton College) for arranging facilities. I thank Mr. T. Sharpe (NMW) and Mrs. S. Newton (OXFUM) for access and permission to quote archival material, and Mr. T. Sharpe, Dr. P. R. Crowther (Bristol), and Mr. Ken Smith and colleagues (NMS) for supplying illustrations. I am especially grateful to Dr. H. S. Torrens (as always), Dr. A. J. Desmond, and Dr. N. Penny for discussion and information, and to Dr. C. McGowan and Professor W. A. S. Sarjeant for helpful referees' reports.

REFERENCES

Allen, D. E. 1976. *The Naturalist in Britain. A Social History*. Allen Lane, Harmondsworth, 292 pp.

Allen, D. E. 1985. The early professionals in British natural history. IN A. Wheeler and J. H. Price (Eds.), *From Linnaeus to Darwin: Commentaries on the History of Biology and Geology*, pp. 1-12. Society for the History of Natural History, London.

Anonymous 1832. [Henry Riley's lecture on "Palaeosaurians"]. *Bristol Mirror*, 26 May, [4], columns 1, 2.

Anonymous 1836. Report from the Select Committee on British Museum; together with the minutes of evidence, appendix and index. Reports from Committees -- (4.) -- British Museum. Session 4 February -- 20 August 1836. Volume X. [Reprinted as Irish University Press Series of British Parliamentary Papers. Education. British Museum 2. Report from the Select Committee on the Condition, Management and Affairs of the British Museum with minutes of evidence, appendix and index. Irish University Press, Shannon.]

Anonymous 1877. *Crystal Palace: A Guide to the Palace and Park by Authority of the Directors 1877 with Illustrations*. Dickens and Evans, London, 32 pp.

Appel, T. A. 1987. *The Cuvier-Geoffroy Debate; French Biology in the Decades Before Darwin*. Oxford University Press, New York and Oxford, 305 pp.

Bell, S. 1987. *IF ... Bounces Back*. Methuen, London, 160 pp.

Benton, M. J. 1990. *The Reign of the Reptiles*. Kingfisher Books, London, 144 pp.

Benton, M. J. and M. A. Taylor. 1984. Marine reptiles from the Upper Lias (Lower Toarcian, Lower Jurassic) of the Yorkshire coast. *Proceedings of the Yorkshire Geological Society* 44:399-429.

Bowen, J. 1854. *A Brief Memoir of the Life and Character of William Baker, F.G.S., ... Prepared Principally from his Diary and Correspondence. May, Taunton, Somerset*. Longman, London, 128 pp.

Broderip, W. J. 1849. *Zoological Recreations.* New edition with additions. Henry Colburn, London, 384 pp.
Buckland, W. 1824. Notice on the *Megalosaurus* or great fossil lizard of Stonesfield. *Transactions of the Geological Society of London* (2)1:390-396.
Buckland, W. 1836. *The Bridgewater Treatises on the Power, Wisdom and Goodness of God as Manifested in His Creation. Treatise VI. Geology and Mineralogy Considered with Reference to Natural Theology.* 2 volumes. Pickering, London, Vol. 1, 599 pp., Vol. 2, 123 pp.
Buffetaut, E. 1983. La paléontologie des vertébrés mésozoiques en Normandie du 18e siècle à nos jours: un essai historique. *Actes du Muséum de Rouen* 2:39-59.
Buffetaut, E. 1987. *A Short History of Vertebrate Palaeontology.* Croom Helm, London, 223 pp.
Chambers, R. 1844. *Vestiges of the Natural History of Creation.* London.
Cleevely, R. J. 1983. *World Palaeontological Collections.* British Museum (Natural History), and Mansell Publishing, London, 365 pp.
Cleevely, R. J. and S. D. Chapman. 1992. The accumulation and dispersal of Gideon Mantell's fossil collections and their role in the history of British palaeontology. *Archives of Natural History* 19:307-364.
Conybeare, W. D. 1822. Additional notices on the fossil genera *Ichthyosaurus* and *Plesiosaurus. Transactions of the Geological Society of London* (2)1:103-123.
Conybeare, W. D. 1824. On the discovery of an almost perfect skeleton of the *Plesiosaurus. Transactions of the Geological Society of London* (2)1:382-389.
Corsi, P. 1988. *The Age of Lamarck. Evolutionary Theories in France 1790-1830.* Revised and updated edition, translated by J. Mandelbaum. University of California Press, Berkeley, 360 pp.
Cumberland, G. 1829. Some account of the order in which the fossil saurians were discovered. *Quarterly Journal of Literature, Science and the Arts* 27:345-349.
De la Beche, H. T. and W. D. Conybeare. 1821. Notice of the discovery of a new fossil animal, forming a link between the *Ichthyosaurus* and the crocodile, together with general remarks on the osteology of *Ichthyosaurus. Transactions of the Geological Society of London* 5:559-594.
Delair, J. B. 1969. A history of the early discoveries of Liassic ichthyosaurs in Dorset and Somerset (1779-1835). *Proceedings of the Dorset Natural History and Archaeological Society* 90:115-127.
Delair, J. B. and W. A. S. Sarjeant. 1978. Joseph Pentland: a forgotten pioneer in the osteology of fossil marine reptiles. *Proceedings of the Dorset Natural History and Archaeological Society* 97:12-16.
Desmond, A. 1982. *Archetypes and Ancestors: Palaeontology in Victorian London 1850-1875.* Blond and Briggs, London, 287 pp.
Desmond, A. 1987. Artisan resistance and evolution in Britain, 1819-1848. *Osiris* 3:77-110.
Desmond, A. 1989. *The Politics of Evolution.* University of Chicago Press, Chicago and London, 503 pp.
Dobson, J. 1954. *William Clift.* Heinemann, London, 144 pp.
Doyle, P. and E. Robinson. 1993. The Victorian "Geological Illustrations" of Crystal Palace

Park. *Proceedings of the Geologists' Association* 104:181-194.
Doyle, P. and E. Robinson. 1995. Report of a field meeting to Crystal Palace Park and West Norwood Cemetery, 11 December, 1993. *Proceedings of the Geologists' Association* 106:71-78.
Edmonds, J. M. 1978. Patronage and privilege in education: a Devon boy goes to school, 1798. *Transactions of the Devonshire Association for the Advancement of Science* 110:95-111.
Fowles, J. 1982. *A Short History of Lyme Regis.* Dovecote Press, Wimborne, Dorset, 53 pp.
Geoffroy Saint-Hilaire, É. 1825. Recherches sur l'organisation des Gavials; sur leurs affinités naturelles, desquelles résulte la nécessité d'une autre distribution générique, *Gavialis, Teleosaurus*, et *Steneosaurus* etc. *Mémoires de la Muséum d'Histoire Naturelle de Paris* 12:97-155.
Geoffroy Saint-Hilaire, É. 1828. Mémoire où l'on se propose de rechercher dans quels rapports de structure organique et de parenté sont entre eux les animaux des âges historiques, et vivant actuellement, et les espèces antédiluviennes et perdues. *Mémoires de la Muséum d'Histoire Naturelle de Paris* 17:209-229.
Gibson, J. A. and D. Heppell. 1988 (Eds.). International Society of Cryptozoology/Society for the History of Natural History Symposium on the Loch Ness Monster. *The Scottish Naturalist* 1988 (2-3):38-214.
Gould, S. J. 1992. Red in tooth and claw. *Natural History* 101(11):14-23.
Halstead, L. B. and J. Halstead. 1984. *A Sea Serpent. The Story of a Nothosaur.* Collins, London, 32 pp.
Haste, H. 1993. Dinosaur as metaphor. *Modern Geology* 18:349-370.
Hawkins, T. 1834. *Memoirs on Ichthyosauri and Plesiosauri, Extinct Monsters of the Ancient Earth.* Relfe and Fletcher, London, 57 pp.
Hawkins, T. 1840. *The Book of the Great Sea Dragons, Gedolim Taninim of Moses.* Pickering, London, 27 pp.
Howe, S. R., T. Sharpe, and H. S. Torrens. 1981. *Ichthyosaurs: A History of Fossil 'Sea-dragons.'* National Museum of Wales, Cardiff, 32 pp.
Howie, F. M. P. 1986. Conserving and mounting fossils: a historical review. *Curator* 29:5-24.
King, G. M. 1991. The aquatic *Lystrosaurus*: a palaeontological myth. *Historical Biology* 4:285-321.
Lang, W. D. 1939. Mary Anning (1799-1847), and the pioneer geologists of Lyme. *Proceedings of the Dorset Natural History and Archaeological Society* 60:142-164.
Leeds, E. T. 1956. *The Leeds Collection of Fossil Reptiles from the Oxford Clay of Peterborough.* (Ed. W. E. Swinton). Blackwell, Oxford, 104 pp.
Lingham-Soliar, T. 1995. Anatomy and functional morphology of the largest marine reptile known, *Mosasaurus hoffmanni* (Mosasauridae, Reptilia) from the Upper Cretaceous, Upper Maastrichtian of the Netherlands. *Philosophical Transactions of the Royal Society of London* B347:155-180.
Lyell, C. 1830. *Principles of Geology.* Volume 1. Murray, London, 511 pp.
McCarthy, S. and M. Gilbert. 1994. *The Crystal Palace Dinosaurs.* Crystal Palace

Foundation, London, 99 pp.

McCartney, P. J. 1977. *Henry De la Beche: Observations on an Observer*. National Museum of Wales, Cardiff, 77 pp.

McGowan, C. 1989. *Leptopterygius tenuirostris* and other long-snouted ichthyosaurs from the English Lower Lias. *Palaeontology* 32:409-427.

McGowan, C. 1990. Problematic ichthyosaurs from southwest England: A question of authenticity. *Journal of Vertebrate Paleontology* 10:72-79.

McGowan, C. 1991. *Dinosaurs, Spitfires and Sea Dragons.* Harvard University Press, Cambridge, Massachusetts, 365 pp.

Murchison, R. I. 1842. Anniversary address of the President. *Proceedings of the Geological Society of London* 3(2):637-687.

North, F. J. 1935. Dean Conybeare, geologist. *Transactions of the Cardiff Naturalists' Society* 66:15-68.

North, F. J. 1957. W. D. Conybeare, his geological contemporaries and Bristol associations. *Proceedings of the Bristol Naturalists' Society* 29:133-146.

Nosotti, S. and G. Pinna. 1989. Storia delle ricerche e degli studi sui rettili placodonti. Parte prima 1830-1902. *Memorie della Società Italiana di Scienze Naturali e del Museo Civico di Storia Naturale di Milano* 24:31-86.

Owen, R. 1840. Report on British fossil reptiles. Part I. *Annual Report of the British Association for the Advancement of Science, for 1839, Reports*:143-216.

Owen, R. 1842. Report on British fossil reptiles. Part II. *Annual Report of the British Association for the Advancement of Science, for 1841, Reports*:60-204.

Owen, R. S. 1895. *The life of Richard Owen by his grandson the Rev. Richard Owen M.A.* 2 volumes. Revised edition. Murray, London, Vol. 1, 409 pp., Vol. 2, 392 pp.

Pearce, J. C. 1846. Notice of what appears to be the embryo of an *Ichthyosaurus* in the pelvic cavity of *Ichthyosaurus (communis?)*. *Annals and Magazine of Natural History* 17:44-46.

Penny, N. 1992. *Catalogue of European Sculpture in the Ashmolean Museum, 1540 to the Present Day. Volume III: British.* Clarendon Press, Oxford, 269 pp.

Porter, R. R. 1977. *The Making of Geology. Earth Science in Britain 1680-1815.* Cambridge University Press, Cambridge, 288 pp.

Porter, R. R. 1978. Gentlemen and geology: the emergence of a scientific career 1660-1920. *Historical Journal* 21:809-836.

Potts, A. 1980. *Sir Francis Chantrey 1781-1841. Sculptor of the Great.* National Portrait Gallery, London, 36 pp.

Read, B. 1982. *Victorian Sculpture.* Yale University Press, New Haven and London, 414 pp.

Rolfe, W. D. I. 1985. William and John Hunter: breaking the Great Chain of Being. IN W. F. Bynum and R. Porter (Eds.), *William Hunter and the Eighteenth-Century Medical World*, pp. 297-319. Cambridge University Press, Cambridge.

Romer, A. S. and A. D. Lewis. 1959. A mounted skeleton of the giant plesiosaur *Kronosaurus*. *Breviora* 112:1-15.

Rudwick, M. J. S. 1975. Caricature as a source for the history of science: De la Beche's anti-Lyellian sketches of 1831. *Isis* 66:534-560.

Rudwick, M. J. S. 1976a. *The Meaning of Fossils. Episodes in the History of Palaeontology*. Second edition. Science History Publications, New York, 287 pp.

Rudwick, M. J. S. 1976b. The emergence of a visual language for geological science 1760-1840. *History of Science* 14:149-195.

Rudwick, M. J. S. 1985. *The Great Devonian Controversy. The Shaping of Scientific Knowledge Among Gentlemanly Specialists*. University of Chicago Press, Chicago, 494 pp.

Rudwick, M. J. S. 1989. Encounters with Adam, or at least the hyaenas: nineteenth-century visual representations of the deep past. IN J. R. Moore (Ed.), *History, Humanity and Evolution. Essays for John C. Greene*, pp. 231-251. Cambridge University Press, Cambridge.

Rudwick, M. J. S. 1992. *Scenes from Deep Time. Early Pictorial Representations of the Prehistoric World*. University of Chicago Press, Chicago, 280 pp.

Rudwick, M. J. S. 1993. Historical origins of the Geological Society's Journal. *Journal of the Geological Society, London* 150:3-6.

Rupke, N. A. 1983. *The Great Chain of History. William Buckland and the English School of Geology (1814-1849)*. Clarendon Press, Oxford, 322 pp.

Rupke, N. A. 1994. *Richard Owen. Victorian Naturalist*. Yale University Press, New Haven and London, 462 pp.

Sarjeant, W. A. S. and J. B. Delair. 1980. An Irish naturalist in Cuvier's laboratory. The letters of Joseph Pentland 1820-1832. *Bulletin of the British Museum (Natural History)*, Historical Series 6:245-319.

Scott, P. and R. Rines. 1975. Naming the Loch Ness monster. *Nature* 258:466-467.

Storrs, G. W. 1984. *Elasmosaurus platyurus* and a page from the Cope-Marsh war. *Discovery* 17(2):25-27.

Storrs, G. W. and M. A. Taylor. 1996. Cranial anatomy of a new plesiosaur genus from the lowermost Lias (Rhaetian/Hettangian) of Street, Somerset, England. *Journal of Vertebrate Paleontology* 16:403-420.

Stukeley, W. 1719. An account of the impression of the almost entire skeleton of a large animal in a very hard stone, lately presented the Royal Society, from Nottinghamshire. *Philosophical Transactions of the Royal Society* 30:963-968, Plate 12.

Taylor, M. A. 1989a. Thomas Hawkins FGS (22 July 1810-15 October 1889). *The Geological Curator* 5:112-114.

Taylor, M. A. 1989b. The other dinosaurs. *New Scientist* 121(1655), 65.

Taylor, M. A. 1994. The plesiosaur's birthplace: the Bristol Institution and its contribution to vertebrate palaeontology. *Zoological Journal of the Linnean Society of London* 112:179-196.

Taylor, M. A. and H. S. Torrens. 1987. Saleswoman to a new science: Mary Anning and the fossil fish *Squaloraja*. *Proceedings of the Dorset Natural History and Archaeological Society* 108:135-148.

Taylor, M. A. and H. S. Torrens. 1995. Fossils by the sea. *Natural History* 104(10):66-71.

Torrens, H. S. 1990. A Wiltshire pioneer and his legacy -- Henry Shorto III (1778-1864), cutler and fossil collector of Salisbury. *Wiltshire Archaeological and Natural History Magazine* 83:170-189.

Torrens, H. S. 1992. When did the dinosaur get its name? *New Scientist* 134(1815):40-44.

Torrens, H. S. 1993. The dinosaurs and dinomania over 150 years. *Modern Geology* 18:257-286.

Torrens, H. S. 1995. Mary Anning (1799-1847) of Lyme: "the greatest fossilist the world ever knew." *British Journal for the History of Science* 28:257-284.

Watson, D. M. S. 1924. The elasmosaurid shoulder girdle and forelimb. *Proceedings of the Zoological Society of London* (1924):885-917.

Watson, D. M. S. 1951. *Paleontology and Modern Biology*. Oxford University Press, Oxford, 216 pp.

Wellnhofer, P. 1991. *The Illustrated Encyclopaedia of Pterosaurs*. Salamander, London, 192 pp.

Whybrow, P. J. 1985. A history of fossil collecting and preservation techniques. *Curator* 28:5-26.

Williston, S. W. 1914. *Water Reptiles of the Past and Present*. University of Chicago Press, Chicago, 251 pp.

Wilson, L. G. 1972. *Charles Lyell. The Years to 1841: The Revolution in Geology*. Yale University Press, New Haven and London, 553 pp.

PART I
Ichthyosauria

Part I: Ichthyosauria

INTRODUCTION

JACK M. CALLAWAY

Order Ichthyosauria, as currently defined, includes only the ichthyosaurs. These very distinctive and highly specialized marine reptiles are known from Mesozoic strata ranging in age from Lower Triassic Smithian (Cox and Smith, 1973; Callaway and Brinkman, 1989) to Cenomanian (Bardet, 1994). Even the oldest known ichthyosaurs are completely adapted to marine life and have no close, gross morphological resemblances to any other reptiles. This uniqueness has firmly defined the group ever since the recognition of ichthyosaurs as reptiles in the nineteenth century and, with the lack of any transitional forms, has kept to a minimum the problem of what to include or exclude from the order. This same uniqueness, however, has posed perplexing problems regarding the ancestry, early evolutionary history, and phylogeny of the group.

The unusual morphology of ichthyosaurs compared to all other reptiles was responsible for well over 100 years of failure to establish their status as reptiles by early naturalists. The first figured remains of ichthyosaurs are probably those produced by the Welch naturalist Lhwyd (1699), who believed the remains were those of fish. A later discovery of an ichthyosaur fossil was made by Scheuchzer (1708) in the vicinity of Altorf, Germany, although he did not recognize the remains as such. Scheuchzer attributed the vertebrae to a human drowned in the biblical Noachan flood. Baier (1708) made finds of similar vertebrae near Altorf but described them as belonging to fish. It was more than a century later that Cuvier (1814) correctly described Scheuchzer's material as belonging to a marine reptile. Hawker (1807) published a brief magazine article describing as a crocodile a nicely articulated ichthyosaur skeleton found near Bath, England. The English anatomist Home described several good ichthyosaur specimens found in the Lyme Regis area in Dorset. He first united them with fish (Home, 1814, 1816), then compared them to the duck-billed platypus and with aquatic birds (Home, 1818), and later allied them with amphibians (Home, 1819a, b), although he was aware of their reptile-like qualities. His acceptance of an amphibian relationship is revealed by the name he assigned the specimens, *Proteosaurus*, which was based on the living salamander, *Proteus* (Home, 1819b). However, in 1818 a British Museum curator named König had proposed the name *Ichthyosaurus* for the same kinds of skeletons as Home's fossils on the assumption that the animals were intermediate between fish and reptiles (Delair, 1969). The general term *ichthyosaur*, which was derived from the proposed name, quickly came to be applied to all known "fish-lizards." De la Beche and Conybeare (1821) and Conybeare (1822) felt that the similar material they were describing did not resemble the salamander and fully recognized the reptilian character of the skeleton. Young (1821) described an ichthyosaur found on the Yorkshire coast at Whitby as being closely related to the Cetacea, especially to *Delphinus*. Within the next few years many ichthyosaur specimens were described by various workers from England, France, and Germany. Recognition of ichthyosaurs as reptiles rapidly became widespread, but Blainville (1835) concluded that they did not fit comfortably into any existing vertebrate group and created a separate and distinct class, Ichthyosauria, to contain them. As noted by Williston (1914), however, Owen (1860) "rather arbitrarily changed Blainville's name Ichthyosauria to Ichthyopterygia, a name which is often, though incorrectly, used to designate this order of reptiles." The use of the term Ichthyopterygia has been a source of confusion from its inception. It has been used at different times to designate various taxonomic categories, including class, order, and suborder. The most common usage has been the application of Ichthyopterygia as the name of the subclass, with Ichthyosauria used for the order (for example, see Camp and Vanderhoof, 1940). This arrangement was further reinforced by Romer (1956). Using cladistic methods, Callaway (1989) and Massare and Callaway (1990) demonstrated that ichthyosaurs do not require their own special subclass and are diapsid reptiles with a derived or modified diapsid skull. As Callaway (1989) pointed out, the subclass Ichthyopterygia is obsolete, and use of that name for any ichthyosaurian taxonomic category should be abandoned.

Once ichthyosaurs were no longer confused with fish, amphibians, mammals, or birds, the problem of their classification and relationships with other reptilian groups engaged the energies of numerous paleontologists over the years. Some of the more interesting hypotheses and classification schemes will illustrate this point. Owen (1860) recognized six extinct vertebrate orders, with the ichthyosaurs as the sole members of the order Ichthyopterygia. Baur (1887) had considered the ichthyosaurs to be specialized *Sphenodon*-like reptiles. The classification of Zittel (1890), one of the most comprehensive and advanced for its time, continued to keep the ichthyosaurs to themselves with no presumption of affinities with any other reptilian taxa. Woodward's (1898) heavy reliance on the use of temporal arches, rather than the fenestrae themselves, resulted in the unlikely grouping of the single-arched ichthyosaurs with the single-arched Anomondontia, Sauropterygia, and Chelonia (the closed cheek region of anapsids was formerly considered to constitute a single arch). Osborn's (1903) classification also relied heavily on skull fenestration, or, more appropriately, temporal arcades. It was this classification that firmly established the use of temporal structure as a basis for defining major subdivisions of the reptiles. Osborn placed all reptiles into two subclasses, the single-arched Synapsida and the double-arched Diapsida. For reasons unexplained by Osborn, the single-arched ichthyosaurs were placed with the Diapsida while the other single-arched marine reptiles, the Sauropterygia and the Placodontia, were placed with the Synapsida. He suggested a close relationship between ichthyosaurs and rhynchocephalians, but made no pronouncements on their origin. Although Osborn assigned ichthyosaurs to the diapsids, he did so for the wrong reasons. McGregor (1906) expressed the opinion that ichthyosaurs had their origin in a rhynchocephalian-like ancestor and postulated that the diapsid Phytosauria were the nearest known relatives of the ichthyosaurs.

Huene (1922, 1923) long believed that *Mesosaurus* from the Lower Permian rocks of South America and Africa was related to the ancestry of the ichthyosaurs and classified them together in the order Ichthyopterygia. He did not place *Mesosaurus* on the direct line of ichthyosaur ancestry, but believed they had a common origin. *Mesosaurus* was thought to be a short side branch of a hypothetical main line leading to ichthyosaurs. This main line formed a parallel branch with the Cotylosauria, both having originated from Carboniferous embolomerous amphibians. Huene (1937) later rejected his classification and phylogenetic hypothesis, but clung to the idea of an embolomerous ancestry for the ichthyosaurs. Huene (1944) again advocated derivation of ichthyosaurs from amphibians, this time from loxommids, but gave few reasons for doing so.

Williston's (1925) classification of reptiles brought together many modifications that had followed Osborn's (1903) proposal, especially those of Williston (1917) and Broom (1924). He erected a separate subclass, Parapsida, for reptiles with an upper temporal opening, other than sauropterygians and placodonts. The four orders included were Proganosauria (Mesosauria), Ichthyosauria, Protorosauria, and Squamata. The Protorosauria were placed here as presumed ancestors of the ichthyosaurs. Romer (1948) attempted to show a relationship between ophiacodont synapsids and ichthyosaurs based on a comparison of *Ophiacodon* with his own restoration of the Jurassic ichthyosaur *Ophthalmosaurus* suggesting an anapsid "pre-ophiacodont" as an ideal ichthyosaur ancestor. However, his use of characters from a highly derived post-Triassic ichthyosaur rather than the available data from more primitive Triassic ichthyosaurs led to a number of false conclusions, and there is no longer any reason to seek out some hypothetical "pre-ophiacodont" ancestor.

The Squamata eventually were determined to be modified diapsids, as initially advocated by Broom (1924), and were removed from the parapsids by Romer (1956). Romer (1956, 1966, 1971) also removed the protorosaurs from the Parapsida and placed them into classifications of his own. Showing little evidence of eligibility as ichthyosaur ancestors, the Proganosauria (Mesosauria) were removed from the Parapsida by Romer (1956) as an uncertain subclass. Romer's (1956, 1966) classifications included some fundamental changes from Williston's (1925) classification, but they clearly show a Willistonian influence. With the dissolution of the Parapsida, largely through his own efforts, Romer (1956) returned the ichthyosaurs to their isolated subclass, Ichthyopterygia.

Colbert (1945, 1955), however, essentially modified the Parapsida into the Euryapsida, which was created to include reptiles with a single upper temporal opening, including the ichthyosaurs. It now seems evident that the Euryapsida as established by Colbert, was largely an artificial assemblage. Kuhn-Schnyder (1980) went so far as to declare the Euryapsida obsolete. Its use as a taxon continues, however. Colbert (1980) retained order Ichthyosauria within the subclass Euryapsida. Mazin (1982) included the superorder Ichthyopterygia within the Euryapsida along with the sauropterygians and placodonts in an effort to defend close phylogenetic ties within the three groups, but his arguments were refuted by Sues (1987). Colbert and Morales (1991) assigned ichthyosaurs to superorder Ichthyosauria within the infraclass Euryapsida and subclass Diapsida. Rieppel (1993), however, restricts the usage of Euryapsida to the Pachypleurosauroidea and Sauropterygia with no mention of ichthyosaurs.

Appleby (1959, 1961) suggested affinities between chelonians and ichthyosaurs based primarily on the unusual course of the ninth cranial nerve as it emerged separately from the braincase to pass through the inner part of the ear, a character he postulated to be shared by turtles and ichthyosaurs. Tarsitano (1982) concluded that ichthyosaurs were diapsids and suggested an affinity with lizards, and later (Tarsitano, 1983) hypothesized a relationship between eosuchians and ichthyosaurs. Carroll (1985, 1988) proposed ichthyosaur affinities with the Archosauria, possibly the thecodonts. This view was founded on comparison of primitive ichthyosaurs with ichthyosaur-like reptiles from China, *Nanchangasaurus* (Wang, 1959) and *Hupehsuchus* (Yang and Dong, 1972), which were thought to be thecodontians. Carroll (1988) included these genera within order Ichthyosauria of subclass Ichthyopterygia. This was later revised when Carroll (1991) reassigned both genera to a new diapsid order, Hupehsuchia, with the suggestion that ichthyosaurs and the Hupehsuchia might share a common aquatic ancestry.

The difficulties of determining ichthyosaurian relationships are clearly evident; however, the possible diapsid nature of ichthyosaurs has been recognized for many years (for example, Callaway, 1989; Carroll, 1985, 1988, 1991; Colbert, 1980;

Colbert and Morales, 1991; McGregor, 1906; Massare and Callaway, 1990; Mazin, 1982; Osborn, 1903; Tarsitano, 1982, 1983; Williston, 1917, 1925). Some workers have offered little or no substantial evidence in support of their phylogenetic scenarios. Others have formed their conclusions based on excellent theoretical grounds, but without any analysis of character states. It is the latter that must be examined if ichthyosaurian relationships are to be discovered. Several in-depth analyses of diapsid synapomorphies, based on derived traits, have been presented in recent years (Benton, 1985; Carroll and Currie, 1991; Evans, 1984, 1988; Gaffney, 1980; Laurin, 1991; Reisz, 1981; Reisz et al., 1984). Based on synapomorphies from the most primitive ichthyosaurs, and using a framework of established diapsid relationships, Callaway (1989) and Massare and Callaway (1990) concluded that ichthyosaurs are diapsids with a modified or derived diapsid skull. They also were able to assign ichthyosaurs to the Neodiapsida (Benton, 1985), which, in a modification by Laurin (1991), is defined as the most recent common ancestor of younginiforms and living diapsids, and all its descendants. Massare and Callaway (1990) proposed a close relationship with the Lepidosauromorpha, especially the Younginiformes. However, with the removal of younginiforms from the Lepidosauromorpha, and thus from the Sauria as well (Laurin, 1991), a reassessment of the position of ichthyosaurs within the Neodiapsida is needed. The Sauria were defined by Gauthier (1984) as the crown group of diapsids, including younginiforms, lepidosaurs, and archosauromorphs. Laurin (1991) determined that the younginiforms were outside the crown group. A suggested, yet untested, phylogenetic position for the ichthyosaurs is shown in Figure 1.

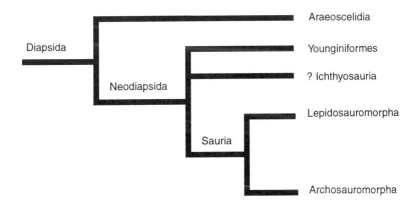

Figure 1. Hypothetical phylogenetic position of the Ichthyosauria within the Neodiapsida. See text for discussion.

Problems have persisted in the understanding of ichthyosaurian phylogeny for several reasons. Most previous hypotheses have been based on the morphology of the better known, better preserved Jurassic species, yet it is the morphology of Triassic species that is critical for understanding ichthyosaurian ancestry and relationships with other reptiles. Also, phylogenetic studies of ichthyosaurs must depend mainly on the morphology of the skull and axial skeleton because the limb and limb girdle features of ichthyosaurs are so highly modified that their use in character analyses is essentially nullified except for use in defining taxa within the order. Additionally, with few exceptions, preservation of Triassic ichthyosaur material is notoriously poor. Many species are defined on the partial, often scrappy, remains of a single specimen. Callaway and Massare (1989a) concluded that species diversity among Triassic ichthyosaurs was as high as during the Jurassic; however, McGowan (1994) believes this is an exaggerated diversity because many Triassic species were erected on inadequate material, especially the five species of *Shastasaurus* described by Merriam (1895, 1902, 1908) and the *Shastasaurus* species of Huene (1925). Certainly an in-depth review and revision of the Shastasauridae is long overdue, and for all other Triassic ichthyosaurs as well. Until then, a full understanding of the early evolution of ichthyosaurs will be hampered. A final difficulty in understanding the phylogeny of the ichthyosaur clade has been the apparent morphological gap between Triassic and post-Triassic forms. Ichthyosaurs are traditionally thought of as "Triassic types" and "Jurassic types" with no hints of possible transitional forms to bridge the evolutionary gap. Some of these problems are beginning to be resolved. A growing awareness of the necessity for data from the more primitive ichthyosaurs has led to a virtual renaissance in the study of Triassic ichthyosaurs over the past two decades. For example, three of the four chapters included in this section are concerned entirely with Triassic ichthyosaurs and the fourth is largely concerned with them as well.

Ichthyosaur dentition is a character that has received scant attention from workers in the past, either because it was thought to have little phylogenetic significance or because of an inability to recognize correctly the nature of the dentition present. An important pioneering effort regarding ichthyosaurian dentitions was that of Mazin (1983), but this work has been found wanting in several respects by Motani (Chapter 4). A major problem has been one of proper nomenclature: the morass of conflicting terminology used in the literature by various workers through the years to describe ichthyosaur dentitions has led to considerable confusion and many difficulties of interpretation. Motani critically examines the types of dental implantation and replacement seen in all ichthyosaurs along with the evolutionary significance implied by this information. Perhaps the most important aspect of Motani's work is the standardization of nomenclature used to describe the full range of dental types seen in ichthyosaurs. Application of this standardized terminology would vastly improve the ability to communicate ichthyosaurian dental information unambiguously in the scientific literature.

Newly described Triassic species, and a new look at earlier described species, are providing new insights into the morphology, paleoecology, early evolution, and phylogeny of ichthyosaurs. For example, the supratemporal bone, once thought to be present in all ichthyosaurs, was shown to be absent by Romer (1968) and McGowan (1973), both working with post-Triassic material, but the supratemporal is present in some of the oldest Triassic ichthyosaurs, however, and is considered here to be a retained primitive diapsid condition in ichthyosaurs. Callaway (1989 and Chapter 2) reaffirms the presence of a supratemporal for *Mixosaurus cornalianus*. A large supratemporal bone is described for *Phalarodon nordenskioeldii*, now considered to belong to the Mixosauridae (Brinkman et al., 1992a; Nicholls et al., in review). Nicholls and Brinkman (1995) describe a large supratemporal in their new Triassic species, *Parvinatator wapitiensis*. The bone Mazin (1982) identified as the quadratojugal for the primitive ichthyosaur *Grippia* in an effort to conform to the skull homologies promulgated by Romer (1968) and McGowan (1973) is excavated ventrally and is forced aberrantly into the border of the supratemporal fenestra. Callaway (1989), Massare and Callaway (1990), and Nicholls and Brinkman (1995) are all in agreement that this bone is the squamosal. No bone that would conform to the quadratojugal is known from *Grippia* and it seems likely that it has been lost. The supratemporal is often thought of as an insignificant bone that is lost in most reptiles (Romer, 1968, for example). As discussed by Nicholls and Brinkman (1995), however, the supratemporal is known in a number of reptiles, often forming one of the most prominent bones of the temporal region. Some sphenodontidans (Wu, 1991), weigeltisaurids (Evans and Haubold, 1987), snakes (Carroll, 1988; Estes et al., 1988), thalattosaurs (Kuhn, 1952; Nicholls and Brinkman, 1995), and many lizards (Estes et al., 1988; Rieppel, 1980) provide examples. In a radical departure from currently accepted views, Nicholls and Brinkman (1995) propose a revision to the homologies of the temporal dermal elements of ichthyosaurs. Based on the relationships among the quadrate foramen, the quadratojugal, the squamosal, and the quadrate, they conclude that the two large bones in the temporal region of all ichthyosaurs are the supratemporal and the squamosal, and that the quadratojugal has been lost in all but a few primitive ichthyosaurs, resulting in a quadrate foramen between the quadrate and squamosal as in primitive diapsids.

An example of new information gleaned from the reexamination of a previously described species is provided by Motani, et al. (1996). *Chensaurus chaoxianensis* (Chen, 1985), possibly the oldest known ichthyosaur, is characterized as having had an anquilliform swimming mode and may represent an evolutionary intermediate between the shorter-bodied terrestrial stock from which ichthyosaurs evolved, and advanced thunniform ichthyosaurs.

New records of old and new species of ichthyosaurs are adding to our knowledge of the paleobiogeography of these marine reptiles. Recent studies on the paleobiogeography of Triassic ichthyosaurs are those of Mazin and Sander (1993) and Sander and Mazin (1993). The range of *Cymbospondylus* has been expanded with reports of specimens from northern Italy (Sander, 1989); Spitsbergen, Norway (Sander, 1992); and Idaho (Massare and Callaway, 1994). New material of *Shastasaurus* from Mexico was reported by Callaway and Massare (1989b) and from British Columbia, Canada, by McGowan (1991, 1994). Within recent years, the Wapiti Lake region of British Columbia has been an especially productive area for new Early and Middle Triassic material, while the Williston Lake locality of British Columbia is becoming increasingly important as a site for new Late Triassic ichthyosaur material (McGowan, Chapter 3). Additional material of *Shastasaurus* and an overview of the paleobiogeography of the genus is provided by Sander (Chapter 1) and by McGowan (Chapter 3). Remains reported from British Columbia by Callaway and Brinkman (1989) extended the ranges of *Phalarodon, Mixosaurus,* and *Pessosaurus*. The ranges of *Grippia* and *Utatsusaurus* now include British Columbia (Brinkman et al., 1992b; Nicholls and Brinkman, 1993).

The gap between "Triassic-type" and "Jurassic-type" ichthyosaurs is beginning to narrow, and the view that there was an abrupt faunal change across the Triassic-Jurassic boundary with no ichthyosaur species crossing that boundary, rather than a period of transition, has been shown to be untrue. McGowan (1989) reported that *Leptopterygius tenuirostris*, a common long-snouted species from the English lower Liassic, is now known from the Rhaetian to the Sinemurian. The Williston Lake fauna contains elements that are typically Triassic and others that are more typical of the Jurassic. A new ichthyosaur species, *Hudsonelpidia brevirostris*, appears to have a mixture of characters shared with *Mixosaurus* of the Early and Middle Triassic and the common Lower Jurassic genus *Ichthyosaurus* (McGowan, 1995 and Chapter 3). The presence of *Ichthyosaurus* itself at Williston Lake is reported also by McGowan (1996 and Chapter 3). Although work on this fauna is still under way, and interpretation of collected data is not yet complete, it seems that the old Triassic-Jurassic wall is being breached even further.

Ichthyosaurian relationships with other reptiles and interrelationships among themselves clearly require additional critical analysis, but much progress has been made in the past few years. An increased interest in the study of the primitive members of this clade is encouraging and is providing a better and more realistic understanding of ichthyosaurian evolution. Continued new finds and a reassessment of previously described material on the basis of new evidence will alter many of our current views of ichthyosaurian morphology, evolution, paleobiogeography, and paleoecology. Considerable effort has been expended in attempting to determine ichthyosaur origins, but little has been directed toward possible causes of their extinction. This is a fertile area for future research and will require an in-depth study of the overall physical and biological Mesozoic marine milieu.

ACKNOWLEDGMENTS

I thank Texas A & M International University and Laredo Community College for their support and my colleagues who read and corrected earlier drafts of this introduction. I also appreciate the review and editorial help of J. Freeman and an anonymous reviewer. This is a Rochester Institute of Vertebrate Paleontology contribution.

REFERENCES

Appleby, R. M. 1959. The origins of the ichthyosaurs. *The New Scientist* 6:758-760.
Appleby, R. M. 1961. On the cranial morphology of ichthyosaurs. *Proceedings of the Zoological Society of London* 137:333-370.
Baier, J. J. 1708. *Oryctographia Norica*. Nuremberg.
Bardet, N. 1994. Extinction events among Mesozoic marine reptiles. *Historical Biology* 7:313-324.
Baur, G. 1887. On the morphology and origin of the Ichthyopterygia. *American Naturalist* 21:837-840.
Benton, M. J. 1985. Classification and phylogeny of the diapsid reptiles. *Zoological Journal of the Linnean Society* 84:97-164.
Blainville, H. M. D. de. 1835. Description de quelqes espèces de reptiles de la Californie. *Nouvelles Annales du Muséum d'Histoire Naturelle* 4:233-296.
Brinkman, D. B., E. L. Nicholls, and J. M. Callaway. 1992a. New material of the ichthyosaur *Mixosaurus nordenskioeldii* from the Triassic of British Columbia, and the interspecific relationships of *Mixosaurus*. *North American Paleontological Convention V, Field Museum of Natural History, Chicago, Abstracts*:37.
Brinkman, D. B., X. Zhao, and E. L. Nicholls. 1992b. A primitive ichthyosaur from the Lower Triassic of British Columbia, Canada. *Palaeontology* 35:465-474.
Broom, R. 1924. The classification of the reptiles. *American Museum of Natural History Bulletin* 51(2):39-65.
Callaway, J. M. 1989. *Systematics, Phylogeny, and Ancestry of Triassic Ichthyosaurs (Reptilia, Ichthyosauria)* Unpublished Ph. D. dissertation, University of Rochester, Rochester, New York.
Callaway, J. M. and D. R. Brinkman. 1989. Ichthyosaurs (Reptilia, Ichthyosauria) from the Lower and Middle Triassic Sulphur Mountain Formation, Wapiti Lake area, British Columbia. *Canadian Journal of Earth Sciences* 26:1491-1500.

Callaway, J. M. and J. A. Massare. 1989a. Geographic and stratigraphic distribution of the Triassic Ichthyosauria (Reptilia; Diapsida). *Neues Jahrbuch für Geologie und Paläontologie*, Abhandlungen 178:37-58.

Callaway, J. M. and J. A. Massare. 1989b. *Shastasaurus altispinus* (Ichthyosauria, Shastasauridae) from the Upper Triassic of the El Antimonio District, northwestern Sonora, Mexico. *Journal of Paleontology* 63(6):930-939.

Camp, C. L. and V. L. Vanderhoof. 1940. Bibliography of Fossil Vertebrates 1928-1933. *Geological Society of America Special Papers 27*.

Carroll, R. L. 1985. Evolutionary constraints in aquatic diapsid reptiles. *Special Papers in Palaeontology* 33:145-155.

Carroll, R. L. 1988. *Vertebrate Paleontology and Evolution*. W. H. Freeman and Company, New York.

Carroll, R. L. 1991. *Hupehsuchus*, an enigmatic aquatic reptile from the Triassic of China, and the problem of establishing relationships. *Philosophical Transactions of the Royal Society of London* B 331:131-153.

Carroll, R. L. and P. J. Currie. 1991. The early radiation of diapsid reptiles. IN H.-P. Schultze and L. Trueb (Eds.), *Origins of the Higher Groups of Tetrapods*, pp. 354-424. Comstock Publishing Associates, Ithaca.

Chen, L. 1985. Ichthyosaurs from the Lower Triassic of Chao County, Anhui. *Regional Geology of China* 15:139-146 (in Chinese).

Colbert, E. H. 1945. The dinosaur book. *The American Museum of Natural History Handbook Series*, No. 14.

Colbert, E. H. 1955. *Evolution of the Vertebrates*. John Wiley and Sons, New York.

Colbert, E. H. 1980. *Evolution of the Vertebrates* (third edition). John Wiley and Sons, New York.

Colbert, E. H. and M. Morales. 1991. *Evolution of the Vertebrates* (fourth edition). Wiley-Liss, New York.

Conybeare, W. D. 1822. Additional notices on the fossil genera *Ichthyosaurus* and *Plesiosaurus*. *Transactions of the Geological Society of London*, Second Series, 1(1):103-123.

Cox, C. B. and D. G. Smith. 1973. A review of the Triassic vertebrate faunas of Svalbard. *Geological Magazine* 110:405-418.

Cuvier, M. G. 1814. Nouvelles observations sur le prétendu homme témoin du déluge de Scheuzer (sic). *Bulletin de la Société Philomatique de Paris*, Series 3, 1:22-23.

De la Beche, H. T. and W. D. Conybeare. 1821. Notice of the discovery of a new fossil animal, forming a link between the *Ichthyosaurus* and the crocodile, together with general remarks on the osteology of the *Ichthyosaurus*. *Transactions of the Geological Society of London* 5:559-594.

Delair, J. B. 1969. A history of the early discoveries of Liassic ichthyosaurs in Dorset and Somerset (1779-1835) and the first record of the occurrence of ichthyosaurs in the Purbeck. *Proceedings of the Dorset Natural History and Archaeological Society* 90:115-132.

Estes, R., K. de Queiroz, and J. Gauthier. 1988. Phylogenetic relationships within Squamata.

IN R. Estes and G. Pregill (Eds.), *Phylogenetic Relationships of the Lizard Families*, pp. 119-281. Stanford University Press, Stanford.
Evans, S. E. 1984. The classification of the Lepidosauria. *Zoological Journal of the Linnean Society* 82:87-100.
Evans, S. E. 1988. The early history and relationships of the Diapsida. IN M. J. Benton (Ed.), *The Phylogeny and Classification of the Tetrapods*, Volume 1: Amphibians, Reptiles, Birds, pp. 221-260. Clarendon Press, Oxford.
Evans, S. E. and H. Haubold. 1987. A review of the Upper Permian genera *Coelurosauravus, Weigeltisaurus*, and *Gracilisaurus* (Reptilia: Diapsida). *Zoological Journal of the Linnean Society* 90:275-303.
Gaffney, E. S. 1980. Phylogenetic relationships of the major groups of amniotes. IN A. L. Panchen (Ed.), *The Terrestrial Environment and the Origin of Land Vertebrates*, pp. 593-610. Academic Press, London.
Gauthier, J. A. 1984. *A Cladistic Analysis of the Higher Systematic Categories of the Diapsida*. Unpublished Ph. D. dissertation, University of California, Berkeley.
Hawker, J. 1807. Fossil crocodile found near Bath. *Gentleman's Magazine* 77:7-8.
Home, E. 1814. Some account of the fossil remains more nearly allied to fishes than to any of the other classes of animals. *Philosophical Transactions of the Royal Society of London* 104:571-577.
Home, E. 1816. Some further account of the fossil remains of an animal, of which a description was given to the Society in 1814. *Philosophical Transactions of the Royal Society of London* 106:318-321.
Home, E. 1818. Additional facts respecting the fossil remains of an animal, on the subject of which two papers have been printed in the *Philosophical Transactions*, showing that the bones of the sternum resemble those of the *Ornithorhynchus paradoxus*. *Philosophical Transactions of the Royal Society of London* 108:24-32.
Home, E. 1819a. An account of the fossil skeleton of the *Proteosaurus*. *Philosophical Transactions of the Royal Society of London* 109:209-211.
Home, E. 1819b. Reasons for giving the name *Proteosaurus* to the fossil skeleton which has been described. *Philosophical Transactions of the Royal Society of London* 109:212-216.
Huene, F. von. 1922. *Die Ichthyosaurier des Lias und ihre Zusammenhänge*. Berlin.
Huene, F. von. 1923. Lines of phyletic and biological development of the Ichthyopterygia. *Geological Society of America Bulletin* 34:463-468.
Huene, F. von. 1925. *Shastasaurus*-Reste in der alpinen Trias. *Centralblatt für Mineralogie, Geologie und Paläontologie*, Abteilung B, 1925:412-417.
Huene, F. von. 1937. Die Frage nach der Herkunft der Ichthyosaurier. *Geological Institutions of the University of Uppsala Bulletin* 27:1-9.
Huene, F. von. 1944. Die Zweiteilung des Reptilstammes. *Neues Jahrbuch für Mineralogie, Geologie und Paläontologie*, Abteilung B, 88(3):427-440.
Kuhn, E. 1952. Die Triasfauna der Tessiner Kalkalpen. XVII. *Askeptosaurus italicus* Nopcsa. *Schweizerische Paläontologische*, Abhandlungen 69:1-73.
Kuhn-Schnyder, E. 1980. Observations on temporal openings of reptilian skulls and the

classification of reptiles. IN L. L. Jacobs (Ed.), *Aspects of Vertebrate History*, pp. 153-175. Museum of Northern Arizona Press, Flagstaff.
Laurin, M. 1991. The osteology of a Lower Permian eosuchian from Texas and a review of diapsid phylogeny. *Zoological Journal of the Linnean Society* 101:59-95.
Lhwyd, E. 1699. Eduardi Luidii apud oxonienses cimeliarchae ashmoleani lithophylacii britannici ichnographia. London.
Massare, J. A. and J. M. Callaway. 1990. The affinities and ecology of Triassic ichthyosaurs. *Geological Society of America Bulletin* 102:409-416.
Massare, J. A. and J. M. Callaway. 1994. *Cymbospondylus* (Ichthyosauria: Shastasauridae) from the Lower Triassic Thaynes Formation of southeastern Idaho. *Journal of Vertebrate Paleontology* 14(1):139-141.
Mazin, J.-M. 1982. Affinités et phylogénie des Ichthyopterygia. IN E. Buffetaut, P. Janvier, J.-C. Rage, and P. Tassy (Eds.), Phylogénie et paléobiogéographie, pp. 85-98. *Geobios, Mémoire Spécial* 6.
Mazin, J.-M. 1983. L'implantation dentaire chez les Ichthyopterygia (Reptilia). *Neues Jahrbuch für Geologie und Paläontologie*, Monatshefte 7:406-418.
Mazin, J.-M. and P. M. Sander. 1993. Palaeobiogeography of the Early and Late Triassic Ichthyopterygia. IN Evolution, ecology, and biogeography of the Triassic reptiles, pp. 93-108. *Paleontologia Lombarda della Società Italiana di Scienze Naturali e del Museo Civico di Storia Naturale di Milano Nuova Serie*, volume II - 1993.
McGowan, C. 1973. The cranial morphology of the Lower Liassic latipinnate ichthyosaurs of England. *Bulletin of the British Museum (Natural History)*, Geology 24(1):1-109.
McGowan, C. 1989. *Leptopterygius tenuirostris* and other long-snouted ichthyosaurs from the English Lower Lias. *Palaeontology* 32:409-427.
McGowan, C. 1991. An ichthyosaur forefin from the Triassic of British Columbia exemplifying Jurassic features. *Canadian Journal of Earth Sciences* 28(10):1553-1560.
McGowan, C. 1994. A new species of *Shastasaurus* (Reptilia: Ichthyosauria) from the Triassic of British Columbia: the most complete exemplar of the genus. *Journal of Vertebrate Paleontology* 14:168-179.
McGowan, C. 1995. A remarkable small ichthyosaur from the Upper Triassic of British Columbia, representing a new genus and species. *Canadian Journal of Earth Sciences* 32:292-303.
McGowan, C. 1996. A new and typically Jurassic ichthyosaur from the Upper Triassic of British Columbia. *Canadian Journal of Earth Sciences* 33:24-32.
McGregor, J. H. 1906. The Phytosauria, with especial reference to *Mystriosuchus* and *Rhytidodon*. *American Museum of Natural History Memoir* 9(2):29-101.
Merriam, J. C. 1895. On some reptilian remains from the Triassic of Northern California. *American Journal of Science* 50:55-57.
Merriam, J. C. 1902. Triassic Ichthyopterygia from California and Nevada. *University of California Publications, Bulletin of the Department of Geology* 3:63-108.
Merriam, J. C. 1908. Triassic Ichthyosauria with special reference to the American forms. *University of California Memoir* 1(1).
Motani, R., H. You, and C. McGowan. 1996. Eel-like swimming in the earliest

ichthyosaurs. *Nature* 382:347-348.
Nicholls, E. L. and D. B. Brinkman. 1993. A new specimen of *Utatsusaurus* (Reptilia: Ichthyosauria) from the Lower Triassic Sulphur Mountain Formation of British Columbia. *Canadian Journal of Earth Sciences* 30:486-490.
Nicholls, E. L. and D. B. Brinkman. 1995. A new ichthyosaur from the Triassic Sulphur Mountain Formation of British Columbia. IN W. A. S. Sarjeant (Ed.), *Vertebrate Fossils and the Evolution of Scientific Concepts*, pp. 521-535. Gordon and Breach Publishers, Amsterdam.
Nicholls, E. L., D. B. Brinkman, and J. M. Callaway. In review. New material of *Phalarodon* (Reptilia: Ichthyosauria) from the Triassic of British Columbia and its bearing on the interrelationships of mixosaurs. *Palaeontographica*.
Osborn, H. F. 1903. The reptilian subclasses Diapsida and Synapsida and the early history of the Diaptosauria. *American Museum of Natural History Memoir* 1:449-507.
Owen, R. 1860. On the orders of fossil Reptilia and their distribution in time. *Report of the British Association for the Advancement of Science*, 29th meeting, Aberdeen, 1859, pp. 153-166.
Reisz, R. R. 1981. A diapsid reptile from the Pennsylvanian of Kansas. *Museum of Natural History, University of Kansas, Special Publications* 7:1-74.
Reisz, R. R., D. S. Berman, and D. Scott. 1984. The anatomy and relationships of the Lower Permian reptile *Araeoscelis*. *Journal of Vertebrate Paleontology* 4:57-67.
Rieppel, O. 1980. The phylogeny of the anguinomorph lizards. *Schweizerische Naturforschende Gesellschaft Denkschriften* 94:1-85.
Rieppel, O. 1993. Euryapsid relationships: a preliminary analysis. *Neues Jahrbuch für Geologie und Paläontologie*, Abhandlungen 188(2):242-264.
Romer, A. S. 1948. Ichthyosaur ancestors. *American Journal of Science* 246(2):109-121.
Romer, A. S. 1956. *Osteology of the Reptiles*. University of Chicago Press, Chicago.
Romer, A. S. 1966. *Vertebrate Paleontology*. University of Chicago Press, Chicago.
Romer, A. S. 1968. An ichthyosaur skull from the Cretaceous of Wyoming. *Contributions to Geology, University of Wyoming* 7(1):27-39.
Romer, A. S. 1971. Unorthodoxies in reptilian phylogeny. *Evolution* 25:103-112.
Sander, P. M. 1989. The large ichthyosaur *Cymbospondylus buchseri, sp. nov.*, from the Middle Triassic of Monte San Giorgio (Switzerland), with a survey of the genus in Europe. *Journal of Vertebrate Paleontology* 9:163-173.
Sander, P. M. 1992. *Cymbospondylus* (Shastasauridae: Ichthyosauria) from the Middle Triassic of Spitsbergen: filling a paleobiogeographic gap. *Journal of Paleontology* 66:332-337.
Sander, P. M. and J.-M. Mazin. 1993. The paleobiogeography of Middle Triassic ichthyosaurs: the five major faunas. IN Evolution, ecology and biogeography of the Triassic reptiles, pp. 145-152. *Paleontologia Lombarda della Società Italiana di Scienze Naturali e del Museo Civico di Storia Naturale di Milano Nuova Serie*, volume II - 1993.
Scheuchzer, J. J. 1708. *Piscium Querelae et Vindiciae*. Zürich.
Sues, H.-D. 1987. On the skull of *Placodus gigas* and the relationships of Placodontia.

Journal of Vertebrate Paleontology 7:138-144.

Tarsitano, S. 1982. A model for the origin of ichthyosaurs. *Neues Jahrbuch für Geologie und Paläontologie*, Abhandlungen 164:143-145.

Tarsitano, S. 1983. A case for the diapsid origin of ichthyosaurs. *Neues Jahrbuch für Geologie und Paläontologie*, Monatshefte 1983(1):59-64.

Wang, K. 1959. Über eine neue fossile Reptilform von Provinz Hupeh, China. *Acta Palaeontologica Sinica* 7(5):373-378.

Williston, S. W. 1914. *Water Reptiles of the Past and Present.* University of Chicago Press, Chicago.

Williston, S. W. 1917. The phylogeny and classification of reptiles. *Journal of Geology* 25:411-421.

Williston, S. W. 1925. *Osteology of the Reptiles.* Harvard University Press, Cambridge, Massachusetts.

Woodward, A. S. 1898. *Outlines of Vertebrate Palaeontology for Students of Zoology.* Cambridge.

Wu, X.-C. 1991. *The Comparative Anatomy and Systematics of Mesozoic Sphenodontidans.* Unpublished Ph. D. dissertation, McGill University.

Yang, Z.-J. and Z.-M. Dong. 1972. Aquatic reptiles from the Triassic of China. *Academia Sinica, Institute of Vertebrate Palaeontology and Palaeoanthropology Memoir* 9 (in Chinese).

Young, G. 1821. Account of a singular fossil skeleton, discovered at Whitby in February 1819. *Wernerian Natural History Society Memoir* 3:450-457.

Zittel, K. A. 1890. *Handbuch der Paläontologie.* I. Abteilung. Palaeozoologie. III. Band. Vertebrata (Pices, Amphibia, Reptilia, Aves). Munich and Leipzig.

Chapter 1

THE PALEOBIOGEOGRAPHY OF *SHASTASAURUS*

P. MARTIN SANDER

INTRODUCTION

The study of Triassic ichthyosaurs is experiencing an unprecedented renaissance that began in the early 1980s. The focus of research has shifted away from the familiar Jurassic forms to the not so well known and commonly more enigmatic representatives of the group that inhabited the Triassic seas. This shift is best illustrated by the topics of presentations on ichthyosaurs given at a recent symposium. Of the six talks and posters exclusively dealing with ichthyosaurs, three were wholly concerned with the Triassic forms, and none was restricted to Jurassic representatives of the group. These trends led to a reflection, catalyzed by two very recent finds of Triassic ichthyosaurs (Sander, et al., 1994a), on the state of the field and the possible directions in which it will progress.

The purpose of this chapter then, is twofold: first, to introduce one of the finds, a new species of *Shastasaurus*, into the literature, and second, to explore its implications for the paleobiogeography of the genus. These sections, however, must be preceded by a clarification of the use of vertebral morphology in Triassic ichthyosaur systematics.

Institutional Abbreviations

BSP = Bayerische Staatssammlung für Paläontologie und historische Geologie, Munich, Germany; GPIBO = Institut für Paläontologie, Universität Bonn, Germany; PIMUZ = Paläontologisches Institut und Museum, Universität Zürich, Switzerland;

PMU = Paleontologiska Museet, Uppsala University, Sweden; UCMP = Museum of Paleontology, University of California at Berkeley.

Rib Articular Facets and Systematics

The position and morphology of rib articular facets have long played an important role in the systematics of Triassic ichthyosaurs. Recently, however, McGowan (1994) has seriously questioned the utility of this character complex. Past and present work leads me to disagree and to make a strong case for rib articulations as overridingly useful characters at higher systematic levels. This case would best be made in the framework of a cladistic analysis of ichthyosaurian relationships. At present, a few key observations will have to suffice.

The rib articulation pattern in the Jurassic *Ophthalmosaurus* and *Ichthyosaurus* (Ichthyosauridae) and *Stenopterygius* (Stenopterygiidae) is very different from that of the Triassic Shastasauridae, but very uniform: anteriorly on the column the round diapophysis connects to the neural arch, then moves down the side of the centrum to fuse with the equally round parapophysis in the sacral region. From anterior to posterior, the parapophysis also moves down the side of the centrum, but not as fast, leading to the fusion of the two facets low on the sacral vertebrae. The caudal vertebrae have only a single articular facet. This pattern was described in detail by Andrews (1910) for *Ophthalmosaurus*, by McGowan (1994) and Owen (1881) for *Ichthyosaurus*, and by Fraas (1891) for *Stenopterygius* and is noted, with variations, for all Liassic genera by Huene (1922). The pattern was easily verified on a mounted cast of *Ophthalmosaurus icenicus* and several specimens of *Stenopterygius* in the GPIBO collections. This pattern of **fusion** of parapophysis and diapophysis in the sacral or posterior dorsal region is the more generalized condition and is found in primitive terrestrial reptiles (Romer, 1956:277).

In Shastasauridae, on the other hand, a fundamentally different pattern involving the **loss** of the parapophysis is encountered. In the cervical vertebrae, a parapophysis is present as in the Jurassic forms, and the elongate diapophysis connects to the neural arch. Going back along the cervical series, the parapophysis gradually diminishes in size and finally is lost around the cervical-dorsal transition. From there posteriorly, only the continually elongating diapophysis is retained, which, in the anterior or middle dorsals, separates from the neural arch and gradually moves down the side of the centrum. In addition, the diapophysis decreases markedly in height, so that at least in the middle caudal region, the diapophysis is more or less round as in other ichthyosaurs. Middle and posterior caudal vertebrae of shastasaurids thus cannot be distinguished from those of other ichthyosaurs by rib articulation patterns.

The pattern of loss of the parapophysis in the cervical region in combination with an elongate diapophysis is clearly autapomorphic for the Shastasauridae and has been documented for *Cymbospondylus* (Sander, 1989:168; Merriam, 1908:47-49), *Shastasaurus* (Merriam, 1908:47-49), and *Californosaurus* (Merriam, 1908:47-49, 132, 133).

Only in *Shonisaurus* (Camp, 1980:169-170) does the situation appear to be somewhat more complex: there is a loss of the parapophysis at the cervical-dorsal transition. However, in a few vertebrae behind the region of loss, the remaining diapophysis is anteroposteriorly constricted and in some instances splits into two separate facets of about equal size. In the mid-dorsal region, these facets fuse again, forming the familiar elongate, slanted diapophysis.

In the remaining Triassic ichthyosaurs, customarily assigned to the Mixosauridae and Omphalosauridae, the pattern of rib articulation is incompletely known, double-headed as well as single-headed ribs being reported. One important difference between the Shastasauridae and these forms is that they lack an elongation of the diapophysis (with the possible exception of *Omphalosaurus*; see below). It is particularly regrettable that there is no detailed published account of rib articulation patterns in *Mixosaurus* despite the wealth of material in the collection of the PIMUZ. In the literature, there are conflicting reports about the condition in *Mixosaurus*: Wiman (1910:131), among others, described the anterior dorsal rib articulations as double and the posterior dorsal ones as single, their shape being round to oval throughout. Huene (1916:10-11), on the other hand, claims that the reverse pattern is present, i.e., double-headed posterior dorsal ribs and single-headed anterior ones. This pattern is confirmed by Callaway (Chapter 2).

After having dealt with the significance of rib articular facets at the family and higher levels, it is necessary to consider their utility for distinguishing different genera of shastasaurids. In particular, can *Cymbospondylus* dorsal vertebrae be distinguished from those of other shastasaurids (*Shastasaurus, Californosaurus, Merriamia, Shonisaurus*), as practiced, among others, by Massare and Callaway (1994), Sander (1989, 1992), and Merriam (1908)? This distinction was made on the basis that the elongate diapophyses of the dorsal vertebrae of *Cymbospondylus* slant forward and make contact with --- or even are truncated by --- the anterior margins of the centra.

McGowan (1994) takes a contrary viewpoint, listing for *Shastasaurus* and *Cymbospondylus* exceptions to this rule figured by Merriam (1908) in which the diapophysis does not reach the anterior margin of the centrum. However, one of McGowan's examples involves cervical vertebrae of *Cymbospondylus* and not dorsals. Another example involves an abnormally developed diapophysis of a *Shastasaurus* vertebra, and the last one is a very posterior dorsal of *Cymbospondylus* in which the anterior truncation may lose its prominence. Reexamination of the type material of *Shastasaurus* and *Californosaurus* in the UCMP collections confirmed that Merriam's (1908) descriptions are remarkably accurate. *Shonisaurus* has posterior dorsals (as described by Camp, 1980:169-170) with a similar arrangement of rib articular facets as in *Cymbospondylus*, but the two genera are easily separated by the much greater foreshortening of the vertebrae in the former (and by a considerable amount of time). In conclusion, the traditionally employed distinction between *Cymbospondylus* and other Shastasauridae holds up. Strings of vertebrae and even isolated dorsal vertebrae of the general shastasaurid type in which the diapophysis makes contact with or is truncated by the anterior margin can be assigned to *Cymbospondylus* with reasonable confidence.

SHASTASAURUS NEUBIGI N. SP. FROM THE GERMAN MUSCHELKALK

The Muschelkalk beds of central Europe have yielded a diverse marine reptile fauna, in particular nothosaurs. However, Muschelkalk vertebrate fossils are generally found as isolated elements, and associated or articulated material is the great exception. This is particularly true for the uncommon finds of ichthyosaurs, despite considerable ichthyosaurian diversity in the fauna (Sander and Mazin, 1993). It is against this background that the partially articulated skeleton of a small shastasaurid is of some interest, as it represents by far the most complete ichthyosaur specimen from the entire Muschelkalk recovered to date.

The specimen was exposed in June 1985 during railroad construction near Karlstadt/Main in Franconia (Figure 1; German coordinates: TK 25 Karlstadt, sheet no. 6024, r. 53500/h. 31700). A significant portion of the presumably once complete skeleton was destroyed before its discovery, but much of the skull, two sections of the vertebral column, and some pelvic girdle and hindlimb elements remain. The specimen is now distributed over eight small to medium-sized slabs which do not make contact with each other. Preparation was done by a combination of acid and mechanical techniques. The orientation of the slabs in the rock was not recorded. The specimen is now housed in the BSP, bearing the accession number 1992 I 39.

Geology and Stratigraphy

The Muschelkalk beds are well exposed along the Main River and its tributaries in Franconia. The new specimen of *Shastasaurus* came from the carbonates of the middle *pulcher/robustus* zone. This zone is situated in the lower upper

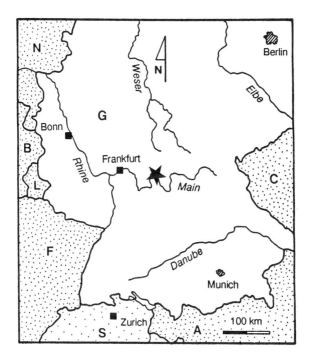

Figure 1. Locality (star) of *Shastasaurus neubigi* n. sp. from the lower upper Muschelkalk (Anisian) of Karlstadt/Main, Germany.

Muschelkalk (mo_1) and is of late Anisian age (Hagdorn, 1991).

The epicontinental Muschelkalk sea was shallow and warm and, particularly in Anisian times, hypersaline (Hagdorn, 1991). Thus, primarily carbonates and some evaporites form the rock record. At least indirect connections existed with the Tethys via the Burgundy Gate in the southwest and the Silesian-Moravian and East Carpathian gates in the east. The Muschelkalk sea had no outlet to the north, however. Even in the basin center, land was never farther away than a few hundred kilometers.

This paleogeographic situation is in accordance with the vertebrate faunal record, which is very poor in pelagic elements such as large ichthyosaurs. Ichthyosaurs in general are very rare, and the smaller, presumably coast-dwelling *Mixosaurus* (Sander et al., 1994b) is distinctly more common than the larger forms such as *Shastasaurus* and *Cymbospondylus* (Huene, 1916). This occurrence of a complete skeleton is best explained by the individual's having accidentally entered the basin or by its being washed in after death. The seeming completeness would argue against this latter possibility.

22   P. Martin Sander

SYSTEMATIC PALEONTOLOGY

Class REPTILIA Linnaeus, 1758
Subclass DIAPSIDA Osborn, 1903
Order ICHTHYOSAURIA Blainville, 1835
Family SHASTASAURIDAE Merriam, 1902

Emended diagnosis: Medium-sized to large ichthyosaurs with isodont dentition. Differ from all other ichthyosaurs in their single elongate diapophysis in the anterior and middle dorsal vertebrae for the articulation of the vertically expanded single rib head. Differ from Mixosauridae in their digital reduction to less than four complete digits.

Genus *SHASTASAURUS* Merriam, 1895

Emended diagnosis: Medium-sized to large ichthyosaurs with isodont dentition. *Shastasaurus* differs from all other shastasaurids except *Merriamia* in having a small notch in the leading edge of the humerus. It differs from *Merriamia* in the much foreshortened humerus, radius, and ulna and also in that the hindlimbs are of roughly equal size as the forelimbs. *Shastasaurus* differs from other shastasaurids except *Shonisaurus* in that the radius is much larger than the ulna. It also differs from *Shonisaurus* in the relatively longer vertebrae. *Shastasaurus* differs from *Cymbospondylus* in the convex occipital condyle and in the lack of diapophysial contact with the anterior margin of the dorsal vertebral centra. Differences with *Californosaurus* appear least significant and consist of shape differences of scapula and ilium.

SHASTASAURUS NEUBIGI n. sp.

Holotype and only known specimen: BSP 1992 I 39, a partial skeleton preserved on eight isolated rock slabs, some isolated phalanges.
Locality: Karlstadt/Main, Franconia, Germany (German coordinates: TK 25 Karlstadt, sheet no. 6024, r. 53500/h. 31700).
Horizon: The specimen was collected from the *pulcher/robustus* zone of the lower upper Muschelkalk (mo_1) and thus is of late Anisian age.
Diagnosis: Medium-sized (about 3-m-long) species of *Shastasaurus*. The shape of the pubis is more rounded and flattened than in other species of the genus. The new species differs from the other species of the genus in its well-developed

zygapophyses in the cervical and dorsal vertebrae, which are nearly horizontal in lateral view. Unlike the condition in some other species, the teeth are set in distinct alveoli. At the level of the genus, *Shastasaurus neubigi* n. sp. exhibits the following autapomorphies: a distinct dorsal process just anterior to the articular facet of the lower jaw, a slender retroarticular process, and a regular alternation of teeth and empty alveoli in the jaws.

Etymology: Named in honor of the discoverer of the holotype, Mr. Bernd Neubig of Euerdorf, Franconia.

DESCRIPTION

Skull

Much of the skull is preserved, albeit in a partially disarticulated and crushed state. The postorbital region is exposed in ventral view on one slab (slab 1, Figure 2); another (slab 2, Figure 3) contains the posterior regions of the lower jaws. Tooth-bearing bones of the rostrum can be found on these two slabs as well as on slab 3 (Figure 4) and slab 6.

The posterior skull region (Figure 2) is truncated anteriorly by an oblique fracture. Of the palate, the basioccipital, the basisphenoid, and the posterior parts of the pterygoids are preserved. Anterior to these bones, the skull roof is exposed from below. The basioccipital bears a distinctly set-off occipital condyle that is strongly convex with a small circular depression in the center. Extending posterolaterally from the basioccipital is a bony process that appears to be fused to the braincase. This tapering bone most likely represents the stapes, which, in ichthyosaurs, buttress the quadrate against the basioccipital and opisthotic. Ichthyosaurian stapes were discussed or illustrated primarily by Camp (1980), McGowan (1973), Romer (1968), Sollas (1916), Andrews (1910), and Merriam (1908). Looking at the occipital condyle from behind, the two exoccipitals can be seen in articulation with the basioccipital. The basisphenoid is situated anterior to this bone. No suture is discernible but a probable internal carotid foramen (Figure 2) is present. The right pterygoid is still connected with the basisphenoid while the left has separated, exposing the basicranial articulation.

Widely separated from the basicranium, the left quadrate, including the articular surface, is seen in ventral view (Figure 2). This wide separation is due to the strong dorsoventral crushing which splayed the bones of the skull roof apart. The quadrate is slender and tapers toward the expanded articular surface. On the right side of the skull, the posteroventral corner of the skull is missing, but the upper temporal opening is visible in ventral view (Figure 2). It is rather long and wide. The bones bordering the upper temporal opening cannot be identified with certainty, as there are no distinct suture lines. The interorbital part of the skull roof shows ventrally

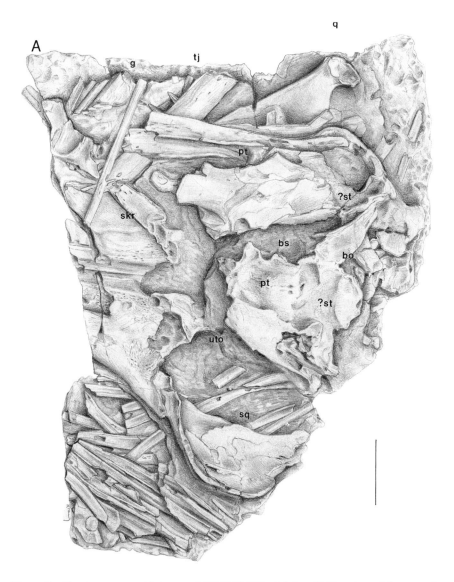

Figure 2. *Shastasaurus neubigi* n. sp., BSP 1992 I 39, slab 1. **A)** Drawing of the posterior skull region in ventral view. Shading indicates skull bones still covered by matrix. Abbreviations: **bo** = basioccipital, **bs** = basisphenoid, **g** = gastral rib, **icf** = internal carotid foramen, **pt** = pterygoid, **q** = quadrate, **skr** = skull roof, **sq** = squamosal, **? st** = probable stapes, **tj** = indeterminate tooth-bearing jaw bone, **uto** = right upper temporal opening. Scale bar = 3 cm.

two longitudinal grooves separated by a distinct keel which marks the skull midline. The pineal foramen is not exposed, but set off from the right groove is a deep foramen which has no counterpart on the left.

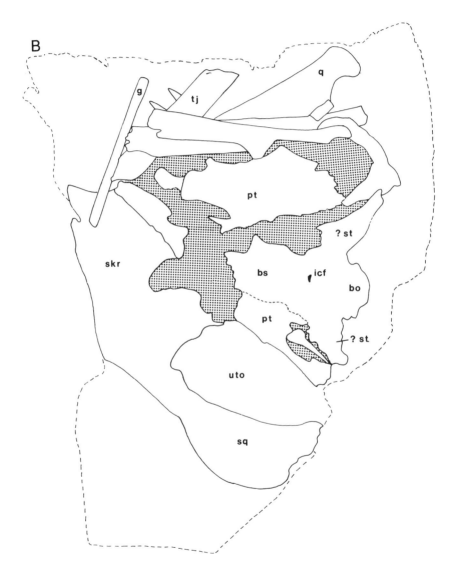

Figure 2 (continued). *Shastasaurus neubigi* n. sp., BSP 1992 I 39, slab 1. **B)** explanatory sketch of the posterior skull region in ventral view. Shading indicates skull bones still covered by matrix. Abbreviations as in legend to Figure 2A.

Isolated tooth-bearing bones of the rostrum are found on several slabs (Figures 2 through 4) but it is difficult to distinguish between premaxillaries and dentaries. This is because the bones are very low and slender and are almost mirror images of each other, as is generally the case in piscivorous ichthyosaurs with elongated rostra.

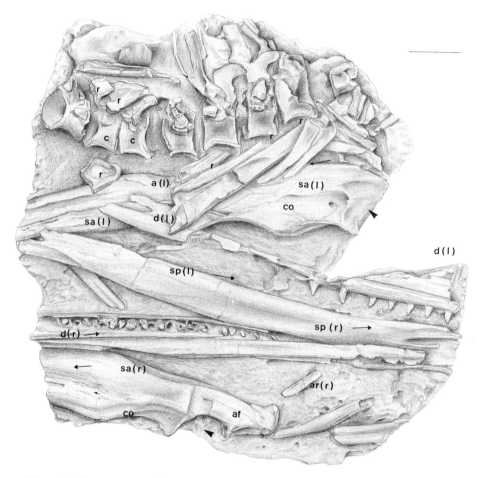

Figure 3. *Shastasaurus neubigi* n. sp., BSP 1992 I 39, slab 2. The disarticulated lower jaws and eight cervical vertebrae. Left surangular and angular seen in medial view, the right ones seen in lateral view. Note distinctive dorsal process (arrowheads) of the surangular anterior to the articular facet and the double rib articular facet in the posterior cervical vertebrae. Slender arrows point toward anterior. Anterior is unknown in the left dentary. The small arrow indicates the foramen on the outside of the surangular. Abbreviations: **af** = articular facet of right lower jaw, **a(l)** = left angular, **ar(r)** = right articular, **c** = vertebral centrum, **co** = coronoid process, **d(l)** = left dentary, **d(r)** = right dentary, **r** = rib, **sa(l)** = left surangular, **sa(r)** = right surangular, **sp(l)** = left splenial, **sp(r)** = right splenial. Scale bar = 3 cm.

Three additional flattened bones of uncertain affinities are found on slab 3 (Figure 4), and one is found on slab 6. These bones are best interpreted as elements of the skull roof, as they are clearly neither girdle nor limb elements.

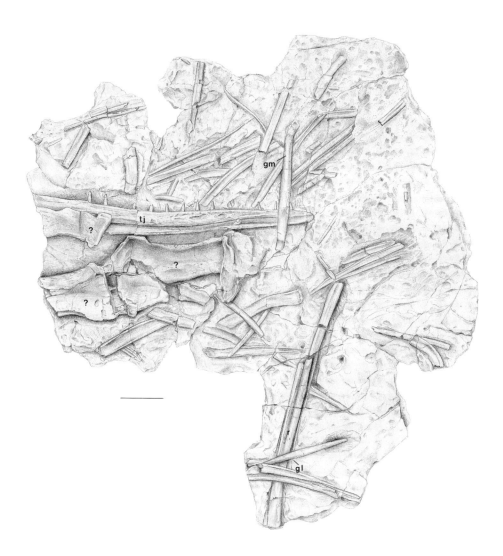

Figure 4. *Shastasaurus neubigi* n. sp., BSP 1992 I 39, slab 3. Large slab with jaw and skull bones, many gastralia, and a rib. The tooth-bearing jaw bone is either the anterior part of the left dentary or, more likely, the right premaxillary. Abbreviations: **gl** = lateral gastral rib, **gm** = medial gastral rib, **r** = rib, **tj** = indeterminate tooth-bearing jaw bone, **?** = indeterminate skull bone. Scale bar = 3 cm.

Lower Jaw

The posterior edentulous portions of both lower jaws are well exposed on slab 2 (Figure 3). The right lower jaw is seen in lateral view and preserved over a length of 15 cm from the posterior end of the retroarticular process forward. The preserved portion of the left lower jaw is 20 cm long and consists of the angular and surangular seen in medial view. There is a distinct but slender, 3-cm-long retroarticular process. At the articular facets, the lower jaw still is low but increases in height significantly below the anterior margin of the articular facet. A characteristic feature of the lower jaw is the high, laterally flattened dorsal process of the surangular (Figure 3), which is not to be confused with the long and low coronoid process of the same bone 4 cm more anteriorly. The coronoid process is also situated more laterally than the flattened process in front of the articular facet. Combined information from the two posterior jaw portions and the dentary indicates that the jaw was rather low and slender, reaching its greatest dorsoventral height between the coronoid process and the articular facet. Just in front of the right coronoid process, about one third of the way down the side of the jaw, a foramen is seen to emerge which is continued forward as a surficial groove (Figure 3). The same foramen can be observed on the inside of the left surangular. The surficial groove can be seen to continue onto the tooth-bearing jaw bones.

In lateral view, the right articular forms the tip of the retroarticular process but does not extend much onto the lateral face of the lower jaw. The inner surface and ventral margin of the lower jaw are formed to a significant extent by the splenials, both of which are preserved for most of their length on slab 2 (Figure 3). Their anterior ends point in the opposite direction from the posterior parts of the lower jaws. As seen in *Ophthalmosaurus* (Andrews, 1910, and a cast in the GPIBO collection) and *Ichthyosaurus* (Sollas, 1916), the anterior end of the splenial becomes two-pronged shortly after it enters into the symphysis. As in the two other forms, the lower prong is the longer one. The noteworthy feature of the splenial of *Shastasaurus neubigi* n. sp., however, is the concave medial surface of that bone in its central region (Figure 3). In other ichthyosaurs this region is flat.

Skull Length Estimate

Based on comparisons with more complete long-snouted ichthyosaur skulls, splenial length can provide a rough estimate of skull length (Table 1). The length of the nearly complete left splenial is 265 mm. Only a few centimeters of the ventral anterior process are missing. Overall length of the splenial, which is useful for estimating lower jaw length, must have been around 300 mm (Table 1). If

Table 1. Comparative data for skull length estimates of *Shastasaurus neubigi* n. sp. Lengths in mm.

Species	Splenial	Lower Jaw	Skull	Skull/Spl	Spl/Skull
Ophthalmosaurus icenicus	435	1090	990	2.27	44%
Cymbospondylus petrinus	992	1190	1105	1.11	90%
Ichthyosaurus communis	380		520	1.37	73%
Shastasaurus neubigi n. sp.	300	-	-	-	-
Skull length estimates based on		*Ophthalmosaurus*		681 mm	
		Cymbospondylus		333 mm	
		Ichthyosaurus		411 mm	

Data from Andrews (1910), Merriam (1908), and Sollas (1916).

Shastasaurus neubigi n. sp. was proportioned similarly as the skull of *Ichthyosaurus* studied by Sollas (1916), splenial length would be three fourths of the skull length and the skull of the new specimen would have been 411 mm long (Table 1). In the cast of *Ophthalmosaurus icenicus* in the GPIBO collection, the splenial equals 44% of the skull length and 40% of the lower jaw length. Using this specimen for an estimate, the skull of the new specimen would have been 681 mm long (Table 1). Using specimen UCMP 9950 of *Cymbospondylus petrinus* (data from Merriam, 1908) for comparison, a skull length of only 333 mm would result. However, the splenial in UCMP 9950 is relatively much longer (90% of lower jaw length) than in the other forms. The shape of the splenial of the new species conforms much closer to *Ophthalmosaurus* than to *Cymbospondylus*. Based on splenial length, the most realistic estimate for skull length would then be around 500-600 mm.

Dentition

Of the dentaries, the right one shows the posterior tooth-bearing region in dorsal view while the left one exposes the middle region in lateral view (Figure 3). As can be seen in the right dentary, the tooth row begins at the medial margin of the bone but gradually assumes a more lateral position as one moves forward. In the posterior part of the tooth row (the posteriormost 8 cm), the alveoli are closely spaced and most of them bear teeth in various stages of development, giving the tooth row in this region an irregular appearance (Figure 3). In front, the tooth row consists of remarkably regularly spaced teeth that alternate with empty alveoli (Figures 3 and 4). The distance between teeth (which is equal to twice the alveolar distance) is very constant at 10 or 11 mm, although in two cases it is 15 and 17 mm, respectively. Tooth height is also rather constant, varying between 6 and 7 mm at a diameter of 3 mm. The regular spacing is in marked contrast to other medium-sized isodont ichthyosaurs but reminiscent of marine crocodiles and sauropterygians in which the lower and upper dentitions interlock precisely. This interlocking cannot be proven for the new specimen because of the disarticulation of the rostrum.

The teeth of *Shastasaurus neubigi* n. sp. are of the acute conical type (Figures 3 and 4; tooth crown type 3 of Massare, 1987) typical for piscivorous ichthyosaurs. The root shows the characteristic infolding of the dentine. The infolding pattern is continued as longitudinal ridges onto the base of the crown. The apical half of the crown is covered by low discontinuous ridges.

Skull Shape

Despite the crushing, some significant information can be extracted about the shape of the skull. The postorbital region is rather long in comparison with other post-Triassic ichthyosaurs and large-eyed Triassic forms such as *Mixosaurus*. The gracile nature of the suspensorium (Figure 2) is in contrast with that of other shastasaurids such as *Cymbospondylus* and *Shonisaurus*. As preserved, the minimum width of the palate, as indicated by the pterygoid, is much smaller than the preserved distances of the two articular surfaces of the quadrate. This suggests that the skull was rather high, as in other ichthyosaurs, and not dorsoventrally flattened.

The relatively gracile lower jaw with its slender retroarticular process (Figures 3 and 4) is in accordance with the gracile suspensorium and the elongate but thin tooth-bearing jaw bones. The overall appearance then of the skull of the new species would have been that of a lightly built, if not gracile, ichthyosaur skull.

Vertebrae and Ribs

Two strings of vertebrae in articulation are available for study. While the seven vertebrae and the scanty remains of an eighth on slab 2 (Figure 3) presumably represent much of the cervical series, the four vertebrae on slab 4 are dorsals (Figure 5). The cervical vertebrae are seen in right lateral view and the dorsal ones in left lateral view. Compared to the disarticulated state of the skull elements, the vertebral column is remarkably well articulated, with the neural arches in connection

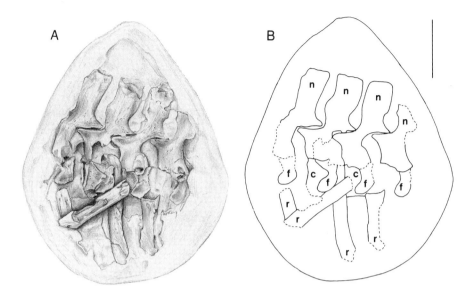

Figure 5. *Shastasaurus neubigi* n. sp., BSP 1992 I 39, slab 4. Four anterior dorsal vertebrae. Abbreviations: **c** = vertebral centrum, **f** = rib articular facet, **n** = neural spine, **r** = rib. Scale bar = 3 cm.

with the centra and the rib heads adjacent to the diapophyses.

The cervical vertebral centra are relatively long and low (average length 17 mm, average height 27 mm) with a pronounced ventral keel which was enhanced by crushing of the specimen. The vertebrae are deeply amphicoelous, with the articular faces evenly sloping toward the center. The vertebrae are notochordal or at least nearly so. As is generally the case in ichthyosaurian cervicals, there is a double rib articulation. The relatively small and oval parapophysis is located halfway up the centrum. In the more anterior vertebrae, the parapophysis is located toward the anterior margin, but further back it moves into the geometric center of the lateral face. The position and shape of the diapophysis are difficult to discern except in the posteriormost vertebra. There it is rather high and narrow, its dorsal part making contact with the neural arch.

The neural arches of the cervical vertebrae are rather large with very well-developed and tightly interlocking zygapophyses (Figure 3). The zygapophyses are nearly horizontal in lateral aspect. In contrast to the neural arches, the laterally flattened neural spines are thin and short, also in contrast to other Triassic ichthyosaurs (Merriam, 1908:35). The neural spines occupy a very posterior position on the arch, rising above the postzygapophysis. This way the neural spines are largely situated above the intervertebral articulations. The cervical ribs are incompletely preserved but the second to posteriormost clearly is double-headed

(Figure 3).

The dorsal vertebrae are similar in shape and proportion to the cervical ones. The length of the dorsal centra is also about 17 mm, while their height is 29-30 mm as measured on an x-ray of slab 4. The most significant feature of the dorsal vertebrae is the nature of the rib articulation. There is only a single elongate articular facet which extends from the base of the neural arch to about the middle of the lateral face of the centrum (Figure 5). Although the facet slants forward, its anteroventral margin is separated by a wide gap from the anterior face of the centrum. The high position of the articular facet, including a small part of the neural arch, suggests that the four vertebra on slab 4 belong to the anterior dorsal region. In shastasaurids in general, the single articular facet separates from the neural arch before the middle dorsal region (Sander, 1989, 1992; Camp, 1980; Merriam, 1908).

As in the cervicals, the nearly horizontal (4°-7°) and well-developed zygapophyses are a distinctive feature (Figure 5). Such zygapophyses differ from those of other Triassic ichthyosaurs, which have anteriorly slanted and often weakly developed zygapophyses in the dorsal vertebrae. The dorsal neutral spines are laterally flattened and may nearly contact each other along the midline (Figure 5). Their anteroposterior width is up to 18 mm. The cross section of the neural spines resembles the shape of a flattened oval. The height of the neural arch, including the spine, is 47 mm in the three cases in which it could be measured. The spines themselves rise 31 mm above the plane of the zygapophyses. The preserved proximal parts of the ribs are single-headed and appear flattened with a blade-like posterodorsal margin (Figure 5). Further out, the ribs have a figure-eight cross section, which then changes to a circular cross section. This change can be observed in dorsal rib fragments on slab 7. Rib diameter is 6-8 mm.

Gastralia

As in all ichthyosaurs, a well-developed gastral apparatus is present in *Shastasaurus neubigi* n. sp., albeit in complete disarticulation (Figures 2 and 4). It consists of at least three elements per row with V-shaped median elements and flattened or rod-like lateral elements. One median element is completely preserved on slab 6 (Figure 4). This element is 175 mm wide and encloses an angle of 155°. For other elements, this angle is 145°. The point of the "V" of the median element lacks an anterior process as seen in some other ichthyosaurs but has a knob that is either dorsally or ventrally oriented. The "wings" of the median element are flattened with a shallow groove on the same side as the median knob. The exact nature of articulation between the median and the lateral elements is unclear.

Appendicular Skeleton

The appendicular skeleton is represented by elements of the pelvic girdle and hindlimb only, most of which are found on slab 5 (Figures 6 and 7). Of the pelvic bones, a pubis and an ?ilium are present, whereas of the hindlimb bones only the tibia can be identified with some confidence. In all these cases, an assignment to the left or the right half of the body is not possible because of the flattened morphology of the bones and their complete disarticulation.

The pubis (Figures 6 and 7) is a flat, somewhat semicircular bone with a convex anteroventral margin with a well-defined obturator notch. The dorsal margin of the bone is slightly concave (Figures 6 and 7). The greatest (anteroposterior) length of the bone is 42 mm. The pubis is similar in shape, and in the location of the obturator notch, to that of *Shastasaurus osmonti* (Merriam, 1908:Plate 16) and to a bone from the lower Muschelkalk of Gogolin figured by Huene (1916:Figure 51). The lower Muschelkalk bone was identified as Shastasauridae *incertae sedis* by Callaway and Massare (1989).

The pubis of *S. pacificus* is also somewhat similar, except that the acetabular area is greatly thickened and expanded and that the obturator notch is approaching closure. The pubis of *Californosaurus*, on the other hand, has a wide-open obturator bay, and that of *Cymbospondylus* has a completely closed obturator foramen.

A squarish flattened bone (Figure 6) near the pubis very probably represents the ilium. This bone has a thickened acetabular area and anterior margin. Its height is 29 mm. The posterior margin bears a peculiar "spur" (Figure 6) somewhat reminiscent of the posterior tubercle of *Californosaurus* as described by Merriam (1908). However, the ilium in this genus is a stout structure without lateral flattening. In view of the not unequivocal identification of the ilium of *Shastasaurus neubigi* n. sp., its spur will not be used for diagnosis of the species nor phylogenetic considerations.

Right next to the pubis on slab 5 lies a flattened subround bone (Figure 7) that is incompletely preserved. Its most conspicuous feature is a deep notch on one margin. Such a notch is either the obturator notch of a pubis or to be found in elements forming the anterior or posterior margins of shastasaurid paddles. Despite the notch and the suggestive location on the slab (Figure 7), the bone is not a pubis, as it is much smaller and more delicate than the adjacent pubis. The flattened nature of the bone in question excludes it from being a humerus. On the other hand, its size (30 mm long) excludes it from being one of the foot bones. The tibia remains as the most likely candidate, as it is more transversely expanded in shastasaurids than the fibula. However, notching is rather unusual for an ichthyosaur tibia (Callaway, personal communication).

Several flattened circular to subcircular bones (Figure 7) varying in diameter from 6 to 20 mm were found scattered over the surface of slabs 5 and 1. They show a distinct radiating pattern of bony ridges which is also seen in the pubis and the tibia. The circular elements presumably represents bones of the hind paddle and, in the case of the two specimens on slab 1, the front paddle. From what is known (McGowan, 1994; Camp, 1980; Merriam, 1908), circular paddle bones with or without anterior or posterior notches are characteristic of many shastasaurids (*Cymbospondylus*, *Shastasaurus*, *Californosaurus*, and *Shonisaurus*, but not *Shastasaurus neoscapularis*, *Toretocnemus*, and *Merriamia*).

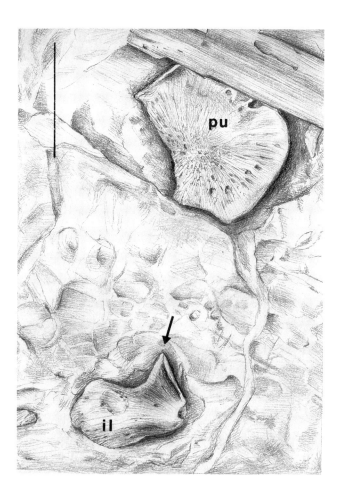

Figure 6. *Shastasaurus neubigi* n. sp., BSP 1992 I 39, front of slab 5. Pubis and probable ilium. Note the posterior "spur" on the ilium (arrow). Abbreviations: **il** = ilium, **pu** = pubis. Scale bar = 3 cm.

DISCUSSION

Assignment to Genus

With the above description in mind, it is now necessary to support the generic assignment of the new species to the Shastasauridae and to *Shastasaurus*. The dorsal vertebrae of the new species are typically shastasaurid in their single, elongate, and anteriorly slanted diapophyses (Figure 5). Such diapophyses in the dorsal vertebrae are an autapomorphy of the Shastasauridae and exclude the Mixosauridae as well as the Californian Late Triassic genus *Toretocnemus* from consideration. In fact, because of this, *Toretocnemus* is best considered family *incertae sedis* at present (Callaway, personal communication).

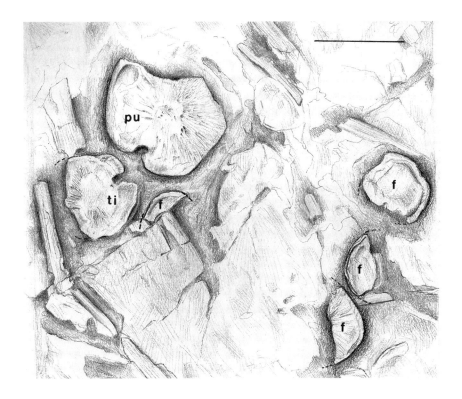

Figure 7. *Shastasaurus neubigi* n. sp., BSP 1992 I 39, back of slab 5. Pubis, tibia, and foot bones. Note the well-defined obturator notch in the pubis. All bones but the pubis are incompletely preserved. Their probable continuation is marked in ink. Abbreviations: **f** = indeterminate foot bones, **pu** = pubis, **ti** = tibia. Scale bar = 3 cm.

Among the Shastasauridae, a number of genera can also easily be rejected. Reference to *Cymbospondylus* (Sander, 1989; Merriam, 1908) is contradicted by the convex occipital condyle (Figure 2), the slender lower jaw, the posterior position of the diapophysis of the dorsal vertebrae, and the presence of an obturator notch in the pubis (Figures 6 and 7) of the new species. *Shonisaurus* from the Upper Triassic of Nevada (Camp, 1980) can also be excluded because it has a much deeper jaw and relatively much shorter and higher vertebrae, apart from its much greater body size.

The Late Triassic *Merriamia* from California (Merriam, 1908) is characterized by teeth set in grooves and by much smaller hindlimbs than forelimbs. The teeth of the new species have well-defined alveoli (Figure 3), and the size of the pubis suggests that the hindlimbs were not particularly small. Arguing against affinities with *Californosaurus* (Merriam, 1908), also from the Upper Triassic of California, are a number of vertebral features. The neural spines of this genus have distinct lateral ridges and anteroposteriorly inclined zygapophyses as opposed to the ridge-less neural spines and horizontal zygapophyses of the new species (Figure 5). As seen in Merriam's (1908) illustrations and supported by examination of the type material, the diapophyses of the dorsal vertebrae of *Californosaurus* are similar in shape to those of the new species but are located more anteriorly on the lateral face of the centrum, almost touching its anterior margin. The pubis of the new species is not very similar in shape to that of *Californosaurus*.

There is only one difference between the new species and the known species of *Shastasaurus*: the relatively long dorsal vertebrae (height/length ratio about 1.6; Figures 3 and 5). Those of *Shastasaurus* generally have a height/length ratio of over 2. However, there is great variability in this ratio which appears to be size-related, with the small species, such as *S. neoscapularis*, having relatively long vertebrae and the large species, such as *S. careyi*, having relatively very short vertebrae. Pending a formal revision of the Shastasauridae, it seems best at present to assign the new species to the genus *Shastasaurus* because of the few differences between the new species and other species of the genus. In addition, the pubis of the new species is most similar to that of the specimens called *S. osmonti* by Merriam (1908).

Upon further study of the type material and new discovery, an assignment to *Californosaurus* might become necessary because the differences between the new species and *Californosaurus* are not very great and may be significant only at the species level. On the other hand, *Californosaurus* may yet prove to be a junior synonym of *Shastasaurus*.

Comparisons

After assigning the new species to *Shastasaurus*, it must be compared to a number of other species and specimens of *Shastasaurus* and shastasaurids.

Other Muschelkalk Material---Most notable among shastasaurid fossils from the Muschelkalk is the isolated left lower jaw from the uppermost Muschelkalk of Bayreuth described by Huene (1916). The jaw is incomplete, lacking the anterior and posterior tips, but the preserved length of 78 cm, the slender shape, and the pointed conical teeth suggest that it came from a large shastasaurid. In a figure caption, the specimen was assigned to ?*Cymbospondylus* by Huene (1916:32), but in the text he noted that it was generically indeterminate. An affinity with *Cymbospondylus* is unlikely because this animal has rather deep and somewhat upwardly curved jaws (Merriam, 1908; Sander, 1989). The jaw of the new species of *Shastasaurus* shows the same slender form but differs in its distinct dorsal process in front of the articular surface (Figure 3) and the regular spacing of the teeth. The Bayreuth jaws thus belonged to a different species than *Shastasaurus neubigi* n. sp. Affinities with the genus *Shastasaurus* are not unlikely because it has been previously recorded from the Muschelkalk (this report; Callaway and Massare, 1989; Huene, 1916).

Despite records of isolated bones, mainly vertebral centra, of *Shastasaurus* from the Muschelkalk (Callaway and Massare, 1989; Huene, 1916), a comparison with the new species is difficult. The isolated bones either were not illustrated (those recorded by Callaway and Massare, 1989) or are of doubtful affinities as well as from different regions of the skeleton (those illustrated by Huene, 1916). In addition, ichthyosaurs are even rarer in the upper Muschelkalk than in the lower Muschelkalk, and none have been recorded from the *pulcher/robustus* zone of the lower upper Muschelkalk before (Huene, 1916).

"*Shastasaurus carinthiacus*"---The only Upper Triassic European shastasaurid material, described as *Shastasaurus carinthiacus* by Huene (1925) is worth comparing to the new species because of similarities in vertebral morphology. *Shastasaurus carinthiacus* is based on a string of four dorsal vertebrae and three isolated caudals from the Carnian of the Carinthia region (Austrian Alps). As already noted by McGowan (1994), the material is insufficient for specific identification but an assignment to *Shastasaurus* sp. appears justified by the nature of the rib articulation. Based on Huene's (1925) description, the Carinthian *Shastasaurus* is, in one respect, similar to the new species: the nearly horizontal orientation of the zygapophyses in lateral view. In its anteroposteriorly compressed centra and the shape of the rib articular surfaces, "*Shastasaurus carinthiacus*" much more resembles the Californian Upper Triassic material of *Shastasaurus* (Merriam, 1908).

The California *Shastasaurus*---*Shastasaurus* is best known from the Upper Triassic Hosselkus Limestone of California that yielded the genotypic material as well as remains of several other individuals. Although described as five different species by Merriam (1908 and earlier papers cited therein), a recent examination of the type material of the Californian *Shastasaurus* species held at the UCMP, Berkeley, suggests that it pertains to only one species, *Shastasaurus pacificus*. A probable exception is *S. careyi*, which appears to be a much larger, *Shonisaurus*-like form. This conclusion was already reached by McGowan (1994), except that he considered *S. careyi* a junior synonym of *S. pacificus*. *Shastasaurus pacificus* clearly differs from the new species in its much more foreshortened vertebrae, the robust neural spines with reduced zygapophyses, its larger size, and details of the morphology of the pubis.

***Shastasaurus neoscapularis*---**McGowan (1994) recently described a new species of *Shastasaurus* from the Norian of British Columbia (Canada) as *S. neoscapularis*. This animal with a skull length of 453 mm is of roughly the same size as the new species (see also Table 1). *Shastasaurus neubigi* n. sp. shares with *S. neoscapularis* a slender, elongate skull but differs in a number of characters, such as the relatively longer vertebrae (they seem to have been of the same height as those of *S. neoscapularis*) and the specific autapomorphies enumerated in the diagnosis. The fact that the teeth are set in distinct alveoli in the new species as opposed to being set in grooves in *S. neoscapularis* underscores the variability of this character in the genus *Shastasaurus* which already was pointed out by McGowan (1994).

Befitting its great stratigraphic age (late Anisian), the new species of *Shastasaurus* appears more generalized than the Late Triassic forms from North America in a number of characters: the animal was smaller than most other representatives of the genus, the teeth are set in alveoli and not in a groove, and the vertebral centra are the relatively longest and lowest in the genus. The well-developed zygapophyses with their horizontal orientation are seemingly generalized as well but are not found in *Mixosaurus*, which is usually considered more primitive than the shastasaurids (Mazin, 1988; Massare and Callaway, 1994). The exact relationships of the new species must await a formal revision of the genus and detailed phylogenetic analysis.

PALEOBIOGEOGRAPHY OF *SHASTASAURUS*

With the find of a new species of *Shastasaurus* in the Muschelkalk beds, a reappraisal of the paleobiogeography of this genus becomes necessary. Clearly, *Shastasaurus* was present in Middle Triassic times in the western Tethys (Figure 8), as had been suggested based on isolated vertebrae (Callaway and Massare, 1989). Possibly *Shastasaurus* is also represented in the undescribed material of large ichthyosaurs from the Middle Triassic Monte San Giorgio fauna. In the other faunal regions from which we have good records for the Middle Triassic, the genus is notably absent during this time span (Figure 8), as it is not known from the Nevada or Wapiti Lake faunas (see also Sander and Mazin, 1993). Its presence also cannot be established in the Spitsbergen fauna. Despite the risks inherent in using negative evidence, the absence of *Shastasaurus* in Spitsbergen and Nevada seems to be real, considering the long history of study and large amount of material from these faunas.

In Late Triassic times, the genus spread to the eastern Pacific Ocean (Figure 8), where sediments containing *Shastasaurus* remains are now found all along the western margin of the North American craton, notably in the Hosselkus Limestone of California (Figure 8; Callaway and Massare, 1989). A new species of the genus, *S. neoscapularis*, was also recently reported "up the coast" from the region of Williston Lake, Canada (Figure 8; McGowan, 1994). *Shastasaurus* still occurred in the western Tethys region at this time, as evidenced by the records from Carinthia, Austria (Figure 8; Huene, 1925). The scarcity of finds from the western Tethys may reflect the lack of suitable marine vertebrate localities.

The distribution of *Shastasaurus* remains in time and space thus records a dispersal event, from the western Tethys to the eastern Pacific (Figure 8). However, the fossil record is too incomplete to permit the tracing of dispersal routes or details of timing.

CONCLUSIONS

A new find of a partial ichthyosaur skeleton from the Muschelkalk beds of Germany is described as *Shastasaurus neubigi* n. sp. The new species represents the first good record of *Shastasaurus* from the Middle Triassic of Europe. The paleobiogeographic pattern thus emerging points to a dispersal of *Shastasaurus* from the western Tethys in Middle Triassic times to the eastern Pacific in Late Triassic times.

The new find of *Shastasaurus* repeats a pattern in recent research that had been pointed out before (Sander 1992; Sander and Bucher; 1990, Sander and Mazin,

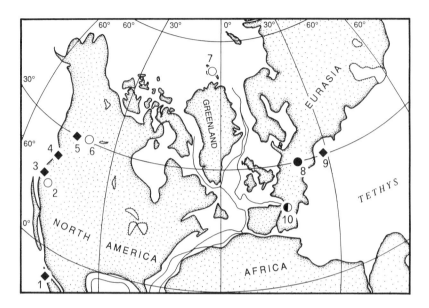

Figure 8. Paleobiogeography of *Shastasaurus*. Circles denote Middle Triassic localities; diamonds denote Upper Triassic localities. Filled symbols indicate the presence of *Shastasaurus* in the respective locality. Note that the genus in Middle Triassic times is only found in the western Tethys region but has spread to the eastern Pacific in Late Triassic times. The rich Middle Triassic localities in western North America are notably devoid of *Shastasaurus*. The half-filled circle at #10 indicates the possible presence of the genus at Monte San Giorgio, Switzerland. **1** = El Antimonio District, Mexico; **2** = north-central Nevada; **3** = Shasta County, California; **4** = Wallowa Mountains, Oregon; **5** = Williston Lake, British Columbia, Canada; **6** = Wapiti Lake, British Columbia; **7** = Spitsbergen, Norway; **8** = Muschelkalk Basin, central Europe; **9** = Carinthia, Austria; **10** = Monte San Giorgio.

1993). Triassic ichthyosaurs are well known only from a core area consisting of North America, Europe, and intervening Spitsbergen, while the record for the rest of the globe is poor. Only the Lower Triassic of East Asia has also yielded a significant ichthyosaur record. In most any instance of finds from new localities or new records from already known localities, the material belongs to already known genera or even species. In this way, many genera of Middle Triassic ichthyosaurs and some of Upper and Lower Triassic ichthyosaurs are now known from all paleogeographic regions, if not all localities from this core area (see also Sander and Mazin, 1993). The reduction in diversity of Triassic ichthyosaurs by synonymy (Motani, Chapter 4; Sander et al., in preparation) also adds to the pattern of a limited number of very widespread forms.

Current research trends are thus clear: Despite a flurry of new discoveries, very few new genera are described, but the record of established ones is continually being expanded by new occurrences and synonymy. These trends, of course, have implications for the utility of studying Triassic ichthyosaur paleobiogeography and for the future direction of research in the field. Obviously, the very widespread and roughly contemporaneous occurrence of individual records of most genera in the core area of North America, Spitsbergen, and Europe leaves little room for the study of dispersal. If this kind of study is to be successful, two conditions have to be met: First, determinable specimens of Triassic ichthyosaurs have to be collected from other parts of the shoreline and shallow marine sediments of Triassic Pangaea, and second, we need robust hypotheses of relationships within the Ichthyosauria as well as of the entire group.

SUMMARY

Triassic ichthyosaurs are as diverse and widespread as the Jurassic forms but, due to the vagaries of preservation, are not so well known. Associated or articulated skeletons are rare, and new finds of this kind provide us with much needed anatomical information. I described here a partial, associated skeleton of a new species of *Shastasaurus* from the German Muschelkalk, the most complete ichthyosaur specimen from this formation.

Shastasaurus neubigi n. sp. is assigned to this genus based on the structure of the vertebrae, in particular the nature of the rib articulations, and of the pubis. The diagnostic characters of the new species are a distinct dorsal process just anterior to the mandibular articular facet, a slender but long retroarticular process, and a peculiar pattern of tooth replacement. The specimen comes from the late Anisian part of the Muschelkalk, making *Shastasaurus neubigi* n. sp. the geologically oldest nominal species of the genus.

The paleobiogeography of *Shastasaurus* records a range expansion from the western Tethys region (Muschelkalk and Alpine Triassic) in Middle Triassic times to the eastern Pacific region (western cratonic margin of North America) in Late Triassic times.

ACKNOWLEDGMENTS

The new *Shastasaurus* specimen was discovered by Mr. Bernd Neubig and jointly collected and initially prepared by him and the other members of the Euerdorf collectors group (Mr. Michael Henz, Mr. Horst Mahler, and Mr. Jürgen Sell). They provided important geological and historical information and generously donated the specimen to the Bayerische Staatssammlung für Paläontologie und historische Geologie.

The material was kindly brought to my attention by Dr. Hans Hagdorn. Dr. Peter Wellnhofer arranged for the preparation and curation of the specimen. I thank Ms. Dorothea Kranz for the preparation of the excellent halftone drawings and Mr. Georg Oleschinski for photographic work.

Special thanks are due to Dr. Rupert Wild for critically reading the entire manuscript as well as for help with the transferral of the find to a public collection. The thorough formal reviews by Drs. Jack Callaway and Hans-Dieter Sues are greatly appreciated, as they considerably improved the manuscript. Dr. Carole Gee offered much needed advice in stylistic matters.

Last, but not least, I would like to thank the editors of this volume, Drs. Elizabeth Nicholls and Jack Callaway, for the invitation to contribute to this book.

REFERENCES

Andrews, C. W. 1910. *A Descriptive Catalogue of the Marine Reptiles of the Oxford Clay.* British Museum (Natural History), London, 202 pp.

Blainville, H. M. D. de. 1835. Systéme d'Herpetologie. *Nouvelles Annales du Muséum (National) d'Histoire Naturelle*, Paris 4:233-296.

Callaway, J. M. and J. A. Massare. 1989. Geographic and stratigraphic distribution of the Triassic Ichthyosauria (Reptilia; Diapsida). *Neues Jahrbuch für Geologie und Paläontologie*, Abhandlungen 178:37-58.

Camp, C. L. 1980. Large ichthyosaurs from the Upper Triassic of Nevada. *Palaeontographica* A170:139-200.

Fraas, E. 1891. *Die Ichthyosaurier der süddeutschen Trias-und Jura-Ablagerungen.* H. Laupp, Tübingen, 81 pp.

Hagdorn, H. 1991. *Muschelkalk. A Field Guide.* Goldschneck-Verlag Werner K. Weidert, Korb, 80 pp.

Huene, F. von. 1916. Beiträge zur Kenntnis der Ichthyosaurier im deutschen Muschelkalk. *Palaeontographica* 62:1-68.

Huene, F. von. 1922. *Die Ichthyosaurier des Lias und ihre Zusammenhänge.* Gebrüder Borntraeger, Berlin, 114 pp.

Huene, F. von. 1925. *Shastasaurus*-Reste in der alpinen Trias. *Centralblatt für Mineralogie, Geologie und Paläontologie*, Abteilung B:412-417.

Linnaeus, C. von. 1758. *Systema Naturae. Regnum Animale.* L. Salvius, Stockholm, 824 pp.

Massare, J. A. 1987. Tooth morphology and prey preference of Mesozoic marine reptiles. *Journal of Vertebrate Paleontology* 7:121-137.

Massare, J. A. and J. M. Callaway. 1994. *Cymbospondylus* (Ichthyosauria: Shastasauridae) from the Lower Triassic Thaynes Formation of southeastern Idaho. *Journal of*

Vertebrate Paleontology 14:139-141.
Mazin, J.-M. 1988. Paléobiogéographie des Reptiles marins du Trias: phylogénie, systématique, écologie et implications paléobiogéographiques. *Memoires des Sciences de la Terre, Université Pierre et Marie Curie* 08/88:1-313.
McGowan, C. 1973. The cranial morphology of the Lower Liassic latipinnate ichthyosaurs of England. *Bulletin of the British Museum (Natural History)*, Geology 93:1-109.
McGowan, C. 1994. A new species of *Shastasaurus* (Reptilia: Ichthyosauria) from the Triassic of British Columbia: the most complete exemplar of the genus. *Journal of Vertebrate Paleontology* 14:168-179.
Merriam, J. C. 1895. On some reptilian remains from the Triassic of Northern California. *American Journal of Science* 50:55-57.
Merriam, J. C. 1902. Triassic Ichthyopterygia from California and Nevada. *University of California Publications, Bulletin of the Department of Geology* 3:63-108.
Merriam, J. C. 1908. Triassic Ichthyosauria, with special reference to the American forms. *Memoirs of the University of California* 1:1-155.
Osborn, H. F. 1903. The reptilian subclasses Diapsida and Synapsida and the early history of the Diaptosauria. *Memoirs of the American Museum of Natural History* 1:449-507.
Owen, R. 1881. A monograph on the fossil Reptilia of the Liassic formations. *Plesiosaurus, Dimorphodon*, and *Ichthyosaurus*. Part III: Order Ichthyopterygia, *Ichthyosaurus*. *Palaeontographical Society Monographs* 35:83-134.
Romer, A. S. 1956. *Osteology of the Reptiles*. The University of Chicago Press, Chicago, 772 pp.
Romer, A. S. 1968. An ichthyosaur skull from the Cretaceous of Wyoming. *Contributions to Geology, University of Wyoming* 7:27-41.
Sander, P. M. 1989. The large ichthyosaur *Cymbospondylus buchseri*, sp. nov., from the Middle Triassic of Monte San Giorgio (Switzerland), with a survey of the genus in Europe. *Journal of Vertebrate Paleontology* 9:163-173.
Sander, P. M. 1992. *Cymbospondylus* (Shastasauridae: Ichthyosauria) from the Middle Triassic of Spitsbergen: filling a paleobiogeographic gap. *Journal of Paleontology* 66:332-337.
Sander, P. M. and H. Bucher. 1990. On the presence of *Mixosaurus* (Ichthyopterygia: Reptilia) in the Middle Triassic of Nevada. *Journal of Paleontology* 64:161-164.
Sander, P. M., C. Faber, and G. Tichy. 1994a. New Middle Triassic ichthyosaur finds and their paleobiogeographical implications. *Journal of Vertebrate Paleontology* 14:44A (abstract).
Sander, P. M. and J.-M. Mazin. 1993. The paleobiogeography of Middle Triassic ichthyosaurs: The five major faunas. *Paleontologia Lombarda, Nuova serie* II:145-152.
Sander, P. M., O. C. Rieppel, and H. Bucher. 1994b. New marine vertebrate fauna from the Middle Triassic of Nevada. *Journal of Paleontology* 68:676-680.
Sollas, W. J. 1916. The skull of *Ichthyosaurus* studied in serial sections. *Philosophical Transactions of the Royal Society of London* B 208:63-126.
Wiman, C. 1910. Ichthyosaurier aus der Trias Spitzbergens. *Bulletin of the Geological Institution of Upsala* 10:124-148.

Chapter 2

A NEW LOOK AT *MIXOSAURUS*

JACK M. CALLAWAY

INTRODUCTION

Ichthyosaurus atavus was described by Quenstedt (1852), and *Ichthyosaurus cornalianus* was described, but not figured, by Bassani (1886). *Ichthyosaurus cornalianus* was placed in a new genus, *Mixosaurus*, and a distinct family, Mixosauridae, by Baur (1887) primarily on differences in limb structure. With the discovery of additional material similar to *I. atavus*, Fraas (1891) determined that species should be assigned to Baur's *Mixosaurus* as well. A redescription of *Mixosaurus cornalianus* was done by Repossi (1902) which included a number of figures. This work is the foundation on which all subsequent works relating to *Mixosaurus* are based. Much of the *M. cornalianus* material collected from Basano, Lombardy, Italy (near Lake Lugano), and belonging to the collections of the Civic Museum of Natural History in Milan, was destroyed by bombing during the Second World War. A neotype for the species was designated by Pinna (1967). The best and most numerous specimens of *M. cornalianus* now available are those located in Zürich which were collected from the Monte San Giorgio region of Ticino, Switzerland (also near Lake Lugano). Unfortunately, this material remains undescribed to the present date, and detailed knowledge of morphology, morphological variation, possible taxic diversity, and ontological data are still lacking because access to researchers has been denied or severely limited. Thus, a wealth of information is unavailable that should form the core of our knowledge of *Mixosaurus*. The only mixosaur described from strata of equivalent age is the specimen of Repossi (1902) which was recovered from the nearby Basano area. Fortuitously, I was allowed a brief examination of the Zürich specimens a few years past. The time available for this examination did not permit extensive, detailed, quantitative data collection, but was sufficient to reveal emphatically that a new look at *Mixosaurus* was needed. Until a detailed study of the Zürich material is conducted, it is hoped that this chapter will serve as an interim update on various aspects of *Mixosaurus*.

SPECIES AND DISTRIBUTION

Mixosaurus has a wide distribution including Canada, China, France, Germany, Indonesia, Italy, Poland, Russia, Svalbard, Switzerland, Turkey, and the United States (Callaway and Massare, 1989; McGowan, 1978; Mazin, 1983, 1988; Sander and Mazin, 1993). All mixosaurs are of Middle Triassic age, with the exception of one specimen of *Mixosaurus* sp. reported from the Early Triassic of Canada (Callaway and Brinkman, 1989). Apart from recognized species, much of the material is indeterminate and assigned to Mixosauridae *incertae sedis* or *Mixosaurus* sp. (Callaway and Massare, 1989; Mazin, 1983). Mazin (1983) recognized five species: *M. atavus* (Quenstedt, 1852), *M. natans* (Merriam, 1908), *M. nordenskioeldii* (Hulke, 1873), *M. cornalianus* (Bassani, 1886), and *M. maotaiensis* Young, 1964. This report recognizes only two valid species: *M. atavus* and *M. cornalianus*.

Mixosaurus atavus (Fraas, 1891; Huene, 1916, 1935, 1943; Mazin, 1983; Quenstedt, 1852) is known from Germany, France, and possibly Poland. It thus appears to be an endemic species restricted to the Germanic Muschelkalk Basin. Although a valid species, it is represented mainly by small amounts of poorly preserved material collected from a number of widely separated localities. Reconstructions by Huene (1916, 1935, 1943) are largely imaginative, leading to false conclusions that *M. atavus* is a well-understood species. Most of the remains are located in Tübingen and Stuttgart, Baden-Württemberg, Germany.

Mixosaurus maotaiensis is known from a single incomplete specimen from China. Young's (1964) Figure 1 is of material essentially identical to *M. cornalianus* in both appearance and size. Young believed he was justified in erecting a new species on the basis of differences perceived between the Chinese material and *M. cornalianus,* including differences in the interclavicle, clavicle, coracoid, and the humerus. Variations noted in the specimens of *M. cornalianus* in Zürich are more than adequate in range to account for the differences observed by Young. It is concluded that *Mixosaurus maotaiensis* is the junior synonym of *Mixosaurus cornalianus*. No known climatic or physical barriers existed that would have prevented free east-west communication between China and the Alpine region of Europe via the Tethys Sea of Middle Triassic Pangaea. Ichthyosaurs living in this region would tend to be cosmopolitan: disjunct ichthyosaur faunas would not be expected.

Mixosaurus natans and *M. nordenskioeldii* are synonymized with *Phalarodon nordenskioeldii* (Nicholls et al., in review). *Mixosaurus nordenskioeldii* (Hulke, 1873) was first described in detail by Wiman (1910) from new finds in Spitsbergen, largest of the Svalbard Islands in the Arctic Ocean. Wiman's material consisted mainly of postcrania with a few isolated skull elements. In the same year, Merriam (1910) described *Phalarodon fraasi*, a new genus and species from the Prida Formation of Nevada in the United States. There was no postcranial material and the referred specimen consisted of a partial skull and mandible. A possible relationship between *M. nordenskioeldii* and *P. fraasi* has been disputed for many years because of the similarity of the unusual posterior dentition found in both species. Merriam (1911) recognized the validity of both species and was of the opinion that both were present in the Spitsbergen material. Wiman (1916) eventually concurred with this opinion. However, Cox and Smith (1973), Dechaseaux (1955), Huene (1916), McGowan (1972a), and Romer (1956, 1966) all synonymized *Phalarodon* with *Mixosaurus*. Kuhn (1934) and Mazin (1983, 1984) kept the genera separate and this lead was followed by Callaway and Brinkman (1989), Carroll (1988), and Sander and Bucher (1990). Everyone, except Kuhn (1934), placed the two genera in different families. Significant new finds of ichthyosaurs from the Wapiti Lake area of eastern British Columbia, Canada, during the past few years have included skulls and jaw elements with associated postcrania. Skulls with a *Phalarodon*-like dentition are associated with *Mixosaurus*-type postcrania. Nicholls et al. (in review) compared the upper and lower dentitions of the new material with the dentigerous material described by Wiman (1910) that he had referred to *M. nordenskioeldii* and concluded that all the material belongs to a single species. The separation of the Spitsbergen material into *P. fraasi* and *M. nordenskioeldii* (Merriam, 1911; Wiman, 1916) represented a separation of upper and lower dentitions. Brinkman et al. (1992a) initially referred the Wapiti Lake material to *M. nordenskioeldii* (Hulke, 1873), which has priority. Additional study, however, led Nicholls et al. (in review) to refer this species to *Phalarodon*. Although the postcranial skeleton is clearly mixosaurian, the skull and dentition are significantly different from either *M. atavus* or the type species, *M. cornalianus*. Based on the postcranial skeleton, *Phalarodon nordenskioeldii* (Hulke, 1873) is assigned to the Mixosauridae (Nicholls et al., in review) and includes all material formerly assigned to *P. fraasi*, *M. nordenskioeldii*, and *M. natans*. The type material of *Mixosaurus natans* (Merriam, 1908) and postcranial material from Nevada referred to *M. natans* by Sander and Bucher (1990) is comparable to the corresponding material of *P. nordenskioeldii* and is synonymized with the latter species, confirming the suspicions of Huene (1916), McGowan (1972a), and Wiman (1916) that the skull of *Phalarodon* belonged with the postcrania of *M. natans* (Nicholls et al., in review).

THE CRANIUM

Skulls of *M. cornalianus* (Figure 1) observed and measured by the author range in length from 185 mm to 270 mm. The skull is primitive in the relatively large size of the maxilla, the subequal lengths of the frontal and nasal bones, and the retention of the quadratojugal. Derived features are the large orbit, short cheek region, lacrimal excluded from the border of the external naris, very small upper temporal fenestra, elongated facial region, and exclusion of the postorbital from the temporal fenestra. The dentition is also derived as discussed below. Derived characters considerably outnumber plesiomorphic traits; thus, any assertions that this is a primitive ichthyosaur skull must be reconsidered. A complete skull reconstruction of *M. cornalianus* by Repossi (1902) was not possible because of insufficient material. With the discovery of additional material, later interpretations of the skull of *M. cornalianus* by Huene (1925, 1935, 1949) were made. These are remarkably accurate, as verified by examination of some of the better specimens in Zürich. The shape and proportions of the skull in Figure 1 are based primarily on observations and measurements by the author of an unnumbered Paläontologisches Institut, Zürich, specimen collected in 1937 from Cava Tre Fontane, Monte San Giorgio. The shape and arrangement of the skull elements were determined mainly from the same specimen and from specimens T ("Tessin") 2418 and T 2405. Several other specimens were examined as well. Figure 1 is a very close approximation of Huene's (1949) reconstruction. The skull of *M. atavus* is known from a limited number of fragments, most of which are located in the Staatliches

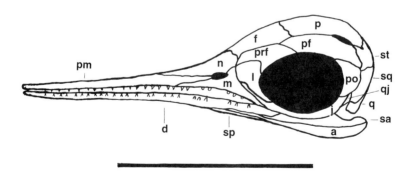

Figure 1. Skull of *Mixosaurus cornalianus* (modified after Huene, 1949). See text for discussion. Abbreviations: **a** = angular, **d** = dentary, **f** = frontal, **j** = jugal, **l** = lacrimal, **m** = maxilla, **n** = nasal, **p** = parietal, **pf** = postfrontal, **pm** = premaxilla, **po** = postorbital, **prf** = prefrontal, **q** = quadrate, **qj** = quadratojugal, **sa** = surangular, **sp** = splenial, **sq** = squamosal, **st** = supratemporal. Bar scale = 100 mm.

Museum für Naturkunde, Stuttgart (SMNS). The largest is SMNS 15378, figured by Huene (1916:Plates I and II). I have examined this specimen in some detail. This skull fragment extends from the occiput to just anterior of the external nares. The palate is quite well preserved and many of the skull elements are present, although determination of sutures is a challenge. The palate confirms the loss of the transverse flange of the pterygoid in mixosaurs, because this is a derived character also belonging to *M. cornalianus*. The flange is present in the mixosaurid *Phalarodon*. No evidence of the frontal or parietal bones could be found, yet Huene (1916) produced a reconstruction showing the upper part of the skull as though it were known, with no dashed lines such as he used for the other unknown regions. In a later publication Huene (1935) discontinues the use of any dashed lines, creating the illusion that the skull of *M. atavus* is known in toto. The identifiable skull elements are comparable in shape and position to those in *M. cornalianus* but, discounting possible distortion, the cheek region is somewhat broader.

A common assumption has been that the dentition of *M. cornalianus* is thecodont, and that the thecodont condition is plesiomorphic for ichthyosaurs (Besmer, 1947; Mazin ,1983; Merriam, 1908, 1910; Peyer, 1968; Romer, 1956). Motani (Chapter 4) shows this not to be true. He concludes that subthecodonty is probably plesiomorphic for ichthyosaurs, and that no ichthyosaurs have true thecodont dentitions. Although the case for the poorly known *M. atavus* is less clear than for *M. cornalianus*, Motani classifies the type of implantation in both species as ankylosed thecodont. The dentition of mixosaurs is thus considered to be a derived condition.

The small temporal opening of *Mixosaurus* (Figure 1) has been cited as a primitive trait by some authors (e.g., Romer, 1956; Stahl, 1974). The assumption that ichthyosaurs had evolved directly from amphibians (Huene, 1937, 1944) or some anapsid reptilian stock (Cope, 1896; Olson, 1971; Romer, 1948) probably led to an interpretation of the very small temporal fenestra as an incipient development (i.e., a primitive character). The superior temporal fenestrae of the specimens of *Mixosaurus cornalianus* observed at Zürich are indeed very small. In fact, they could not be found with any certainty in the majority of specimens examined (despite the general overall excellent preservation of the Zürich mixosaurian material, the skulls are usually badly disarticulated). The smallness of the opening must be interpreted as a highly derived trait rather than a primitive one, however, because ichthyosaurs are now generally thought to be diapsid reptiles (Massare and Callaway, 1990), and primitive diapsids, such as *Petrolacosaurus* and *Youngina* (Carroll, 1988), have much larger openings. Ichthyosaurs determined to be more primitive than *Mixosaurus* also have larger openings, such as *Parvinatator wapitiensis* n. gen., n. sp. Nicholls and Brinkman, 1996; *Grippia longirostris* Wiman, 1929 (Mazin, 1981); and *Thaisaurus chonglakmanii* n. gen., n. sp. Mazin et al., 1991. Reconstructions of the skull of *M. atavus* (Huene, 1916, 1935, 1943) showing upper temporal fenestrae cannot provide any useful information, for reasons already cited.

The postorbital does not enter into the border of the supratemporal fenestra as it does in early diapsids and the primitive ichthyosaur genera *Grippia* (Mazin, 1981), *Thaisaurus* (Mazin et al., 1991), and *Parvinatator* (Nicholls and Brinkman, 1996). This is a derived character shared with all other ichthyosaurs for which adequate knowledge of the skull is available.

Romer (1966) stated that the short face of *Mixosaurus* was a primitive character. Merriam (1908) and Appleby (1979) also commented on the relatively short facial region. It is assumed that "face" length is the equivalent of snout length which, McGowan (1994) defines as the distance between the tip of the snout and the anterior (internal) margin of the orbit. Simply measuring snout length is inadequate, however, for comparison with other species. A ratio is needed. McGowan (1994) uses a snout ratio which is snout length divided by jaw length, with jaw length being the distance between the tip of the dentary and the posterior edge of the angular. Time constraints during my limited access to the Zürich collections allowed for few measurements, but the parameters needed for a snout ratio were obtained for two well-preserved skulls. A ratio of 0.65 was obtained for both. Whether this is representative of the total variation possible in the available specimens of *M. cornalianus* is not known, of course, but at least it is a figure that can be used for rough comparisons. This ratio falls within the range of variation for all species of the Late Liassic ichthyosaur, *Stenopterygius*, and exceeds that of the Middle Triassic ichthyosaur, *Cymbospondylus petrinus*, according to ratios published by McGowan (1972a). It equals or exceeds the ratios of the Early Liassic species of *Ichthyosaurus* reported by McGowan (1974), but falls well short of the ratios for the Cretaceous ichthyosaur, *Platypterygius* (McGowan, 1972a). *Mixosaurus* does not have a short face and is not primitive in that respect.

All ichthyosaurs have large eyes, but there is the implication that the orbits of Triassic genera are smaller, and thus more primitive than later forms (Romer, 1956). *Orbital ratio* is defined as the orbital diameter divided by jaw length (McGowan, 1994). Orbital diameter is the internal diameter of the orbit, measured along its longitudinal axis. Using the same skulls of *M. cornalianus* as were used for snout ratios, orbital ratios of 0.22 and 0.24 were obtained. These ratios are within the ranges published for *Stenopterygius* (McGowan, 1972a). The 0.22 ratio is slightly less than those published for *Ichthyosaurus* (McGowan, 1974), but the 0.24 ratio is within the range for that genus. Both ratios are much greater than for *Cymbospondylus petrinus* and species of *Platyterygius*. *Mixosaurus* has a large orbit and clearly is derived in that respect.

Working with post-Triassic species, Romer (1968) and McGowan (1973) demonstrated that there are only two bones in the posterior temporal region of ichthyosaurs, which they interpreted as being the squamosal and quadratojugal. Earlier interpretations included a supratemporal; however, these interpretations were based on specimens in which the ventral element was broken and was thought to represent two bones instead of one (Romer, 1968). Three bones were observed in the Zürich specimens of *M. cornalianus*. These are interpreted here to be the supratemporal, squamosal, and quadratojugal (Figure 1). The latter is very small, when it can be found at all. Following the lead of Romer and McGowan, Mazin (1982) considered the third bone seen in some Triassic ichthyosaurs to be a neomorph, located between what he believed to be the squamosal and quadratojugal. However, as pointed out by Callaway (1989), the trend throughout vertebrate evolution has been toward the reduction of temporal elements, not the addition of new ones, and it is more parsimonious to accept these bones as the supratemporal, squamosal, and quadratojugal. The supratemporal is present also in the mixosaurid, *Phalarodon* (Nicholls, et al., in review). The supratemporal may be a plesiomorphic character present in all ichthyosaurs (Nicholls and Brinkman, 1996), in which case the two bones in the posterior temporal region of advanced ichthyosaurs are the supratemporal and squamosal, not the squamosal and quadratojugal. The retention of the quadratojugal in the Mixosauridae would thus represent a primitive condition for ichthyosaurs.

The lacrimal is excluded from the border of the external naris in all mixosaurids (*Mixosaurus* and *Phalarodon*), a derived trait shared with *Grippia* (Mazin, 1981), *Thaisaurus* (Mazin et al., 1991), *Omphalosaurus* (Mazin, 1986a), and *Parvinatator* (Nicholls and Brinkman, 1996) and with some other derived diapsids, such as *Youngina*. Primitive tetrapods, and the most primitive diapsids such as *Petrolacosaurus*, have lacrimals that take part in both orbital and nasal openings, a condition retained in all other ichthyosaurs for which this element is known.

Although not shown in Figure 1, mixosaurs possessed sclerotic rings, a trait that may be present in all ichthyosaurs. These are rarely found in place and are usually partially disarticulated. The plates meet edge to edge with no overlapping. The rings curve inward and apparently must have encased a considerable segment of the eyeball. Presumably, the ring offered mechanical support for the gelatinous mass of the very large eye.

Skull reconstructions of mixosaurs often show the quadrate as being largely concealed by the squamosal (Huene, 1949, for example). Examination of Zürich *M. cornalianus* skulls revealed that the posterior portion of the quadrate is considerably exposed in lateral view (Figure 1). The quadrate is emarginated posteriorly. The same condition is evident from my examination of the partial skull of *M. atavus* (SMNS 15378) in Stuttgart.

APPENDICULAR SKELETON

The pelvic girdle of *M. cornalianus* has the primitive, broad, plate-like pubis and ischium typical of most Triassic species. The pubis as figured by Repossi (1902) is in error in its lack of an obturator foramen. Merriam (1908) and Appleby (1979) also indicated lack of an obturator foramen or notch in mixosaurs. Many pubes of *M. cornalianus* were observed in Zürich and all had well-defined foramina. The ilium is relatively short and somewhat widened distally, a feature common to most Triassic ichthyosaurs.

The elements of the pectoral arch examined by the author are in close agreement with those figured for *M. cornalianus* by Repossi (1902). The coracoid and scapula are almost identical in form and it is sometimes difficult to differentiate the two. Both are somewhat fan-shaped, but the coracoid is generally less symmetrical than the scapula and commonly has a sickle-shaped extension.

Many excellent fins of *M. cornalianus* are present in Zürich. Repossi's (1902: Plate IX, Figure 2) figure of the forefin is essentially correct, but comment is required pertinent to the humerus. Most of the humeri I observed have a broad, semicircular flange which forms the anterior edge of the bone, but some have an emarginated anterior border. Repossi's Figure 2 shows an emarginated anterior border, but in his Plate VIII, Figure 1, and Plate IX, Figure 1, are illustrations of humeri with an anterior flange. Mazin (1986b) considered the flange to be a primitive character for ichthyosaurs because such a flange is present on some Early Triassic genera: *Utatsusaurus* Shikama et al., 1978 (Mazin, 1986b), *Omphalosaurus* Merriam, 1906 (Mazin and Bucher, 1987), and *Grippia longirostris* Wiman, 1929 (Mazin, 1981). Nicholls et al. (in review) disagree, because it is not a character widespread among primitive ichthyosaurs. It is absent in *Thaisaurus* Mazin et al., 1991; *Chaohusaurus* Yang and Dong, 1972; and *Chensaurus* (Chen, 1985). Also, the character is unknown in other primitive diapsids. Some publications illustrate the forefin of *M. cornalianus* with the flange (McGowan, 1972b, for example) and others without (Merriam, 1905, for example). The flange is present also in *Phalarodon*, and is shown in a forefin reconstruction for *M. atavus* by Huene (1916), but this reconstruction is suspect because it is based on a few poorly preserved fragments. These two shapes of the humerus could be an expression of morphological variation, but this seems unlikely because there are no intermediate shapes. Humeri are either emarginated or have a well-developed flange. The differences may reflect some degree of sexual dimorphism, but this also seems unlikely as sexual dimorphism is presently unknown in ichthyosaurs. Another possibility is that the shape is a reflection of various stages of ontological development; however, either shape may occur on specimens of any size. Finally, more than one species may be concealed under the name *M. cornalianus*, but one character such as this is inadequate to erect a new species. Clearly, this emphasizes the need for a detailed examination of all the available material. For now, it is probably best to assume the difference is the result of morphological variation.

Callaway (1989) identified the two large proximal elements in the mixosaur tarsus as the astragalus and calcaneum and noted that the centrale is lost. He also suggested that the five elements distal to the astragalus and calcaneum were distal tarsals 1-5; however, the presence of a large fifth distal tarsal is a feature that contradicts the hypothesis that ichthyosaurs were derived from diapsid reptiles, a hypothesis that otherwise seems consistent with the morphology of early ichthyosaurs and primitive diapsid reptiles (Callaway, 1989; Massare and Callaway, 1990; Tarsitano, 1983). The basal element of the fifth digit was reidentified as a fifth metatarsal by Brinkman et al. (1992b). I concur with their reidentification based on my examinations of excellent hindlimb material from *M. cornalianus*.

AXIAL SKELETON

Neural spines of *M. cornalianus* are very high and straight-sided with dorsal tips sharply truncated. Insufficient material is known from *M. atavus* for comparison; however, the same type of spines are also found in the mixosaurid *Phalarodon*. No neural spines I examined show the curved sides and rounded tips figured by Repossi (1902:Plate IX, Figure 9a). Repossi's Figure 9c, represented as a neural arch, is actually a chevron. Anterior caudal spines stand vertically as they do throughout the presacral region, but have a slight posterior curve, giving the back edge of the spine a concave shape. A gentle upward arch in the vertebral column occurs in the middle caudal region and it is here that the neural spines become slightly taller and develop a sharp anterior slope (Repossi:1902, Plate IX, Figure 9). The spines decrease rapidly in height posterior to this region.

Chevrons depicted by dashed lines beneath the middle caudal vertebrae by Repossi (1902:Plate IX, Figure 8) do not exist for any specimens of *M. cornalianus*. Their depiction as factual in the skeletal reconstruction of *M. cornalianus* by Kuhn-Schnyder (1963), which is reproduced also in Carroll (1988) and Kuhn-Schnyder and Rieber (1986), is incorrect. The chevrons are united ventrally to form long stems.

Ribs of *M. cornalianus* are holocephalous throughout the vertebral column, except for the last two or three posterior dorsal vertebrae and the first two or three anterior caudal vertebrae, which are dichocephalous. The same arrangement occurs in *Phalarodon*. Ribs are not adequately known from *M. atavus*. This pattern of rib articulation among ichthyosaurs is known only for the Mixosauridae.

Gastralia are well developed in mixosaurs, forming a tightly knit plastron.

These structures are found in other ichthyosaurs as well and in a variety of other ancient marine reptiles. What function gastralia performed in marine reptiles is uncertain. They may simply represent a feature retained from terrestrial ancestors, although they seem very well developed for mere vestigial structures. Gastralia are particularly well preserved in a large, belly-up, Zürich specimen of *M. cornalianus* (T 2405), a photograph of which was used by Dechaseaux (1955:Figure 1) in a general review of all ichthyosaurs.

Repossi (1902) reported that the vertebral column of *M. cornalianus* consists of approximately 100-105 vertebrae. None of the specimens I examined had a complete column, but an approximate count of 110-122 was determined by interpolation from several specimens. The number of presacrals was between 45 and 50. Presacral vertebrae are subcircular to hexagonal in shape, the latter shape tending to become more common posteriorly along the column with a gradual increase in size as well. All the vertebrae are higher than they are long, but no data for specific height/length ratios were collected. Centra within most of the caudal region are strongly compressed laterally and are very tall. The laterally flattened centra, plus the very tall anterior and middle caudal neural spines and elongated chevrons, all serve to broaden the tail and produce an effective sculling organ.

CONCLUSIONS

Mixosaurus, with its many derived characters, cannot be considered the most primitive of ichthyosaurs as formerly thought by many authors (e.g., Kuhn-Schnyder, 1964; Romer, 1956). As are all ichthyosaurs, *Mixosaurus* is a mosaic of primitive and derived characters, but it has more than enough derived characters to remove it from the purely primitive status to which it has long been relegated. We simply don't know as much about *Mixosaurus* as we should, considering the amount of material that has been recovered. A detailed study of the superb Monte San Giorgio material especially would help to answer many questions about this small, fascinating ichthyosaur. The amount of material available offers an opportunity to statistically analyze morphological variation, and reach some conclusions regarding ontogeny and evolution. Whether or not *Mixosaurus* could have been a potential ancestor to any Late Triassic and post-Triassic species, or whether it was an evolutionary dead end, are among the types of questions that would be worthy of examination.

SUMMARY

Following the description of *Mixosaurus atavus* in 1852, and descriptions of *Mixosaurus cornalianus* in 1886 and 1902, *Mixosaurus* became the quintessential Triassic ichthyosaur used to illustrate both scientific and popular literature. Since publication of those initial reports, however, additional remains of *Mixosaurus* have been collected from widely separated regions of the world. Particularly valuable and numerous specimens were collected from the famous Besano-Monte San Giorgio region of Italy and Switzerland, but little or nothing has been published about this material, which has resulted in the unavailability of a wealth of valuable data. Triassic ichthyosaurs are generally notorious for their poor quality and scarcity of remains relative to those of younger age; however, the well-preserved Besano-Monte San Giorgio mixosaur fauna is a notable exception. Very little is left of the Italian Besano material, but many undescribed, excellent specimens of *Mixosaurus cornalianus* from Monte San Giorgio exist that have the potential to answer many unresolved questions relevant to *Mixosaurus* paleobiology and evolution, but availability of these specimens for research, comparative studies, and publication has been selectively or completely restricted for many years. A fortuitous, but brief, access to this material indicated that a new look at *Mixosaurus* was needed. Based on that access, this chapter provided an interim update of various aspects of the genus. Mixosaurs are not the most primitive of ichthyosaurs, as often depicted in much of the older literature, but are in some respects as derived as or more derived than Late Triassic and post-Triassic forms. Only two valid species are recognized: *Mixosaurus atavus* and *Mixosaurus cornalianus*.

ACKNOWLEDGMENTS

I thank Texas A & M International University and Laredo Community College for the support they have given me in the preparation of this chapter and those colleagues who read and corrected earlier drafts. I very much appreciate the review and editorial assistance of J. Freeman and an anonymous reviewer. I am particularly indebted to Olivier Rieppel, formerly of the Paläontologisches Institut, Zürich, and now at the Field Museum of Natural History, Chicago, who made it possible for me to gain a precious, but all too brief, access to the Monte San Giorgio ichthyosaur collection in Zürich. This is a Rochester Institute of Vertebrate Paleontology contribution.

REFERENCES

Appleby, R. M. 1979. The affinities of Liassic and later ichthyosaurs. *Palaeontology* 22(4):921-946.

Bassani, F. 1886. Sui fossili e sull'età degli schisti bituminosi triasici di Besano in Lombardia. Communicazione preliminare. *Atti della Società Italiana di Scienze Naturali e del Museo Civili di Storia Naturale* 29:15-72.

Baur, G. 1887. On the morphology and origin of the Ichthyopterygia. *American Naturalist* 21:837-840.

Besmer, A. 1947. Beiträge zur Kenntnis des Ichthyosauriergebisses. *Schweizerische Palaeontologische*, Abhandlungen 65:1-21.

Brinkman, D. B., E. L. Nicholls, and J. M. Callaway. 1992a. New material of the ichthyosaur *Mixosaurus nordenskioeldii* from the Triassic of British Columbia, and the interspecific relationships of *Mixosaurus*. *North American Paleontological Convention V, Field Museum of Natural History, Chicago, Abstracts*:37.

Brinkman, D. B., Z. Xijin, and E. L. Nicholls. 1992b. A primitive ichthyosaur from the Lower Triassic of British Columbia, Canada. *Palaeontology* 35(2):465-474.

Callaway, J. M. 1989. *Systematics, Phylogeny, and Ancestry of Triassic Ichthyosaurs (Reptilia, Ichthyosauria)*. Unpublished Ph. D. dissertation, University of Rochester, Rochester, New York.

Callaway, J. M. and D. B. Brinkman. 1989. Ichthyosaurs (Reptilia, Ichthyosauria) from the Lower and Middle Triassic Sulphur Mountain Formation, Wapiti Lake area, British Columbia, Canada. *Canadian Journal of Earth Sciences* 26:1491-1500.

Callaway, J. M. and J. A. Massare. 1989. Geographic and stratigraphic distribution of the Triassic Ichthyosauria (Reptilia; Diapsida). *Neues Jahrbuch für Geologie und Paläontologie*, Abhandlungen 178:37-58.

Carroll, R. L. 1988. *Vertebrate Paleontology and Evolution*. W. H. Freeman and Co., New York.

Chen, L. 1985. Ichthyosaurs from the Lower Triassic of Chao County, Anhui. *Regional Geology of China* 15:139-146 (in Chinese).

Cope, E. D. 1896. *Primary Factors of Organic Evolution*, p. 115.

Cox, C. B. and D. G. Smith. 1973. A review of the Triassic vertebrate faunas of Svalbard. *Geological Magazine* 110(5):405-418.

Dechaseaux, C. 1955. Ichthyopterygia. IN J. Piveteau (Ed.), *Traite de Paleontologie*, pp. 376-408. Paris.

Fraas, E. 1891. *Die Ichthyosaurier der Süddeutschen Trias --- und Jura-Ablagerungen*. Tübingen.

Huene, F. von. 1916. Beiträge zur Kenntnis der Ichthyosaurier im deutschen Muschelkalk. *Palaeontographica* 62:1-68.

Huene, F. von. 1925. Einige Beobachtungen an *Mixosaurus* (Bassani). *Centralblatt für Mineralogie, Geologie und Paläontologie*, Abteilung B 1925:289-295.

Huene, F. von. 1935. Neue Beobachtungen an *Mixosaurus*. *Palaeontologische Zeitschrift* 17:159-162.

Huene, F. von. 1937. Die Frage nach der Herkunft der Ichthyosaurier. *Geological Institutions of the University of Uppsala Bulletin* 27:1-9.
Huene, F. von. 1943. Bemerkungen über primitive Ichthyosaurier. *Neues Jahrbuch für Mineralogie, Geologie und Paläontologie*, Monatshefte, Abteilung B 1943:154-156.
Huene, F. von. 1944. Die Zweiteilung des Reptilstammes. *Neues Jahrbuch für Mineralogie, Geologie und Paläontologie*, Abteilung B 88(3):427-440.
Huene, F. von. 1949. Ein Schädel von *Mixosaurus* und die Verwandtschaft der Ichthyosaurier. *Neues Jahrbuch für Mineralogie, Geologie und Paläontologie*, Monatshefte, Abteilung B 1949:88-95.
Hulke, J. W. 1873. Memorandum on some fossil vertebrate remains collected by the Swedish expedition to Spitzbergen in 1864 and 1868. *Bihang till K. Svenska Vetenskapsakademiens Handlingar* 1, Afdelning IV(9):1-11.
Kuhn, O. 1934. Ichthyosauria. *Fossilium Catalogus. I: Animalia* 63:1-75.
Kuhn-Schnyder, E. 1963. I Sauri del Monte San Giorgio. *Comunicazioni dell 'Instituto di Paleontologia dell 'Universita di Zurigo* 20:811-854.
Kuhn-Schnyder, E. 1964. Die Wirbeltierfauna der Trias der Tessiner Kalkalpen. *Geologische Rundschau* 53:393-412.
Kuhn-Schnyder, E. and H. Rieber. 1986. *Handbook of Paleozoology* (translated into English by E. Kucera). The Johns Hopkins Press, Baltimore and London.
Massare, J. A. and J. M. Callaway. 1990. The affinities and ecology of Triassic ichthyosaurs. *Bulletin of the American Geological Society* 102:409-416.
Mazin, J.-M. 1981. *Grippia longirostris* Wiman, 1929, un Ichthyopterygia primitif du Trias inférieur du Spitsberg. *Bulletin du Muséum National d'Histoire Naturelle à Paris*, Série IV 3(C):317-340.
Mazin, J.-M. 1982. Affinités et phylogénie des Ichthyopterygia. IN E. Buffetaut, P. Janvier, J.-C. Rage, and P. Tassy (Eds.), Phylogénie et paléobiogéographie, pp. 85-98. *Geobios, Mémoire Spécial* 6.
Mazin, J.-M. 1983. Répartition stratigraphique et géographique des Mixosauria (Ichthyopterygia). Provincialité marine au Trias moyen. IN E. Buffetaut, J.-M. Mazin, and E. Salmon (Eds.), *Actes du Symposium Paléontologique Georges Cuvier, 1982, Montbéliard, France*, pp. 375-387.
Mazin, J.-M. 1984. Les Ichthyopterygia du Trias du Spitsberg. Descriptions complémentaires a partir d'un nouveau matériel. *Bulletin du Muséum National d'Histoire Naturelle à Paris*, Série IV 6(C):309-320.
Mazin, J.-M. 1986a. A new interpretation of the type specimen of *Omphalosaurus nevadanus* Merriam, 1906. *Palaeontographica*, Abteilung A 195:19-27.
Mazin, J.-M. 1986b. A new interpretation of the fore-fin of *Utatsusaurus hataii* (Reptilia, Ichthyopterygia). *Palaeontologische Zeitschrift* 60:313-318.
Mazin, J.-M. 1988. Paléobiogéographie des reptiles marins du Trias. *Mémoires des Sciences de la Terre, Université Paris VI*, 8/88.
Mazin, J.-M. and H. Bucher. 1987. *Omphalosaurus nettarhynchus*, une nouvelle espèce d'Omphalosauridé (Reptilia, Ichthyopterygia) du Spathien de la Humboldt Range (Nevada, U. S. A.). *Comptes Rendus de l'Académie des Sciences*, Série II 305:823-

828.

Mazin, J.-M., V. Suteethorn, E. Buffetaut, J.-J. Jaeger, and R. Helmcke-Ingavat. 1991. Preliminary description of *Thaisaurus chonglakmanii* n. g., n. sp., a new ichthyopterygian (Reptilia) from the Early Triassic of Thailand. *Comptes Rendus de l'Académie des Sciences*, Série II 313:1207-1212.

McGowan, C. 1972a. Evolutionary trends in longipinnate ichthyosaurs with particular reference to the skull and fore fin. *Life Sciences Contributions, Royal Ontario Museum* 83:1-38.

McGowan, C. 1972b. The distinction between latipinnate and longipinnate ichthyosaurs. *Life Sciences Occasional Papers, Royal Ontario Museum* 20:1-8.

McGowan, C. 1973. The cranial morphology of the Lower Liassic latipinnate ichthyosaurs of England. *Bulletin of the British Museum (Natural History)*, Geology 24:1-109.

McGowan, C. 1974. A revision of the latipinnate ichthyosaurs of the Lower Jurassic of England (Reptilia: Ichthyosauria). *Life Sciences Contributions, Royal Ontario Museum* 100:1-30.

McGowan, C. 1978. Further evidence for the wide geographical distribution of ichthyosaur taxa (Reptilia, Ichthyosauria). *Journal of Paleontology* 52:1155-1162.

McGowan, C. 1994. A new species of *Shastasaurus* (Reptilia: Ichthyosauria) from the Triassic of British Columbia: the most complete exemplar of the genus. *Journal of Vertebrate Paleontology* 14(2):168-179.

Merriam, J. C. 1905. The types of limb structure in the Triassic Ichthyosauria. *American Journal of Science*, Series 4 19:23-30.

Merriam, J. C. 1906. Preliminary note on a new marine reptile from the Middle Triassic of Nevada. *University of California Publications, Bulletin of the Department of Geology* 5(5):71-79.

Merriam, J. C. 1908. Triassic Ichthyosauria with special reference to the American forms. *University of California Memoir* 1(1).

Merriam, J. C. 1910. The skull and dentition of a primitive ichthyosaurian from the Middle Triassic. *University of California Publications, Bulletin of the Department of Geology* 5(24):381-390.

Merriam, J. C. 1911. Notes on the relationships of the marine saurian fauna described from the Triassic of Spitzbergen by Wiman. *University of California Publications, Bulletin of the Department of Geology* 6(13):317-327.

Nicholls, E. L. and D. B. Brinkman. 1996. A new ichthyosaur from the Triassic Sulphur Mountain Formation of British Columbia. IN W. A. S. Sarjeant (Ed.), *Vertebrate Fossils and the Evolution of Scientific Concepts*, pp. 521-535. Gordon and Breach Publishers, Amsterdam.

Nicholls, E. L., D. B. Brinkman, and J. M. Callaway. In review. New material of *Phalarodon* (Reptilia: Ichthyosauria) from the Triassic of British Columbia and its bearing on the interrelationships of mixosaurs. *Palaeontographica*.

Olson, E. C. 1971. *Vertebrate Paleozoology*. John Wiley and Sons, New York.

Peyer, B. 1968. *Comparative Odontology*. University of Chicago Press, Chicago and London.

Pinna, G. 1967. La collezione di rettili triassici di Besano (Varese) del Museo Civico di Storia Naturale di Milano. *Natura* 58:177-192.
Quenstedt, F. A. 1852. *Handbuch der Petrefactenkunde*. Tübingen.
Repossi, E. 1902. Il mixosauro degli strati triassici di Besano in Lombardia. *Atti della Società Italiana di Scienze Naturali e del Museo Civili di Storia Naturale* 41:361-372.
Romer, A. S. 1948. Ichthyosaur ancestors. *American Journal of Science* 246(2):109-121.
Romer, A. S. 1956. *Osteology of the Reptiles*. University of Chicago Press, Chicago.
Romer, A. S. 1966. *Vertebrate Paleontology*. University of Chicago Press, Chicago.
Romer, A. S. 1968. An ichthyosaur skull from the Cretaceous of Wyoming. *Contributions to Geology, University of Wyoming* 7(1):27-39.
Sander, P. M. and H. Bucher. 1990. On the presence of *Mixosaurus* (Ichthyopterygia: Reptilia) in the Middle Triassic of Nevada. *Journal of Paleontology* 64:161-164.
Sander, P. M. and J.-M. Mazin. 1993. The paleobiogeography of Middle Triassic ichthyosaurs: the five major faunas. IN Evolution, ecology and biogeography of the Triassic reptiles, pp. 145-152. *Paleontologia Lombarda della Società Italiana di Scienze Naturali e del Museo Civico di Storia Naturale di Milano Nuova Serie*, volume II--- 1993.
Shikama, T., T. Kamei, and M. Murata. 1978. Early Triassic ichthyosaurus, *Utatsusaurus hataii* gen. et sp. nov., from the Kitakami Massif, northeast Japan. *Tohoku University, Science Reports*, Series 2 48:77-97.
Stahl, B. J. 1974. *Vertebrate History: Problems in Evolution*. McGraw-Hill, New York.
Tarsitano, S. F. 1983. A case for the diapsid origin of ichthyosaurs. *Neues Jahrbuch für Geologie und Paläontologie, Monatshefte* 1983:59-64.
Wiman, C. 1910. Ichthyosaurier aus der Trias Spitzbergens. *Bulletin of the Geological Institutions of the University of Upsala* 10:124-148.
Wiman, C. 1916. Notes on the marine Triassic reptile fauna of Spitzbergen. *University of California Publications, Bulletin of the Department of Geology* 10(5):63-73.
Wiman, C. 1929. Eine neue Reptilien-Ordnung aus der Trias Spitzbergens. *Bulletin of the Geological Institutions of the University of Upsala* 22:183-196.
Yang, Z.-J. and Z.-M. Dong. 1972. *Chaohusaurus geishanensis* from Anhui Province. IN Z.-J. Yang and Z.-M. Dong (Eds.), Aquatic reptiles from the Triassic of China, pp. 11-14. *Academia Sinica, Institute of Vertebrate Palaeontology and Palaeoanthropology Memoir* 9 (in Chinese).
Young, C. C. 1964. On a revised determination of a fossil reptile from Jenhui, Kweichou with note on a new ichthyosaur probably from China. *Vertebrata PalAsiatica* 9(4):368-375.

Chapter 3

A TRANSITIONAL ICHTHYOSAUR FAUNA

CHRIS MCGOWAN

INTRODUCTION

Williston Lake, in northeastern British Columbia, is an immense artificial body of water, formed when the W. A. C. Bennett Dam was built across the Peace River in 1968 (Figure 1). Some of the rocks forming its shores belong to the Pardonet Formation, a Late Triassic marine horizon that extends from the late Carnian to the close of the Norian (Harland et al., 1990; McLearn, 1960). Ichthyosaurs are abundant, and are represented by a diverse array, from salmon-sized to whale-sized forms. Indeed, it is the diversity of form that sets Williston Lake apart from many other Triassic ichthyosaur localities, which tend to predominate in one species. The richly fossiliferous Middle Triassic locality of Monte San Giorgio, Switzerland, for example, predominates in *Mixosaurus* (Sander, 1989), while the Upper Triassic Luning Formation of Nevada is dominated by *Shonisaurus* (Camp, 1976, 1980). In this regard Williston Lake is more like the rich Lower Jurassic localities of southern England and Germany, where a wide diversity of species has been described.

The youngest Williston Lake ichthyosaurs are late Norian in age (Orchard, personal communication). They are therefore separated from the diverse Lower Jurassic assemblage of southern England by as little as only about 4 Ma (Harland et al., 1990), though most are probably more than about 10 Ma older. The available evidence suggests that ichthyosaurs had wide geographical ranges (Sander, 1989; McGowan, 1978); consequently,there would have been no geographical barriers between European and North American ichthyosaur assemblages. Given the small temporal separation between them, it is not surprising that the Williston Lake ichthyosaurs exemplify features similar to those of the English Lower Jurassic (McGowan, 1991b, 1995). This Late Triassic prelude to their Early Jurassic

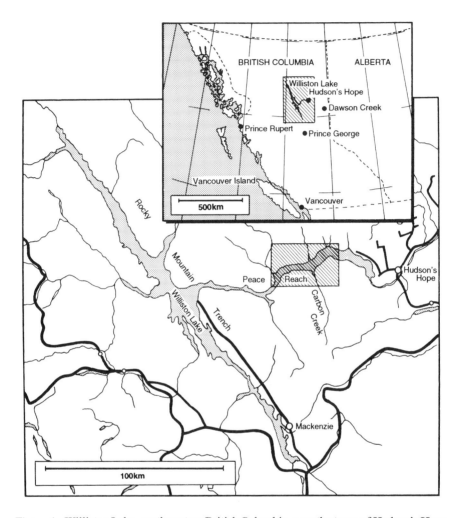

Figure 1. Williston Lake, northwestern British Columbia, near the town of Hudson's Hope.

radiation thus promises to provide some insights into the evolution of the Ichthyosauria at a critical time in their geological history.

The rate at which new Williston Lake specimens are being prepared is much lower than the rate they are being collected, so it will be many years before a complete picture emerges. This chapter is therefore intended as an interim report, to give some idea of the extent of the finds and of their diversity.

MATERIALS AND METHODS

The abbreviations used for the institutions referred to in the text are as follows: BMNH, The Natural History Museum (formerly the British Museum [Natural History]); GSC, Geological Survey of Canada; RBCM, Royal British Columbian Museum, Victoria; ROM, Royal Ontario Museum, Toronto; and SMNS, Staatliches Museum für Naturkunde, Stuttgart, Germany. Specimen numbers with the prefix of the year are field numbers and are included for completeness since they correspond to numbers included in the specimen log, located in the Department of Vertebrate Palaeontology, ROM. In accordance with guidelines for collecting paleontological material in British Columbia, specimens have been given RBCM accession numbers in addition to their ROM numbers.

Half-kilogram matrix samples were collected for each of the specimens for conodont analysis. Where possible, five samples were taken: at the level of the specimen, 50 cm and 100 cm above and 50 cm and 100 cm below. Conodont identifications for the more important ichthyosaurs have generously been made by M. J. Orchard of the GSC. Once the conodonts have been identified, the specimen locality is given a unique GSC locality number.

For descriptive purposes two types of digits are recognized, described as primary (or major) digits and accessory digits. A primary digit is one that takes proximal origin from the distal margin of the distal carpal series, whereas an accessory digit does not (McGowan, 1972). Accessory digits are often shorter than primary digits, frequently comprise small elements, often lie distal to the level of the distal carpal series, and are usually postaxial rather than preaxial in position. The total digital count is the sum of primary and accessory digits. The degree of hyperphalangy of a fin is given by counting the number of elements in the longest digit (McGowan, 1972). For convenience the count commences at the level of the epipodials, and continues to the distalmost element, even though this may be small and well separated from the rest. The count is therefore four elements higher than the number of phalanges in the longest digit. Forefin length is measured from the distal end of the humerus to the distal margin of the most distal phalanx (for details see McGowan, 1974a, and Table 1 of the same reference for definitions and other measurements used).

WILLISTON LAKE

The fossil-producing exposures at Williston Lake (Figures 1 and 2) thus far extend over a distance of about 16 miles (26 km). Lithologies vary from limestones and calcareous shales to siltstones and sandstones. There is also some variation in

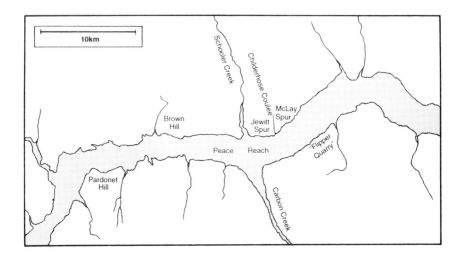

Figure 2. Peace Reach, a side branch of Williston Lake. The primary collecting localities are all Norian in age (Upper Triassic). From west to east these localities are: Pardonet Hill, Brown Hill, Jewitt Spur, McLay Spur, and "Flipper Quarry."

the quality of preservation of the enclosed fossils, from essentially uncrushed three-dimensional bones with excellent preservation to extremely crushed ones that readily crumble during collection. The collecting sites, which are on both sides of the lake, can be reached only by water, and all supplies have to be ferried in from the marina, a distance of about 20 miles (32 km). The water levels fluctuate by as much as about 49 ft (15 m) from a low in April to a high in July or August. The outcrops are therefore regularly scoured by the action of water and ice, as well as being abraded by the wind, so the rate of weathering in the most exposed areas is fairly high. The field season begins in early May, as soon as the lake is ice-free, and since water levels can rise by as much as 1 ft (30 cm) each day, the season ends in late June or early July, when most of the exposures have been flooded.

Most of the specimens are found in calcareous shales and siltstones. These usually not only are very hard, but also lack clean bedding planes, making collecting all the more difficult. The usual collecting procedure is to cut out blocks by drilling holes around the specimen with a jackhammer, using feathers and wedges to split the rock.

The ROM began prospecting at Williston Lake in 1987, and the first important specimen, an almost complete skeleton of *Shastasaurus*, was collected the following year (McGowan, 1994a). The initial reconnaissance trip and first three collecting seasons were led by Peter May, who deserves full credit for the Museum's presence at Williston Lake. Subsequent collecting trips have been ably led by Ian Morrison. The museum completed its fifth field season in 1994.

MATERIAL COLLECTED

Forty-five ichthyosaurs have been collected, ranging from isolated elements and associations of elements to complete or near-complete skeletons. In addition to these ichthyosaurs, ten fishes, four thalattosaurs, and a few elements of a fairly large sauropterygian have been collected. The first partial skeleton to be described from Williston Lake was that of a new species of *Shastasaurus, S. neoscapularis* (McGowan, 1994a; Figure 3). The specimen (ROM 41993) was collected at "Flipper Quarry" on the south shore, opposite McLay Spur (GSC locality number C-176546). The specimen is geologically slightly younger than the Californian *Shastasaurus* material (Merriam, 1908), being early Norian rather than late Carnian. The skeleton possesses features indicative of immaturity (McGowan, 1994a; Johnson, 1977), including the possession of a skull that appears to be fairly large compared with the length of the forefin (McGowan, 1994a). The entire animal would probably not have exceeded a total length of 2 m, and is probably best described as a subadult.

The skull has all the appearances of a typical Jurassic ichthyosaur like *Ichthyosaurus*, with a large orbit, set well back to give a narrow postorbital region, and with a small maxilla. Because of the incompleteness of the Californian material, it is not known whether these features are typical of the genus or of the species. The new species certainly differs from the other *Shastasaurus* material in its scapula, which is more like that of a Jurassic ichthyosaur than the hatchet shape that Merriam figured for *Shastasaurus* (Merriam, 1902:Plate 10, Figure 4). The pelvic girdle is also similar to that of Jurassic ichthyosaurs, the pubis and ischium being rod-like, in contrast to the plate-like elements that are characteristic of Triassic taxa. Although the skull is typically Jurassic, and the limb girdles are more Jurassic than Triassic in appearance, the forefin is quite unlike that of any post-Triassic ichthyosaur. Thus the humerus is short and broad, with a small anterior notch, and there is a prominent foramen enclosed between the radius and ulna, the latter feature being common to all Triassic ichthyosaurs. In marked contrast to the post-Triassic trend toward hyperdactyly, *Shastasaurus* is characterized by a reduction in the number of digits, and Callaway and Massare (1989) noted a tendency toward didactyly. Their contention is borne out by this geologically youngest representative of the genus, which has two digits with only traces of a third.

Although the small size of *Shastasaurus neoscapularis* may be due to its immaturity, the same is not true for an unusual small ichthyosaur recently described as a new genus and species, *Hudsonelpidia brevirostris* (McGowan, 1995). The holotype (ROM 44629), a near-complete skeleton (Figure 4), would have been less than 3 ft (1 m) long in life, but there are no indications of immaturity. A second,

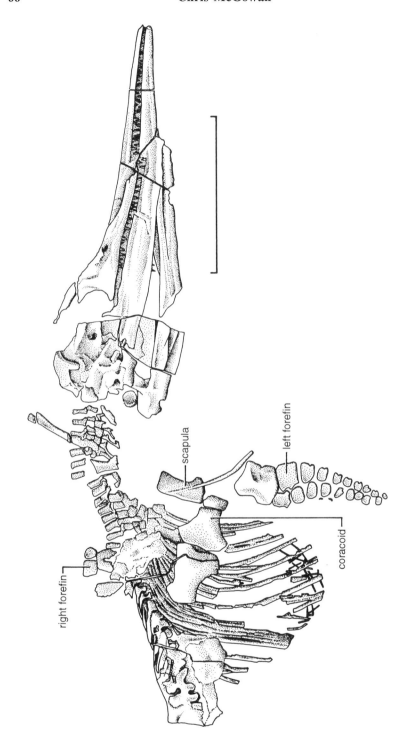

Figure 3. The holotype of *Shastasaurus neoscapularis* (ROM 41993) was the first partial skeleton to be described from Williston Lake. The specimen was collected from "Flipper Quarry" in 1988. Entire specimen. Scale bar = 200 m.

Figure 4. This small mature skeleton (ROM 44629) was collected in 1988 at Brown Hill and described as a new genus and species (McGowan, 1995). Scale bar =100 mm.

fragmentary, skeleton was described as the referred specimen (ROM 44633). It, too, is small, and was found in the same bedding plane as the holotype, at Brown Hill on the north shore (GSC locality number C-301269). In contrast to the other localities, Brown Hill has well-laminated strata that readily cleave along their bedding planes, greatly simplifying the collecting procedure. However, most of the specimens have undergone considerable compression, reducing them to thin wafers of bone. Furthermore, the strata are nearly vertical and the bedding planes are often invaded by tree roots, sometimes to a considerable depth, resulting in poor preservation. Aside from being compressed and weathered, the holotype is quite well preserved, revealing much of its anatomy.

The initial impression is that *Hudsonelpidia* resembles *Mixosaurus*, partly because of its small size. Closer inspection reveals a mixture of characters shared with *Mixosaurus* and with *Ichthyosaurus*. Like *Mixosaurus*, the skull has a relatively large orbit for a Triassic ichthyosaur, but this is even larger than in *Mixosaurus*, and in this regard it is similar to *Ichthyosaurus*. The snout is strikingly short, similar to that of *I. breviceps*, and this contributes to the relatively large orbit, which again is similar to that of *I. breviceps*. The forefin is incomplete, but that which has been preserved corresponds to the condition seen in *Ichthyosaurus*. Thus, the humerus is slender rather than short and broad, and the epipodials are rectangular rather than elongate. However, there is evidence of notching in the last digit, a feature shared with *Mixosaurus*. The hindfin is clearly more like the mixosaurian condition in that the tibia is elongate and broadly notched, rather than being compact and rectangular as in *Ichthyosaurus* and other post-Triassic taxa. Some of the more distal elements also have notches on their preaxial and postaxial margins, again a feature shared with *Mixosaurus* and other Triassic taxa, but not with post-Triassic ones. In contrast to *Mixosaurus*, the pelvic girdle lacks the broad, plate-like pubis and ischium that typifies Triassic species, these elements being more slender, like those seen in the post-Triassic. One of the most striking features of the hindfin is the large size of the femur, which is almost 75% as long as the rest of the hindfin.

At the time of description *Hudsonelpidia* was represented by two specimens, but a third was collected during the 1994 season (ROM 1994-3/ROM 44663/RBCM EH 91.2.10). The new material, which is still in its field jacket, possesses an articulated series of 43 vertebrae, and a humerus with the same elongate erosional feature as that seen in the holotype and referred specimen (McGowan, 1995). It is estimated that there would have been about ten vertebrae anterior to the level of the humerus, making the most posterior vertebra in the series the 53rd. Since there are less than 50 presacral vertebrae in the holotype, it is conceivable that the pelvic region has been preserved.

Very large ichthyosaurs are commonly encountered at Williston Lake, often as strings of vertebrae, ribs, and massive limb and girdle elements, comparable in size to the gigantic *Shonisaurus* from the late Carnian of Nevada (Camp, 1976, 1980). A large partial skeleton (Figure 5) was collected in 1991 from the base of Pardonet Hill (ROM 1991-1/ROM 44295/RCBM EH.91.2.6; GSC locality number C-301271). It comprises the rostral portion of the skull, the distal end of one fin and the proximal part of another, several series of articulated vertebrae totaling 67, numerous ribs, and some unidentified elements that probably belong to the girdles. The specimen was disposed in several blocks on a boulder-strewn section of the shore, and it was impossible to determine from where it had weathered. The specimen has been undergoing acid preparation, but more work is needed, especially on one of the fins. Detailed comparisons have to be made with *Shonisaurus* and with *Temnodontosaurus* and *Leptopterygius* before anything definitive can be said. Nevertheless, some preliminary remarks can be made, based largely on comparisons with Camp's figures and descriptions for *Shonisaurus*.

The distal fin segment is associated with the skull, and therefore probably represents a forefin. It appears to have three primary digits and at least one accessory digit. The absence of accessory digits on the other fin margin identifies it as being preaxial rather than postaxial. The other fin (Figure 6) is tentatively identified as a hindfin because of the similarity between its propodial and the femur of *Shonisaurus* (Camp, 1980:Figures 57b and 59a). The tibia also corresponds with that figured by Camp (1980) (compare Figure 6 with Camp's Figure 59b). Thus, the tibia has a wide emargination on its leading edge, and a marginally smaller one on its trailing edge. Only the anterior portion of the fibula has been preserved, and this has a preaxial emargination, so the epipodials enclose a foramen between them.

Camp (1980:Figures 50 and 65) described how notching occurred in one of the digits of the forefin, but was barely developed in the hindfin. The notched digit was said to be the last rather than the first, but this is contrary to the condition in all post-Triassic ichthyosaurs. It is also contrary to the forefin of *Shastasaurus* (McGowan, 1994a), and for an unnamed forefin described from Williston Lake (McGowan, 1991b). While it is true that notching does occur in the last digit in some Triassic taxa, including *Merriamia* and *Mixosaurus*, the first digit is also notched. It would therefore be most unusual if the only digit with notched elements were in the last rather than in the first digit. In both fins of the present material it is the first digit that is notched, so either Camp was mistaken in the orientation of the fins of *Shonisaurus*, or ROM 44295 cannot be referred to that genus. It also appears that the hindfin is as markedly notched as the forefin, which again is contrary to Camp's description for *Shonisaurus*. This suggests that, in spite of its similarity in size, the new material may not be referable to *Shonisaurus*.

The maximum number of elements preserved in the longest digit of the forefin is 16. The small size of the most distal element, and the fact that it is well spaced

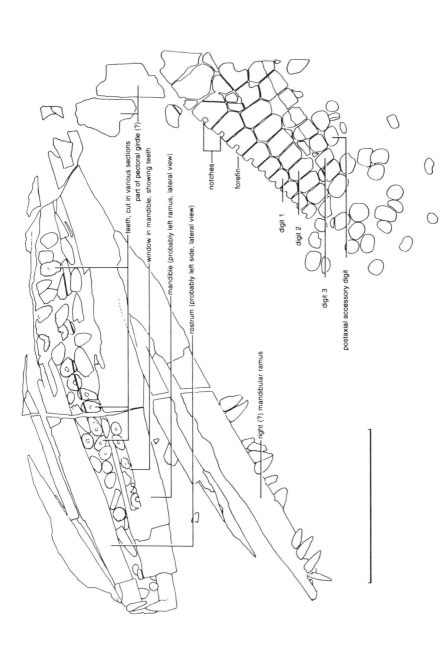

Figure 5. The incomplete skull, mandible, and fin (probably forefin) of a giant ichthyosaur (ROM 44295), collected from Pardonet Hill in 1991. Scale bar =100 mm.

from the rest, suggest that it probably represents the tip of the fin. Camp (1980:182, Figure 50) described 19 phalanges for each primary digit of the forefin of *Shonisaurus*, corresponding to a count of 23 for the number of elements in the longest digit. According to Camp's figure (1980:Figure 50), the 16 most distal phalanges comprise about half of the length of the fin (excluding the humerus). Therefore, if the present material were similar to *Shonisaurus* in its limb proportions, the preserved segment of the forefin, which is approximately 740 mm long, would only represent about half its length.

The cranial material comprises the distal portion of what appears to be the left side of the skull and mandible, seen in lateral view, and the displaced distal portion of the right mandibular ramus, which appears to extend to the tip. The main segment is about 1080 mm long. There are numerous teeth, and these have been weathered into oblique and transverse sections. The teeth are large and robust, the longest ones being about 67 mm long and 26 mm in diameter. Most of the teeth lie in the gap between the upper and lower jaw margins, but some are also exposed in windows that have been eroded through the mandible and snout. The teeth are so numerous that they appear to be contiguous, making it unlikely that they were set in individual sockets. Furthermore, the terminal teeth in the displaced mandibular ramus lie in shallow depressions that appear to lie at the base of a continuous dental groove (Figure 7). These terminal teeth are about 52 mm long

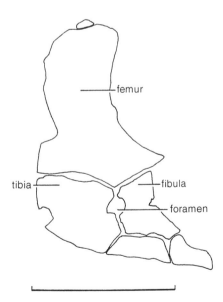

Figure 6. An incomplete fin, probably the hindfin, of the giant ichthyosaur from Pardonet Hill (ROM 44295). Scale bar = 200 mm.

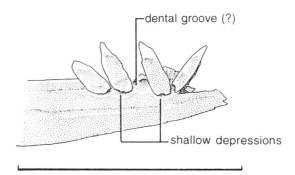

Figure 7. Terminal teeth in the displaced mandibular ramus of the giant ichthyosaur from Pardonet Hill (ROM 44295). Scale = 200 mm.

(root and crown) and about 16 mm in diameter. Camp (1980:164-166, Figure 22) figured and described the teeth of *Shonisaurus* as being well spaced and set in individual alveoli. The terminal mandibular teeth were described as being about half the size as the others---only about 15 mm high---which is less than one third of the length of the present teeth. These marked differences in dental features confirm that the new material cannot be referred to *Shonisaurus*.

Temnodontosaurus, from the English Lower Jurassic (Hettangian and Sinemurian), is another large species, though it is somewhat smaller than *Shonisaurus*. Like the new material, it has numerous large teeth that are set in a groove. However, it differs in having relatively shorter forefins, with fewer elements in the longest digit, and notching is restricted to the radius and succeeding one or two elements (McGowan, 1994b). *Leptopterygius burgundiae*, from the Lower Jurassic (Toarcian) of Germany and France, is a large species that appears to have more features in common with the new material than does *Temnodontosaurus*. The teeth are large and numerous, and the forefins are long and slender, with ≥ 21 elements in the longest digit, and with extensive notching (McGowan, 1979). However, there is no marked disparity in size between the tibia and fibula, nor do the epipodials of the fore- and hindfins enclose a foramen. Furthermore, *L. burgundiae* is somewhat smaller than the new material, the largest individual (SMNS 50,000) having a skull length of about 1480 mm, compared with an estimate of well over 2 m for the present skull. These differences show that this partial skeleton represents a new species.

Two more large specimens were collected at the close of the 1994 season. The first (ROM 1994-24/ROM 4461/RBCM EH.91.2.11), a partial skeleton comprising a string of vertebrae, ribs, and some unidentified elements, extended over a length of almost 8 m. The vertebral centra have diameters of about 273 mm, compared with only about 80 mm in the previous specimen. The second specimen (ROM 1994-27/ROM 44662/RBCM EH.91.2.9), a skull that is approximately 2.4 m long, may be associated with postcranial material which is still in the field. The skull has numerous large teeth that extended back from the tip of the snout for a distance of at least 1.5 m, which represents about 62% of the skull length. It seems likely that both of these specimens will be found to belong to the same taxon as the Pardonet specimen (ROM 44295). Two fragmentary specimens of large ichthyosaurs were also collected, and these appear to represent girdle and possibly limb elements (1994-4 and 1994-21). As there is presently so little information about them, they have not yet been assigned catalog numbers.

The first ichthyosaur described from Williston Lake was a single fin, from McLay Spur on the north shore (McGowan, 1991b; ROM 41991; GSC locality number C-301235). It was concluded that this was probably a forefin (Figure 8), and it appeared that there may have been more material still in the field. Subsequent quarrying has uncovered a series of about 20 vertebrae with associated ribs. Their orientation shows that the published fin was located at the anterior end of the rib cage, supporting the conclusion that it is a forefin. There are also remnants of another fin, this one lying in the pelvic region on the other side of the vertebral column, which is almost certainly a hindfin. A second specimen is associated with this material, comprising a series of postsacral vertebrae and two partial hindfins. It appears to be about the same size as the previous specimen, and although they may both belong to the same species, there are insufficient common features for comparison.

The forefin of ROM 41991 has much in common with the Lower Jurassic species *Ichthyosaurus communis*, *Leptopterygius tenuirostris*, and *Stenopterygius quadriscissus*, but, in the absence of a complete humerus, it was not possible to comment any further on its affinities (McGowan, 1991b). The same year that the forefin was found, a skull and partial forefin were collected at Jewitt Spur, an adjacent and contemporaneous locality on the north shore (ROM 1990-7/ROM 41992; RBCM EH.91.2.5; GSC locality number C-301270). The forefin has now been prepared, and is so similar to that of the previous specimen that they could well belong to the same species (Figure 9). Unfortunately, the humerus is missing, so it is not possible to determine which Jurassic species it resembles most. As in ROM 41991, there are four primary digits and one postaxial accessory digit. The maximum number of elements in the longest digit appears to be 16, though there could be one or two elements missing, and the maximum length is at least 220 mm. A prominent foramen is enclosed between the radius and ulna, a feature only hinted at in ROM 41991. The free margin of the radius and of the next six elements are notched, as in ROM 41991.

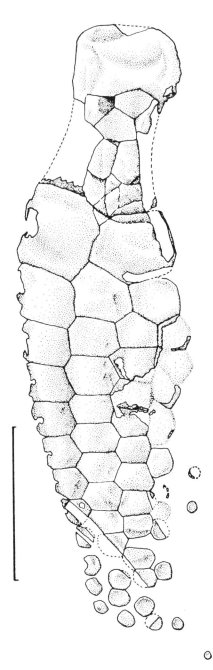

Figure 8. The first ichthyosaur described from Williston Lake was a single fin (ROM 41991), collected from McLay Spur in 1990 (McGowan, 1991b). Further remains of the specimen have subsequently been found. Scale bar = 100 mm.

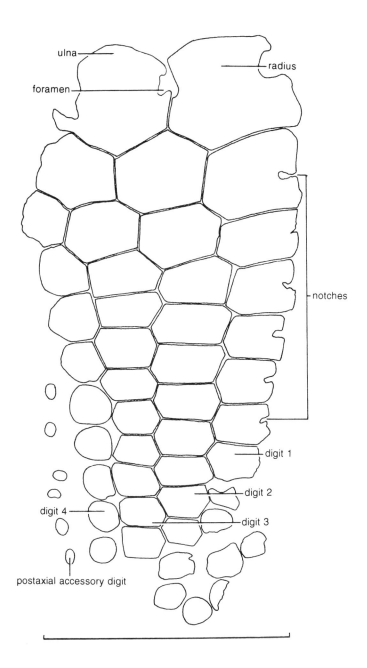

Figure 9. Partial forefin of a new short-snouted ichthyosaur (ROM 41992), collected from Jewitt Spur in 1990. Scale bar measures = 100 mm.

The skull was exposed from the right side, and although rather badly weathered, it was possible to obtain some measurements (Figure 10). This right side has now been embedded in clear epoxy resin, and the left side, which promises to be better preserved, is being acid-prepared. Preparation is not yet complete so the following preliminary account will be based on measurements obtained from the right side. The skull is remarkable for the brevity of the snout and the large size of the orbit. In both of these regards it compares closely with *Ichthyosaurus breviceps*, from the Lower Jurassic of England, the values for the snout and orbital ratios falling within the observed ranges for that species. However, *I. breviceps* is a much smaller species, the largest individual having a skull length of 239 mm (BMNH 43006) compared with an estimated length of 493 mm for ROM 41992. *Ichthyosaurus breviceps* also has a relatively longer and wider forefin, with at least five primary digits, a total digital count as high as nine, and up to 28 elements in the longest digit (McGowan, 1974a). The new material clearly cannot be referred to *I. breviceps*. Nor can it be referred to the large short-snouted species *Temnodontosaurus eurycephalus*, which has a relatively much smaller orbit (McGowan, 1974a). The new material clearly represents a new taxon, but its formal description will be postponed until the left side of the skull has been prepared. At this point it appears that almost all of its features are consistent with the emended diagnosis for *Ichthyosaurus* (McGowan, 1974b). Thus, the forefin has no fewer than four primary digits, the total digital count is not less than five, and the orbital ratio is greater than 0.20. The only discrepancy is in the

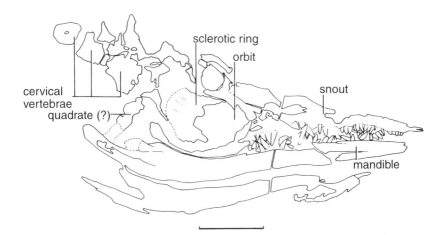

Figure 10. The right side of the skull of the new short-snouted ichthyosaur (ROM 41992). The skull is being acid-prepared from the other side, and will be described as a new species. Scale bar = 100 mm.

possession of a foramen enclosed between the radius and ulna, a feature that is characteristic of all Triassic ichthyosaurs. This foramen is an almost constant feature of *Leptopterygius tenuirostris*, an English Liassic species that extends across the Triassic-Jurassic boundary (McGowan, 1989). If *Ichthyosaurus* similarly crossed the boundary, it would be expected that the Triassic forms would possess this foramen too. Resolution of the generic status of this new taxon awaits completion of preparation, but at this point it seems that it may be referable to *Ichthyosaurus*. This would extend the geological range of that genus from the middle Norian to the Sinemurian, a temporal extension of roughly 8 Ma (based on Harland et al., 1990).

DISCUSSION

The new Williston Lake ichthyosaur fauna possesses unmistakably Triassic elements, as exemplified by *Shastasaurus*, and by the mixosaurian features of the recently described small new genus (McGowan, 1995). However, the fauna is predominately Jurassic in character, and many of the specimens would not look out of place if found in one of the Lower Jurassic localities of England or Germany. As already noted, it is possible that *Ichthyosaurus*, the commonest lower Liassic genus, may be represented at Williston Lake.

These new findings necessitate a change in our thinking of the tempo of ichthyosaurian evolution. The typical post-Triassic ichthyosaur, with its large orbit, narrow postorbital region, teeth set in grooves, and fins marked by their hyperphalangy, often hyperdactyly too, obviously appeared long before the close of the Triassic. These Triassic antecedents probably also had a fusiform body, with a markedly angled tailbend supporting a lunate tail, as seen in certain Holzmaden specimens in which the body outline has been preserved as a carbonaceous film. It is also likely that the more typically Triassic forms, like *Shonisaurus*, *Shastasaurus*, and *Cymbospondylus*, had tailbends; there is evidence for this in the occurrence of wedge-shaped centra in the caudal regions of *Shonisaurus* and *Cymbospondylus* (McGowan, 1991a). The thunniform body shape normally associated with post-Triassic ichthyosaurs must therefore have evolved very much earlier than the Early Jurassic, perhaps even before the Late Triassic. The close of the Triassic was therefore probably not marked by a major turnover in ichthyosaurian species. Instead, there was probably a more gradual period of transition, that may have lasted for most of the Late Triassic.

SUMMARY

Triassic ichthyosaurs are generally poorly known. A relatively new series of Upper Triassic localities in British Columbia is proving to be immensely productive, both in terms of the numbers of ichthyosaurs found and in their diversity. Two new taxa have already been described (McGowan, 1994a, 1995) and two more specimens, discussed here, probably represent two additional new forms. While some elements of this new ichthyosaur fauna are typically Triassic, as exemplified by *Shastasaurus*, others are more typical of the Jurassic. Indeed, there is evidence that the common Lower Jurassic genus *Ichthyosaurus* is represented. The close of the Triassic was therefore probably not marked by a mass turnover in ichthyosaurian species. Instead, there was probably a more gradual period of transition that may have lasted for most of the Upper Triassic.

ACKNOWLEDGMENTS

I thank Jan McCarthy and the Hudson's Hope Museum and Historical Society for their continued interest and support of our work. My thanks also to the BC Ministry of Crown Lands for collecting permission, and to Richard Hebda (RBCM).

Thanks to Ian Morrison for his efficient coordination of field operations, and to Raoul Bain, Ted Ecclestone, Tim Fedak, Brian Iwama, and Ryosuke Motani for their work in the field. Thanks also to Messrs. Bain, Ecclestone, Iwama, and Morrison (all from ROM) for preparation of the specimens. I am very grateful to Michael Orchard (GSC) for conodont identifications and discussions, and thank Peter Von Bitter, Kathy David (both from ROM), and Ted Ecclestone for conodont preparations. Julian Mulock drew all the figures, for which I thank him. I also thank Brian Boyle for photography, Hans-Dieter Sues for reading the manuscript and making many helpful suggestions, and Catherine Skrabec for clerical assistance (all from ROM). My thanks go to Don Brinkman and Jack Callaway for reviewing the manuscript, making many good suggestions for improvements.

This research was supported by Grant A 9550 from the Natural Sciences and Engineering Research Council of Canada.

REFERENCES

Callaway, J. M. and J. A. Massare. 1989. *Shastasaurus altispinus* (Ichthyosauria, Shastasauridae) from the Upper Triassic of the El Antimonio district, northwestern Sonora, Mexico. *Journal of Paleontology* 63:930-939.

Camp, C. L. 1976. Vorläufige Mitteilung über grosse Ichthyosaurier aus der oberen Trias von Nevada. *Sitzungsberichte der Österreichischen Akademie der Wissenschaften, Mathematisch-naturwissenshaftliche Klasse*, Abteilung 1 185:125-134.

Camp, C. L. 1980. Large ichthyosaurs from the Upper Triassic of Nevada. *Palaeontographica*, Abteilung A 170: 139-200.

Harland, W. B., R. L. Armstrong, A. V. Cox, L. E. Craig, A. G. Smith, and D. G. Smith. 1990. *A Geological Time Scale 1989*. Cambridge University Press, Cambridge, 263 pp.

Johnson, R. 1977. Size independent criteria for estimating relative age and the relationships among growth parameters in a group of fossil reptiles (Reptilia: Ichthyosauria). *Canadian Journal of Earth Sciences* 14:1916-1924.

McGowan, C. 1972. The distinction between latipinnate and longipinnate ichthyosaurs. *Life Sciences Occasional Papers, Royal Ontario Museum* 20:1-12.

McGowan, C. 1974a. A revision of the longipinnate ichthyosaurs of the Lower Jurassic of England, with descriptions of two new species (Reptilia: Ichthyosauria). *Life Sciences Contributions, Royal Ontario Museum* 97:1-37.

McGowan, C. 1974b. A revision of the latipinnate ichthyosaurs of the Lower Jurassic of England (Reptilia: Ichthyosauria). *Life Sciences Contributions, Royal Ontario Museum* 100:1-30.

McGowan, C. 1978. Further evidence for the wide geographical distribution of ichthyosaur taxa (Reptilia: Ichthyosauria). *Journal of Paleontology* 52:1155-1162.

McGowan, C. 1979. A revision of the Lower Jurassic ichthyosaurs of Germany with the description of two new species. *Palaeontographica*, Abteilung A 166:93-135.

McGowan, C. 1989. *Leptopterygius tenuirostris* and other long-snouted ichthyosaurs from the English Lower Lias. *Palaeontographica*, Abteilung A 166:93-135.

McGowan, C. 1991a. *Dinosaurs, Spitfires, and Sea Dragons*. Harvard University Press, Cambridge, Massachusetts, 365 pp.

McGowan, C. 1991b. An ichthyosaur forefin from the Triassic of British Columbia exemplifying Jurassic features. *Canadian Journal of Earth Sciences* 28:1553-1560.

McGowan, C. 1994a. A new species of *Shastasaurus* (Reptilia: Ichthyosauria) from the Triassic of British Columbia: the most complete exemplar of the genus. *Journal of Vertebrate Paleontology* 14:168-179.

McGowan, C. 1994b. *Temnodontosaurus risor* is a juvenile form of *T. platyodon* (Reptilia, Ichthyosauria). *Journal of Vertebrate Paleontology* 14:472-479.

McGowan, C. 1995. A remarkable small ichthyosaur from the Upper Triassic of British Columbia, representing a new genus and species. *Canadian Journal of Earth Sciences* 32:292-303.

McLearn, F. H. 1960. Ammonoid faunas of the Upper Triassic Pardonet Formation, Peace River Foothills, British Columbia. *Geological Survey of Canada Memoir* 311, 118 pp.

Merriam, J. C. 1902. Triassic Ichthyopterygia from California and Nevada. *University of California Publications, Bulletin of the Department of Geology* 3:63-108.

Merriam, J. C. 1908. Triassic Ichthyosauria, with special reference to the American forms. *Memoirs of the University of California* 1:1-196.

Sander, P. M. 1989. The large ichthyosaur *Cymbospondylus buchseri* sp. nov., from the Middle Triassic of Monte San Giorgio (Switzerland), with a survey of the genus in Europe. *Journal of Vertebrate Paleontology* 9:163-173.

Chapter 4

TEMPORAL AND SPATIAL DISTRIBUTION OF TOOTH IMPLANTATIONS IN ICHTHYOSAURS

RYOSUKE MOTANI

INTRODUCTION

One of the common features of post-Triassic ichthyosaurs is that their teeth are set in a longitudinal groove, which is often referred to as a "dental groove," in the upper and lower jaw margins. It has been suggested that the dental groove evolved from the thecodont condition found in some of the Triassic forms, such as *Mixosaurus cornalianus* (Merriam, 1908; Peyer, 1968; Mazin, 1983). Mazin (1983) coined the term *aulacodont* for the dental implantation in post-Triassic ichthyosaurs, defining it as a derivative of thecodont implantation. While post-Triassic ichthyosaurs are quite abundant and well preserved, especially those from the Jurassic, Triassic ones are comparably rare and poorly preserved. Consequently, our knowledge of ichthyosaurian dental biology is largely biased toward Jurassic forms, and little is known about the evolution of ichthyosaurian dentitions during the Triassic. However, a recent study of the mandibular dentitions of *Utatsusaurus hataii* and *Grippia longirostris*---among the most primitive ichthyosaurs from the Early Triassic---established the occurrence of a dental groove (Motani, in press, 1996). Thus, a reconsideration of the evolution of the ichthyosaurian dental groove is required.

When discussing dental implantation in ichthyosaurs, it is important to recognize that two kinds of implantation may occur within the same jaw ramus, depending on position. Therefore, it is useful to divide the dentition into four parts when describing tooth implantation, namely, the maxillary, premaxillary, anterior dentary, and posterior dentary dentitions. However, the published descriptions are scarcely available for all four parts, not only because complete jaw materials are rare, but also because most authors did not pay attention to the change in dental implantation within a given jaw ramus. Before discussing the evolution of the dental groove, a review of the geological history of various dental implantations among ichthyosaur species is necessary. Although a similar review has already been undertaken by Mazin (1983), he did not consider the occurrence of more than one kind of dental implantation within an individual, leading to some misunderstanding of the true situation.

MATERIALS AND METHODS

The abbreviations used for institutions are as follows: BMNH, Natural History Museum, London (formerly the British Museum [Natural History]); IGPS, Institute of Geology and Paleontology, Tohoku University, Sendai, Japan; PMU, Paleontologiska Museet, Uppsala Universitet, Sweden; UCMP, University of California, Museum of Paleontology, Berkeley. The specimens with the prefix SVT are stored in MNHN (Institut de Paléontologie du Muséum National d'Histoire Naturelle, Paris).

Tooth implantation and replacement in ichthyosaurs were reviewed based on a literature survey and on examination of some specimens. Specimens examined are: *Utatsusaurus hataii*---IGPS 95941 and 95942; *Grippia longirostris*---PMU R445 and R449, SVT 201 and 202; *Mixosaurus cornalianus*---BMNH R5702. Described specimens of *Pessosaurus polaris, Mixosaurus nordenskioeldii, Himalayasaurus tibetensis*, and *Shastasaurus neoscapularis* were also observed to confirm the descriptions in the literature. The summary of the review is given in Table 1.

TERMINOLOGY

The terminology for tooth implantation in amniotes varies among authors, leading to confusion; therefore, it is important to clarify the usage in this chapter. I essentially follow Romer (1956) and Edmund (1969), and recognize five basic types, namely, acrodonty, pleurodonty, subthecodonty, ankylosed thecodonty, and thecodonty (Figure 1). The term *aulacodonty* (Mazin, 1983) is employed for the implantation in post-Triassic ichthyosaurs, but without evolutionary implication as originally defined. Other terms are either synonymous with, or variations on, these five basic patterns.

Tooth Implantations in Ichthyosaurs

Table 1. Described tooth implantations of various ichthyosaurs. Descriptions in the literature are compiled for tooth implantation of ichthyosaurs, and are interpreted according to the terminology defined in the text.

EPOCH	TAXON	REFERENCE	Upper Jaw Premaxillary	Upper Jaw Maxillary	Lower Jaw Anterior	Lower Jaw Posterior
	Jurassic and Cretaceous ichthyosaurs general	see Text	dental groove	dental groove	dental groove	dental groove
U. Jurassic	Ichthyosaur from Normandy	Mazin, 1988	?	socket		?
U. Triassic	*Himalayasaurus tibetensis*	Dong, 1972	subpleurodont	subpleurodont	pleurodont	pleurodont
	Merriamia zitteli	Merriam, 1903		groove		
	UCPM 27141 (*?Shonisaurus*)	Callaway and Massare, 1989		socket		
	Shastasaurus neoscapularis	McGowan, 1994	groove	groove	groove	groove
	Shonisaurus popularis	Camp, 1980		socket		
		Figures 22, 23	socket	?	socket	socket
M. Triassic	*Cymbospondylus petrinus*	Merriam, 1908	?	groove and shallow socket	socket	socket
	Mixosaurus atavus	Fraas, 1891	groove and rudimentary socket	groove and rudimentary socket		
	Mixosaurus cornalianus	Repossi, 1902	groove	socket	groove	groove
		Besmer, 1947	socket	socket	groove	groove
	Mixosaurus nordenskioeldii	Wiman, 1910	more or less separated alveoli (= subthecodont?)			
		Table 5, Figure 8	subthecodont			
	Phalarodon fraasi	Merriam, 1910	subthecodont?	socket	?	socket
	Pessosaurus polaris	Wiman, 1910		groove		
	Large ichthyosaur from Tessin	Besmer, 1947	?	?	pleurodont	pleurodont
L. Triassic	*Crippia longirostris*	Wiman, 1929	pleurodont	pleurodont		
		Wiman, 1933	shallow alveoli (= subthecodont?)	shallow alveoli (= subthecodont?)		
		Mazin, 1981	fused to the bone (= subthecodont?)	fused to the bone (= subthecodont?)		
		Mazin, 1983	thecodont	thecodont		
		Motani, in press b	?	?	?	subthecodont
	Utatsusaurus hataii	Motani, in press a	?	?	pleurodont	subthecodont

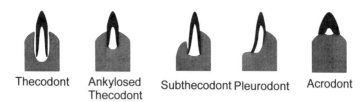

Figure 1. Five major types of tooth implantation, and several variations of them, are recognized in amniotes. After Romer (1956) and Edmund (1960).

Acrodonty

Definition: Teeth are ankylosed to the jaw bone (Miles and Poole, 1967). A dental groove or socket is absent, and the teeth are fixed to the margin of the jaw.
Example: Some lizards, and *Sphenodon*. Not known in ichthyosaurs.

Ankylosed Thecodonty

Definition: Teeth are set in sockets, which can be deep, up to a depth of about the height of the crown (Edmund, 1969). The surrounding bones of sockets are ankylosed to the teeth. Edmund (1969) pointed out that ankylosed thecodonty merges with subthecodonty because they can be distinguished only by the relative depth of the socket. Although it is difficult to set a clear line between the two, ankylosed thecodonty can be distinguished from subthecodonty by the absence of the dental groove in the former.
Example: Maxillary teeth in *Mixosaurus cornalianus*.

Aulacodonty

Definition: Mazin (1983) described the tooth implantation seen in post-Triassic ichthyosaurs as "aulacodonty," defining this as a derivative of thecodonty. However, as will be discussed later, the ichthyosaurian dental groove is not necessarily derived from the thecodont condition; therefore, this term is used here without evolutionary implication. The teeth are set in a longitudinal dental groove along the jaw margin, and there is no proper socket. Whether the teeth are ankylosed to the jaw bone is not well established in the literature. Mazin (1983) noted that the fixation is of a nonmineralized type, but did not provide any evidence. Should the fixation prove

to be of an ankylosis type with a shallow socket, aulacodonty would become a junior synonym for subthecodonty.
Example: Post-Triassic ichthyosaurs such as *Ichthyosaurus*.

Labial Pleurodonty

Rieppel (1978) explained that this term is used for pleurodont implantation where only the labial side of the tooth is in contact with the labial wall of the jaw; the lingual part of the tooth rests on a horizontal bony shelf that extends lingually from the bottom of the labial wall. He used this term to distinguish the complete pleurodonty seen in platynoan lizards from the implantation of many other lizards. However, in many cases the labial wall of the dental groove gradually shifts to the horizontal shelf toward the bottom; hence, it is very difficult to set a clear line between the labial wall and the shelf. Because pleurodonty and labial pleurodonty cannot always be distinguished, I follow Romer (1956) in considering labial pleurodonty as a variation of pleurodonty. Pleuroacrodonty (subacrodonty) of Wild (1973) may be similar to labial pleurodonty.

Pleuroacrodonty

See labial pleurodonty.

Pleurodonty

Definition: There is no proper socket, and the teeth are ankylosed to the surface of the jaw bones. A longitudinal dental groove with a high labial and low lingual wall may exist (Romer, 1956), or the lingual wall is lost (Edmund, 1969). I follow Romer's definition and consider that the loss of the lingual wall is not a necessary condition for pleurodonty. The teeth are mainly attached to the lingual side of the labial wall, while they may also be attached to the bottom of the dental groove.
Example: Varanid and iguanid lizards.

Pleurothecodonty

See subthecodonty.

Prothecodonty

See subthecodonty.

Protothecodonty

See subthecodonty.

Subacrodonty

See labial pleurodonty.

Subpleurodonty

The definition of *subpleurodonty* is not well established. Smith (1958) used this term to describe pleurodont implantation with the varanid type of tooth replacement (*sensu* Edmund, 1960), where replacement teeth occur in the interdental positions. Dong (1972) used this term for the implantation in a Late Triassic ichthyosaur, *Himalayasaurus tibetensis*, which is essentially pleurodont but "the jaw bone joins the roots of the teeth [translation]." Presch (1974) used this term for the implantation in some teiid lizards, such as *Dracaena guianensis*, where the pleurodont teeth are covered with extensive bone of attachment at their bases. Smith's (1958) subpleurodont is synonymous with pleurodont, because the only difference is the mode of replacement, which should not be considered in the terminology for tooth implantation. The usages of Dong (1972) and Presch (1974) are similar: in both cases, pleurodont implantation is strengthened by the well-developed bone of attachment. However, it is difficult to set a clear line between pleurodonty and subpleurodonty, because there is an intermediate state. For example, in the rhinoceros iguana, *Cyclura cornuta* (ROM R1154), the bone of attachment is better developed than in typical pleurodonty in varanid lizards, but it covers the roots only partially. Therefore, subpleurodonty is considered to be a variation of pleurodonty.

Subthecodonty

Definition: Teeth are set in shallow sockets arranged at the bottom of a longitudital dental groove with high labial and low lingual walls (Romer, 1956). Reduction of the lingual wall would result in the formation of a lingual shelf, as described for *Petrolacosaurus* (Reisz, 1981), but this is interpreted here as a variation of a dental groove. *Prothecodont* (Peyer, 1968), *protothecodont* (Edmund, 1969), and *pleurothecodont* (Wild, 1973) are synonymous with *subthecodont*, and are accordingly not used here to avoid confusion. The word *protothecodont* was also used by Wild (1973) in a sense which resembles that for *ankylosed thecodont* of Edmund (1969), but because this usage is confusing, it is not employed here.

Example: Most early amniotes, including *Paleothyris*, and *Petrolacosaurus*. Other tetrapods, such as *Seymouria*, also have this implantation.

Thecodonty

Definition: Teeth are set in sockets which are deeper than the height of the tooth crowns. There is no ankylosis between the teeth and the jaw bone, and the teeth are fixed to the jaw bone by fibrous organic connective tissue. The roots of the teeth are cylindrical.

Example: Crocodilians and many other archosaurs; mammals.

TOOTH IMPLANTATION AND REPLACEMENT IN ICHTHYOSAURS

Early Triassic Forms

Grippia longirostris **Wiman, 1929**

Known Parts: Maxillary, posterior part of the mandibular, and posteriormost part of the premaxillary dentitions.

Implantation: Wiman (1929) stated that dental implantation in *Grippia longirostris* is pleurodont, as in Recent varanid lizards. However, Wiman (1933) revised his previous description, and redescribed the teeth as being set in shallow sockets. Mazin (1983) described the dental implantation of the species as ankylosed thecodont, without specifying the part of the dentition. I have shown that implantation is subthecodont, at least in the posterior part of the mandible (Motani, in press).

Replacement: The arrangement of the maxillary teeth suggests that a replacement tooth occurs disto-lingual of a functional tooth, and migrates toward the latter to replace it (Motani, in press).

Utatsusaurus hataii Shikama, Kamei, and Murata, 1978

Known Parts: Mandibular dentition.

Implantation: The mandibular dentition is pleurodont at the anterior tip, gradually changing to subthecodont posteriorly (Motani, 1996). The teeth are well fused to the labial wall of the dental groove all along the jaw (Figure 2).

Replacement: A replacement tooth appears disto-lingual of a functional tooth, and migrates toward the latter to replace it (Motani, 1996). Formation of the resorption cavity in the functional tooth is not known.

Utatsusaurus sp. Nicholls and Brinkman (1993)

Known Parts: Maxillary, premaxillary, and partial mandibular dentitions.

Implantation: Nicholls and Brinkman (1993) reported that implantation is subthecodont, without specifying the position.

Replacement: Unknown.

Middle Triassic Forms

Cymbospondylus petrinus Leidy, 1886

Known Parts: All parts of the dentition.

Implantation: Merriam (1908) described the teeth of *Cymbospondylus petrinus* as set in distinct pits, at least in part of the mandible. He figured three cross sections of the mandibular teeth in horizontal, transverse, or disto-mesial section (Merriam, 1908:Figures 10-12). In the transverse section, it is seen that the bottom of the tooth is ankylosed to the bottom of the socket, while the upper part of the tooth is clearly free from the wall of the socket. This ankylosis at the bottom of the root is confirmed in the horizontal section, where the folded root of a mature tooth (i.e., the one associated with a replacement tooth) is shown. This implantation might be called thecodont if it were not for the ankylosis at the bottom of the socket and for the folding of the root. It also differs from ankylosed thecodont in that the ankylosis is restricted to the bottom of the socket. Therefore, tooth implantation in *C. petrinus* is tentatively described as ichthyosaurian thecodonty.

Replacement: Judging from Merriam (1908:Figure 10), tooth replacement in *C. petrinus* is similar to that in sauropterygians, such as nothosaurs (Edmund, 1960): a replacement tooth occurs in its own crypt, located lingual of a functional tooth.

Mixosaurus atavus (Quenstedt, 1852)

Known Parts: Maxillary and mandibular dentitions.

Implantation: Fraas (1891:38) noted that the teeth of *M. atavus* are fixed to the bone, although they are arranged in a common groove. Huene (1916:4)

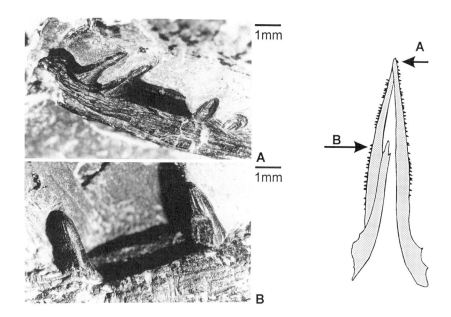

Figure 2. Dental groove of *Utatsusaurus hataii* (IGPS 95941, the holotype). The tooth implantation is pleurodont anteriorly, and subthecodont posteriorly. **A** and **B**) Photographs of the teeth and dental groove at the respective positions indicated in the drawing on the right.

redescribed the dentition of the species, and stated that the teeth are set in a groove, with rudimentary bony septa between the bottom portions of the roots. Tooth implantation is therefore probably subthecodont. Huene (1916:Plate 3, Figure 7) illustrated a cross section of an isolated jaw fragment, with a tooth which is unusual for having a long and tapering root that curves as it tapers. However, no such root morphology is known in other ichthyosaurs, and the jaw rami of other specimens of *M. atavus* are too shallow compared to the crown height to accommodate such a long root.
Replacement: Unknown.

Mixosaurus cornalianus (Bassani, 1886)

Known Parts: All parts of the dentition.
Implantation: Repossi (1902) gave a good description of the dental implantation of *M. cornalianus*: maxillary teeth are set in alveoli, while a dental groove is present elsewhere. He noted that the groove becomes very narrow between the teeth in the premaxilla and dentary, and gave a clear figure. However,

Besmer (1947) could not find the dental groove in the premaxilla, and stated that the teeth are set in distinct pits in the upper jaw while they are set in a groove in the mandible. The dental groove is known at least in the premaxillae of some mixosaur specimens from Tessin. Therefore, tooth implantation is probably subthecodont in the mandible and in the anterior part of the upper jaw, but becomes ankylosed thecodont posteriorly in the upper jaw, through the reduction of the dental groove.

Merriam (1908) briefly noted that the teeth are set in distinct pits in *M. cornalianus*, but did not specify the relative position in the dentition. Because he examined the same specimen as Repossi's (1902), I interpret that his comment was based on the maxillary teeth. Merriam (1910) described tooth implantation in early ichthyosaurs, including *M. cornalianus*, as being thecodont but his usage of the term is incorrect. Peyer (1968), referring to Besmer (1947), noted that the teeth of *M. cornalianus* are set in distinct alveoli. He provided a radiograph quoted from Besmer (1947) to show the presence of sockets (Peyer, 1968:Plate 62b). However, this radiograph was taken through the mandible of a large ichthyosaur from Tessin which does not belong to the genus *Mixosaurus*, and this ichthyosaur has a pleurodont implantation (Besmer, 1947). Therefore, Peyer's comment is irrelevant to *Mixosaurus*.

Replacement: Unknown.

Mixosaurus nordenskioeldii (Hulke, 1873)

Known Parts: Maxillary and premaxillary dentitions.

Implantation: Wiman (1910:130) noted that the teeth are set in "more or less separated alveoli." However, judging from his figure (Wiman, 1910:Plate 5, Figure 8), there seems to be a dental groove anteriorly in the upper jaw; therefore, tooth implantation is probably subthecodont in this region. It is possible that tooth implantation in the upper jaw is similar to that in *M. cornalianus*, where the subthecodont implantation in the anterior region becomes ankylosed thecodont posteriorly, through the reduction of the dental groove.

Replacement: Unknown.

Comment: I was informed that this species is being assigned to *Phalarodon* in a paper in review which I have yet to consult (Nicholls et al., in review).

Pessosaurus polaris (Hulke, 1873)

Known Parts: Jaw fragments of uncertain position.

Implantation: Wiman (1910) described that the teeth are set loosely in a groove. Judging from his figures (Wiman, 1910:Plate 7, Figure 7, and Plate 10,

Figure 28), the root is expanded and folded, as Mazin (1983) mentioned; therefore, subthecodont implantation is likely.

Replacement: Wiman (1910:Plate 10, Figure 28) shows a small tooth between a pair of mature teeth shifted to one side of the dental groove. This small tooth is probably a replacement tooth, and this side of the groove (top in his figure) is possibly lingual, because the dental lamina is located lingually in amniotes. A replacement tooth occurs outside the pulp cavity, possibly disto-lingual of each functional tooth: replacement teeth may move from distal to mesial in other ichthyosaurs, but never from mesial to distal.

Phalarodon fraasi Merriam, 1910

Known Parts: Maxillary and posterior mandibular dentitions.

Implantation: According to Merriam (1910), the posterior teeth are set in sockets in both upper and lower jaws, and the bones of the sockets surround the roots very closely. He noted that the sockets are located at the bottom of a shallow longitudinal groove, at least anteriorly in the upper jaw. The roots are associated with vertical grooves, and the presence of cementum is not established. The sockets cannot be very deep, judging from the depth of the jaws. Although bony fixation is not described, tight sockets and the vertical grooves of the roots are indicative of fixation by bony tissue. The presence of shallow sockets and of a dental groove suggests subthecodont implantation, at least anteriorly in the upper jaw. A dental groove is absent posteriorly, and implantation is therefore ankylosed thecodont. This arrangement of tooth implantation in the upper jaw is similar to that in *M. cornalianus*. Implantation is certainly not thecodont *sensu stricto*.

Replacement: Merriam (1910) described two tooth rows per maxilla, one row comprising smaller teeth than the other. *Phalarodon fraasi* has very robust posterior teeth, for which a durophagous diet has been proposed (Merriam, 1910). In the Recent durophagous lizard *Dracaena guianensis*, which similarly has robust posterior teeth, two tooth rows per maxilla are reported for young individuals (Dalrymple, 1979). The lingual tooth row of the maxilla of *D. guianensis* comprises replacement teeth for the labial row, and these are larger than the functional teeth: in this lizard species, tooth size, rather than tooth number, increases as it grows. A similar replacement pattern is possible for *P. fraasi*, or replacement never occurred and the tooth rows were continuously added as in *Captorhinus aguti*.

Comment: I was informed that this species is being synonymized with *P. nordenskioeldii* in a paper in review (Nicholls et al., in review).

Late Triassic Forms

Himalayasaurus tibetensis Dong, 1972

Known Parts: Almost all parts of the dentition.
Implantation: Dong (1972) stated that the upper and lower jaws of *Himalayasaurus tibetensis* have different dental implantations. He called the condition in the lower jaw pleurodont, and that in the upper jaw subpleurodont, a variation of pleurodont with an extensive bone of attachment covering the root.
Replacement: Unknown.

Merriamia zitteli (Merriam, 1903)

Known Parts: Maxillary and posterior mandibular dentitions.
Implantation: Merriam (1903) described the teeth as set in an open groove, with no evidence of bony partitions between the teeth. The teeth are numerous and closely packed, at least in the posterior portion of the jaw. Merriam figured a transverse section of the jaw, where a loose dental groove is seen. Therefore, tooth implantation seems to be similar to that described for post-Triassic ichthyosaurs.
Replacement: Unknown.

Shastasaurus neoscapularis McGowan, 1994

Known Parts: All parts of the dentition.
Implantation: McGowan (1994) described the teeth of *Shastasaurus neoscapularis* as being set in dental grooves. The holotype of the species has the only complete dentition of the genus. Although Merriam (1908) mentioned that *Shastasaurus* had a dental groove, there was no substantial support for this statement, as Callaway and Massare (1989) noted. *Shastasaurus neoscapularis* establishes the presence of dental grooves in *Shastasaurus* for the first time. The teeth are numerous and closely packed posteriorly, and the implantation seems to be similar to that described for post-Triassic ichthyosaurs (aulacodonty).
Replacement: Unknown.

Shonisaurus popularis Camp, 1976

Known Parts: All parts of the dentition.
Implantation: Camp (1980) figured and described deep alveoli for the mandibular dentition of *Shonisaurus popularis*: implantation in the upper jaw was figured for the premaxilla but was not mentioned in the text. The root, which is

much longer than the crown, is folded and there is a gap between the teeth and the wall of the sockets. According to Camp, the cementum covers the roots as a thin coating and fills the necks of the alveoli. Judging from his figure (Camp, 1980: Figure 23), there is possibly a bony fixation between the bottom of the root and the socket. This implantation would be described as thecodont if it were not for the folding of the root, the possible bony fixation at the bottom, and the cementum filling the neck of the alveolus. Because tooth implantation in *Cymbospondylus petrinus* is similar to this condition, tooth implantation in *S. popularis* is tentatively described as ichthyosaurian thecodonty. The teeth are well spaced.

Replacement: Camp noted that replacement teeth occur in pockets lying against the roots of old teeth. Camp (1980:Figure 2A) depicts two small crypts just beside alveoli partially connected to the latter. Their occurrence is similar to that of the replacement teeth in *Cymbospondylus petrinus*, and to those of sauropterygians. These two crypts are on the same side of the dental row, and although Camp did not specify the lingual direction in his figure, this side is likely to be lingual, considering the position of the dental lamina.

UCMP 27141 (? *Shonisaurus*) Callaway and Massare (1989)

Known Parts: Posterior part of the upper and lower dentitions.

Implantation: Callaway and Massare (1989) described how the teeth of UCMP 27141 are set in clearly defined sockets. They did not discuss whether the fixation was bony or not. The teeth are well spaced.

Replacement: Unknown.

Comment: Callaway and Massare (1989) assigned UCMP 27141 to *Shastasaurus altispinus* Merriam, 1902, based on features of the dorsal vertebrae and podial elements. However, these features are also known for *Shonisaurus*, a contemporary of *Shastasaurus* (compare Callaway and Massare, 1989:Figures 5 and 6, to Camp, 1980:Figures 29-30 and 50, respectively). Other features of this specimen, namely, the teeth set in deep sockets and the premaxilla excluding the nasal from the external naris, have yet to be confirmed in other specimens of *Shastasaurus*: the former character is known for both *Shonisaurus* and *Cymbospondylus*, and the latter for *Cymbospondylus*. According to Camp's (1980) description, the nasal region of *Shonisaurus* is represented by inadequate materials, so the reconstruction of this region is speculative. The identification of the specimen as *Shastasaurus altispinus* is therefore not well established, and it is possible that the specimen may be referable to *Shonisaurus* or related forms.

Triassic *Incertae Sedis*

Thaisaurus chonglakmanii Mazin, Suteethorn, Buffetaut, Jaeger, and Helmcke-Ingavat, 1991

Known Parts: Partial premaxillary and partial mandibular dentitions.
Implantation: Mazin et al. (1991) stated that the teeth are set in incomplete alveoli without ossified transverse septa, and tightly fused to the bone. The root is smooth-walled, which is exceptional for ichthyosaurs.
Replacement: Unknown.
Comment: Mazin et al. (1991) stated that the specimens are from the Lower Triassic, based on the elongated podial elements. However, elongated podial elements are known from other levels of the Triassic too. Moreover, the age should be determined by evidence independent of the material that is being described. Because associated ammonoids were identified as being Triassic (Mazin et al., 1991), the age of this species is regarded here as Triassic without further subdivision.

Post-Triassic Forms

The presence of a dental groove is well established for post-Triassic ichthyosaurs. For example, a dental groove has been described for *Platypterygius compylodon* from the Cretaceous (Owen, 1851), *Ophthalmosaurus icenicus* from the Middle and Late Jurassic (Andrews, 1910), *Ichthyosaurus quadriscissus* and *I. acutirostris* from the upper Liassic (Besmer, 1947), and *Ichthyosaurus* sp. from the lower Liassic (Sollas, 1916). However, tooth implantation is not described for many other post-Triassic ichthyosaurs.

One exception to this mode of implantation has been depicted by Mazin (1988), who described the presence of bony partitions between the posterior maxillary teeth of a partial ichthyosaurian skull from the Toarcian of France. Although this skull was previously identified as *Ichthyosaurus tenuirostris*, it lacks the diagnostic features, such as the long, slender snout and a well-constricted humeral shaft; hence, the identification is questionable (McGowan, personal communication). Moreover, *I. tenuirostris* is so far known from the Rhaetian to Sinemurian, and not from the Toarcian. Because the anterior part of the skull is missing, only the most posterior two tooth positions are preserved in the figured maxilla, and these are located in a very shallow longitudinal groove (Mazin, 1988:Plate 3a). This longitudinal groove seems to be identical to the dental groove described for the maxilla of *Ichthyosaurus* (McGowan, 1973), which becomes shallow and wide posteriorly. In Mazin's (1988) specimen, the bony partitions between the tooth positions seem to be incomplete, forming a bar rather than a wall (Mazin, 1983). Tooth implantation in the maxilla of post-Triassic ichthyosaurs has not been well documented; therefore, it is possible that the teeth are located in pits at the bottom of the dental groove in the most posterior part where the dental groove is shallow. Sollas (1916), who made serial cross sections of the skull of *Ichthyosaurus* sp. for every 1 mm, did not describe a bony partition between each pair of maxillary teeth. However, it is possible that 1 mm is not fine enough to detect a thin partition.

Tooth replacement in Jurassic ichthyosaurs was described by Edmund (1960). The replacement tooth emerges disto-lingual of each functional tooth, outside the pulp cavity, then moves mesio-labially to replace the functional tooth. A resorption cavity is formed in each functional tooth, allowing the replacement tooth to enter the pulp cavity.

DISCUSSION

Many authors, such as Merriam (1908, 1910), Edinger (1934), Peyer (1968), and Mazin (1983), have stressed that the teeth of Triassic ichthyosaurs are set in sockets, and Merriam and Mazin both applied the term *thecodont* to this condition. Their arguments are largely based on the condition in *Mixosaurus*. However, as I have shown, tooth implantation in *Mixosaurus* has been incorrectly interpreted by these authors, and there is no thecodonty in this genus. My compilation of described tooth implantations of ichthyosaurs shows that dental grooves are more dominant in Triassic ichthyosaurs than deep alveoli (Table 2). The only ichthyosaurs without dental grooves are those with ichthyosaurian thecodonty, which slightly differs from the thecodonty of archosaurs for having folded roots and possible bony fixation with the bottom of the sockets. Ichthyosaurian thecodonty is reported only for some of the large species from the Middle to Late Triassic, namely, *Cymbospondylus petrinus* and *Shonisaurus popularis*, and for UCMP 27141, which possibly belongs to *Shonisaurus*, as discussed earlier in this report.

It is likely that the subthecodont implantation of the oldest ichthyosaurs, such as *Utatsusaurus hataii*, is ancestral for the group, because subthecodonty is common among early amniotes. Accordingly, the presence of a dental groove, a shallow socket, and bony fixation seems to be plesiomorphic for the Ichthyosauria. The absence of a dental groove in *Cymbospondylus* and in *Shonisaurus* is therefore probably a derived character, assuming that ichthyosaurs are monophyletic. This shared derived character may establish the monophyly of the subfamily Cymbospondylinae Callaway, 1989, which was originally designated as a paraphyletic group.

Tooth implantation in ichthyosaurs has three essential elements, namely, a dental groove, sockets, and bony fixation. Depending on how the character states for these three are combined, four types of tooth implantation, which I refer to as the subthecodont, ankylosed thecodont, aulacodont, and ichthyosaurian thecodont types, are recognizable. A brief summary of the features of each type is given in Table 2 and Figure 3, while stratigraphical distribution is summarized in Table 3.

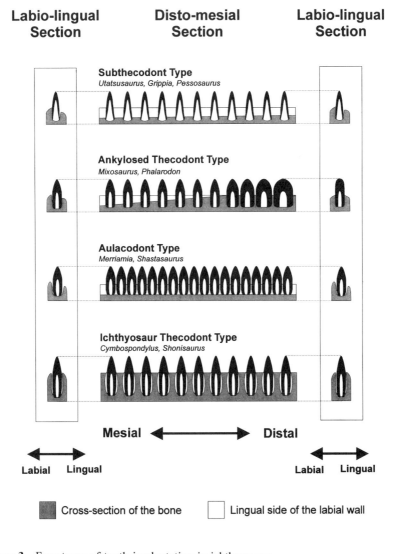

Figure 3. Four types of tooth implantation in ichthyosaurs.

Tooth Implantations in Ichthyosaurs 97

	Dental Groove		Sockets		Root	Taxonomic Distribution
	Anterior	Posterior	Anterior	Posterior		
Subthecodont Type	Y	Y	Shallow	Shallow	Expanded	*Grippia, Utatsusaurus, Pessosaurus*
Ankylosed Thecodont Type	Y	N	Shallow	Moderate	Straight	*Mixosaurus, Phalarodon*
Ichthyosaurian Thecodont Type	N	N	Deep	Deep	Straight	*Cymbospondylus, Shonisaurus*
Aulacodont Type	Y	Y	None	None	Straight	*Merriamia, Shastasaurus, post-Triassic species*

Table 2. Features of the four types of tooth implantation in ichthyosaurs.

	L. Triassic		M. Triassic		Stratigraphical Distribution			
	...Spathian	Anisian	Ladinian	Carnian	U. Triassic	Norian	Rhaet.	post-Triassic
Subthecodont Type	▓	▓						
Ankylosed Thecodont Type		▓	▓					
Aulacodont Type				▓	▓	▓	▓	
Ichthyosaurian Thecodont Type								▓

Table 3. Stratigraphical distribution of the four types of tooth implantation in ichthyosaurs.

Subthecodont Type

Taxonomic Distribution: *Utatsusaurus, Grippia*, and *Pessosaurus.*
Stratigraphical Distribution: Lower to Middle Triassic.
Description: Dental groove is present for the entire tooth-bearing portion of the jaw margin. The teeth are set in shallow sockets located at the bottom of the dental groove to which they are fused. The root of the tooth is expanded, the pulp cavity is open, and the walls of the root are folded.

Ankylosed Thecodont Type

Taxonomic Distribution: *Mixosaurus* and *Phalarodon.*
Stratigraphical Distribution: Middle Triassic.
Description: Dental groove is present anteriorly, shallowing posteriorly, and may be absent in the most posterior region. The teeth are usually set in shallow sockets, but posteriorly, where the dental groove is absent, the sockets may be as deep as the height of the crowns. The roots are not very much expanded, and are straight in many cases. The sockets fit tightly to the roots, and fixation is probably bony.

Aulacodont Type

Taxonomic Distribution: *Merriamia, Shastasaurus*, possibly *Himalayasaurus*, and post-Triassic ichthyosaurs.
Stratigraphical Distribution: Upper Triassic to Cretaceous.
Description: The teeth are set in a common dental groove which forms the margin of the tooth-bearing portions of the jaws. The teeth may fit inside the groove tightly or loosely, and there is no complete bony partition between the teeth. Bony fixation may be present at the bottom of the groove, but this is not well established in the literature. The roots of the teeth are not expanded but are straight.

Ichthyosaurian Thecodont Type

Taxonomic Distribution: *Cymbospondylus* and *Shonisaurus.*
Stratigraphical Distribution: Middle to Upper Triassic.
Description: Dental groove is absent and the teeth are set in deep sockets which are deeper than the height of the crowns. Teeth seem to be fused to the

bottom of the socket; therefore, implantation is not truly thecodont, although there is a gap between the wall of the socket and the tooth. The roots of the teeth are not expanded. Replacement teeth occur in their own crypts, located lingual of the socket of functional teeth.

Although Mazin (1983) suggested that the dental grooves of post-Triassic ichthyosaurs were derived from the thecodont condition in Triassic ichthyosaurs, this is not necessary because the dental groove is probably plesiomorphic for ichthyosaurs. The aulacodont condition could be derived from either the subthecodont or ankylosed thecodont types. The nonexpanded root of post-Triassic ichthyosaurs is probably a derived character, but whether this character is homologous with that of the ichthyosaurian thecodont condition cannot be established without a comprehensive phylogenetic analysis of the Ichthyosauria.

The ankylosed thecodont type of implantation can easily be derived from the subthecodont condition, through a strengthening of the tooth fixation in the posterior region of the dentition. All ichthyosaurs with the ankylosed thecodont type of implantation have varying degrees of differentiation in their dentition, more robust teeth being located in the posterior region than in the anterior region; therefore, there is a functional advantage for a stronger fixation of the posterior teeth. The tendency of the posterior teeth toward being more strongly fixed to the bone than the anterior teeth is seen also in the upper jaw of an ichthyosaur from the Toarcian of France (Mazin, 1988) and in the mandible of *Utatsusaurus* (Motani, 1994). This is reasonable on functional grounds: the stress resulting from the adduction of the jaw is higher in the posterior portion of the dentition; hence stronger attachment of the teeth is functionally adaptive.

Fraas (1891) noted the presence of both a groove and of sockets in *M. atavus* is very similar to the condition in young crocodilians, but the conditions are not the same. In young individuals of some crocodilians, such as *Alligator mississippiensis*, dental grooves occur posteriorly because ossification of interalveolar bony septa, which starts anteriorly in the jaw ramus, is still incomplete in this region. In *M. atavus*, dental grooves are present anteriorly, while posterior teeth are set in independent alveoli.

Peyer (1968) stated that Triassic ichthyosaurs, which he collectively called "mixosaurs," have slight or no folding of the roots. This statement is largely based on Besmer (1947), who figured cross sections of the teeth of *Mixosaurus cornalianus* without the folding of the dentine wall, but this is the only example of ichthyosaurian teeth lacking deep folding of the dentine wall (plicidentine). However, plicidentine is reported even for Early Triassic ichthyosaurs, e.g., *Grippia longirostris* (Mazin, 1981:Figure 7b) and isolated teeth from Spitsbergen (Wiman, 1910:Plate 10, Figures 24-27). It is possible that Besmer's (1947) cross sections were taken at a higher level than the plicated part of the dentine. However, since plicidentine in *Grippia* is reported only for the bulbous maxillary teeth, it is necessary to cross-section the teeth of *Utatsusaurus*, which shows less functional adaptation than those of *Grippia*, before concluding whether plicidentine is a universal feature among ichthyosaurs. Unfortunately, scarcity of material prevents such a destructive study (Motani, 1996).

All ichthyosaurs have replacement teeth that occur outside the pulp cavity. Although this may be an important synapomorphy for the order, it has to be tested against a reasonable hypothesis for the ichthyosaurian relationship to other amniotes. Formation of a resorption cavity in the functional tooth that is being replaced is reported only for Jurassic ichthyosaurs; therefore, it is probably a derived feature.

SUMMARY

Four essential types of dental implantation are recognized for ichthyosaurs, namely, the subthecodont, ankylosed thecodont, ichthyosaurian thecodont, and aulacodont types. The subthecodont type, as exemplified by the oldest ichthyosaurs such as *Utatsusaurus*, is common among early amniotes; therefore, subthecodonty is probably plesiomorphic for ichthyosaurs. Some authors described *Mixosaurus* and *Phalarodon* as having thecodont implantation, but they actually possess dental grooves anteriorly and ankylosed thecodont implantation posteriorly; therefore, they are categorized as being of the ankylosed thecodont type. The ichthyosaurian thecodont type is rare and has only been described for *Shonisaurus* and *Cymbospondylus*. The aulacodont type, which is the commonest, is dominant among post-Triassic ichthyosaurs, as well as in Late Triassic species. The previously proposed derivation of the aulacodont type from the thecodont is not necessarily so, because the presence of a dental groove is probably plesiomorphic for ichthyosaurs. Replacement teeth always occur outside the pulp cavities in ichthyosaurs, and this feature may be an important synapomorphy of the group.

ACKNOWLEDGMENTS

Dr. C. McGowan supplied generous educational and financial support to the present study. I thank Dr. H.-D. Sues for the discussion of dental implantation in amniotes. Dr. R. MacDougall, known as R. Johnson for her work on ichthyosaurs, patiently corrected my written English in the earlier drafts of this chapter. Dr. G. Deiuliis kindly translated Italian text into English. Access to the specimens was facilitated by Drs. D. J. Goujet (MNHN), I. Hayami (Kanagawa University and University of Tokyo), K. Mori (IGPS), and S. Stuenes (PMU). Dr. J. M. Callaway and an anonymous reviewer gave useful suggestions which improved the quality of the manuscript. Special thanks go to the editors of this volume for their support. This study was financially supported by an NSERC grant provided to C. McGowan (A 9550).

REFERENCES

Andrews, C. W. 1910. A descriptive catalogue of the marine reptiles of the Oxford Clay. Part 1. Printed by Order of the Trustees of the British Museum, London.

Bassani, F. 1886. Sui fossili e sull'età degli schisti bituminosi triasici di Besano in Lombardia. Communicazione preliminare. *Atti della Società Italiana di Scienze Naturali e del Museo Civili di Storia Naturale* 29:15-72.

Besmer, A. 1947. Beiträge zur Kenntnis des Ichthyosauriergebisses. *Schweizerische Palaeontologische*, Abhandlungen 65:1-21.

Callaway, J. M. 1989. *Systematics, Phylogeny, and Ancestry of Triassic Ichthyosaurs (Reptilia, Ichthyosauria)*. Unpublished Ph. D. dissertation, University of Rochester, Rochester, New York.

Callaway, J. M. and J. A. Massare. 1989. *Shastasaurus altispinus* (Ichthyosauria, Shastasauridae) from the Upper Triassic of the El Antimonio district, northwestern Sonora, Mexico. *Journal of Paleontology* 63:930-939.

Camp, C. L. 1976. Vorläufige Mitteilung über grosse Ichthyosaurier aus der oberen Trias von Nevada. *Österreichische Akademie der Wissenschaften, Mathematisch-Naturwissenschaftliche Klasse, Sitzungsberichte*, Abteilung I 185:125-134.

Camp, C. L. 1980. Large ichthyosaurs from the Upper Trassic of Nevada. *Paläontographica*, Abteilung A 170:139-200.

Dalrymple, G. H. 1979. On the jaw mechanism of the snail-crushing lizards, *Dracaena* Daudin 1802 (Reptilia, Lacertilia, Teiidae). *Journal of Herpetology* 13:303-311.

Dong, Z.-M. 1972. An ichthyosaur fossil from the Qomolangma Feng region. IN Z.-J. Yang and Z.-M. Dong (Eds.), Aquatic reptiles from the Triassic of China, pp. 7-10. *Academia Sinica, Institute of Vertebrate Paleontology and Paleoanthropology Memoir* 9 (In Chinese).

Edinger, T. von. 1934. *Mixosaurus*-Schädelrest aus Rüdersdorf. *Jahrbuch der Preußischen Geologischen Landesanstalt* 55:341-347.

Edmund, A. G. 1960. Tooth replacement phenomena in the lower vertebrates. *Life Sciences Contributions, Royal Ontario Museum* 52:1-190.

Edmund, A. G. 1969. Dentition. IN C. Gans, A. d'A. Bellairs, and T. S. Parsons (Eds.), *Biology of the Reptilia* 1, pp. 117-200. Academic Press, London.

Fraas, E. 1891. *Die Ichthyosaurier der Süddeutschen Trias - und Jura-Ablagerungen.* Tübingen.

Huene, F. von. 1916. Beiträge zur Kenntnis der Ichthyosaurier im deutschen Muschelkalk. *Paläontographica* 62:1-68.

Hulke, J. W. 1873. Memorandum on some fossil vertebrate remains collected by the

Swedish expedition to Spitzbergen in 1864 and 1868. *Bihang till K. Svenska Vetenskapsakademiens Handlingar* 1, Afdelning IV 9.

Leidy, J. 1868. Notice of some reptilian remains from Nevada. *Proceedings of the Philadelphia Academy of Science* 20:177-178.

Mazin, J.-M. 1981. *Grippia longirostris* Wiman 1929, un Ichthyopterygia primitif du Trias inférieur du Spitsberg. *Bulletin du Muséum National d'Histoire Naturelle* 4:317-340.

Mazin, J.-M. 1983. L'implantation dentaire chez les Ichthyopterygia (Reptilia). *Neues Jahrbuch für Geologie und Paläontologie*, Monatschefte 1983:406-418.

Mazin, J.-M. 1988. Le crane d'*Ichthyosaurus tenuirostris* Conybeare 1822 (Toarcian, La Caîne, Normandie, France). *Bulletin de la Société Linnéenne de Normandie* 112-113:121-132.

Mazin, J.-M., V. Suteethorn, E. Buffetaut, J.-J. Jaeger, and R. Helmcke-Ingavat. 1991. Preliminary description of *Thaisaurus chonglakmanii* n. g., n. sp., a new ichthyopterygian (Reptilia) from the Early Triassic of Thailand. *Comptes Rendus de l'Académie des Sciences*, Série II 313:1207-1212.

McGowan, C. 1973. The cranial morphology of the Lower Liassic latipinnate ichthyosaurs of England. *Bulletin of the British Museum (Natural History)*, Geology 24:1-109.

McGowan, C. 1994. A new species of *Shastasaurus* (Reptilia: Ichthyosauria) from the Triassic of British Columbia: the most complete exemplar of the genus. *Journal of Vertebrate Paleontology* 14:168-179.

Merriam, J. C. 1903. New Ichthyosauria from the Upper Triassic of California. *University of California Publications, Bulletin of the Department of Geology* 3:249-263.

Merriam, J. C. 1908. Triassic Ichthyosauria, with special reference to the American forms. *Memoirs of the University of California* 1:1-196.

Merriam, J. C. 1910. The skull and dentition of a primitive ichthyosaurian from the Middle Triassic. *University of California Publications, Bulletin of the Department of Geology* 5:381-390.

Miles, A. E. W. and D. F. G. Poole. 1967. The history and general organization of dentitions. IN A. E. W. Miles (Ed.), *Structural and Chemical Organization of Teeth*, pp. 3-44. Academic Press, New York and London.

Motani, R. 1994. Temporal and spatial distribution of tooth implantations in ichthyosaurs [Abstract]. *Journal of Vertebrate Paleontology* 14, Supplement to Number 3:39A.

Motani, R. 1996. Redescription of the dental features of an Early Triassic ichthyosaur *Utatsusaurus hataii*. *Journal of Vertebrate Paleontology* 16(3):396-402.

Motani, R. In press. Redescription of the dentition of *Grippia longirostris* with a comparison with *Utatsusaurus hataii*. *Journal of Vertebrate Paleontology*.

Nicholls, E. L. and D. Brinkman. 1993. An new specimen of *Utatsusaurus* (Reptilia: Ichthyosauria) from the Lower Triassic Sulphur Mountain Formation of British Columbia. *Canadian Journal of Earth Sciences* 30:486-490.

Nicholls, E. L., D. B. Brinkman, and J. M. Callaway. In review. New material of *Phalarodon* (Reptilia: Ichthyosauria) from the Triassic of British Columbia and its bearing on the interrelationships of mixosaurs. *Palaeontographica*.

Owen, R. 1851. Monograph of the fossil Reptilia of the Cretaceous formations.

Palaeontographical Society Monograph 5:1-118.

Peyer, B. 1968. *Comparative Odontology*. University of Chicago Press, Chicago and London.

Presch, W. 1974. A survey of the dentition of the macroteiid lizards (Teiidae: Lacertilia). *Herpetologica* 30:344-349.

Quenstedt, F. A. 1852. *Handbuch der Petrefactenkunde*. Tübingen.

Reisz, R. R. 1981. A diapsid reptile from the Pennsylvanian of Kansas. *Special Publication of the Museum of Natural History, University of Kansas* 7:1-74.

Repossi, E. 1902. Il mixosauro degli strati Triasici di Besano in Lombardia. *Atti della Società Italiana di Scienze Naturali* 41:61-72.

Rieppel, O. 1978. Tooth replacement in Anguinomorph lizards. *Zoomorphologie* 91:77-90.

Romer, A. S. 1956. *Osteology of the Reptiles*. University of Chicago Press, Chicago and London.

Smith, H. M. 1958. Evolutionary lines in tooth attachment and replacement in reptiles: their possible significance in mammalian dentition. *Transactions of the Kansas Academy of Science* 61:216-225.

Sollas, W. J. 1916. The skull of *Ichthyosaurus* studied in serial sections. *Philosophical Transactions of the Royal Society of London* B 208:63-126.

Wild, R. 1973. *Tanystropheus longobardicus* (Bassani). IN E. Kuhn-Schnyder and B. Peyer (Eds.), Die Triasfauna der Tessiner Kalkalpen 23, pp. 1-162. *Schweizerische Palaeontologische*, Abhandlungen 95.

Wiman, C. 1910. Ichthyosaurier aus der Trias Spitzbergens. *Bulletin of the Geological Institutions of the University of Uppsala* 10:124-148.

Wiman, C. 1929. Eine neue Reptilien-Ordnung aus der Trias Spitzbergens. *Bulletin of the Geological Institutions of the University of Uppsala* 22:183-196.

Wiman, C. 1933. Über *Grippia longirostris*. *Nova Acta Regiae Societatis Scientiarum Upsaliensis* 9:1-19.

PART II
Sauropterygia

Part II: Sauropterygia

INTRODUCTION

OLIVIER RIEPPEL

The Sauropterygia is a monophyletic clade of diapsid reptiles which invaded the Mesozoic seas. It comprises such popular fossils as pachypleurosaurs, nothosaurs, plesiosaurs, and pliosaurs. The clade can be divided into stem-group Sauropterygia from the Triassic, and crown-group taxa from the Jurassic and Cretaceous. The first appearance of stem-group sauropterygians is in the latest Early Triassic of central Europe and of the western United States (perhaps also of China, where the stratigraphic control is less stringent), and a first radiation populated the Middle Triassic epicontinental seas and near shore areas of the eastern (China) and western (Europe) Tethyan Province. Stem-group taxa were superseded by the crown-group plesiosaurs and pliosaurs invading the open marine habitat and achieving a cosmopolitan distribution. The first occurrence of diagnostic plesiosaurs is at the Triassic-Jurassic boundary in England (Storrs and Taylor, 1993). The enigmatic genus *Pistosaurus* from the upper Muschelkalk (lower Ladinian) of Germany is believed to be the sister-taxon to the plesiosaurs (Sues, 1987a). The Sauropterygia became extinct in the Late Cretaceous.

The history of analysis of sauropterygian fossils goes back to the very beginning of the science of paleontology. The initial discovery and study of sauropterygians from the Lower Jurassic and Cretaceous laid the foundation for vertebrate paleontology in England (Taylor, Foreword). Sauropterygian remains from the Germanic Muschelkalk triggered the development of vertebrate paleontology in Germany (Weiss, 1983; Freyberg, 1972), and received a first comprehensive treatment in a beautifully illustrated folio compendium published by H. von Meyer during the years from 1847 to 1855. In Italy, vertebrate paleontology started with the description of fossils from Middle Triassic deposits in the southern Alps, including sauropterygians (Pinna and Teruzzi, 1991). In spite of this widespread early interest in that group, it took a long time and much effort to sort out sauropterygian interrelationships.

A number of basic subgroups of the Sauropterygia were readily recognized, such as the plesiosaurs and pliosaurs with their streamlined skulls, needle-shaped teeth, elongated necks, and limbs transformed to hydrofoils. Nothosaurs and their allies were perceived as a distinct group, but the distinction between the relatively large nothosaurs and the generally smaller pachypleurosaurs of similar habitus was less easily drawn.

Another major problem was the interpretation of the large, black, and shiny crushing tooth plates found in the upper Muschelkalk deposits around Bayreuth in Germany. The first skull bearing such teeth was found in 1824, and described by Münster (1830), who solicited Louis Agassiz' help in its identification. Agassiz attributed the remains to an as yet unknown genus of durophagous fish (Bronn, 1831), which he later named *Placodus* (Agassiz, 1833-1845). It was left to Owen (1858) to recognize the reptilian nature of that genus. Comparing *Placodus* to *Simosaurus*, Owen postulated sauropterygian affinities of placodonts. He later (Owen, 1860) formalized this view by the inclusion of nothosaurs, plesiosaurs, and placodonts, as well as some other enigmatic fossils, in his Sauropterygia.

The stem-group Sauropterygia include three major monophyletic clades (Figure 1), the Placodontia, the Pachypleurosauroidea, and Triassic Eusauropterygia. The history of analysis of the Placodontia has been reviewed in detail by Nosotti and Pinna (1989). The first skull ever collected, and figured by Münster (1830) as "specimen I," is the holotype of *Placodus gigas*, a genus which remains restricted to the lower and upper Muschelkalk of the Germanic Triassic. Together with *Paraplacodus* from the Grenzbitumen horizon (Anisian-Ladinian boundary) of the southern Alps, it constitutes the unarmored placodonts, the Placodontoidea (Mazin and Pinna, 1993). The armored placodonts, or Cyamodontoidea (Mazin and Pinna, 1993), range from the lower Muschelkalk of the Germanic Triassic and deposits of equivalent age in Israel to the Upper Triassic (Rhaetian) of the Alpine Triassic and possibly also England (Storrs, 1994). Although a subdivision of the Placodontia into these two major clades is generally accepted, the three-taxon statement including *Placodus*, *Paraplacodus*, and the Cyamodontoidea still requires critical reevaluation. Placodonts are highly adapted for durophagous feeding, preying on bottom-dwelling shelled invertebrates in nearshore habitats or shallow epicontinental seas. In *Placodus*, strongly procumbent anterior incisors were used to pick up brachiopods (Huene, 1933) from the substrate; anterior teeth are reduced or absent in the narrow and pointed rostrum of cyamodontoids. Dermal ossification is restricted to a row of scutes tipping the neural spines of the vertebrae in *Placodus*. The Cyamodontoidea carry elaborate dermal armoring comparable to the turtle shell. Placodonts have long been recognized as a separate radiation of Triassic marine reptiles, but their relationships with other Sauropterygia have been questioned until recent times (Carroll and Currie, 1991; Sues, 1987b; see below for further

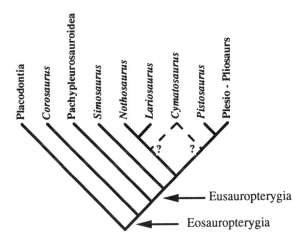

Figure 1. Hypothetical relationships of the Sauropterygia. See text for further discussion.

comments).

A basal dichotomy between the remaining two clades of stem-group Sauropterygia, the pachypleurosaurs and nothosaurs, was first recognized by Nopcsa (1928), who erected the family Pachypleurosauridae (the Pachypleurosauroidea of Huene, 1956). The correct composition of the Pachypleurosauridae, and a diagnosis for the group, were first provided by Peyer (1934). The pachypleurosaurs constitute a second monophyletic clade of the sauropterygian radiation (Storrs, 1991; Rieppel, 1987, 1989; Sues, 1987a) which first appears in the upper Lower Triassic and lowermost Middle Triassic of the eastern Germanic Basin (*Dactylosaurus*: Rieppel and Hagdorn, Chapter 5; Gürich, 1884). Pachypleurosaurs are also recorded from the lower Middle Triassic epicontinental seas of China (*Keichousaurus*: Young, 1958, 1965). The clade is abundantly represented in uppermost Anisian and lower Ladinian (Middle Triassic) deposits at Monte San Giorgio, Switzerland (genera *Neusticosaurus* and *Serpianosaurus*: Rieppel, 1989; Sander, 1989; Carroll and Gaskill, 1985). *Psilotrachelosaurus* is a small pachypleurosaur from the middle Ladinian of the Austrian Alps (Rieppel, 1993). The clade has not been recorded beyond the upper Ladinian (upper Middle Triassic), perhaps lower Carnian (lower Upper Triassic) deposits in the southern Alps (Ca' del Frate, Italy: Tintori et al., 1985). Pachypleurosaurs are of small overall size, and show a number of supposedly plesiomorphic traits such as conservative skull proportions with no significant elongation of the jaws, a relatively short postorbital region of the skull, and an impedance-matching middle ear for the efficient transmission of airborne sound. The vertebral column is characterized by pachyostosis of the neural arches, which bear low neural spines. Pachyostosis of dorsal ribs is absent in the early representatives of the clade, and the limbs preserve a generalized morphology except for modest phalangeal reduction in some late representatives (*Neusticosaurus*). Throughout the clade a distinct sexual dimorphism is expressed primarily in limb proportions and humerus morphology. Pachypleurosaurs are restricted to epicontinental deposits or coastal stretches of the Tethys, and accordingly show a rather patchy geographical distribution (see Rieppel and Hagdorn, Chapter 5, for further discussion on the historical biogeography of the Pachypleurosauroidea).

Sauropterygia other than placodonts and pachypleurosaurs (and possibly *Corosaurus*: Storrs, 1991; see below) are included in the Eusauropterygia. Stem-group Eusauropterygia include a number of genera from the Middle Triassic of China, Israel, and Europe, which are currently under revision. The Eusauropterygia are characterized by a number of derived characters (as compared to pachypleurosaurs) such as elongation of the jaws and of the mandibular symphysis, the development of strongly procumbent rostral (premaxillary and dentary) dentition as well as paired maxillary fangs, and the elongation of the dorsoventrally depressed skull, correlated with a size increase of the upper temporal fenestrae and the development of a complex dual jaw adductor system suited for the capture of fast-moving prey in the marine environment by a rapid lateral snapping bite (Storrs, 1993a; Taylor, 1992). The earliest known eusauropterygian is *Cymatosaurus*, which first appears in the upper Buntsandstein and lower Muschelkalk of the eastern part of the Germanic Basin. The genus remains restricted to the lower Anisian of western Europe. *Cymatosaurus latissimus* (Gürich, 1891) from the lowermost Muschelkalk of Gogolin, Upper Silesia (Gorny Slask, Poland) was referred to a new genus, *Germanosaurus*, by Nopcsa (1928; for *Eurysaurus* Frech [1903-1908]) in recognition of the fact that the species is, in some characters (differentiation of nasal bones, relation of pre- and postfrontals), intermediate between *Cymatosaurus* and *Nothosaurus*. *Proneusticosaurus* Volz, 1902, from the lower Muschelkalk of Upper Silesia (Gorny Slask, Poland) most probably is a junior synonym of *Cymatosaurus* (Rieppel and Hagdorn, Chapter 5; Sues, 1987a). *Micronothosaurus stensioeii* Haas, 1963, from the upper (?) Muschelkalk of Israel was referred to *Cymatosaurus* (Schultze, 1970) on the basis of reduced nasal bones and a forward displacement of the pineal foramen (Sues, 1987a). If correct, this would significantly expand the geological range of occurrence of *Cymatosaurus*.

Simosaurs and nothosaurs are known from the Middle and lower Upper Triassic of Europe, Israel, and China. *Simosaurus gaillardoti* is the only known species of its genus, restricted to the upper Muschelkalk and Lettenkeuper of the Germanic Triassic (Rieppel, 1994a). *Shingyisaurus* has been described as a simosaur from the Middle Triassic of China (Young, 1965). The earliest occurrence of the genus

Nothosaurus is in the lower Muschelkalk of Gogolin (Wysogorski, 1904), and in contemporary deposits of the Muschelkalk of Wadi Ramon in Israel (Swinton, 1952). In Europe, the genus persists up into the Lettenkeuper of the Germanic Triassic, and into the Ladinian of the Alpine Triassic (Furrer et al., 1992; Bürgin et al., 1991). The Chinese nothosaurs (Young, 1959, 1960, 1965, 1978) are difficult to judge without critical revision.

Lariosaurs are closely related to nothosaurs and remain restricted to the Middle and Upper Triassic of the western Tethyan Province. The earliest representative of the clade is *Silvestrosaurus* (Storrs, 1993b; Kuhn-Schnyder, 1990; *Lariosaurus* fide Tschanz, 1989) from the Grenzbitumen horizon (Anisian-Ladinian boundary) of Monte San Giorgio in the southern Alps. *Ceresiosaurus* (Peyer, 1931) appears to be restricted to the lower Ladinian of Monte San Giorgio, but may have occurred in the Germanic Muschelkalk also (Rieppel, 1994a). The genus *Lariosaurus* finally is restricted to the middle and upper Ladinian of the southern Alps (Peyer, 1934) and to the Spanish Muschelkalk (Sanz, 1976, 1983a; for further discussion and references see Rieppel, 1994b), again with a rare occurrence in the Gipskeuper of the Germanic Triassic (Schultze and Wilczewski, 1970).

The only stem-group sauropterygian described so far from the New World is *Corosaurus* from the upper Lower Triassic of Wyoming (Storrs, 1991). New material has been collected in the Middle Triassic of Nevada (Sander et al., 1993), and fragmentary material is also known from the Triassic of British Columbia (E. L. Nicholls, personal communication). *Corosaurus* may represent the sister-taxon to all other Eusauropterygia, but further analysis---particularly of the Chinese material---is required to test that hypothesis.

Pistosaurus is unique in the upper Muschelkalk (Ladinian) of the Germanic Triassic (Sues, 1987a; Meyer, 1847-1855), with a questionable occurrence in Spain (Sanz, 1983b). The genus remains poorly understood, but it shares with *Cymatosaurus* the much-reduced nasals which remain excluded from the external nares. Current phylogenies of sauropterygians place *Pistosaurus* as sister-taxon to the crown-group plesiosaurs and pliosaurs (Storrs, 1991, 1993b; Sues, 1987a).

The Plesiosauria became the dominant marine reptiles of the Jurassic and Cretaceous. Growing to larger body size (up to 50 ft total length) than the Triassic Sauropterygia, these pelagic animals reached a cosmopolitan distribution by the Upper Cretaceous. The fossil record of plesiosaurs is particularly rich in Europe for the Jurassic, and in North America for the Cretaceous. Early taxonomy of plesiosaurs was based on body proportions such as the relative size of the skull and the relative length of the neck. Owen (1841, cited and discussed in Brown, 1981) introduced the name *Pliosaurus* to include plesiosaurs with larger heads and shorter necks, as opposed to *Plesiosaurus* with a longer neck and a smaller head. Today, these two genera stand as types for two superfamilies, the Plesiosauroidea and the

Pliosauroidea. The plesiosauroids are divided into two groups, the more generalized Plesiosauridae with up to 28 cervical vertebrae, and the Elasmosauridae with 32-76 cervical vertebrae. This traditional classification is currently under revision, however, with many more characters (particularly those of the skull) being subject to cladistic analysis. Relative skull size and neck length have obvious implications for feeding strategies (Massare, 1987): plesiosaurs would feed on relatively small fish and cephalopods, with some taxa possibly representing filter or straining feeders on small fish and crustaceans, whereas pliosaurs with large skulls (reaching a length of 10 ft in some taxa) and jaws carrying "canine" teeth were able to tackle larger prey (Taylor, 1992).

All plesiosaurs share a relatively short trunk, elongated pectoral and pelvic girdles, a relatively short tail, an exceptionally well-developed gastral rib cage, and the transformation of the limbs into paddles characterized by a shortening of the proximal long bones and by distinct hyperphalangy. Locomotion patterns of plesiosaurs were reviewed by Godfrey (1984; see also Massare, 1988), who chose the sea lion as a model for comparison. The elongation of the pectoral girdle would have allowed a powerful horizontal retraction of the forelimb, creating most of the propulsive force in addition to some anterior lift. Unlike in true underwater flight demonstrated by sea turtles or penguins, the horizontal recovery stroke would not create propulsion in plesiosaurs which, due to the inertia of their large bodies, would continue to glide forward during the recovery stroke, as do sea lions.

In summary, the Eusauropterygia include a number of Triassic genera distributed in the eastern (China) and western (Europe) Tethyan Province as well as in the eastern Pacific Province (western North America). At the same time, stem-group Eusauropterygia show progressive adaptations to life in the open, if shallow, epicontinental seas, culminating in the highly modified and pelagic crown-group plesiosaurs and pliosaurs (Storrs, 1993a; Sues, 1987a). Stem-group eusauropterygians therefore represent the crucial link within the nearshore to offshore gradient linking pachypleurosaurs with plesiosaurs and pliosaurs (Storrs, 1993a). Adaptations to the marine habitat affected body proportions, locomotion, jaw mechanics, and hearing. Locomotion shifted from axial undulation to paraxial propulsion. A relative (allometric) size increase of the forelimbs, correlated with the development of hyperphalangy (starting in the *Lariosaurus-Ceresiosaurus* clade and culminating in crown-group plesiosaurs and pliosaurs), renders the forelimbs the main propulsive organs. Depression of the skull, correlated with an elongation of the jaws and the development of a strongly procumbent dentition, results in a head shape well adapted for lateral snapping bites in pursuit of fast-moving prey (Taylor, 1987, 1992), but results in problems for the arrangement of the jaw adductor musculature. The low temporal region of the skull requires subdivision of the jaw adductors into two main functional units (Rieppel, 1989, 1994a; Taylor, 1992). The anterior unit with an anterodorsally directed resultant force is well suited for the initiation of rapid jaw closure. The posterior unit with a posterodorsally directed resultant force is capable of delivering the adductive force required to pierce ammonite shells or ganoid fish scales. Changes of skull structure in adaptation to an increasingly marine mode of life also resulted in the loss of an impedance-matching middle ear suitable for the transmission of airborne sound but in danger of injury during rapid dives (Rieppel, 1989, 1994a; Storrs, 1993a; Taylor, 1987, 1992). Ultimately, however, the proper interpretation of the paleobiogeography and of the evolutionary history of aquatic adaptations in the Sauropterygia must relate to a robust hypothesis of phylogenetic interrelationships of the group. The current placement of *Corosaurus* and placodonts as successive sister-groups of all other Sauropterygia indicates that many of the supposedly plesiomorphic characters of pachypleurosaurs (jaw adductor system, middle ear morphology) might, in fact, have resulted from a reversal in character evolution. Current knowledge of the phylogenetic relationships of the Sauropterygia points to either of three alternative interpretations: that pachypleurosaurs secondarily invaded the nearshore environment; that advanced characters which placodonts and, in particular, *Corosaurus* share with Eusauropterygia are convergent; or that sauropterygian interrelationships require further analysis.

As mentioned above, it was Owen (1860) who recognized the relationships of "nothosaurs" and plesiosaurs with placodonts, and included these taxa in his Sauropterygia. A revised concept of the Sauropterygia was proposed by Williston (1925), who included the nothosaurs, plesiosaurs, and placodonts in the Synaptosauria (a term taken from Baur, 1887) on the basis of the presence of a single upper temporal fossa. Williston's (1925) Synaptosauria were renamed as Euryapsida by Colbert (1955), and the Lower Permian genus *Araeoscelis* (Vaughn, 1955; Williston, 1914) was included within the group, as it shares the presence of a single upper temporal fossa (Romer, 1968).

The concept of Euryapsida contrasts with the hypothesis of a diapsid relationships for the Sauropterygia, first proposed by Jaekel (1910, who did not consider placodonts in his analysis), following the investigation of the cranial structure of the pachypleurosaur *Anarosaurus* and of *Simosaurus*. A diapsid derivation of sauropterygians was further supported by Kuhn-Schnyder (1967, 1980), who again used the skull of *Simosaurus* in support of his argument (Kuhn-Schnyder, 1961), and by Carroll (1981), who described the marine diapsid genus *Claudiosaurus* from the Permo-Triassic of Madagascar as a sauropterygian (plesiosaur) ancestor. Both latter authors denied a diapsid status of placodonts, and therewith any close relationship between placodonts and other sauropterygians (Carroll and Currie, 1991; Kuhn-Schnyder, 1967, 1980, 1990).

Initial cladistic analysis of reptile interrelationships (Evans, 1988; Benton, 1985;

Gauthier, 1984; Reisz et al., 1984) showed *Araeoscelis* to be a stem-group diapsid, and modified the diagnosis of the Diapsida to refer (among other characters) to the presence of an upper temporal fossa only. This rendered the Sauropterygia diapsids by definition. The relations of the Sauropterygia within the Diapsida were not addressed in any of these cladistic analyses, and although the Placodontia were recognized as diapsids (Sues, 1987b), a close relationship between them and sauropterygians continued to be rejected.

Further cladistic analysis of sauropterygian interrelationships confirmed the monophyly of Owen's (1860) Sauropterygia, including placodonts, pachypleurosaurs, and eusauropterygians (Rieppel, 1989; Zanon, 1989). The Sauropterygia were furthermore shown to be nested within the Diapsida (Storrs, 1991, 1993b; Sues, 1987b); indeed, they were found to be placed within the crown-group diapsids (Sauria *sensu* Gauthier, 1984) close to the lepidosauromorph clade (Rieppel, 1994a; Sues, 1987a). These results confirm a terrestrial ancestry of the clade. Interrelationships within the Sauropterygia currently remain somewhat controversial and/or unresolved. Traditionally, the Pachypleurosauroidea were viewed as the sister-group to all other Sauropterygia (Sues, 1987a), but inclusion of placodonts in the analysis resulted in two alternative hypotheses. Rieppel (1989) defended a sister-group relationship of the Placodontia and Sauropterygia, resurrecting the Euryapsida as a subgroup of the Diapsida (see also Zanon, 1989; Owen's [1860] Sauropterygia has priority over Euryapsida). In contrast, Storrs' (1991, 1993b) analysis showed the placodonts to be the sister-group of the Eusauropterygia within the Sauropterygia. Storrs (1991) grouped the Placodontia and Eusauropterygia in a new taxon, the Nothosauriformes, with the pachypleurosaurs as its sister-group. A comprehensive review of stem-group Sauropterygia from the Triassic of Europe resulted in an expanded database used to test placodont relationships (Rieppel, 1994a). The results showed the Placodontia to be the sister-group of all other Sauropterygia, termed Eosauropterygia (Rieppel, 1994a). *Corosaurus* from the western United States was found to be the sister-taxon of an unnamed clade comprising the Pachypleurosauroidea and the Eusauropterygia (*sensu* Tschanz, 1989). Within the Eusauropterygia, *Simosaurus* was found to be the sister-taxon of all other genera. A close relationship of *Nothosaurus* (including *"Paranothosaurus"*) and lariosaurs is indicated by braincase characters, and the two lineages were included within a monophyletic taxon, Nothosauridae, by Rieppel (1994c). The lariosaurs constitute a monophyletic lineage, currently comprising the genera *Ceresiosaurus* Peyer, 1931; *Lariosaurus* Curioni, 1847; and *Silvestrosaurus* Kuhn-Schnyder, 1990. *Cymatosaurus* appears to be closely related to the Nothosauridae, and may, in fact, represent their sister-taxon (Rieppel, 1994c), but a relationship of *Cymatosaurus* with *Pistosaurus* represents a possible alternative (Storrs, 1991, 1993b; Sues, 1987a). *Pistosaurus*, finally, has been hypothesized to represent the sister-taxon of the plesiosaurs and pliosaurs (Storrs, 1991, 1993a; Sues, 1987a).

Paleoecological considerations discussed above indicate problems of convergence versus reversals in the interpretation of character evolution within the Sauropterygia. These in turn highlight the fact that sauropterygian interrelationships require further critical analysis. It may well be expected that the redescription of the Chinese material, as well as new finds from the Triassic of the western United States, will modify our future views on sauropterygian evolution. Furthermore, the invasion of the Mesozoic seas by reptiles will not be fully understood unless other groups such as ichthyosaurs and thalattosaurs are added to the phylogenetic analysis. We still have a far way to go toward a more complete understanding of the interrelationships and evolution of Mesozoic marine reptiles and their invasion of the Mesozoic seas.

ACKNOWLEDGMENTS

I thank E. L. Nicholls and J. M. Callaway for their editorial help with this chapter, earlier versions of which were was kindly read by H.-D. Sues and G. W. Storrs. Research for this chapter was supported by NSF grant DEB-9220540.

REFERENCES

Agassiz, L. 1833-1845. *Recherches sur les Poissons Fossiles*, Vol. II. Imprimerie de Petitpierre, Neuchâtel, 310 pp.

Baur, G. 1887. On the phylogenetic arrangement of the Sauropsida. *Journal of Morphology* 1:93-104.

Benton, M. J. 1985. Classification and phylogeny of the diapsid reptiles. *Zoological Journal of the Linnean Society* 84:97-164.

Bronn, G. 1831. G. Graf zu Münster: Über einige ausgezeichnete fossile Fischzähne aus dem Muschelkalk bei Bayreuth. *Jahrbuch für Mineralogie, Geognosie, Geologie und Petrefaktenkunde* 2:470-471.

Brown, D. S. 1981. The English Upper Jurassic Plesiosauroidea (Reptilia) and a review of the phylogeny and classification of the Plesiosauria. *Bulletin of the British Museum (Natural History), Geology* 53:1253-347.

Bürgin, T., U. Eichenberger, H. Furrer, and K. Tschanz. 1991. Die Prosanto-Formation---eine fischreiche Fossil-Lagerstätte in der Mitteltrias der Silvretta-Decke (Kanton Graubünden, Schweiz). *Eclogae Geologicae Helvetiae* 84:921-990.

Carroll, R. L. 1981. Plesiosaur ancestors from the Upper Permian of Madagascar. *Philosophical Transactions of the Royal Society of London* B 293:315-383.

Carroll, R. L. and P. J. Currie. 1991. The early radiation of diapsid reptiles. IN H.-P.

Schultze and L. Trueb (Eds.), *Origins of the Higher Groups of Tetrapods, Controversy and Consensus*, pp. 354-424. Comstock Publishing Association, Ithaca, New York.

Carroll, R. L. and P. Gaskill. 1985. The nothosaur *Pachypleurosaurus* and the origin of plesiosaurs. *Philosophical Transactions of the Royal Society of London* B 309:343-393.

Colbert, E. H. 1955. *Evolution of the Vertebrates*, 1st edition. John Wiley and Sons, New York, 479 pp.

Curioni, G. 1847. Cenni sopra un nuovo saurio fossile dei monti di Perledo sul Lario e sul terreno che la racchiude. *Giornale J. R. Instituto Lombardo delle Science, Lettre ed Arti* 16:159-170.

Evans, S. E. 1988. The early history and relationships of the Diapsida. IN M. J. Benton (Ed.), *The Phylogeny and Classification of the Tetrapods, Volume 1: Amphibians, Reptiles, Birds*, pp. 221-260. Systematics Association Special Publication no. 35A. Clarendon Press, Oxford.

Frech, F. 1903-1908. *Lethaea geognostica. Handbuch der Erdgeschichte. II. Teil. Das Mesozoicum. 1. Band. Trias.* E. Schweizerbart'sche Verlagsbuchhandlung (E. Nägele), Stuttgart.

Freyberg, B. v. 1972. Die erste erdgeschichtliche Erforschungsphase Mittelfrankens (1840-1847). Eine Briefsammlung zur Geschichte der Geologie. *Erlanger Geologische Abhandlungen* 92:1-33.

Furrer, H., U. Eichenberger, N. Froitzheim, and D. Wüster. 1992. Geologie, Stratigraphie und Fossilien der Ducankette und des Landwassergebiets (Silvretta-Decke, Ostalpin). *Eclogae Geologicae Helvetiae* 85:246-256.

Gauthier, J. A. 1984. *A Cladistic Analysis of the Higher Systematic Categories of the Diapsida*. Ph. D. Thesis, University of California, Berkeley. Univ. Microfilm, Int., No. 85-12825, Ann Arbor, Michigan.

Godfrey, S. 1984. Plesiosaur subaqueous locomotion: a reappraisal. *Neues Jahrbuch für Geologie und Paläontologie*, Monatshefte 11:661-672.

Gürich, G. 1884. Über einige Saurier des oberschlesischen Muschelkalkes. *Zeitschrift der Deutschen Geologischen Gesellschaft* 36:125-144.

Gürich, G. 1891. Über einen neuen *Nothosaurus* von Gogolin, Oberschlesien. *Zeitschrift der Deutschen Geologischen Gesellschaft* 43:967-970.

Haas, G. 1963. *Micronothosaurus stensiöii*, ein neuer Nothosauride aus dem Oberen Muschelkalk des Wadi Ramon, Israel. *Paläontologische Zeitschrift* 37:161-178.

Huene, F. v. 1933. Die Placodontier. 4. Zur Lebensweise und Verwandtschaft von *Placodus*. *Abhandlungen der Senckenbergischen Naturforschenden Gesellschaft* 38:365-382.

Huene, F. v. 1956. *Paläontologie und Phylogenie der Niederen Tetrapoden*. G. Fischer, Jena, 716 pp.

Jaekel, O. 1910. Über das System der Reptilien. *Zoologischer Anzeiger* 35:324-341.

Kuhn-Schnyder, E. 1961. Der Schädel von *Simosaurus*. *Paläontologische Zeitschrift* 35:95-113.

Kuhn-Schnyder, E. 1967. Das Problem der Euryapsida. *Colloques Internationaux du Centre National de la Recherche Scientifique, Paris* 163:335-348.

Kuhn-Schnyder, E. 1980. Observations on the temporal openings of reptilian skulls and the classification of reptiles. IN L. Jacobs (Ed.), *Aspects of Vertebrate History*, pp. 153-175. Museum of Northern Arizona Press, Flagstaff, Arizona.

Kuhn-Schnyder, E. 1990. Über Nothosauria (Sauropterygia, Reptilia)---ein Diskussionsbeitrag. *Paläontologische Zeitschrift* 64:313-316.

Massare, J. M. 1987. Tooth morphology and prey preference of Mesozoic marine reptiles. *Journal of Vertebrate Paleontology* 7:121-137.

Massare, J. M. 1988. Swimming capabilities of Mesozoic marine reptiles: implications for method of predation. *Paleobiology* 14:187-205.

Mazin, J.-M. and G. Pinna. 1993. Palaeoecology of the armoured placodonts. IN J.-M. Mazin and G. Pinna (Eds.), *Evolution, Ecology and Biogeography of the Triassic Reptiles*, Paleontologia Lombarda, N.S. 2:109-130.

Meyer, H. v. 1847-1855. *Zur Fauna der Vorwelt. Die Saurier des Muschelkalkes mit Rücksicht auf die Saurier aus buntem Sandstein und Keuper*. Heinrich Keller, Frankfurt am Main, 167 pp.

Münster, G. 1830. *Über einige ausgezeichnete fossile Fischzähne aus dem Muschelkalk bei Bayreuth*. F. C. Birner, Bayreuth.

Nopcsa, F. 1928. Palaeontological notes on reptiles. *Geologica Hungaria, Ser. Palaeontologica* 1:3-84.

Nosotti, S. and G. Pinna. 1989. Storia delle ricerche e degli studi sui rettili placodonti. *Memorie della Società Italiana di Scienze Naturali e del Museo Civico di Storia Naturale di Milano* 24:29-86.

Owen, R. 1858. Description of the skull and teeth of the *Placodus laticeps* Owen, with indications of other new species of *Placodus*, and evidence of the saurian nature of that genus. *Philosophical Transactions of the Royal Society of London* 148:169-184.

Owen, R. 1860. *Palaeontology*. Adam and Charles Black, Edinburgh, 420 pp.

Peyer, B. 1931. Die Triasfauna der Tessiner Kalkalpen. IV. *Ceresiosaurus calcagnii* nov. gen. nov. spec. *Abhandlungen der Schweizerischen Paläontologischen Gesellschaft* 51:1-68.

Peyer, B. 1934. Die Triasfauna der Tessiner Kalkalpen. VII. Neubeschreibung der Saurier von Perledo. *Abhandlungen der Schweizerischen Paläontologischen Gesellschaft* 53-54:1-130.

Pinna, G. and G. Teruzzi. 1991. Il giacimento paleontologico di Besano. *Natura (Milano)* 82:1-55.

Reisz, R. R., D. S. Berman, and D. Scott. 1984. The anatomy and relationships of the Lower Permian reptile *Araeoscelis*. *Journal of Vertebrate Paleontology* 4:57-67.

Rieppel, O. 1987. The Pachypleurosauridae: an annotated bibliography. With comments on some lariosaurs. *Eclogae Geologicae Helvetiae* 80:1105-1118.

Rieppel, O. 1989. A new pachypleurosaur (Reptilia: Sauropterygia) from the Middle Triassic of Monte San Giorgio, Switzerland. *Philosophical Transactions of the Royal Society of London* B 323:1-73.

Rieppel, O. 1993. Status of the pachypleurosauroid *Psilotrachelosaurus toeplitschi* Nopcsa (Reptilia, Sauropterygia), from the Middle Triassic of Austria. *Fieldiana (Geology)*

27:1-17.
Rieppel, O. 1994a. Osteology of *Simosaurus gaillardoti* and the relationships of stem-group Sauropterygia. *Fieldiana (Geology)*: in press.
Rieppel, O. 1994b. *Lariosaurus balsami* Curioni (Reptilia, Sauropterygia) aus den Gailtaler Alpen. *Carithia* II:184 (104):345-356.
Rieppel, O. 1994c. The braincases of *Simosaurus* and *Nothosaurus*: monophyly of the Nothosauridae (Reptilia: Sauropterygia). *Journal of Vertebrate Paleontology* 14:9-23.
Romer, A. S. 1968. *Notes and Comments on Vertebrate Paleontology*. University of Chicago Press, Chicago, 304 pp.
Sander, P. M. 1989. The pachypleurosaurids (Reptilia: Nothosauria) from the Middle Triassic of Monte San Giorgio (Switzerland), with the description of a new species. *Philosophical Transactions of the Royal Society of London* B 325:561-670.
Sander, P. M., O. Rieppel, and H. Bucher. 1993. New marine vertebrtate fauna from the Middle Triassic of Nevada. *Journal of Paleontology* 68:676-680.
Sanz, J. L. 1976. *Lariosaurus balsami* (Sauropterygia, Reptilia) de Estada (Huesca). *Estudios Geoogicos* 32:547-567.
Sanz, J. L. 1983a. Los Nothosaurios (Reptilia, Sauropterygia) Espanoles. *Estudos Geologicos* 39:193-215.
Sanz, J. L. 1983b. Consideraciones sobre el genero *Pistosaurus*. El suborden Pistosauria (Reptilia, Sauropterygia). *Estudos Geologicos* 39:451-458.
Schultze, H.-P. 1970. Über *Nothosaurus*. Neubeschreibung eines Schädels aus dem Keuper. *Senckenbergiana Lethaea* 51:211-237.
Schultze, H. and N. Wilczewski. 1970. Ein Nothosauride aus dem unteren Mittel-Keupers Unterfrankens. *Göttinger Arbeiten in Geologie und Paläontologie, H. Martin-Festschrift*:101-112.
Storrs, G. W. 1991. Anatomy and relationships of *Corosaurus alcovensis* (Diapsida: Sauropterygia) and the Triassic Alcova Limestone of Wyoming. *Bulletin of the Peabody Museum of Natural History* 44:1-151.
Storrs, G. W. 1993a. Function and phylogeny in sauropterygian (Diapsida) evolution. *American Journal of Science* 293-A:63-90.
Storrs, G. W. 1993b. The systematic position of *Silvestrosaurus* and a classification of Triassic sauropterygians. *Paläontologische Zeitschrift* 67:177-191.
Storrs, G. W. 1994. Fossil vertebrate faunas of the British Rhaetian (latest Triassic). *Zoological Journal of the Linnean Society* 112:217-259.
Storrs, G. W. and M. A. Taylor. 1993. Cranial anatomy of a plesiosaur from the Triassic/Jurassic boundary of Street, Somerset, England. *Journal of Vertebrate Paleontology* 13:59A.
Sues, H.-D. 1987a. The skull of *Placodus* and the relationships of the Placodontia. *Journal of Vertebrate Paleontology* 7:138-144.
Sues, H.-D. 1987b. Postcranial skeleton of *Pistosaurus* and interrelationships of the Sauropterygia (Diapsida). *Zoological Journal of the Linnean Society* 90:109-131.
Swinton, W. E. 1952. A nothosaur vertebra from Israel. *Annales and Magazines of Natural History* (12)5:875-876.

Taylor, M. A. 1987. How tetrapods feed in water: a functional analysis by paradigm. *Zoological Journal of the Linnean Society* 91:171-195.

Taylor, M. A. 1992. Functional anatomy of the head of the large aquatic predator *Rhomaleosaurus zetlandicus* (Plesiosauria, Reptilia) from the Toarcian (Lower Jurassic) of Yorkshire, England. *Philosophical Transactions of the Royal Society of London* B 335:247-280.

Tintori, A., G. Musico, and S. Nardon. 1985. The Triassic fossil fishes localities in Italy. *Rivista Italiana di Paleontologie e Stratigrafia* 91:197-210.

Tschanz, K. 1989. *Lariosaurus buzzii* n. sp. from the Middle Triassic of Monte San Giorgio (Switzerland), with comments on the classification of nothosaurs. *Palaeontographica* A 208:153-179.

Vaughn, P. 1955. The Permian reptile *Araeoscelis* restudied. *Bulletin of the Museum of Comparative Zoology* 113:305-467.

Volz, W. 1902. *Proneusticosaurus*, eine neue Sauropterygier-Gattung aus dem untersten Muschelkalk Oberschlesiens. *Palaeontographica* 49:121-162.

Weiss, G. (Ed.). 1983. *Bayreuth als Stätte alter erdgeschichtlicher Entdeckungen.* Druckerei Ellwanger, Bayreuth, 70 pp.

Williston, S. W. 1914. The osteology of some American Permian vertebrates. *Journal of Geology* 22:364-419.

Williston, S. W. 1925. *The Osteology of the Reptiles.* Harvard University Press, Cambridge, Massachusetts, 298 pp.

Wysogorski, J. 1904. Die Trias in Oberschlesien. *Zeitschrift der Deutschen Geologischen Gesellschaft* 56:260-264.

Young, C. C. 1958. On the new Pachypleurosauridae from Keichow, South West China. *Vertebrata PalAsiatica* 2:69-82.

Young, C. C. 1959. On a new Nothosauria from the Lower Triassic Beds of Kwangsi. *Vertebrata PalAsiatica* 3:73-78.

Young, C. C. 1960. New localities of sauropterygians in China. *Vertebrata PalAsiatica* 3:73-78.

Young, C. C. 1965. On the new nothosaurs from Hupeh and Kweichou, China. *Vertebrata PalAsiatica* 9:337-356.

Young, C. C. 1978. A nothosaur from Lu-hsi County, Yunnan Province. *Vertebrata PalAsiatica* 16:222-224.

Zanon, R. T. 1989. *Paraplacodus* and the diapsid origin of Placodontia. *Journal of Vertebrate Paleontology* 9:47A.

Chapter 5

PALEOBIOGEOGRAPHY OF MIDDLE TRIASSIC SAUROPTERYGIA IN CENTRAL AND WESTERN EUROPE

OLIVIER RIEPPEL and HANS HAGDORN

INTRODUCTION

Current taxonomy of vertebrate fossils reflects a tradition bias of considering the Germanic and Alpine Triassic as separate biotas. The literature records little, if any, overlap of vertebrate taxa in these two depositional realms. In contrast, many faunal elements among most invertebrate groups occur in both depositional realms, and indicate close affinities that allow biostratigraphic and chronostratigraphic correlation as well as reconstruction of paleobiogeographical migratory routes. This is especially true for Anisian and Ladinian times when the Germanic Basin was flooded by the Muschelkalk sea. The marine transgressions starting in late Olenekian times and persisting from early Anisian through Ladinian times induced the formation of the Röt (upper Lower Triassic) and Muschelkalk (Middle Triassic) sediments (Hagdorn, 1991), which were deposited in a shallow epicontinental sea and were formerly believed to differ fundamentally from the intraplatform basin facies characteristic of the Middle Triassic of the Alpine region. More recent analyses indicate, however, that the Germanic Basin and the Alpine region were subject to similar sedimentation processes during the upper Scythian, lower Anisian, and early upper Anisian times, and that distinct facies deviations do not become apparent prior to the Illyrian (upper Anisian), in some areas not earlier than Fassanian (lower Ladinian) times (Mostler, 1993). Sequence stratigraphy, with biostratigraphical control, provides a welcome tool for comparing eustatic sea level fluctuations in different peritethys basins that give evidence for faunal migrations.

According to Kozur (1974), the invertebrate fauna characteristic of the earliest Muschelkalk deposits (Gogolin beds, lower Anisian) in the eastern part of the Germanic Basin (Holy Cross Mountains, central Poland) shows strong Asiatic affinities, and appears to have reached the Muschelkalk Basin from the Paleotethys through the East Carpathian Gate. Kovács (1992) believes that during the early Middle Triassic, a marine strait connecting the Germanic Basin with the northern Paleotethys branch did not pass through the east Carpathian region, but rather passed through the pre-Dobrogean zone. Marine faunas populating the Muschelkalk Basin seem to have dispersed to the south into the intraplatform basins of the Alpine Triassic, which itself became connected to the pelagic Tethyan realm much later only, i.e., during Pelsonian times (upper lower Muschelkalk: Mostler, 1993). Based on the analysis of invertebrate fossils and conodonts, these reconstructions of sedimentological and paleogeographical relations between the Germanic and Alpine Triassic raise the question of whether the revision of vertebrate taxa might not also document a closer paleobiogeographic relation between these two regions than is apparent from the current literature.

In the central Germanic Basin, the earliest (although not diagnostic) sauropterygian remains (an isolated ischium, MHI 1299) come from the upper Olenekian (*Gipsresidualbank*, 1.5 m below the *tenuis*-Bank, lower Röt) of Jena, Thuringia. The earliest diagnostic sauropterygian remains are those of the genus *Cymatosaurus* from the upper Buntsandstein (so$_2$, Röt) and lower Muschelkalk (Huene, 1944) of the Germanic Triassic. The recent redescription of "*Anarosaurus multidentatus*" (Huene, 1958) from the lower Anisian of the Austrian Alps showed this taxon to be referable to the genus *Cymatosaurus* (Rieppel, 1995a). The distribution of *Cymatosaurus* in the upper Buntsandstein and lower Muschelkalk of the eastern Germanic Basin, and in the lower Anisian of the northern Alps, fits the scenario of an early population of the Germanic Basin by vertebrates of eastern affinities, and an early subsequent dispersal of at least some faunal elements from the Muschelkalk Basin southwestward into the Alpine Triassic (Mostler, 1993). In this chapter we report the earliest occurrence of sauropterygians from the southern Alpine Triassic and discuss their relation to the Germanic Triassic.

Museum acronyms used in this chapter are the following: GPIT, Institut und Museum für Geologie und Paläontologie, University of Tübingen; Ha, Institut für Geologische Wissenschaften, Martin-Luther-Universität, Halle/Saale; MB, Museum für Naturkunde der Humboldt-Universität, Berlin; MGU, Institute of Geological Sciences, University of Wroclaw; MHI, Muschelkalk Museum Hagdorn, Ingelfingen; and P, Phyletisches Museum, Jena.

SAUROPTERYGIA FROM THE LOWER ANISIAN OF THE VICENTINIAN ALPS

An isolated eusauropterygian vertebral neural arch (MHI 1279) was found in the "*Formazione a gracilis*" (Barbieri et al., 1980; lower Anisian) near Recoaro, Val Camonda, above the classical *Dadocrinus* site Cava di Gesso, in the Vicentinian Alps. The same locality yielded a poorly preserved neural arch of a small pachypleurosaur (MHI 1292), which represents the earliest record of the Pachypleurosauroidea in the Alpine Triassic. The occurrence of *Dadocrinus gracilis* indicates an equivalent age and a closely similar depositional environment for the "*Formazione a gracilis*" in the Vicentinian Alps, and for the upper Gogolin beds in Silesia, a conclusion which is also supported by sedimentological and benthic paleocommunity evidence (Hagdorn, in press).

The eusauropterygian neural arch (MHI 1279; Figure 1) shows a total height of 27.5 mm. The neural spine is rather low, and measures approximately 11.5 mm in height. The transverse processes are broad and stout, and the total width of the neural arch, measured across the transverse processes, is 36.5 mm. Although the neural arch superficially resembles those of *Nothosaurus*, important differences can be noted. *Nothosaurus* vertebrae from the lower Muschelkalk (Figures 2 and 3) generally show a relatively low neural spine, as does MHI 1279, but the latter differs from *Nothosaurus* by the fact that the neural spine is in an upright position and broadens toward its dorsal margin as a result of the diverging anterior and posterior margins. At the base, the length of the neural spine is 15.6 mm; along its dorsal margin, the length is 20.2 mm. In *Nothosaurus*, the neural spine

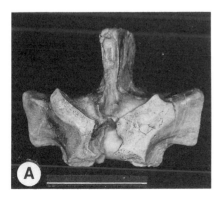 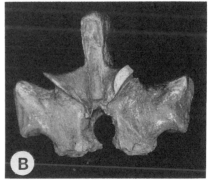

Figure 1. Eusauropterygian neural arch (MHI 1279) from the "*Formazione a gracilis*" (lower Anisian) of the Vicentinian Alps. **A)** Anterior view. **B)** Posterior view. Scale bar = 20 mm.

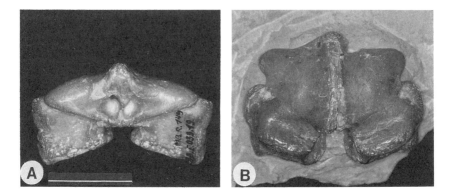

Figure 2. *Nothosaurus* neural arch (MB.R. 149) from the Schaumkalk (upper lower Muschelkalk) of Oberdorla. **A**) Posterior view. **B**) Dorsal view (original of Peyer, 1939: Figure 22). Scale bar = 20 mm.

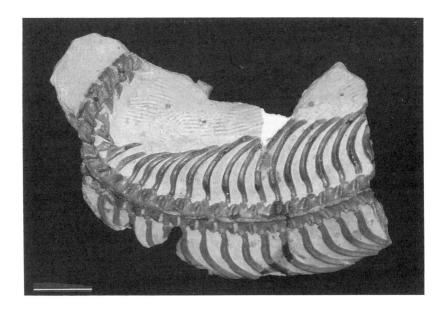

Figure 3. Partial postcranium of *Nothosaurus* (MB.R. 150) from the Schaumkalk (upper lower Muschelkalk) of Oberdorla (original of Peyer, 1939:Figure 23 and 24). Scale bar = 50 mm.

typically slants backward to a slight degree and shows no dorsal expansion. The articulation facets of pre- and postzygapophyses are more or less horizontally oriented throughout the vertebral column in *Nothosaurus*, but strongly obliquely oriented in MHI 1279. Significant differences are also observed in the detailed structure of the zygosphene-zygantrum articulation.

The zygosphene is an accessory articular process located between the prezygapophyses. In *Nothosaurus*, it appears partially subdivided by an anterior groove. In MHI 1279, the zygosphene is fully divided by a deep anterior indentation. The zygantrum is a depression in the base of the neural arch located between the postzygapophyses and receiving the zygosphene from the successive vertebra. In *Nothosaurus*, the zygantrum is subdivided by a thin medial vertical septum. In spite of the deep subdivision of the zygosphene, this vertical septum is absent in the zygantrum in MHI 1279.

Morphological detail allows the ready distinction of MHI 1279 from *Nothosaurus* vertebrae, and its occurrence in the *gracilis* beds (lower Anisian) of the southern Alps raises the question as to whether the same type of vertebra also occurs in the lower Muschelkalk.

COMPARISON WITH FAUNAL ELEMENTS FROM THE LOWER MUSCHELKALK, INCLUDING "*PRONEUSTICOSAURUS*" *SILESIACUS* VOLZ

In spite of superficial similarities, the neural arch from the lower Anisian of the Vicentinian Alps (MHI 1279) is readily distinguished from those of *Nothosaurus* on the basis of several characters discussed in the preceding section. Vertebrae closely resembling MHI 1279, and hence not congeneric with *Nothosaurus*, are also found in the lower Muschelkalk of the Germanic Basin.

A neural arch (MHI 1293/1), associated (but not articulated) with a centrum (MHI 1293/2), comes from the upper lower Muschelkalk (*Spiriferina*-Bank, *decurtata* biozone, Pelsonian) exposures in the MACKERT quarry in Hettingen near Buchen, Badenia (Figure 4). The right transverse process as well as the neural arch show the effect of slight erosion, but otherwise all characters typical for MHI 1279 can be identified: the low neural spine which expands toward its dorsal margin, the inclined articular facets of pre- and postzygapophyses, the deeply divided zygosphene, and the absence of a vertical septum in the zygantrum. The total height of the neural arch is 23.1 mm; total width across the transverse processes (as preserved) is 36 mm. The centrum was found 30 cm away from the neural arch on the same bedding plane. The articular surface is platycoelous, 18.9 mm high, and 17.1 mm wide. The length of the centrum is 20.3 mm. The articular facet which

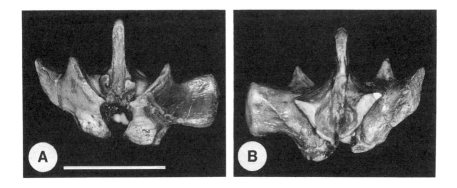

Figure 4. Eusauropterygian neural arch (MHI 1293/1) from the upper lower Muschelkalk (*Spiriferina*-Bank, *decurtata* biozone, Pelsonian) of Hettingen near Buchen, Badenia. **A**) Anterior view. **B**) Posterior view. Scale bar = 20 mm.

receives the pedicels of the neural arch is expanded into the cruciform or butterfly-shaped platform diagnostic of Eosauropterygia (Rieppel, 1994a).

Similar neural arches, different from those of *Nothosaurus* but similar in morphological detail to MHI 1279, can be reported from other lower Muschelkalk localities in the central part of the Germanic Basin. One specimen (Ha, drawer M4/12; Figure 5A) comes from the lower Muschelkalk (mu_1) of Halle/Saale. The neural arch measures 40.5 mm across the transverse processes; its total height is 35 mm. A further occurrence of this type of vertebra is the upper lower Muschelkalk (mu_2, Schaumkalk) from Freyburg/Unstrut. Two specimens have been identified from this locality. The first (P 1795; Figure 5B) is a rather poorly preserved neural arch with a broken neural spine. Nevertheless, the inclination of the articular facets of the pre- and postzygapophyses is clearly identifiable, as is the deeply bifurcated zygosphene and the undivided zygantrum. The neural arch is still associated with the centrum, and the latter contributes to the formation of the transverse processes as is typical for sacral vertebrae. The centrum is platycoelous and non-notochordal, the diameter of its articular surface is 25.5 mm, and its length is 27 mm. Total width of the neural arch across the transverse processes is 37 mm. The second specimen from the Schaumkalk of Freyburg/Unstrut (MB.R. 449) is again a sacral element with a broken neural spine.

In 1902, Volz described two incomplete postcranial skeletons from the lower Muschelkalk of Upper Silesia, which he referred to the new genus "*Proneusticosaurus*." The first specimen, "*Proneusticosaurus*" *silesiacus* Volz, 1902, comes from the Gogolin beds (mu_1) of Gogolin; the second specimen, "*Proneusticosaurus*" *madelungi* Volz, 1902, comes from equivalent beds near Sacrau, situated close to Gogolin. Of these two specimens, only one posterior slab

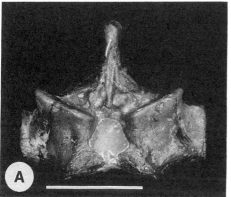 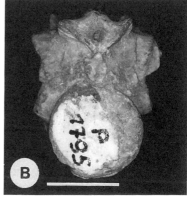

Figure 5. A) Eusauropterygian neural arch from the lower Muschelkalk (mu$_1$) of Halle/Saale (Halle, uncatalogued). **B)** Sacral vertebra from the Schaumkalk (upper lower Muschelkalk, mu$_2$) of Freyburg/Unstrut (Jena, P 1795). Scale bar = 20 mm.

of the holotype of "*Proneusticosaurus*" *silesiacus* survived World War II (Figure 6; MGU Wr. 4438s). The specimen is very incomplete, but it shows a number of characteristics which distinguish it from *Nothosaurus*, including posterior dorsal vertebrae of the type described above.

The remaining part of "*Proneusticosaurus*" (Figure 7) shows fragments of posterior dorsal ribs, parts of the gastral rib cage, remnants of two digits of the left manus, a number of posterior dorsal vertebrae, the complete right pubis and the ventral (medial) part of the left pubis, the proximal head of the right ischium, and the proximal part of the right femur. It differs from *Nothosaurus* in a number of characters.

Each gastral rib is composed of five elements as in the latter genus, but the distal end of the lateral elements is rather broad and rounded as in *Cymatosaurus* (Schrammen, 1899:Plate 26), instead of tapering as in *Nothosaurus*. The obturator foramen in the pubis is open, as in many *Nothosaurus* (specimens of different size are variable with respect to that character: Rieppel, 1994a), but it is rounded and not slit-like as it always is in the latter genus. The pubis of *Nothosaurus* shows a characteristic concavity at its ventral (medial) margin which is absent in "*Proneusticosaurus*." In the latter genus, the symphysis between the two pubes is a closed suture. The height of the right pubis is 64.5 mm, its dorsal width is 39 mm, its minimal width is 38.2 mm, and its ventral width is 54.5 mm. The same morphology of the pubis as in "*Proneusticosaurus*" is also observed in an isolated element, again from the lower Gogolin beds (MGU Wr. 3917s; Figure 8). The femur of "*Proneusticosaurus*" shows a well-developed trochanter which is reduced

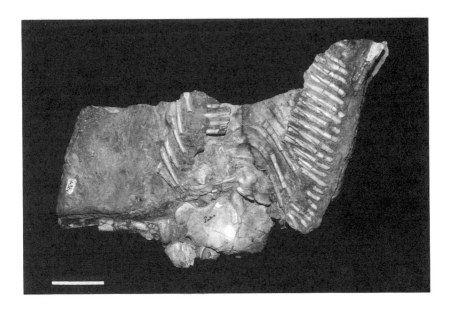

Figure 6. "*Proneusticosaurus silesiacus*" Volz, 1902 (holotype, MGU Wr. 4438s), from the Gogolin beds (lower Muschelkalk, mu_1) of Gogolin, Upper Silesia. Scale bar = 50 mm.

to a greater degree in *Nothosaurus*. The width of the proximal head of the femur is 19.5 mm; width at the mid-diaphysis is approximately 12 mm. An isolated femur (MGU Wr. 3864s; Figure 9) from the lower Muschelkalk of Gogolin shows a similar morphology with a sigmoidally curved shaft and a well developed trochanter (Figure 9).

A total of nine posterior dorsal and sacral vertebrae are at least partially exposed. The centrum of these vertebrae is platycoelous and non-notochordal. Their body is not constricted but shows slightly convex lateral and ventral margins. The length of a centrum of a posterior dorsal vertebra is 19.5 mm; the diameter of its articular surface is 15 mm. The articular surface which receives the pedicels of the neural arch is expanded into the butterfly-shaped platform typical of Eosauropterygia. Two posterior vertebrae lying in front of the left pubis show a low neural spine which slightly expands towards its dorsal margin. The neural spine is broken in a vertebra that lies at the posterior margin of the right pubis. This element shows the strongly inclined articular facets of the pre- and postzygapophyses, the deeply bifurcated zygosphene, and the undivided zygantrum characteristic of the vertebrae discussed above.

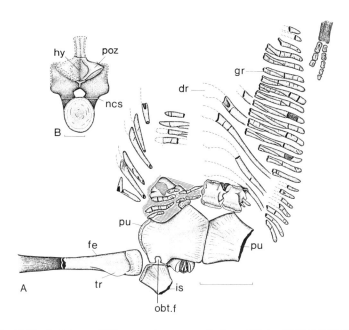

Figure 7. A) The holotype of "*Proneusticosaurus silesiacus*" Volz, 1902 (MGU Wr. 4438s). Scale bar = 50 mm. **B)** Posterior dorsal vertebra of "*Proneusticosaurus silesiacus*" Volz, 1902 (holotype, MGU Wr. 4438s) in posterior view. Scale bar = 20 mm. Abbreviations: **dr** = dorsal rib fragments, **fe** = femur, **gr** = gastral rib fragments, **hy** = hyposphene, **is** = ischium, **ncs** = neurocentral suture, **obt.f** = obturator foramen, **poz** = postzygapophysis, **pu** = pubis, **tr** = trochanter.

TAXONOMIC CONCLUSIONS

"*Proneusticosaurus*" *silesiacus* shares with *Nothosaurus* from the lower Muschelkalk a number of characters such as the slight pachyostosis of the proximal part of the dorsal ribs, the low neural spine, and the broad and "swollen" pre- and postzygapophyses. It differs from *Nothosaurus* by the dorsal expansion of the neural spines, by the greater inclination of the articular surfaces on pre- and postzygapophyses, by the deeply subdivided zygosphene and the undivided zygantrum, by the complete ossification of the ventral margin of the pubis, by the rounded obturator foramen which remains widely open, and by the well-developed trochanter on the femur. The vertebral characters allow the conclusion that the vertebrae discussed above, from the lower Muschelkalk of the central and southwestern Germanic Basin, as well as from the lower Anisian of the southern Alps, may be referred to "*Proneusticosaurus*."

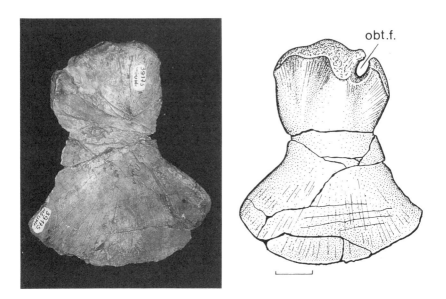

Figure 8. Isolated pubis (MGU Wr. 3917s) from the Gogolin beds (lower Muschelkalk, mu_1) of Gogolin, Upper Silesia. Abbreviation: **obt.f** = obturator foramen. Scale bar = 20 mm.

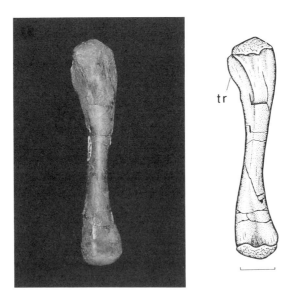

Figure 9. Isolated femur (MGU Wr. 3864s) from the Gogolin beds (lower Muschelkalk, mu_1) of Gogolin, Upper Silesia. Abbreviation: **t** = trochanter. Scale bar = 20 mm.

Unequivocal identification of the generic identity of "*Proneusticosaurus*" is impossible. The platycoelous vertebrae as well as the butterfly-shaped expansion of the platform receiving the pedicels of the neural arch are synapomorphies of the Eosauropterygia (Rieppel, 1994a). The Gogolin beds in Silesia have yielded three eosauropterygian genera represented by skulls: the pachypleurosaur *Dactylosaurus* (Gürich, 1884; GPIT, uncatalogued), the eusauropterygian *Nothosaurus* (Wysogorski, 1904; Kunisch, 1888; and personal observation), and *Cymatosaurus* (Huene, 1944; Wysogorski, 1904; and personal observation). *Lamprosauroides goepperti* (*Lamprosauroides* Schmidt, 1927, replaces *Lamprosaurus* Meyer, 1860) is represented by a right maxilla (Figure 10; MGU Wr. 3871s) from the lowermost Muschelkalk of Krappitz (now Krapkowice) in Upper Silesia which was first described by Meyer (1860). A synonymy of *Lamprosauroides* with *Cymatosaurus* was suggested by Schrammen (1899:408; *Lamprosauroides* would take priority), but is problematic in view of the fragmentary nature of the holotype of *Lamprosauroides*. Because of its incomplete nature, *Lamprosauroides goepperti* must be considered a nomen dubium.

"*Proneusticosaurus*" differs from *Dactylosaurus* in size and morphology, and from *Nothosaurus* in morphology (see above). Sues (1987:129) suggested that

Figure 10. *Lamprosauroides goepperti* Meyer, 1860 (holotype, MGU Wr. 3871s), from the Gogolin beds (lower Muschelkalk, mu_1) of Krappitz (Krapkowice), Upper Silesia. Scale bar = 50 mm.

"*Proneusticosaurus*" Volz, 1902, is referable to *Cymatosaurus* Fritsch, 1894, a conclusion which may be supported by the observation that vertebrae similar to those observed in "*Proneusticosaurus*" occur in other localities that have produced eusauropterygian cranial remains that are all referable either to *Nothosaurus* or to *Cymatosaurus*. This is true of the lower Muschelkalk of Halle/Saale, as well as of the Schaumkalk of Freyburg/Unstrut (Fritsch, 1894; and personal observation). *Cymatosaurus* is also known from the lower Anisian of the northern Alps (Rieppel, 1995a).

A COMPARISON OF SAUROPTERYGIA IN THE GERMANIC AND ALPINE TRIASSIC

Synonymy of "*Proneusticosaurus*" with *Cymatosaurus* is likely, but must remain conjectural. Nevertheless, the vertebrae described in this contribution, along with what remains of the holotype of "*Proneusticosaurus*" *silesiacus*, document unequivocally that the same taxon (most probably *Cymatosaurus*) occurs in the lowermost lower Muschelkalk of the eastern Germanic Basin, in the lower Muschelkalk of the central and southwestern Germanic Basin (MHI 1293/1), and in the lower Anisian of the southern Alps (MHI 1279). *Cymatosaurus* represents the earliest sauropterygian genus to appear in the eastern part of the Germanic Basin (*Cymatosaurus erythreus* from the upper Buntsandstein [so_2, Röt] of Rüdersdorf: Huene, 1944), from where it seems to have spread south as is indicated by its (possible) occurrence in the lower Muschelkalk of the southern part of the Germanic Basin (MHI 1293/1), in the lower Anisian of the northern (*Cymatosaurus multidentatus*: Rieppel, 1995a), and (perhaps) in the southern Alps (MHI 1279). This scenario is based on the assumption of a marine migratory route from the Germanic Basin into the southern Alpine intraplatform realm during early Anisian times, as is also indicated by the *Dadocrinus* benthic paleocommunity (Hagdorn, in press). Since vertebrate paleontologists have traditionally viewed the Germanic and Alpine Triassic as separate faunal provinces (but see Wild, 1972, for further discussion), it seems appropriate at this junction to summarize current knowledge on the distribution of sauropterygian taxa in the Middle Triassic of the Germanic Basin and the Alpine region at the generic level (Figures 11 and 12).

Pachypleurosauria are known from the Middle Triassic of China and Europe. A recent cladistic analysis (Rieppel and Lin Kebang, 1995) shows the Chinese taxon *Keichousaurus* to be the sister-group of all European genera. The hypothesis of an immigration of pachypleurosaurs along the northern border of the Tethys into the Germanic Basin from the east may be supported by reference to Asiatic affinities of lower Muschelkalk invertebrate faunas (Kozur, 1974). The

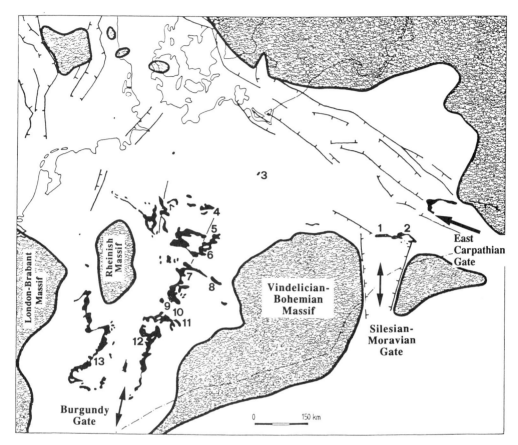

Figure 11. The distribution of fossil-bearing localities in the Germanic Basin (see text for further discussion). Key to fossil sites: **1** = Gogolin, **2** = Tarnowitz, **3** = Rüdersdorf, **4** = Remkersleben, **5** = Halle and Freyburg/Unstrut, **6** = Jena, **7** = Sulzheim, **8** = Bayreuth, **9** = Hettingen, **10** = Eberstadt, **11** = Crailsheim, **12** = Hoheneck, **13** = Luneville.

earliest pachypleurosaur to appear in the Germanic Basin is *Dactylosaurus* from the lowermost Muschelkalk of the eastern Germanic Basin (lower Gogolin beds: Gürich, 1884). Diagnostic material of that taxon has not been recorded outside Upper Silesia. Isolated bones of a pachypleurosaur of *Dactylosaurus* size (MHI 1300) have been collected from the approximately contemporaneous base of the *Myophoria* beds (lower Anisian) of Thuringia, which are included into the upper Röt (so_3).

The pachypleurosaur genus *Anarosaurus* is known from the lower Muschelkalk of the western Germanic Basin (Winterswijk, the Netherlands: Hoojer, 1959; Oosterink, 1986:79-80, Photos 38-40 [in these papers, *Anarosaurus* is misidentified as *Cymatosaurus*]), from the Schaumkalk (upper lower Muschelkalk) of Freyburg/Unstrut (Rieppel and Lin Kebang, in preparation), and from the *orbicularis*

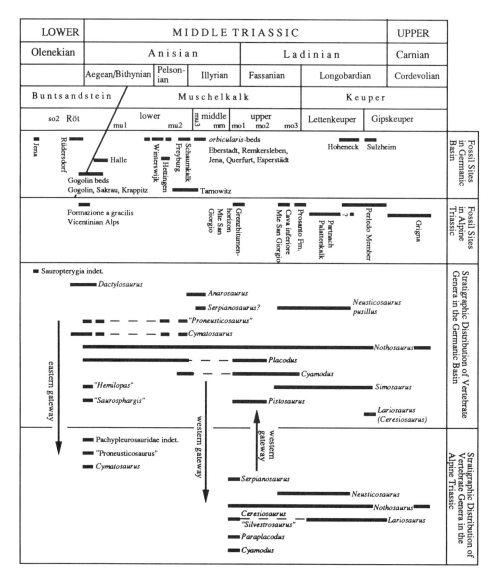

Figure 12. The stratigraphic relationship of Middle Triassic fossil genera in the Germanic Basin and Alpine region (see text for further discussion).

beds (lower middle Muschelkalk) of Remkersleben near Magdeburg (Dames, 1890). The latter two occurrences are in the central part of the Germanic Basin. Diagnostic remains of *Anarosaurus* have not been recorded elsewhere, although isolated postcranial elements from the middle Muschelkalk of Eberstadt (Badenia) might belong to that genus (Hagdorn and Simon, 1993; see below also).

The earliest pachypleurosaur occurrence in the Alpine Triassic is the neural arch (MHI 1292) from the "*Formazione a gracilis*" in the Vicentinian Alps mentioned above. Although not diagnostic, this specimen documents a pachypleurosaur of a size comparable to that of *Dactylosaurus*. Since the Alpine Triassic was not connected to the pelagic Tethyan realm until Pelsonian times (Mostler, 1993), this specimen again indicates a lower Anisian connection between the epicontinental Germanic Basin and southern Alpine intraplatform basins. The earliest diagnostic pachypleurosaur material from the Alpine Triassic is *Serpianosaurus* from the Grenzbitumen horizon (Anisian-Ladinian boundary) of the southern Alps (Rieppel, 1989). According to Kozur (1974), this boundary is between the lower (mo_1) and middle (mo_2) upper Muschelkalk (*compressus* biozone) of the Germanic Basin. Evidence from palynomorph biostratigraphy (Visscher et al., 1993) calibrates the Anisian/Ladinian boundary higher upsection, i.e., with the *enodis* biozone. Diagnostic remains of *Serpianosaurus* have not been recorded from the Germanic Muschelkalk, although it seems possible to refer isolated postcranial elements from the middle Muschelkalk of Eberstadt (Badenia) to that genus rather than to *Anarosaurus* (Hagdorn and Simon, 1993). If so, a persisting exchange of reptile faunas between the Germanic Muschelkalk and the southern Alpine intraplatform basin may be assumed to have existed during Anisian times. Invertebrate faunas indicate Pelsonian through lower Illyrian migration routes via the Silesian-Moravian Gate in the East, and a subsequent breakthrough of a western passage via the Burgundy Gate with the onset of the upper Muschelkalk transgression in late Illyrian times (Kozur, 1974).

The only other diagnosable pachypleurosaur from the Germanic Triassic is *Neusticosaurus pusillus* from the Lettenkeuper of Hoheneck near Ludwigsburg. In the Alpine Triassic, the earliest occurrence of the genus *Neusticosaurus* (*N. pusillus*) is in the Cava Inferiore beds of the lower Meridekalke, Monte San Giorgio (lower Ladinian: Sander, 1989). The species is also known from the Prosanto Formation of the eastern Swiss Alps, believed to be of middle Ladinian age (Bürgin et al., 1991), and from the Perledo Member of the Perledo-Varenna Formation (Rieppel, 1995b) of late Ladinian age (Gaetani et al., 1992). The genus *Neusticosaurus* may also be represented (by very poorly preserved material: Zapfe and König, 1980; Warch, 1984; Rieppel, 1994b) in the upper unit of the Partnach Plattenkalk of the Gailtaler Alps, Austria, of middle or late Ladinian age (see discussion in Rieppel, 1993).

The earliest eusauropterygian genus other than *Cymatosaurus* to appear in the European Triassic is *Nothosaurus*, in the lowermost lower Muschelkalk of the eastern basin (Gogolin beds, Upper Silesia: Kunisch, 1888). Since the lower Muschelkalk deposits in the west are somewhat younger than those in the east

(Kozur, 1974), an early westward dispersal of *Nothosaurus* within the Muschelkalk Basin may be assumed. However, a biozonal scheme for the Muschelkalk of Winterswijk has not been established yet, and a precise age comparison of these faunas therefore is still impossible. In the Germanic Basin, the genus *Nothosaurus* persists throughout the Muschelkalk and into the Lettenkeuper and Gipskeuper (Carnian). Northward expansion of the genus is documented by an isolated neural arch from the English Triassic (Ladinian; Walker, 1969). An isolated nothosaur vertebra has been reported from the lowermost lower Muschelkalk of Israel (Swinton, 1952), equivalent in age to the "transition zone" of Kozur (1974) with *Beneckeia buchi* and *Myophoria vulgaris* (Brotzen, 1957). This occurrence again supports eastern affinities of the genus. In the Alpine Triassic, *Nothosaurus* does not appear until much later, i.e., in the Grenzbitumen horizon of Monte San Giorgio, marking the Anisian-Ladinian boundary equivalent to the transition from mo_1 to mo_2 in the upper Muschelkalk of the Germanic Basin. The Grenzbitumen horizon has yielded as yet undescribed *Nothosaurus* remains, as well as *Paranothosaurus* (Peyer, 1939), a junior synonym of *Nothosaurus* (Rieppel and Wild, 1996). In the Alpine Triassic, the genus persists into the middle Ladinian (Bürgin et al., 1991) and lower Carnian (Boni, 1937), respectively. The stratigraphic and geographic distribution of *Nothosaurus* again indicates a southern connection from the Muschelkalk Basin into the Alpine realm. However, since *Nothosaurus* is lacking in the Alpine Triassic up to the uppermost Anisian, there is no evidence for an early migration of *Nothosaurus* from the Germanic Basin through the eastern (Silesian-Moravian) gateway into the Alpine realm. Instead, the genus must have arrived there through the western connection: in the Alpine Triassic, *Nothosaurus* does not appear before the western passage through the Burgundy Gate had opened.

The genus *Lariosaurus* was described by Curioni (1847) on the basis of specimens from the Perledo Member of the Perledo-Varenna Formation of the southern Alps (see also Ticli, 1984; Peyer, 1933-1934; Mariani, 1923; Boulenger, 1898); additional material has come from the Kalkschieferzone of Monte San Giorgio (Kuhn-Schnyder, 1987) and northern Italy (Renesto, 1993; Tintori and Renesto, 1990), from the upper unit of the Partnach Plattenkalk of the Gailtaler Alps (Warch, 1984; Zapfe and König, 1980), from the calcareous black shales of Amélie-les-Bains in the eastern Pyrenees (Mazin, 1985), and from the Spanish Muschelkalk (Sanz, 1976, 1983a). The age for all these specimens is middle or late Ladinian. Kuhn-Schnyder (1990) named a new genus, *Silvestrosaurus*, for a specimen from the Grenzbitumen horizon of Monte San Giorgio (Anisian-Ladinian boundary) described by Tschanz (1989) as *Lariosaurus buzzi*. Another eusauropterygian from the Monte San Giorgio deposits is *Ceresiosaurus* (Peyer, 1931). It appears from preliminary studies (Rieppel and Tschanz, work in progress), that the genus *Lariosaurus* is paraphyletic if it excludes *Silvestrosaurus* and *Ceresiosaurus*. Potential synapomorphies of this clade are the shape and relative size of the humerus, the broad ulna, the morphology of the ilium, the number of sacral ribs, and hyperphalangy in the manus.

Whether or not this monophyletic clade is considered to include one, two, or three genera, it comprises a total of at least five species (*balsami, buzzi, calcagnii, valceresii,* and an undescribed species), and it consistently comes out as sister-group to the *Nothosaurus-Paranothosaurus* clade in recent cladistic analyses (Rieppel, 1994a; Storrs, 1991, 1993; Tschanz, 1989). The (rare) occurrence of this clade in the Germanic Muschelkalk is indicated by a specimen from the lower middle Keuper (Gipskeuper, Grundgipsschichten, uppermost Ladinian) of Sulzheim near Würzburg (southern Germany), described by Schultze and Wilczewski (1970), who compared the specimen either to *Ceresiosaurus* or to *Lariosaurus*. The occurrence of the *Lariosaurus* clade in the Germanic Muschelkalk is also indicated by an isolated ulna from the upper Lettenkeuper of Hoheneck (SMNS 7175: Rieppel, 1994a:Figure 61). In contrast to *Cymatosaurus* and, perhaps, *Serpianosaurus*, which indicate an early (lower and upper Anisian) southeastern connection of the Muschelkalk Basin with the southern Alpine realm, probably via the Silesian-Moravian Gate, the *Lariosaurus* clade seems to have diversified primarily along the northwestern shoreline of the Tethys, from where rare representatives ventured into the northern peritethys basins, e.g., rarely into the Germanic Basin during Ladinian times. The same may be true of the *Neusticosaurus* clade, which again diversified within Alpine intraplatform basins (Sander, 1989), and reached the Germanic Basin during the Ladinian (*Neusticosaurus pusillus*), at a time (during the Lettenkeuper deposition) when the Germanic Basin became more land-influenced with a small scale cyclic change of lacustrine and marine conditions.

Phylogenetic conclusions may be drawn if sister-group relationships are translated into ancestor-descendant relationships under biostratigraphical control. The *Lariosaurus* clade would then appear as related to, i.e., descended from, the *Nothosaurus* clade, and *Serpianosaurus* is the sister-taxon (Rieppel, in preparation), i.e., ancestor, of the *Neusticosaurus* clade. On that basis it would appear that coming from the Muschelkalk Basin, ancestral taxa (*Nothosaurus, Serpianosaurus*) reached the southern Alpine realm during late Anisian times, i.e., at a time when distinct facies deviations start to become apparent between the two biotas (Mostler, 1993). The Germanic Basin was increasingly influenced by terrestrial influx from the lower Ladinian *enodis* biozone onward (upper part of upper Muschelkalk). This period is characterized by highly endemic invertebrate faunas. In the Alpine realm, strong facies differentiation from uppermost Anisian and earliest Ladinian times onward in carbonate platforms separated two major pelagic carbonate basins, also including anaerobic or dysaerobic intraplatform troughs of Ladinian age like the

Besano (Grenzbitumen horizon), Meride, and Perledo-Varenna Formations.

The descendant clades diversified in the fragmented southern Alpine intraplatform basin facies, and returned to the Muschelkalk Basin during Ladinian times via the Burgundy Gate. This scenario is in accordance with the observation that during the lower Muschelkalk, the sauropterygian fauna is most diverse and numerous in the central and eastern part of the Germanic Basin. By upper Muschelkalk, Lettenkeuper, and Gipskeuper times (Ladinian), marine reptiles are most diverse and most abundant in the southwestern part of the Germanic Basin, which was situated relatively close to the marine strait connecting it with the western Tethys and its marginal seas. During late Ladinian times, marine reptiles, together with invertebrate faunal elements, repeatedly invaded episodic Gipskeuper playa lakes covering the Germanic Basin along its subsidence axes.

The Germanic Triassic has yielded a number of genera which have not been recorded from the Alpine Triassic, and vice versa. The genus *Placodus* is known from the earliest Muschelkalk deposits (Gogolin beds) in Upper Silesia through the middle upper Muschelkalk (mo_2) of Southwest Germany. The earliest occurrence of *Cyamodus* is from the Karchowice Beds of Tarnowitz in Upper Silesia (now Tarnowskie Gory, Poland), which belong to the uppermost lower Muschelkalk (lower Illyrian) (Gürich, 1884). The specimen (lost during World War II) documents the early occurrence of *Cyamodus*, along with *Placodus*, in the eastern Muschelkalk basin. *Placodus* has never been unequivocally identified in the Alpine Triassic, where the earliest placodont records are *Paraplacodus* and *Cyamodus* from the Grenzbitumen horizon (Anisian-Ladinian boundary) of Monte San Giorgio. These are still somewhat later than the first *Cyamodus* occurrence in the eastern Germanic Basin. The significance of this distributional pattern is hard to assess in the absence of a well-corroborated hypothesis of phylogenetic interrelationships of *Placodus*, *Paraplacodus*, and the Cyamodontoidea (Rieppel, in preparation). The situation is further complicated by the fact that *Saurosphargis*, a fragmentary postcranium from the lower Muschelkalk of Upper Silesia (lost during World War II) had been compared with *Paraplacodus* (Huene, 1936).

Eusauropterygians from the Germanic Muschelkalk which have never before been recorded from the Alpine Triassic include the genus *Simosaurus* from the upper Muschelkalk (mo_3), Lettenkeuper, and lowest Gipskeuper (Rieppel, 1994a), and *Pistosaurus* from the lower and middle upper Muschelkalk (mo_1 through mo_2) of southwestern Germany. The only other pistosaur record is from the Ladinian of Montral-Alcover (Sanz, 1983b; Sanz et al., 1993); however, this specimen is difficult to interpret, and the identification may be erroneous. Since the earliest occurrence of *Cymatosaurus* and *Nothosaurus* is in the Gogolin beds of Upper Silesia, and in view of the highly corroborated hypothesis that *Simosaurus* is the sister-taxon of all other Eusauropterygia (Sues, 1987, Tschanz, 1989; Storrs, 1991,

1993; Rieppel, 1994a), the restriction of *Simosaurus* to the upper Muschelkalk and Lettenkeuper of the Germanic Basin is a clear indication of the incompleteness of the fossil record (Norell and Novacek, 1992a, b). There are, however, jaw fragments from the lowermost lower Muschelkalk (Gogolin beds) of Upper Silesia with a highly characteristic dentition, originally described by Meyer (1847:575, 1851:236, Plate 28, Figure 16) as those of a fossil fish named *Hemilopas mentzeli*, and identified by Jaekel (1907) as anterior teeth of the enigmatic genus *Tholodus* of probable ichthyosaur affinities (Sander and Mazin, 1993). Because of similarities in tooth morphology, it could well be that these jaw fragments represent an early simosaurid, filling the stratigraphic gap for this clade indicated by cladistic analysis.

Alpine taxa not recorded from the Germanic Muschelkalk include two incomplete specimens from the middle to upper Ladinian of the northern Alps (Partnachschichten, Vorarlberg, Austria) described as *Microleptosaurus schlosseri* and *Partanosaurus zitteli* by Skuphos (1893a, b, c). A recent reinvestigation of these fossils (Rieppel, work in progress) showed that *Partanosaurus zitteli* is a junior synonym of *Simosaurus gaillardoti*, adding to the evidence of faunal interchange between the Germanic and Alpine Triassic during the late Ladinian.

SUMMARY AND CONCLUSIONS

The distribution of sauropterygian genera in the Germanic and Alpine Lower and Middle Triassic supports a four-step sauropterygian migration scenario in western Tethys and peritethys basins.

1. Late Olenekian (Röt): earliest (not diagnostic) sauropterygians immigrated into the Germanic Basin from the Asiatic faunal province via an eastern marine strait ("Polish East Carpathian Gate") following the northern Paleotethys branch.

2. Early Anisian (uppermost Röt to lower part of lower Muschelkalk): with the Muschelkalk transgression expanding further to the west, a diverse sauropterygian fauna comprising *Dactylosaurus*, *Cymatosaurus*, *Nothosaurus*, *Cyamodus*, and *Placodus* established itself in the Germanic Basin. Abundance and diversity are highest in the eastern and central parts of the Basin. From there, *Cymatosaurus* and a small pachypleurosaur (*Dactylosaurus*?) reached the eastern part of the southern Alpine realm (Vicentinian Alps).

3. Upper Anisian, lower Illyrian (upper part of lower Muschelkalk): *Placodus*, *Cyamodus*, *Nothosaurus*, and *Cymatosaurus* persisted in the Germanic Basin, but there is no evidence for further faunal exchange with the Alpine realm. During the middle Muschelkalk sea level lowstand, *Cymatosaurus* disappeared, and the large pachypleurosaurs *Anarosaurus* and possibly also *Serpianosaurus* appeared in the central and western Germanic Basin. During that time, *Nothosaurus*,

Serpianosaurus, and *Cyamodus* reached the western part of the southern Alpine realm (Grenzbitumen horizon) via the Burgundy Gate, a marine strait along a subsiding graben-like structure which connected the Germanic Basin with the western Tethys (Ziegler, 1982).

4. Early Ladinian: the southern Alpine realm became a speciation center as facies deviations between the Germanic and Alpine Triassic become apparent. *Neusticosaurus* may have originated from *Serpianosaurus*, and the *Ceresiosaurus/ Lariosaurus* clade may be descended from *Nothosaurus*. These originally southern Alpine genera episodically immigrated into the Germanic Basin via Burgundy Strait in late Ladinian times (Lettenkeuper, lowest Gipskeuper). While *Cyamodus* and *Nothosaurus* persisted in the Germanic Basin until the end of upper Muschelkalk (early Ladinian) or the end of Gipskeuper (latest Ladinian/early Carnian), *Placodus* and *Pistosaurus* disappeared at the end of the latest Anisian/earliest Ladinian (*enodis* biozone), when the onset of the highstand induced the beginning of highly endemic development among Muschelkalk invertebrate faunas. *Simosaurus* is an endemic element of the southwestern Germanic Basin. During Ladinian the sauropterygian fauna is most diverse and abundant in the southwestern part of the Germanic Basin.

ACKNOWLEDGMENTS

We thank the editors, E. L. Nicholls and J. M. Callaway, for their efforts in putting this volume together. R. L. Carroll, Montreal, and M. J. Benton, Britsol, kindly read an early draft of the manuscript, offering helpful advice and criticism. We are grateful to a number of colleagues who granted free access to the collections in their care. These include J. Gorczyca-Skala, Institute of Geological Sciences, University of Wraclaw; A. Liebau and F. Westphal, Institut und Museum für Geologie und Paläontologie, University of Tübingen; H. Haubold, Institut für Geowissenschaften, Martin-Luther-Universität, Halle/Saale; W.-D. Heinrich, Humboldt Museum, Berlin; D. v. Knorre, Phyletisches Museum, Jena; H. U. Schlüter, Bundesanstalt für Geowissenschaften und Rohstoffe, Berlin; R. Wild, Staatliches Museum für Naturkunde, Stuttgart. This study was supported by NSF grant DEB-9220540 (to O. R.).

REFERENCES

Barbieri, G., G. De Vecchi, V. De Zanche, E. Di Lallo, P. Frizzo, P. Mietto, and R. Sedea. 1980. Note illustrative della carta geologica dell'area di Recoaro. *Memorie di Scienze*

Geologiche già Memorie degli Istituti di Geologia e Mineralogia dell' Università di Padova 34:23-52.

Boni, A. 1937. Sulla presenza di un nothosauride nel Raibliano della Grigna. *Rivista Italiana di Paleontologia* 43:81-92.

Boulenger, A. G. 1898. On a nothosaurian reptile from the Trias of Lombardy, apparently referable to *Lariosaurus*. *Transactions of the Zoological Society of London* 14:1-10.

Brotzen, F. 1957. Stratigraphical studies on the Triassic vertebrate fossils from Wadi Ramon, Israel. *Arkiv för Mineralogi och Geologi* 2:191-217.

Bürgin, T., U. Eichenberger, H. Furrer, and K. Tschanz. 1991. Die Prosanto-Formation - eine fischreiche Fossil-Lagerstätte in der Mitteltrias der Sivretta-Decke (Kanton Gaubünden, Schweiz). *Eclogae Geologicae Helvetiae* 84:921-990.

Curioni, G. 1847. Cenni sopra un nuovo saurio fossile dei monti di Perledo sul Lario e sul terreno che lo racchiude. *Giornale del' J. R. Instituto Lombardo di Scienze, Lettre ed Arti* 16:159-170.

Dames, W. 1890. *Anarosaurus pumilio* nov. gen. nov. sp. *Zeitschrift der Deutschen Geologischen Gesellschaft* 42:74-85.

Fritsch, K. von. 1894. Beitrag zur Kenntnis der Saurier des Halle'schen unteren Muschelkalkes. *Abhandlungen der Naturforschenden Gesellschaft zu Halle* 20:273-302.

Gaetani, M., M. Gnaccolini, G. Poliasni, D. Grignani, M. Gorza, and L. Matrellini. 1992. An anoxic intraplatform basin in the Middle Triassic of Lombardy (southern Alps, Italy): anatomy of a hydrocarbon source. *Rivista Italiana di Paleontologia e Stratigrafia* 97:329-354.

Gürich, G. J. E. 1884. Über einige Saurier des Oberschlesischen Muschelkalkes. *Zeitschrift der Deutschen Geologischen Gesellschaft* 36:125-144.

Hagdorn, H. 1991. The Muschelkalk in Germany---an introduction. IN H. Hagdorn, T. Simon, and J. Szulc (Eds.), *Muschelkalk---A Field Guide*, pp. 7-21. Goldschneck Verlag, Korb.

Hagdorn, H. In press. Palökologie der Trias-Seelilie *Dadocrinus*. *Geologische und Paläontologische Mitteilungen Innsbruck*.

Hagdorn, H. and T. Simon. 1993. Rinnenbildung und Emersion in den Basisschichten des Mittleren Muschelkalks von Eberstadt (Nordbaden). *Neues Jahrbuch für Geologie und Paläontologie*, Abhandlungen 189:119-145.

Hoojer, D. A. 1959. Records of nothosaurians from the Muschelkalk of Winterswijk, Netherlands. *Geologie en Mijnbouw* 21:37-39.

Huene, E. von. 1944. *Cymatosaurus* und seine Beziehungen zu anderen Sauropterygiern. *Neues Jahrbuch für Mineralogie, Geologie und Paläontologie*, Monatshefte B:192-222.

Huene, F. von. 1936. *Henodus chelyops*, ein neuer Placodontier. *Palaeontographica* A 84:99-148.

Huene, F. von. 1958. Aus den Lechtaler Alpen ein neuer *Anarosaurus*. *Neues Jahrbuch für Geologie und Paläontologie*, Monatshefte (8/9):382-384.

Jaekel, O. 1907. *Placochelys placodonta* aus der Obertrias des Bakony. Resultate der Wissenschaftlichen Erforschung des Balatonsees. I. Band. 1. Teil. *Paläontologie---Anhang*. Victor Hornyánzky, Budapest.

Kovács, S. 1992. Tethys "western ends" during the Late Paleozoic and Triassic and their possible genetic relationships. *Acta Geologica Hungarica* 35:329-369.

Kozur, H. 1974. Probleme der Triasgliederung und Parallelisierung der germanischen und tethyalen Trias. Teil II: Anschluss der germanischen Trias an die internationale Triasgliederung. *Freiberger Forschungshefte. Reihe* (C): *Paläontologie* 2:51-77.

Kuhn-Schnyder, E. 1987. Die Triasfauna der Tessiner Kalkalpen. XXVI. *Lariosaurus lavizzarii* n. sp. (Reptilia, Sauropterygia). *Abhandlungen der Schweizerischen Paläontologischen Gesellschaft* 110:1-24.

Kuhn-Schnyder, E. 1990. Über Nothosauria (Sauropterygia, Reptilia)---ein Diskussionsbeitrag. *Paläontologische Zeitschrift* 64:313-316.

Kunisch, H. 1888. Über eine Saurierplatte aus dem oberschlesischen Muschelkalke. *Zeitschrift der Deutschen Geologischen Gesellschaft* 40:671-693.

Mariani, E. 1923. Su un nuovo esemplare di *Lariosaurus balsami* Cur. trovato negli scisti di Perledo sopra Varenna (Lago di Como). *Atti della Societa Italiana di Scienze Naturali* 62:218-225.

Mazin, J.-M. 1985. A specimen of *Lariosaurus balsami* Curioni 1847, from the Eastern Pyrenees (France). *Palaeontographica* A 189:159-169.

Meyer, H. von. 1847. Mittheilungen an Professor Bronn gerichtet. *Neues Jahrbuch für Mineralogie, Geognosie, Geologie und Petrefaktenkunde* pp. 572-580.

Meyer, H. von. 1851. Fossile Fische aus dem Muschelkalk von Jena, Querfurt und Esperstädt. *Palaeontographica* A 1:195-208.

Meyer, H. von. 1860. *Lamprosaurus Göpperti*, aus dem Muschelkalke von Krappitz in Ober-Schlesien. *Palaeontographica* 7:245-247.

Mostler, H. 1993. Das Germanische Muschelkalkbecken und seine Beziehungen zum tethyalen Muschelkalkmeer. IN H. Hagdorn and A. Seilacher (Eds.), *Muschelkalk. Schöntaler Symposium 1991*, pp. 11-14. Goldschneck Verlag, Korb.

Norell, M. and M. Novacek. 1992a. The fossil record and evolution. Comparing cladistic and paleontological evidence for vertebrate history. *Science* 225:1690-1693.

Norell, M. and M. Novacek. 1992b. Congruence between superpositional and phylogenetic pattern: comparing cladistic patterns with fossil records. *Cladistics* 8:319-337.

Oosterink, H. W. 1986. Winterswijk, geologie deel II. De Trias-periode (geologie, mineralen en fossielen). *Wetenschappelijke Mededeling van de Koninklijke Nederlandse Natuurhistorische Vereniging* 178:1-120.

Peyer, B. 1931. Die Triasfauna der Tessiner Kalkalpen. IV. *Ceresiosaurus calcagnii* nov. gen. nov. spec. *Abhandlungen der Schweizerischen Paläontologischen Gesellschaft* 51:1-68.

Peyer, B. 1933-1934. Die Triasfauna der Tessiner Kalkalpen. VII. Neubeschreibung der Saurier von Perledo. *Abhandlungen der Schweizerischen Paläontologischen Gesellschaft* 53-54:1-130.

Peyer, B. 1939. Die Triasfauna der Tessiner Kalkalpen. XIV. *Paranothosaurus amsleri* nov. gen. nov. spec. *Abhandlungen Schweizerischen Paläontologischen Gesellschaft* 62:1-87.

Renesto, S. 1993. A juvenile *Lariosaurus* (Reptilia, Sauropterygia) from the Kalkschieferzone (uppermost Ladinian) near Viggiù (Varese, Northern Italy). *Rivista*

Italiana di Paleontologia e Stratigrafia 99:199-212.

Rieppel, O. 1989. A new pachypleurosaur (Reptilia: Sauropterygia) from the Middle Triassic of Monte San Giorgio, Switzerland. *Philosophical Transactions of the Royal Society of London* B 323:1-73.

Rieppel, O. 1993. The status of the pachypleurosauroid *Psilotrachelosaurus toeplitschi* Nopcsa (Reptilia, Sauropterygia), from the Middle Triassic of Austria. *Fieldiana* (Geology), N. S. 27:1-17.

Rieppel, O. 1994a. Osteology of *Simosaurus gaillardoti* and the relationships of stem-group Sauropterygia. *Fieldiana* (Geology), N. S. 28:1-85.

Rieppel, O. 1994b. *Lariosaurus balsami* Curioni (Reptilia, Sauropterygia) aus den Gailtaler Alpen. *Carithia II* 184 (104):345-356.

Rieppel, O. 1995a. The status of *Anarosaurus multidentatus* Huene (Reptilia, Sauropterygia), from the Lower Anisian of the Lechtaler Alps (Arlberg, Austria). *Paläontologische Zeitschrift* 69:289-299.

Rieppel, O. 1995b. The pachypleurosaur *Neusticosaurus* (Reptilia, Sauropterygia) from the Middle Triassic of Perledo, northern Italy. *Neues Jahrbuch für Geologie und Paläontologie*, Monatshefte 1995:205-216

Rieppel, O. and L. Kebang. 1995. Pachypleurosaurs (Reptilia: Sauropterygia) from the lower Muschelkalk, and a review of the Pachypleurosauroidea. *Fieldiana* (Geology), N. S. 32:1-44.

Rieppel, O. and R. Wild. 1996. A revision of the genus *Nothosaurus* (Reptilia: Sauropterygia) from the Germanic Triassic, with comments on the status of *Conchiosaurus clavatus*. *Fieldiana* (Geology), N. S. 34:1-82.

Sander, P.M. 1989. The pachypleurosaurids (Reptilia: Nothosauria) from the Middle Triassic of Monte San Giorgio (Switzerland), with the description of a new species. *Philosophical Transactions of the Royal Society of London* B 325:561-670.

Sander, P. and J.-M. Mazin. 1993. The paleobiogeography of Middle Triassic ichthyosaurs: the five major faunas. IN J.-M. Mazin and G. Pinna (Eds.), Evolution, Ecology and Biogeography of the Triassic Reptiles, pp. 145-151. *Paleontologia Lombarda*, N. S. 2.

Sanz, J. L. 1976. *Lariosaurus balsami* (Sauropterygia, Reptilia) de Estada (Huesca). *Estudios Geologicos* 32:547-567.

Sanz, J. L. 1983a. Los Nothosaurios (Reptilia, Sauropterygia) Españoles. *Estudos Geologicos* 39:193-215.

Sanz, J. L. 1983b. Consideraciones sobre el genero *Pistosaurus*. El suborden Pistosauria (Reptilia, Sauropterygia). *Estudos Geologicos* 39:451-458.

Sanz, J. L., L. S. Alafont, and J. J. Moratalla. 1993. Triassic reptile faunas from Spain. IN J.-M. Mazin and G. Pinna (Eds.), Evolution, Ecology and Biogeography of the Triassic Reptiles. *Paleontologia Lombarda*, N. S. 2:153-164.

Schmidt, K. P. 1927. New reptilian generic names. *Copeia* 1927 (163):58-59.

Schrammen, A. 1899. 3. Beitrag zur Kenntnis der Nothosauriden des unteren Muschelkalkes in Oberschlesien. *Zeitschrift der Deutschen Geologischen Gesellschaft* 51:388-408.

Schultze, H.-P. and N. Wilczewski. 1970. Ein Nothosauride aus dem unteren Mittel-Keuper Unterfrankens. *Göttinger Arbeiten zur Geologie und Paläontologie* 5:101-112.
Skuphos, T. 1893a. Vorläufige Mittheilung über *Parthanosaurus Zitteli*, einen neuen Saurier aus der Trias. *Zoologischer Anzeiger* 16:67-69.
Skuphos, T. 1893b. *Partanosaurus Zitteli* (s. No. 413 p. 67). *Zoologischer Anzeiger* 16:96.
Skuphos, T. 1893c. Über *Partanosaurus Zitteli* Skuphos und *Microleptosaurus Schlosseri* nov. gen., nov. spec., aus den Vorarlberger Partnachschichten. *Abhandlungen der Geologischen Reichsanstalt* 15:1-16.
Storrs, G. W. 1991. Anatomy and relationships of *Corosaurus alcovensis* (Diapsida: Sauropterygia) and the Triassic Alcova Limestone of Wyoming. *Bulletin of the Peabody Museum of Natural History* 44:1-151.
Storrs, G. W. 1993. The systematic position of *Silvestrosaurus* and a classification of Triassic sauropterygians. *Paläontologische Zeitschrift* 67:177-191.
Sues, H.-D. 1987. Postcranial skeleton of *Pistosaurus* and interrelationships of the Sauropterygia (Diapsida). *Zoological Journal of the Linnean Society* 90:109-131.
Swinton, W. E. 1952. A nothosaur vertebra from Israel. *Annales and Magazines of Natural History* (12) 5:875-876.
Ticli, B. 1984. Due esemplari di *Lariosaurus balsami* Curioni presenti nei Musei Civici di Lecco. *Natura* (Milano) 75:69-74.
Tintori, A. and S. Renesto. 1990. A new *Lariosaurus* from the Kalkschieferzone (uppermost Ladinian) of Valceresio (Varese, N. Italy). *Bollettino della Società Paleontologica Italiana* 29:309-319.
Tschanz, K. 1989. *Lariosaurus buzzii* n. sp. from the Middle Triassic of Monte San Giorgio (Switzerland), with comments on the classification of nothosaurs. *Palaeontographica* A 208:153-179.
Visscher, H., W. A. Brugman, and M. van Houte 1993. Chronostratigraphical and sequence stratigraphical interpretation of the palynomorph record from the Muschelkalk of the Obernsees Well, South Germany. IN H. Hagdorn and A. Seilacher (Eds.), *Muschelkalk. Schöntaler Symposium 1991*, pp. 145-152. Goldschneck Verlag, Korb.
Volz, W. 1902. *Proneusticosaurus*, eine neue Sauropterygiergattung aus dem untersten Muschelkalk Oberschlesiens. *Palaeontographica* 49:121-164.
Walker, A. D. 1969. The reptile fauna of the 'Lower Keuper' Sandstone. *Geological Magazine* 106:470-476.
Warch, A. 1984. Saurier-Fossilfunde in den Gailtaler Alpen. *Carinthia II* 94:79-90.
Wild, R. 1972. Die Wirbeltierfaunen der fränkischen und südalpinen Mitteltrias (ein Vergleich). *Zeitschrift der Deutschen Geologischen Gesellschaft* 123:229-234.
Wysogorski, J. 1904. Die Trias in Oberschlesien. *Zeitschrift der Deutschen Geologischen Gesellschaft* 56:260-264.
Zapfe, H. and H. König. 1980. Neue Reptilienfunde aus der Mitteltrias der Gailtaler Alpen (Kärnten, Österreich). *Österreichische Akademie der Wissenschaften, Mathematisch-Naturwissenschaftliche Klasse, Sitzungsberichte*, Abteilung I 189:65-82.
Ziegler, P. A. 1982. Triassic rifts and facies patterns in western and central Europe. *Geologische Rundschau* 71:747-772.

Chapter 6

MORPHOLOGICAL AND TAXONOMIC CLARIFICATION OF THE GENUS *PLESIOSAURUS*

GLENN W. STORRS

INTRODUCTION

Plesiosaurs (Diapsida: Sauropterygia: Plesiosauria) are extinct Mesozoic reptiles comprising one of the most successful and widely distributed groups of marine tetrapods. They developed a world wide range early in their history and some representatives of the clade survived into the Maastrichtian, becoming extinct perhaps only at the terminal Cretaceous. The evolutionary and systematic relationships of the Plesiosauria, however, are almost entirely unknown. The group appears as isolated bones and associated partial skeletons in the Middle Triassic (Anisian) of Germany, but the first unambiguous, fully articulated specimens occur in the uppermost Triassic and Lower Jurassic (Liassic) of England. By Liassic times, the plesiosaurs were particularly diverse, already fully marine, and highly modified from the presumed terrestrial condition of their forebears. The adoption of limb-dominated locomotion through two symmetrical sets of hyperphalangic appendages was a unique functional response by the plesiosaurs to a secondary invasion of the sea (Storrs, 1993a). In the Lias also, the group had already begun to exhibit specializations that led to several distinct lineages in the later Mesozoic, although the origins of this lineage diversification are obscure. The Lower Jurassic plesiosaurs, because of their often excellent state of preservation and well-constrained stratigraphic positions, currently provide the best potential for elucidation of early plesiosaur diversification and testing of postulated lineage monophyly.

Among the Lower Jurassic plesiosaurs, dozens of species of which have been described, the archetypal genus is *Plesiosaurus*, created by De la Beche and Conybeare (1821) on the basis of isolated and disparate material from Lyme Regis, Dorset, and the Bristol region of England. The first specific description of a plesiosaur, however, that of the type species *P. dolichodeirus*, was provided by Conybeare (1824) for a complete skeleton found by Mary Anning (1799-1847) in December 1823 in the Lower Lias (Sinemurian) near Lyme Regis (Torrens, 1995). This skeleton (BMNH 22656, Figure 1) is now universally recognized as the genoholotype. Thereafter, the Lower Jurassic of Dorset continued to produce a number of fine fossils of this animal (Andrews, 1896; Buckland, 1837; Lydekker, 1889; Owen, 1840a, 1860, 1865) until the cessation of quarrying activities in the Lias Group, early in this century.

Surprisingly, since Conybeare's (1824) description and additional work by Owen (1840a, 1865), there have been few attempts to clarify the anatomy and status of the type species of *Plesiosaurus* or, indeed, of the genus itself. As a result, a historical lack of understanding of the range of variation in plesiosaurs and a poor knowledge of their generic characteristics has reduced *Plesiosaurus* to the status of a wastebasket taxon that now includes poor and problematic material from the Rhaetian to the Maastrichtian. Literally hundreds of species of *Plesiosaurus* have been created worldwide, often without regard for sample size or quality of specimen. Solving the problem of the specific composition of *Plesiosaurus* by identifying diagnostic characters of the genus is therefore critical to the further interpretation of plesiosaurian relationships.

Examination of all existing Liassic plesiosaur holotypes has revealed that *Plesiosaurus* may be monotypic. Certainly for the English specimens, *Plesiosaurus dolichodeirus* is the only species of the genus that is based on proper, diagnostic material. Other English taxa are frequently nomina dubia or, if founded on sufficient specimens, must be referred to distinct genera. Therefore, an examination of lower Liassic *Plesiosaurus* need focus only on the holotype and quality referred specimens of *Plesiosaurus dolichodeirus*. *Plesiosaurus* is not yet known to occur in the English Upper Liassic, but is present in the Posidoniaschiefer (Toacian) of Germany and France. *Plesiosaurus guilelmiiperatoris* Dames, 1895 (emend. Brown, 1981) is potentially distinguished by subtle proportional differences, a more delicate skull, and a more robust humerus. Sciau et al. (1990) argue that *Plesiosaurus tournemirensis* is also a valid species, but it is conceivably another example of the German taxon.

Material

English coastal exposures of, and quarries in, the predominantly Lower Jurassic

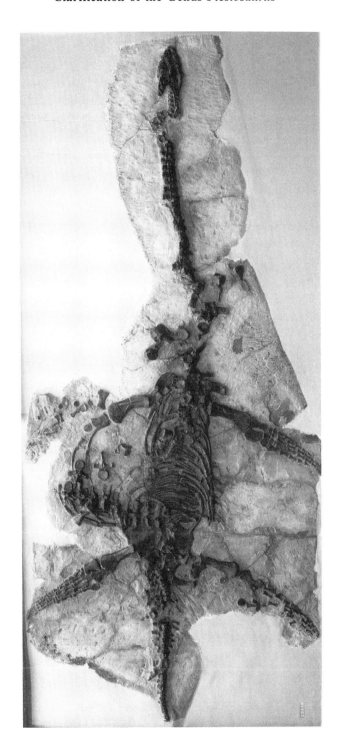

Figure 1. The holotype specimen of *Plesiosaurus dolichodeirus*, BMNH 22656, from the lower Lias of the Lyme Regis area, Dorset, England. Exposed in dorsal view.

Lias Group have produced many of the most important plesiosaur specimens known. Perhaps the most famous such locality is the area of Lyme Regis (and neighboring Charmouth), Dorset, whence came the world's first major collections of fossil reptiles (largely marine). Among these finds were numerous specimens of *Plesiosaurus dolichodeirus*; unequivocal examples remain restricted to the environs of Lyme Regis. While historical descriptions of these and other skeletons frequently provided illustrations of high aesthetic quality (e.g., Andrews, 1896; Hawkins, 1834, 1840; Owen, 1840b, 1865; Sollas, 1881; Stutchbury, 1846), their lack of detail often does not satisfy current needs. Some important specimens have never been scientifically reported, while others have been inaccessible or only partially prepared. Recent cleaning, preparation, and conservation of much of this classic material have facilitated new anatomical study and provided a basis for the description of *Plesiosaurus* presented here.

The most significant specimens of *Plesiosaurus dolichodeirus* remain in England, although relevant exported fossils exist also in Ireland, France, the Netherlands, and the United States. The best are in the Natural History Museum in London (BMNH) and the University Museum at Oxford (OXFUM). A good specimen (NMING F:8758) is in the National Museum of Ireland, having been sent to Dublin when that collection formed part of the British national museum system, and another exists at University College Galway (JMM FC M 032). In the United States, an articulated juvenile skeleton was acquired in the nineteenth century by Princeton University (YPM-PU 3352, now on indefinite loan to NJSM) and the Teylers Museum, the Netherlands, purchased a fine specimen (TM 13286) through the offices of Henry Woodward. A partial skeleton in the Muséum National d'Histoire Naturelle in Paris (NMHN A.-C. 8592) is of historical interest because of its reputed discovery by Mary Anning.

Repository Abbreviations

The abbreviations used for the institutions referred to in the text are as follows: ANSP, Academy of Natural Sciences, Philadelphia; BGS, British Geological Survey, Keyworth, Nottingham; BM, Museum für Naturkunde der Humbolt Universitet, Berlin; BMNH, the Natural History Museum, London; BRSMG, City of Bristol Museum and Art Gallery; CAMSM, Sedgwick Museum, Cambridge; DBYMU, Derby Museum and Art Galleries; GPIT, Geologische und Paläontologische Institut, Tübingen; JMM, James Mitchell Museum, Department of Geology, University College Galway; NJSM, New Jersey State Museum, Trenton; NMHN, Muséum National d'Histoire Naturelle, Paris; NMING, National Museum of Ireland, Dublin; OXFUM, University Museum, Oxford; SMNS, Staatliches Museum für Naturkunde,

Stuttgart; TM, Teylers Museum, Haarlem, the Netherlands; and YPM, Yale Peabody Museum of Natural History, New Haven, Connecticut.

SYSTEMATIC PALEONTOLOGY

DIAPSIDA Osborn, 1903
SAUROPTERYGIA Owen, 1860
EUSAUROPTERYGIA Tschanz, 1989
PLESIOSAURIA Blainville, 1835

PLESIOSAURUS De la Beche and Conybeare, 1821

Type species: *Plesiosaurus dolichodeirus* Conybeare, 1824.
Type locality: Lyme Regis, Charmouth coastal region, Dorset, England.
Range: Uppermost Sinemurian *Echioceras raricostatum* zone, Black Ven Marl Formation, Lower Lias.
Diagnosis: Moderately sized plesiosaur up to 3.5 m in length, differing from other genera by its small head with relatively short antorbital and temporal regions; subcircular supratemporal fenestrae; narrow temporal bar but only weak cheek emargination; broad yet distally pointed rostrum; prominent pineal foramen; relatively narrow and sharp sagittal crest; non spatulate mandibular symphysis; homodont dentition lacking carinae; elongate neck comprising approximately 40 cervical vertebrae; cervical rib heads pierced by elongate longitudinal foramen resulting in distinctly paired facets on the vertebral centrum; slightly to markedly rugose edges to the articular faces of virtually all mature vertebral centra; prominent anterior "U" notch in clavicular arch; stout longitudinal pectoral bar between ovate fenestrae; slender dorsal blade of scapula; coracoids of moderate breadth; acute posterolateral expansion to the coracoid plate in adult condition; humerus with prominent shaft curvature and marked posterodistal expansion but weak anterodistal corner; robust, pillar-like anterior epipodials offset to extend distally beyond posterior epipodials; broad, crescentic posterior epipodials; large, lozenge-shaped spatia interossea; pubis slightly longer anteroposteriorly than ischium; generally convex anterior margin of pubis; small, ovate to circular, puboischiatic fenestrae with unfused longitudinal pelvic bar; ilium with little twist to shaft; very elongate yet narrow limbs; forelimbs slightly longer than hindlimbs in adult; up to nine phalanges in the middle digit; digit IV the longest.

PLESIOSAURUS DOLICHODEIRUS Conybeare, 1824

? *Plesiosaurus* De la Beche and Conybeare, 1821, p. 560.
? *Plesiosaurus priscus* Parkinson, 1822, p. 294.
Plesiosaurus Dolichodeirus Conybeare, 1824, p. 389.
Plesiosaurus dolichodeirus Conybeare, Owen, 1840, p. 50.
Plesiosaurus macromus Owen, 1840, p. 72.
Plesiosaurus macrourus Owen, Stutchbury, 1846, p. 417 (Erore).
Plesiosaurus dolichodirus Conybeare, Lydekker, 1889, p. 255 (Erore).

Holotype: BMNH 22656.
Referred specimens: BMNH 28332, 33287 (Owen, 1865:Plate 4, Figures 1 and 2), 36183 (Owen, 1865:Plates 1 and 2), 39490 (Owen, 1865:Plate 3, Figures 1 and 3), 39491 (Owen, 1865:Plate 3, Figure 2), 41101 (Andrews, 1896:Figure 1), R.255 (a complete jaw purchased from the Edgerton collection in 1882), R.1313 (collected by Mary Anning, 1829; Buckland's Geology and Mineralogy, "The Bridgewater Treatise," 1837:Plate 16, Figure 2), R.1314 (a well-preserved right forelimb), ?R.1315, R.1316 (Owen, 1865:Plate 3, Figures 4 and 6), R.1316b (Owen, 1865:Plate 4, Figures 3 and 8), ?R.1330, R.1756 (a partial skeleton with good right limbs; Lydekker, 1890:277), ?BRSMG Ce17972 (the smallest potential *Plesiosaurus* skeleton known; Storrs, in press), JMM FC M 032, NMHN A.-C. 8592, NMING F:8758, OXFUM J.10304, J.13809, J.28586 (typical dentaries from the Philpott collection, measured by Owen, 1840a:61), J.28587, TM13286 (Winkler, 1873:Plate 7), YPM 1654 (an isolated pair of dentaries with well-preserved teeth), YPM-PU 3352 (a juvenile specimen), and numerous isolated, but frequently nondiagnostic, bones assigned to *P. dolichodeirus* are contained in virtually all collections of Lias material.
Diagnosis: Generally as for genus; may be distinguished from Toarcian *P. guilelmiimperatoris* Dames, 1895, by a relatively more robust skull, a very much larger pineal foramen, a greater anterior edge curvature to the humerus with a shallow but obvious interepipodial groove on its ventral surface, and subtle proportional differences that must await renewed study of the German material for modern characterization.
Remarks: Although Conybeare (1824) implicitly assumed specific identity of his new skeleton (BMNH 22656) with the material described and figured by De la Beche and Conybeare (1821), the latter fossils came from various localities and stratigraphic horizons. The description of these original specimens is not sufficient for generic diagnosis, nor are they preserved for comparison, and De la Beche and Conybeare (1821) designated no specific name for their material. As noted by Lydekker (1890), Parkinson's (1822) designation of *Plesiosaurus priscus* for one of

these specimens is a nomen dubium. Therefore, the first valid species holotype (BMNH 22656) is considered by all subsequent authors the de facto genoholotype as well. BMNH 22656 was purchased by the trustees of the British Museum in 1848 from the estate of Richard Grenville (1776-1839), the first Duke of Buckingham, having been described by Conybeare (1824) while on loan from His Grace's collection to the Reverend Professor William Buckland of Oxford. Conybeare (1824, explanation to Plate 18) stated that a second specimen from the type locality, collected contemporaneously, was presented by Buckland to Oxford. This is presumably OXFUM J.10304 (Figure 2A), which preserves an example of the Sinemurian ammonite, *Echioceras raricostatum*, in its matrix (P. Powell, personal communication). Once thought to have originated in the *Psiloceras planorbis* zone (Hettangian) of the Blue Lias Formation (e.g., Mansel-Pleydell, 1888; Delair, 1959; Persson, 1963), the holotype must therefore be Sinemurian as well, and this is confirmed by a contemporary newspaper account of the fossil's discovery below Black Ven (Torrens, 1995).

DESCRIPTION

The holotype of *Plesiosaurus dolichodeirus* (Figure 1) is a moderately sized, young adult exposed in dorsal aspect. Only the dorsal and posterior cervical vertebral series are disrupted in what is otherwise a complete and articulated skeleton (figured by Conybeare, 1824, and Cuvier, 1824, redrawn from Conybeare). Like all Lyme Regis/Charmouth specimens, the bone is black/brown in a dark gray matrix, in this case limestone, but frequently shale. A certain amount of compression is evident, again as in all Lias material. Notable dimensions of the holotype and other specimens are given in Appendix 1.

Skull

The skull is rather difficult to interpret. It is dorsoventrally crushed and somewhat distorted by virtue of its preservation in dewatered marine sediments. Exposed in dorsal aspect, it is slightly obscured by matrix, and a small amount of the tip of the snout is missing. Few sutures other than those of the straight midline are visible. There is, however, a large and prominent pineal foramen lying just behind the orbits. A slight surface texture or stippling of the maxillae, presumably an artifact, is present. The nares are not obvious and were apparently small. The orbital margins are also difficult to distinguish, but a key hole-shaped extension to the anterior corner of the orbit possibly represents the position of the prefrontal,

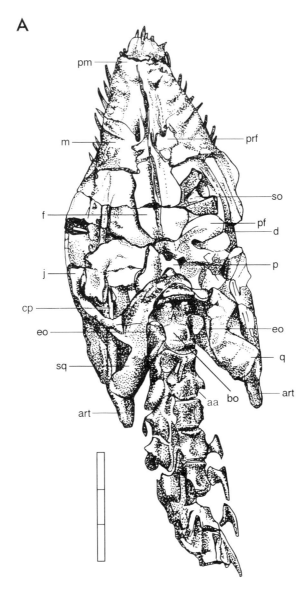

Figure 2. Skull of *Plesiosaurus dolichodeirus*, in dorsal aspect. **A)** OXFUM J.10304. For list of abbreviations, see Appendix 2. Scale bar = 3 cm.

presumed to be displaced. There is no diastema or constriction of the short, bluntly pointed, rostrum. The anterior edges of the supratemporal fenestrae are clear but because the squamosals have been lost, the back edges of the fenestrae are missing.

A sharp, longitudinal parietal ridge is evident, however, and the supratemporal fenestrae are thought to have been slightly longer than wide. The postorbital bar is relatively narrow compared with that of "*P.*" *hawkinsi* (Storrs and Taylor, 1996), but the apparent postorbital abuts against the parietal in the same fashion. The basioccipital is exposed from the dorsal side because both exoccipitals are disarticulated, revealing ovate facets. A distinct lip is present around the basal edge of the occipital condyle. The epipterygoids remain upright and in place as small, but robust, triangular prongs anterolateral to the basisphenoid; a basisphenoid/basiocciptal suture is not apparent. The narrow, ridge-like, basioccipital rami of the pterygoids are just visible. A few teeth only (about eight total) are to be seen on the right side of the skull. Little other detail is apparent.

Comparable specimens with well-preserved skulls, however, give virtually complete coverage of the cranial osteology of *Plesiosaurus dolichodeirus*. OXFUM J.10304, a putative topotype, is a generally good skeleton with a very well-preserved skull exposed in dorsal aspect (Figure 2A). Although crushed, it is otherwise only slightly distorted and the braincase is well exposed and intact, save that the three-dimensional supraoccipital is displaced into the right orbit. The supraoccipital is the typical U-shaped element of plesiosaurs with its wide hemicircular notch forming the dorsal part of the foramen magnum. The dorsal portion of the supraoccipital slopes forward as it does in "*P.*" *hawkinsi* (Storrs and Taylor, 1996). The ventrolateral corners of the supraoccipital are fashioned into stout, subtriangular exoccipital facets. These correspond to the robust columnar exoccipitals that remain upright lateral to the basioccipital. The paroccipital processes taper toward the medial faces of the suspensoria, which they contact. The occipital face of the skull has been forced forward onto the anterior basicranium, clearly exposing the posterior surfaces and flared articular ends of the suspensoria. The large, triradiate nature of the squamosal is evident, although their quadrate rami are bent medially. The quadrate's contact with the mandibular cotylus is clear, and both retroarticular processes are dorsally exposed. These are straight, with little or no median curvature, narrow and blunt. The position of the jaw below the skull is visible through the fenestrae and orbits, the palate suffering compression damage.

The frontals are broad and joined by a straight longitudinal suture. They are overridden at their anterior end by the premaxillae and by clearly distint prefrontals whose notched anterior ends form the posterior margin of the nares. The right prefrontal appears to be somewhat larger and more regularly shaped than the left in this individual. The posterior rami of the maxillae are transversely thin and form the lateral borders of the orbits. In this example, they are medially exposed by postmortem crushing. The posterior borders of the orbits and the shapes of the temporal fenestrae are somewhat obscured by deformation. The procumbent, needle-like teeth of both the premaxillae and maxillae are intact. Some dentary teeth are also preserved. The premaxilla contains five teeth as in most plesiosaurs.

Perhaps the best preserved skull, however, is BMNH 39490 (Figure 2B), as it is virtually undistorted and well displays the characteristics of adult members of the

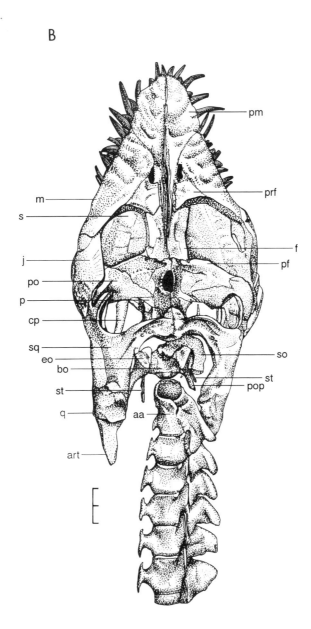

Figure 2 (continued). *Plesiosaurus dolichodeirus*, in dorsal aspect. **B)** BMNH 39490. For list of abbreviations, see Appendix 2. Scale bar = 2 cm.

species. It is also very similar in size and shape to OXFUM J.10304. BMNH 39490 is a complete, fully adult skull and jaws in dorsal aspect, along with the first 7.5 cervical vertebrae in articulation (including atlas/axis). This Charmouth specimen was purchased in 1865 and described by Owen (1865:Plate 3, Figure 1). The mandible is in place and the left retroarticular process is well exposed, although the right is hidden. Similarly to but more clearly than OXFUM J.10304, the crushing of the palate onto the mandible gives a raised-relief indication of the position of the jaw and its alveoli underneath, especially in the right orbit. The tips of the coronoid processes project through the temporal fenestrae and lie along the medial side of each temporal bar.

The seemingly textured snout of this individual, like the holotype, is perhaps artificial and the result of intense fracturing, also causing the base of a lower tooth to be forced up through the left premaxilla. The straight median suture is bounded by reinforced bone, resulting in a slight dorsal keel after crushing (a similar phenomenon surrounds the frontals of OXFUM J.10304). The premaxillary/maxillary suture is so tight that its position is obscure; its position is estimated based on the standard plesiosaurian condition, also seen in OXFUM J.10304. The typical posterior extensions of the premaxillae (nasals of Owen, 1865) override and interdigitate with the anterior half of the frontals, thereby separating them in dosal view for much of their length. The posterior ramus of the maxilla overlaps the jugal (malar of Owen, 1865) caudad---the posterior tip of the left maxilla has broken away to expose this relationship well. The prefrontals (regular and subequal in this example) are firmly joined, nearly fused, to the maxillae with a straight suture line visible between them, and override both the frontals posteriorly and the premaxillae anteriorly. The small, retracted, external nares again lie before a notch at the anterior edge of the prefrontals. This position was noted by Owen (1865:9), but he believed that part of the prefrontal represented an undefinable lachrymal. No separate lachrymal is present in any example of *Plesiosaurus* and is probably absent from the group (Storrs, 1991, 1993b; Storrs and Taylor, 1996; cf. Cruickshank, 1994a; Taylor, 1992; Taylor and Cruickshank, 1993).

The orbit has a rather sharp posterolateral corner at the junction of the jugal and postorbital (somewhat exaggerated on the right side by the loss of the posterodorsal edge of the jugal). The jugal is tallest at the back corner of the orbit and some external surface texturing, in the form of slight grooves running posteroventrally from the orbit, is present on it. Four sclerotic plates, the medial one with a natural crenulated edge as is seen in a specimen of "*P.*" *hawkinsi* (Storrs and Taylor, 1996), lie on the left orbital floor appressed to the medial surface of the palate. These flat sheets of bone are extremely thin and delicate.

The broad frontals are paired but the parietals are potentially fused. There is a slight depression or fossa in the center of each frontal, possibly created or accentuated by crushing. The frontal is excluded from the temporal fenestra by the postfrontal. The parietal/frontal suture is quite narrow, as the postfrontals firmly join the frontals at their posterolateral corners. As in the holotype, the pineal foramen is very large, occupying much of the dorsal surface of the parietals, and the parietal crest is sharp---rather more so than in specimens of "*P.*" *hawkinsi*, particularly CAMSM J.46986 (Storrs and Taylor, 1996).

The postorbital possesses a strong posterior prong that fits between the posteromedial edge of the jugal and an anterior fork of the squamosal. The dorsal surface of this posterior prong is keeled. The suture between the postorbital and postfrontal forms a dorsolateral to ventromedial line, the postorbital overlapping the posterolateral edge of the postfrontal. The postorbital thus forms the anterolateral edge of the temporal fenestra and both the lateral and ventral parts of its anterior wall. The postfrontal, however, forms the dorsomedial part of this wall as well as most of the posterior rim of the orbit. The medial ends of both the postfrontal and postorbital overlie the lateral wall of the parietal. The anterior wall of the upper temporal fenestra is crushed flat, resulting in an unnaturally broad postorbital bar, and artificially broad temporal fenestra. A subcircular supratemporal fenestra was found in life.

The squamosals meet, overlapping the parietals and supraoccipital, in a digitate median suture that is also the highest point of the skull. The anteromedial ends of their dorsal rami (mastoids of Owen, 1865) also overlap the posterior ends of the parietals around the lateral wall of the braincase. The posterior edges of the dorsal rami are very rugose for the attachment of the nuchal ligature; this rugosity is limited to a long crescent tapering lateroventrally. The remainder of the squamosal, including its dorsal surface, is very smooth. The anterior ramus of the squamosal underlaps both the jugal and postorbital; this relationship, noted above, is seen clearly on both sides of the skull. The quadrate appears to overlap the squamosal on the posterior face of a suspensorium "box-beam," although the dorsal tip of the left quadrate may be broken away. This deeply cupped box construction of the suspensorium is common to all plesiosaurs (Storrs and Taylor, 1996; Taylor, 1992). As in all known plesiosaurs, there is no quadratojugal (cf. Welles, 1943, 1952, 1962).

The supraoccipital is displaced slightly forward and to the right and the dorsal squamosal bar is pushed forward. The U-shaped supraoccipital is robust as in OXFUM J.10304. The exoccipitals and their paroccipital processses remain in near articulation to the basioccipital and the suspensorium, but are displaced forward slightly. The tops of the exoccipitals are pitted, rugose, and potentially vascularized and contain portions of the semicircular canals (Cruickshank, 1994b; Storrs and Taylor, 1996). The occipital condyle is constructed almost exclusively from the basioccipital, although there is an extremely slight contribution of the exoccipitals to its formation. Both stapes are present, but at least the left has come free of the suspensorium; each is quite thin. The left stapes extends about 1.5 cm beyond the braincase and must still be more or less firmly attached to the exoccipital (Storrs and Taylor, 1996; Taylor, 1992), but its posterior end is missing (unossified?). Stapes are rarely preserved in plesiosaurs.

The epipterygoids at the back medial edge of the postorbital bar form a relatively strong pillar-like support, judging by the uncrushed nature of the left element. Nevertheless, these are much narrower than the dorsal pillars of the exoccipitals. Portions of what may be the posterior pterygoid rami are just visible in the floor of each temporal fenestra.

OXFUM J.13809 (Figure 3) is an isolated, largely matrix-free skull of a relatively young animal from the Philpot collection (Edmonds, 1978), and is perhaps that noted by Conybeare (1824:382). Again dorsoventrally crushed and incomplete, it importantly shows the prefrontals as separate elements with little fusion to the maxillae. The prefrontals override the anterolateral corners of the frontals, also clearly unfused. The "box-beam" sloping construction of the suspensorium, the notochordal occipital condyle with a moderate lip and slight exoccipital contributions, and the overlap of the dorsal rami of the squamosals onto the posterodorsal parietals are also noteworthy. The true shape of the subcircular supratemporal fenestra is suggested by the undistorted left window. The occipital condyle contains a notochordal pit as do most plesiosaur specimens. Such a pit is presumed to be present in all *P. dolichodeirus* specimens.

Much of the palate of *P. dolichodeirus* can also be observed in OXFUM J.13809 (Figure 3B), although some breakage has occurred. Nevertheless, a narrow, slit-like, anterior interpterygoid vacuity and elongate posterior interpterygoid vacuities are clearly displayed. So, too, are the posterior pterygoid rami, which clearly do not meet to floor the ventral surface of the basioccipital, a condition opposite that reported in some "pliosaurs" (Cruickshank, 1994a; Taylor, 1992; Taylor and Cruickshank, 1993). The constricted, hourglass-shaped, articular surface of the quadrate typical of plesiosaurs, an apparent fusion of the parasphenoid to the basisphenoid, and tall but slim quadrate rami of the pterygoids are evident. The internal nares are situated at the same approximate level as the external nares and not anterior to them as reported for *Rhomaleosaurus* (Cruickshank et al., 1991).

BMNH 41101 (Andrews, 1896:Figure 1) from Lyme Regis, however, is a better palate for description as most elements remain intact (Figure 4). The tip of the snout is rather bluntly pointed. The premaxillary/maxillary suture (especially the left) is clear and again lies behind the fifth tooth. This suture is directed anterolaterally-posteromedially in standard fashion. The maxilla is broadest on the palate at the level of the third maxillary tooth; the internal nares lie just behind the level of this point. Only about 13 distinct alveoli are obviously present in the

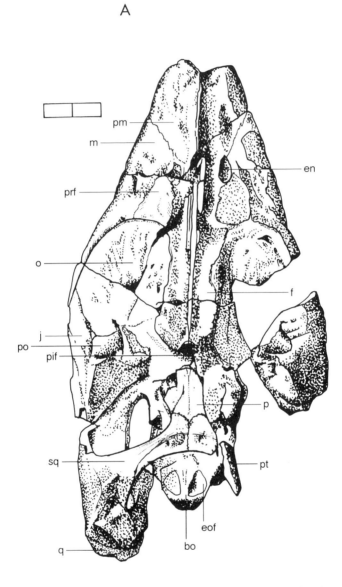

Figure 3. Isolated skull of *Plesiosaurus dolichodeirus*, OXFUM J.13809. **A)** Dorsal aspect. For list of abbreviations, see Appendix 2. Scale bar = 2 cm.

maxilla; posteriorly, slight indications of very small pits suggest the presence of at least 18 alveoli in life. Only the base of one primary tooth remains.

Small, replacement alveoli are obvious medial to much of the tooth row; the

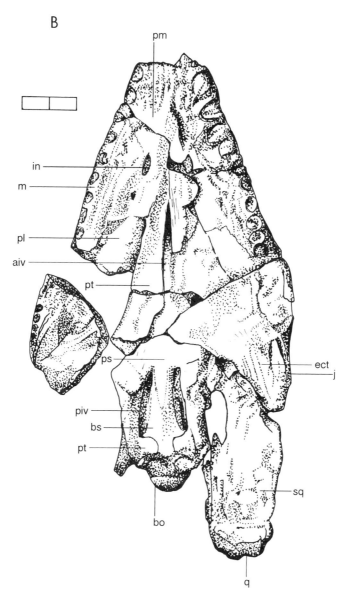

Figure 3 (continued). Isolated skull of *Plesiosaurus dolichodeirus*, OXFUM J.13809. **B)** Ventral aspect. For abbreviations, see Appendix 2. Scale bar = 2 cm.

usual groove containing replacement teeth is not apparent in this individual save at the anteriormost end of the premaxillae. Plesiosaur dentitions differ from other reptiles in that the new teeth, instead of being formed in the pulp cavity of the old,

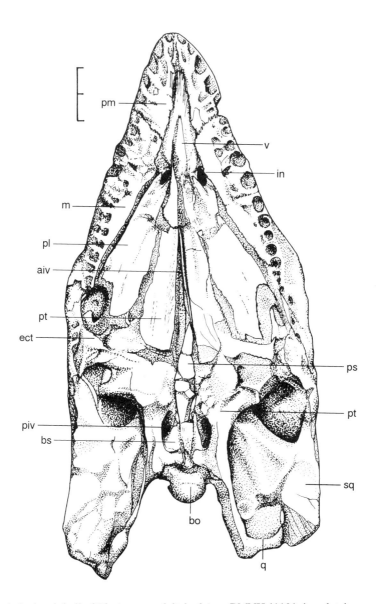

Figure 4. Isolated skull of *Plesiosaurus dolichodeirus*, BMNH 41101, in palatal aspect. For list of abbreviations, see Appendix 2. Scale bar = 2 cm.

emerge through distinct tracks situated medial to the sockets of the old teeth. Replacement teeth entered the functional alveoli by bone reformation (resorption). Two replacement teeth are preserved here---one lying medial to the base of the functional tooth, the second in a replacement alveolus nearly opposite. The lateral surfaces of the maxillae have been crushed outward at their posterior ends.

In the midline, the fused vomers overlie, or lie in a depression between, the premaxillae, where their central portion presents a raised diamond-shaped area (possibly enhanced by crushing). A small foramen enters this "diamond" at its anterior end, while at its posterior, a vomerine ridge is formed between the internal nares. The posterior end of the vomers overlies the anteromedial edge of the palatine and the anterior end of the pterygoid. In their midsection, a lateral notch within each vomer borders the naris both anteriorly and anteromedially, thus forming its anterior wall. The palatines form the narial border posterolaterally and, by means of a raised medial prong, posteromedially. The palatines are probably not fused to the pterygoids, as a slight preservational offset in elevation exists between them, but the critical sutural areas are obscured by breakage. An apparent fenestra at the back of the right palatine (noted by Lydekker, 1889, and called a suborbital vacuity by Andrews, 1896:248) is actually a preparational artifact; a second on the left side is far less pronounced and some bone shows through under the matrix there.

The pterygoids are not fused at the midline of the palate and a long, narrow, anterior interpterygoid vacuity lies between them as in the previous specimen. The medial edge of the pterygoid palatal ramus is roughened by shallow longitudinal grooves, while the lateral portions are smooth; a few deeper grooves are also to be found on the palatines. The lateral, ectopterygoid, ramus of the pterygoid is overlain anteriorly by the posterior edge of the palatine. The posterior edges of this ramus are poorly preserved. The posterior interpterygoid vacuities lie alongside the basisphenoid between the two posterior (quadrate) rami of the pterygoids. Behind these vacuities, the pterygoids only partially cover the basicranium as in OXFUM J.13809; they do not meet, but expose a parallel-sided gap between the accessory basioccipital tuber articulations of the palate that are unique to plesiosaurs. The long quadrate ramus of the pterygoid is very narrow but rather deep, again as typical for plesiosaurs. There is no pterygoid flange. Both ectopterygoid areas are damaged or covered and their margins indefinite, but the ectopterygoid (transverse bone or transpalatine of Andrews, 1896) itself is stout and firmly joins the palate to the tooth row at the posterior end of the maxilla.

The basisphenoid is visible between the interpterygoid vacuities and dorsal to the parasphenoid for the entire vacuity length, tapering anteriorly. The parasphenoid is a long, narrow, diamond-shaped bone; it is fused to the palatal surface of the basisphenoid and extends anteriorly between the palatal rami of the pterygoids to fill the posterior part of the anterior interpterygoid vacuity. The basioccipital is unremarkable and its small occipital condyle also contains a prominent notochordal pit.

The suspensoria (squamosals/quadrates) are laterally flared at their posterior ends. Crushing has played a role here, although the left quadrate is about naturally positioned. On the other hand, part of the "box-beam" construction of the left suspensorium has come apart; the appearance of the squamosal, although compressed, is more natural on the right side. The anterior ramus of the squamosal broadly underlaps the jugal and seems to contact the ectopterygoid and the posterior tip of the maxilla. The quadrate condyles lie ventral to the constricted midsection of the articular surface; the lateral condyle lies anterior to the medial one.

The excellent, palatally exposed, skull of NMING F:8758 (Figure 5) displays proportions and relationships in accord with those above, notably the fused, diamond-shaped basisphenoid and parasphenoid, and the open basioccipital between the pterygoids. Here, however, the full extent of the ectopterygoids is for the first time clear. They join the pterygoids medially in a stout V-shaped suture and anteriorly, lie deep to the thin bony sheet of each palatine. It is also clear from this specimen that no suborbital fenestration exists. Laterally, the ectopterygoid firmly abuts the posterior ramus of the maxilla and the ventral edge of the jugal/squamosal. The posterior edge of the ectopterygoid, forming the back of the palate, is raised in a slight boss where it joins the pterygoid. Such a boss has been described previously, in, e.g., *Rhomaleosaurus* (Cruickshank, 1994a; Taylor, 1992), as a "pterygoid boss." Once again, there is no true pterygoid flange. The basioccipital is obviously displayed between the separated pterygoids at the back of the posterior interpterygoid vacuities.

BMNH R.1313 (Buckland, 1836:Plates 16 and 18) from Lyme Regis (collected by Mary Anning; Torrens, 1995) is one of the articulated skeletons referred to by Owen (1865). It is exposed in dorsal view, but the skull is twisted on the neck so as to be viewed also from its palatal side. The entire mandible except for the tip of the right retroarticular process (once seemingly present but now lost) is also visible (Figure 6). This is slightly displaced from its natural articulation, thus obscuring some palatal details, but these latter do not generally differ from those discussed above. The elongate posterior palatal vacuities, however, extend slightly anterior to the posterior, concave, edge of the pterygoid palatal ramus, indicating some variability in the size of these openings. The basioccipital is well shown and yet again, the pterygoids do not entirely underlap it. Unlike the case in BMNH 41101, there is no obvious lip rimming the occipital condyle, but this seems to be an artifact of preparation. Many long teeth extrude from both the mandible and the skull, showing that there are over 20 teeth in each maxilla.

In BMNH 36183, another articulated partial skeleton from Lyme Regis (Owen, 1865:Plates 1 and 2, in reverse image), the skull and mandible are uniquely crushed to present their profiles, but few sutures are visible. The snout is low and approximately half the length of the skull. The nares are not apparent. There

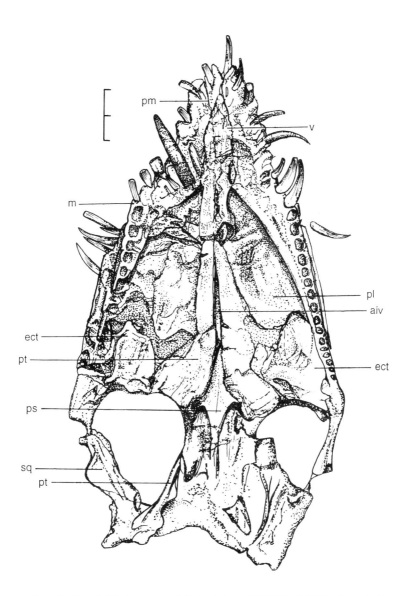

Figure 5. Skull of *Plesiosaurus dolichodeirus*, NMING F:8758, in palatal aspect. Abbreviations in Appendix 2. Scale bar = 2 cm.

superficially is little cheek emargination and a relatively low skull table, but the skull roof is artificially depressed; in reality, the sagittal crest was high. As the suspensorium is in natural articulation with the mandible, disarticulation here is not

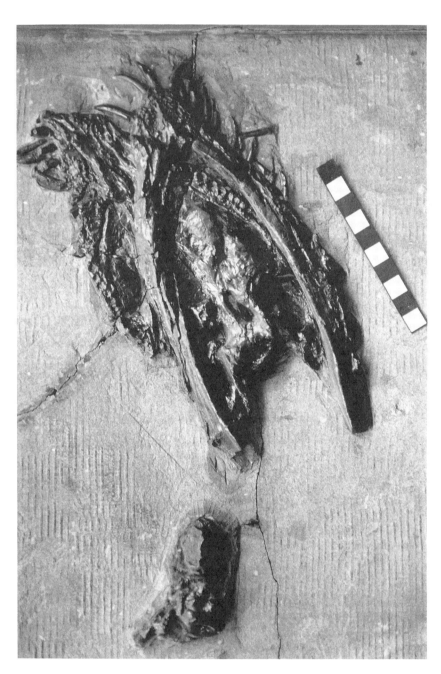

Figure 6. Skull of *Plesiosaurus dolichodeirus*, BMNH R.1313, in palatal aspect. Bar scale = 10 cm.

responsible. Rather, the skull table (comprising parietal and supraoccipital without an obvious suture) is displaced into the temporal region; the dorsal ramus of each squamosal (mastoid of Owen, 1865), because of lateral compression, is unnaturally elevated. The true shape of the temporal emargination is difficult to delimit, but it seems that the posterior edge was slightly deeper than the anterior; the emargination may have been slight, even though the temporal bar was rather narrow. The mandibular articulation lies at the average level of the tooth row, although well below its posterior end. As expected, the supraoccipital is robust and uncrushed; its ventral end exhibits the exoccipital facet, the suture between them having been opened by the skull table displacement. The top of the exoccipital presently lies just ventral and slightly anterior to this facet. The blunt, hook-shaped top of the epipterygoid (directed caudad) can be seen in this specimen, lying just anterior to the exoccipital. The pterygoid contacts the quadrate near the jaw articulation. Again, approximately 20 teeth are visible in the upper jaw and as pointed out by Owen (1865:8), two teeth are sometimes inserted between adjoining teeth of the opposing tooth row.

The totality of cranial material of *Plesiosaurus dolichodeirus* discussed above allows the first confident reconstruction of the complete skull of this animal (Figure 7). The orbit is placed approximately midway along the skull, the rostrum is low and bluntly triangular, and the skull table is high with a relatively narrow sagittal crest. The broadest part of the skull is formed by the postorbital bars; the large pineal foramen occupies the center point between these bars in the middle of the skull table. The orbits and temporal fenestrae are subcircular and subequal in size. The suspensoria are posteriorly and slightly laterally flared. The external nostrils overlie the internal nares and are not explicitly suggestive of underwater olfaction as suggested for *Rhomaleosaurus* by Cruickshank et al. (1991). Approximately 20-25 teeth formed each upper tooth row. Although the standard plesiosaurian posterior union of the pterygoids and basioccipital is present, the incomplete flooring of the basioccipital and the long anterior interpterygoid vacuity indicate an incompletely buttressed palate/cranium connection; a less robust connection than in, for example, the pliosaurs *Rhomaleosaurus* and *Pliosaurus* (Cruickshank, 1994a; Taylor, 1992; Taylor and Cruickshank, 1993). A weak "pterygoid boss" is present at the posterior of the plate-like ectopterygoid. The palatal bones are thin, but there is no suborbital fenestra.

Mandible

The mandible of the holotype is largely hidden. Nevertheless, it shows that the coronoid process was rather thin and low and that the tip of the retroarticular

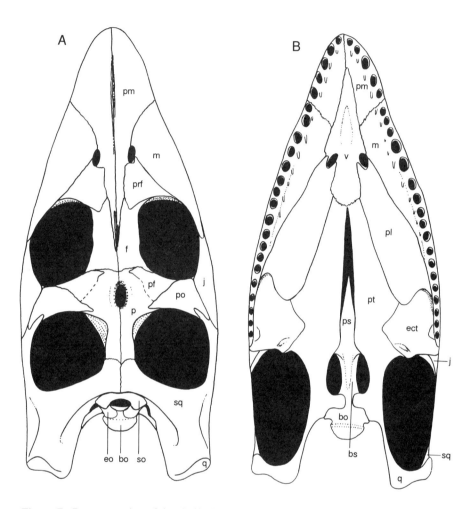

Figure 7. Reconstruction of the skull of *Plesiosaurus dolichodeirus*. **A)** Dorsal aspect. **B)** Ventral aspect. For list of abbreviations, see Appendix 2.

process was blunt. Several additional examples of *P. dolichodeirus* lower jaws complete the picture (Figures 6, 8, and 9). BMNH 39491 (Figure 8), for example, is a set of fused dentaries from Lyme Regis (Owen, 1865:Plate 3, Figure 2). The rami form a nearly perfect "V" at an angle of about 45° with only a slight medial curvature. Dorsoventral compression imparts a small amount of lateral torsion to the left ramus. Both dentaries are entire and contain 24 (primary) alveoli. The first alveolus in each ramus retains a tooth, the remainder are empty and posterior to the third, alveoli gradually decrease in size caudad. The longitudinal replacement tooth

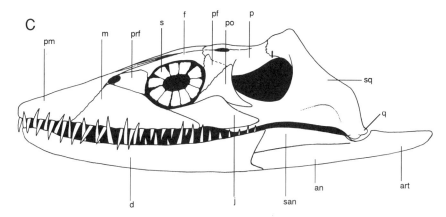

Figure 7 (continued). Reconstruction of the skull of *Plesiosaurus dolichodeirus*. **C**) Lateral aspect. For list of abbreviations, see Appendix 2.

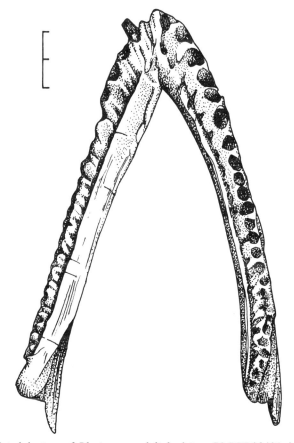

Figure 8. Isolated dentary of *Plesiosaurus dolichodeirus*, BMNH 39491, in dorsal aspect. Scale bar = 2 cm.

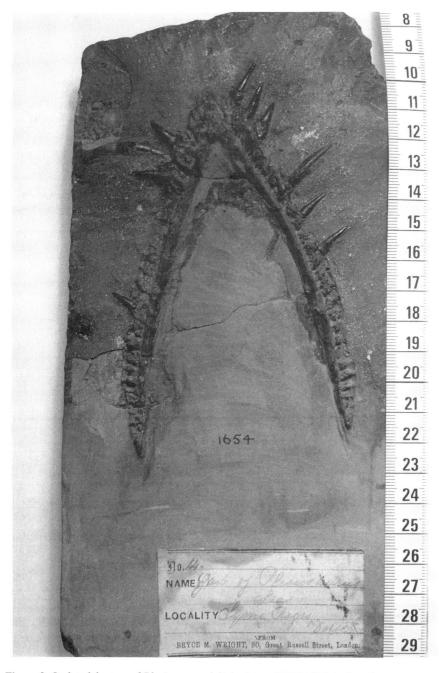

Figure 9. Isolated dentary of *Plesiosaurus dolichodeirus*, YPM 1654, in dorsal aspect. Scale in centimeters.

troughs medial to the outer tooth rows are well displayed and the nearby resorption paths are visible as diagonal grooves. The areas of these grooves and the primary alveoli they enter are rough and unfinished. Medial to the replacement tooth trough, the bone surface is smooth.

The anterior end of each dentary is broad and robust. A strong, fused symphysis, pointed like a shallow scoop, incorporates 2.5 alveoli in each ramus. This is slightly in excess of the single symphysial tooth traditionally allotted to *Plesiosaurus* and other long-necked genera (Brown, 1981; Cruickshank, 1994b). In BMNH R.1313 (Figure 6) the robust symphysis, although not particularly spatulate, may contain as many as four teeth; its straight median suture is confined within a robust, sculptured ridge. The posterior end of each mandibular ramus of BMNH 39491 rises slightly as it forms the anterior edge of the coronoid process. A shallow, medial depression or groove for reception of the splenial lies below this process, narrowing anteriorly. Externally, the dentary is similarly, though not as markedly, depressed for reception of the angular and surangular; these elements are missing. The exterior surface is lightly sculptured with grooves and nutrient sulci, whereas the lingual surface is smooth. Most of the thin posterior edge is oriented subvertically, but a sharp (1-cm) prong protrudes caudad along its ventral portion. In the obliquely exposed mandible of TM 13286, the posterior edge of the dentary forms a more V-shaped notch; the notch is reduced to a near-foramen in BMNH 36183 in another example of individual variation.

Until well-prepared examples of postdentary bones other than the articular are known, it may be presumed that they are similar to those of "*P.*" *hawkinsi* (Storrs and Taylor, 1996). Certainly these bones are frequently separated from, and therefore must be loosely attached to, the fused dentaries.

Dentition

The teeth of *Plesiosaurus dolichodeirus* are simple, needle-like, cones---slightly curved and circular in transverse section. They are slender and sharp with fine, longitudinal striae and no obvious carinae. In better preserved examples, the striae run from the top of the alveolus nearly to the tip of the tooth; there are approximately three striae per mm. All striae begin at the base of the tooth; the root, gradually increasing in diameter below the alveous rim, is smooth in typical fashion. Some striae are lost toward the tip, thus a more or less even spacing is maintained as the tooth narrows. In BMNH 39490, a slight crenulation of the tooth striae is visible under low magnification. No wear facets are present on normal crowns. All teeth are procumbent (Figures 2, 5, 6, and 9), but especially those most anteriorly situated, lying approximately 10-15° above horizontal. The upper teeth occlude with the lowers in a generally, but not exclusively, alternating pattern, the first premaxillary tooth lying medial to the first dentary tooth (Owen, 1865:Plate 3, Figure 3). Various examples suggest a full complement of between 20 and 25 teeth in each upper and lower jaw.

Vertebral Column

The holotype skeleton (Figure 1) preserves approximately 41 cervical vertebrae (OXFUM J.10304 has an estimated 42 cervicals and BMNH R.1313 about 38 or 39). The atlas/axis complex, as shown in both BMNH 39490 and OXFUM J.10304, is fused in the fully adult condition (Figure 2). Indeed, all ribs and arches are fused to the centra, with almost no indication of suturing, in BMNH 39490. This is a typical feature of late ontogeny. The atlas forms a deeply cupped, hemispherical bowl. A small, posteriorly projecting cervical rib with little or no anterior prong is situated low on the axial centrum. The axis neural spine slopes anteriorly as in "*P.*" *hawkinsi* (Storrs and Taylor, 1996), and while there is no axial prezygapophysis, this is perhaps an artifact.

The cervical centra of *P. dolichodeirus*, except those most proximal, are relatively elongate; their length is normally slightly greater than their height, but equals or is exceeded by their breadth. Their articular faces are slightly concave and kidney-shaped, with rounded, slightly rugose, edges. Such rugosity may be variably developed as there is little rugosity on the edges of the centra of BMNH 39490. The middle of each centrum is constricted. The foramina subcentralia are small and elongate; there is a slight ridge or bulge between each. The neural arches are fused to the centra through barely visible U-shaped sutures, a mid to late ontogenetic development. Typical anterior cervical ribs are hatchet-shaped with subequal anterior and posterior processes (Figure 2); cervical neural spines have rounded, progressively erect, tips; the spines of the proximal cervicals are subquadrate. The cervical ribs are obviously double-headed; their facets are elongate and closely spaced anteriorly, with a longitudinal passage or foramen between them. The posterior facets are round and pedicelate. A typical *P. dolichodeirus* cervical vertebra is illustrated by Owen (1865:Plate 3; BMNH R.1316).

A juvenile specimen, YPM-PU 3352, exhibits cervical arches and ribs which are, as expected, not fused to and usually separate from the centra. A distinct, V-shaped neurocentral suture is obvious on each of these vertebra. The suture is confluent with the upper rib facet through an open and unfinished vertical "canal"; this obviously closes with age. The cervicals, however, retain rugose edges.

The holotype posesses four or five "pectoral" vertebrae, i.e., those at the base of the neck in which the transverse process is carried by both the centrum and the neural arch (Owen, 1865; Welles, 1943). About 21 vertebrae are dorsals, and there are at least three sacrals. Approximately 28 caudals are present, but a few are likely missing at both ends of the tail. All of these types are well described by Owen (1865:Plate 4). Some dorsal vertebrae, like the cervicals, have an articular edge rugosity. The neural spines of the thorax are tall and subrectangular in profile; their tips are transversely thickened. The dorsal ribs are single-headed, thick, and "pachyostotic" in typical plesiosaurian fashion. The sacral ribs are short, robust, and blunt or knob-like on both ends. The caudal vertebrae are unremarkable; the centra generally lack rugosity, their arches are fused in adults, but the ribs and chevrons are free. These are the last elements to coossify during ontogeny. The caudal centra are subquadrate with ovate to circular rib facets as in other plesiosaurs. The short, squarish, caudal spines taper little, if at all. The chevrons thicken distally; their facets are shared by adjacent centra. The anterior caudal ribs are pointed; the remainder are broad and blunt.

The interarch space from the back edge of the coracoid to the pubis is filled with pachyostotic gastralia. The V-shaped, anteriorly pointing median elements are matched with three lateral elements per side. All are pointed save for the distal ends of the most lateral elements; these are flat and spatulate (recurving caudad), although possibly somewhat crushed. One lateral element is double-forked at its proximal end, high lighting individual variation in gastralia (Storrs, 1991). At least nine sets of gastralia are present.

Forelimb

Seeley (1874) reviewed the history and early confusion surrounding the interpretation of the plesiosaurian pectoral girdle. Most of this confusion is a result of difficult to interpret, incompletely preserved specimens. The pectoral girdle of BMNH 22656, as in most other specimens, is largely hidden by matrix or obscured by overlying vertebrae and ribs. What is visible, however, appears typical for plesiosaurs. The fused clavicular arch lies medial to the scapulae as in all sauropterygians. The scapular dorsal blade is slender and subrectangular, meeting the ventral plate at a marked angle, and the glenoid ramus is stout. The glenoid, as in all plesiosaurs, is shared by the scapula and coracoid. The coracoid is much longer than the scapula. A referred specimen, NMHN A.-C. 8592, clearly displays ovate pectoral fenestrae at the medial junction of the scapulae and coracoids; a prominent longitudinal bar is therefore created between the fenestrae. Hulke (1883: Figure 15) illustrates these fenestrae and the relationships and shapes of the scapula, (partial) coracoid, and articulated humerus based on BGS GSM 118412, a cast of

NMHN A.-C. 8592 (portions of which have been lost; a second cast is ANSP 17427). Hulke's (1883:Figures 2 and 15) figures, in which he designates part of the scapula a precoracoid and the clavicular arch an omosternum, are modified in Figure 10 on the basis of the known material.

Most of the pectoral girdle mystery is resolved by the fine shoulder girdle preserved as BMNH R.1315, "*Plesiosaurus*," from Keynsham near Bath in County Avon. By virtue of its characteristic humerus and epipodials (see below), this specimen almost certainly represents a large adult *P. dolichodeirus*. A prominent, U-shaped notch in the clavicular arch, situated between the anteromedial edges of the scapulae, ovate and diagonally directed pectoral fenestrae adjacent to the stout longitudinal bar, and the slightly flared posterolateral corners of the coracoids in this mature specimen are notable. The flared corners are also obvious in the left coracoid of TM 13286 (Winkler, 1873:Plate 7, Figure 2). These corners grew and became more acute with age, as revealed by their very well-rounded condition in the juvenile YPM-PU 3352, and in other plesiosaur juveniles, e.g., *Cryptoclidus* and "*Leurospondylus*" (Andrews, 1910; Brown, 1913; Welles, 1952, 1962). The

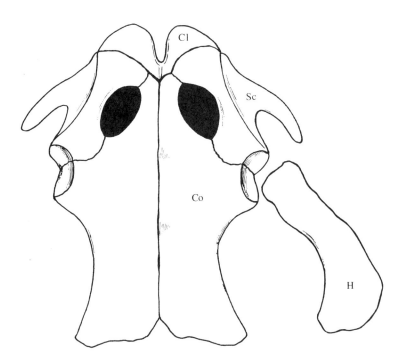

Figure 10. Reconstructed pectoral girdle and humerus of *Plesiosaurus dolichodeirus* in ventral aspect. Based in part on BMNH R.1315, NMHN A.-C. 8592, and BGS GSM 118412, partially redrawn from Hulke (1883). Dorsal blades of scapulae artificially flared for clarity. For list of abbreviations see Appendix 2.

posterior plates of the coracoids, other than the posterolateral corners, are only moderately broad. The posterior edge usually is not crenulated in the manner seen in, e.g., *Cryptoclidus* (Brown, 1981).

The right forelimb is especially well preserved and completely articulated in the holotype (Figures 1, 11A). It is a very elongate and relatively narrow structure, far more so than in most plesiosaurian taxa, particularly those of the later Mesozoic. This narrow limb morphology is presumed to be the plesiomorphic condition. Most notably, *Plesiosaurus* exhibits a characteristic humerus with a profound curvature to the propodial shaft not generally found in other plesiosaurs (Figure 1; Dames, 1895; Owen, 1865:11, Plate 1, Figures 1 and 2; Watson, 1924:Figure 4), but retained from stem-group Sauropterygia (Storrs, 1991, 1993a, b). *Plesiosaurus dolichodeirus* exhibits this curvature to an even greater extent than does the German *P. guilelmiimperatoris*. This bowed shaft may be slightly less pronounced in larger and older individuals (e.g., OXFUM J.10304). The breadth of the posterodistal expansion of the humerus increases with age, but there is little or no anterodistal corner in any but the oldest individual (e.g., BMNH 36183; Owen, 1865:Plate 1, Figure 1). Proximally, the humeri are slightly rugose (stippled), but there are no obvious muscle scars (although a bold one is found proximally in the large NMHN A.-C. 8592), and the tuberosity is small and anteriorly located. The humeral head has a lipped edge. Distally, a shallow, but marked, longitudinal groove or "interepipodial trough" exists on the ventral surface of the humerus in mature individuals (e.g., NMING F:8758). This trough is particularly characteristic of the species. Prominent epipodial facets also are developed in full adults (BMNH 36183, NMHN A.-C. 8592, NMING F:8758).

The plate-like, flat ulnae are semilunate (crescent moon shaped) and very broad, whereas the radius is robust and pillar-like. Although the same length as the ulna, the radius is offset to extend beyond it distally; both are tightly articulated to the humerus and there is a marked lozenge-shaped spatium interosseum. The radii, at least, are subject to morphological variation. For example, in BMNH 36183, a distinct notch is found in the middle of its anterior edge (Owen, 1865:Plate 1, Figure 1); this may be a feature of advanced age and/or large size. Similarly, a stepped anterior border exists in NMHN A.-C. 8592.

Of the six carpals, the ulnare and intermedium are the largest, with the others subequal in size. The anteriormost proximal carpal (radiale) is subrectangular; the posterior proximal (ulnare), circular. The intermedium is subquadrate with clear ulnar and radial facets; these are separated by a shallow groove or notch where this bone contributes to or enters the spatium interosseum (well shown in BMNH R.1314). The three distal carpals are subcircular. There is little intervening space between any of the carpals, a consequence of full adult ossification. A small supernumerary carpal illustrated by Conybeare (1824; and Owen, 1865:Plate 1,

Figure 11. BMNH 22656, holotype of *Plesiosaurus dolichodeirus*. **A)** Right forelimb. Bar scale = 10 cm.

Figure 11 (continued). BMNH 22656, holotype of *Plesiosaurus dolichodeirus*. **B)** Left hindlimb. Bar scale = 10 cm.

Figure 2) between and behind the ulna and ulnare is no longer present. Such a supernumary, however, is found in an isolated young adult right forelimb, BMNH R.1314. In OXFUM J.10304, the complete right carpus has a very small ($\approx$1-cm-diameter) anterior supernumerary.

The phalanges and most of the metacarpals, although flattened by compression, are spool-like and constricted in the middle. They decrease in size distally (metapodial IV is longest); their ends are flat to slightly convex. Only metacarpal V is moderately asymmetrical, having a more concave posterior edge. Digit I consists of three phalanges (the distalmost missing?), digit II of seven, III of seven (perhaps originally more), IV of six (at least one or two more), and V of four (certainly one or two more). The digit V is retracted as in all plesiosaurs. There may be variation in the number of phalanges, for in the large individual, OXFUM J.10304, digit I of the right forelimb has four (although the last is a nubbin), II has eight, III has nine (again with a tiny circular bone at the end), IV has eight, and V has at least six.

The forelimbs are much larger than the hindlimbs in full adults (e.g., OXFUM J.10304), but this relationship is less obvious in smaller individuals (Appendix 1). Seemingly, the proportions of fore- and hindlimbs change through ontogeny as they do, for example, in stem-group sauropterygians such as pachypleurosaurs (Sander, 1989). In the juvenile limbs of YPM-PU 3352, the bones are not well formed and are well spaced. Propodial tuberosities are almost nonexistent and there are no epipodial facets. Fewer mesopodials (circular and subequal in size) are ossified. Relatively greater spacing between all bones, more rounded edges, and less well-developed facets are seen in the limbs of the juvenile stages of all plesiosaurs.

Hindlimb

Most of the holotype pelvis is visible due to disruption of the thoracic vertebrae. The pubes are subquadrate in form, and while their anterior borders are hidden, the right lateral corner suggests a slightly concave anterolateral margin. However, the complete, well-formed pelvis of OUM J.10304 (Figure 12) displays anterior pubic edges that are rounded and convex; the bones are almost semilunate (as they are also in NMHN A.-C. 8592 and in the juvenile YPM-PU 3352; TM 13286, too, has a convex anterior border). There are small, ovate (circular in OXFUM J.10304), puboischiatic fenestrae, with slight median bar contact between the pubis and ischium, although no fusion. The ischium is typical of virtually all plesiosaur taxa, save for some stippled rugosity (a potential pathology) at its contribution to the acetabulum. The juvenile ischium of YPM-PU 3352 is extremely short with little posterior extension; such extension as is present is widely

Figure 12. Pelvis of *Plesiosaurus dolichodeirus*, OXFUM J.10304, exposed in medial aspect. Scale bar = 10 cm.

separated from its mate. The adult ilium is blade-like, in keeping with other Lower Jurassic forms, and there is little twist to its shaft, in which regard it is relatively plesiomorphic. The flattened sacral end is relatively broad (4 cm), whereas the acetabular end articulating with the pelvic plate is rod-like (3 cm across).

The femur is almost longitudinally symmetrical; the posterodistal corner is very slightly larger than the anterior (Figures 1 and 11B). The femoral shaft is straight and both the anterior and posterior edges are concave. The tuberosity is a little more pronounced than that of the humerus; the lipped femoral head is apparent (a large, raised, ventral muscle rugosity is present in NMHN A.-C. 8592). The tibia and fibula are similar in size to one another, but the robust tibia has a wider proximal than distal end, whereas the fibula is semilunate. As in the forelimb, the anterior epipodial reaches distally beyond its flatter epipodial partner. A substantial, lozenge-shaped spatium interosseum is also present. As another indication of ontogenetic or individual variation, the right fibula of TM 13286 displays a prominent notch in its posterior edge.

Mirroring the condition in the forelimb, there are six subcircular tarsals, the largest two being the intermedium and fibulare. The tibiale and the anterior and middle distal tarsals are quite small. Although not present in the holotype, the tarsal intermedium of the larger OXFUM J.10304 has a well-developed anteroproximal notch separating the tibial and fibular facets and forming the distal end of the spatium interosseum. This is akin to the carpal intermedium condition in the holotype. In the holotype, digit I has three phalanges (perhaps at one time four, as the last preserved is blunt), II has seven, III has nine, IV has eight, and the retracted digit V has seven (probably another one is missing). Unlike the forelimb, there are no supernumerary bones. Exhibiting variation, BMNH R.1313 has eight, eight, and six phalanges in the last three digits, respectively. Like the forelimb, the hindlimb is quite long and narrow.

The left femur of BMNH 22656 has a rugose distal end, although the right does not. In fact, many elements of the right hindlimb, including both epipodials, most tarsals, and most phalanges, have such stippling on their ends and some shafts. This condition is especially pronounced in the distal phalanges of the middle three digits, yet the distal phalanges of digit V are not at all rugose, while the proximal ones are. Could this be a pathology? One distal phalanx of digit II may be fused to digit I by a bony growth. Such variation within plesiosaur limbs appears to be quite common.

DISCUSSION

Plesiosaurus dolichodeirus is apparently the most common plesiosaur species from the English Liassic, or at least is that known from the most material. It is apparent also that most Liassic plesiosaur species named subsequent to Conybeare's (1824) description either do not belong to the genus *Plesiosaurus* or are nomina dubia (e.g., Storrs, 1994). For example, "*Plesiosaurus*" *homalospondylus* Owen, 1865, is now referred to as *Microcleidus* (Watson, 1909); "*P.*" *rostratus* Owen, 1865, is known by the generic name *Archaeonectrus* (Novozhilov, 1964); "*P.*" *conybeare* Sollas, 1881, has been renamed *Attenborosaurus* (Bakker, 1993); and "*P.*" *hawkinsi* Owen, 1840a, also represents a distinct genus (renamed by Storrs and Taylor, 1996). None is equivalent to *Pleisosaurus dolichodeirus*.

Of the hundreds of Upper Jurassic and Cretaceous plesiosaur species referred to *Plesiosaurus* (e.g., Bogolubov, 1911; Persson, 1963; Sauvage, 1882, 1898; Welles, 1952, 1962; White, 1940) none are correctly assigned, and it appears that only *P. guilelmiimperatoris* Dames, 1895, from the Toarcian Holzmaden area Lagerstätten of Baden-Württemberg, and possibly *P. tournemirensis* Sciau et al., 1990, from the upper Toarcian of the Causse du Lazarc, France, may be valid species of *Plesiosaurus*. Both of these latter taxa are represented by good material and undoubtedly are *Plesiosaurus*, but the French taxon's distinctness from *P. guilelmiimperatoris* has yet to be clearly demonstrated. *Plesiosaurus guilelmiimperatoris* may have a more delicate skull and a more robust, straighter humerus than *P. dolichodeirus*, among other characters. Most of the remainder are certainly nomina dubia, based on very fragmentary and nondiagnostic materials. *Plesiosaurus brachypterygius* Huene, 1923, which is known from an excellent holotype (GPIT HUENE 1923 T. 1-2), is undoubtedly a junior synonym of *P. guilelmiimperatoris*, as its only features of note are limb proportions that differ minorly from the holotype of *P. guilelmiimperatoris* (MB.1893/94.1961 [Holotype] O). It is clear from description of *P. dolichodeirus* that these proportions vary during ontogeny. Similarly, *Seeleyosaurus holzmadensis* White, 1940 (holotype SMNS 12039), is merely an individual variant of *P. guilelmiimperatoris* (cf. Bakker, 1993), based on perceived differences in the pectoral girdle that are also subject to ontogenetic variation.

Plesiosaurus dolichodeirus is known primarily from the type area of Lyme Regis and Charmouth, Dorset. Certainly, the only complete specimens of this taxon originated here. It is probable, however, that some isolated remains from the Lias of Somerset, Gloucestershire, and the Bristol area pertain to *Plesiosaurus*. In particular, the tentatively referred partial skeleton from Keynsham, BMNH R.1315, discussed above, may represent *P. dolichodeirus*. Interestingly, Stukeley (1719) announced the discovery near Fulbeck, Nottinghamshire, of a partial skeleton of a Lower Lias "crocodile or porpoise" (BMNH R.1330)---actually a plesiosaur cf. *Plesiosaurus*, and possibly *P. dolichodeirus*. Dames (1895) referred an isolated humerus from the Lower Liassic of Germany to *P. dolichodeirus*, but this requires confirmation, even though its stratigraphic position is suggestive.

The classic ammonite zonation of the English Lias allows precise stratigraphic placement of many plesiosaur specimens. Most complete Dorset marine reptile fossils are thought to have come from the upper part of the Blue Lias, the "Shales with Beef," and the lower Black Ven Marls, corresponding to the lower Sinemurian (*Arietites bucklandi, Arnioceras semicostatum,* and *Caenisites turneri* ammonite zones) (Arkell, 1933; Cope et al., 1980; Delair, 1959; Macfayden, 1970; Storrs and Taylor, 1996; Woodward, 1893). This has been the presumed stratigraphic distribution of most *Plesiosaurus dolichodeirus* individuals, although Mansel-Pleydell (1888), Delair (1959), and Persson (1963) placed *P. dolichodeirus* in the *Psiloceras planorbis* zone. However, OXFUM J.10304 (and thus seemingly Conybeare's holotype) is apparently from the *Echioceras raricostatum* zone and thus the uppermost Sinemurian top of the Black Ven Marls, probably below Stonebarrow Hill. Lydekker (1889:255-256) and Delair (1959:54) synonymized "*P.*" *cliduchus* Seeley, 1865 (CAMSM J.35180), from the basal Lias (Rhaetian/Hettangian boundary) of Street, Somerset, with *P. dolichodeirus*, but the material is too poor to demonstrate equivalence and "*P.*" *cliduchus* is more likely a poor example of "*P.*" *hawkinsi* (Storrs and Taylor, 1996). However, an unprepared and damaged skull (DBYMU 355-1903), seemingly from the lower Hettangian or uppermost Rhaetian Hydraulic Limestone of Cropwell Bishop, Nottinghamshire, may conceivably represent the earliest occurrence of *Plesiosaurus*, if identity and horizon are verified.

The holotype of *Plesiosaurus macromus* Owen, 1840 (OXFUM J.28587, part of the Misses Philpott collection presented to Oxford in 1880; Powell and Edmonds, 1978), collected by Mary Anning from the Lower Lias of Lyme Regis, is here synonymized with *P. dolichodeirus*. It was distinguished originally (Owen, 1840: 72-74) on the basis of a disproportion in size between the fore- and hindlimbs and on very slight variations of the vertebral centra. The relative limb sizes are again readily explained by allometric changes during ontogeny, and the vertebral differences may be ascribed to individual variation. Owen's specimen comprises three-dimensional, matrix-free bones from a partial skeleton: all four propodials, most epipodials, and 47 vertebral centra. The characteristically curved humeri of *P. dolichodeirus* with their shallow interepipodial grooves on their ventral surfaces are evident in this fossil. Indeed, Owen (1840:73) noted the resemblance of the humerus to that of *P. dolichodeirus*, describing both the shaft curvature and distal trough, but did not recognize their importance.

In addition to its characteristic humerus, *Plesiosaurus dolichodeirus* is identified by its small head with broadly similar pre- and postorbital lengths, moderately sized subcircular temporal fenestrae, and large pineal opening. It has broad parietals, a rather shallowly emarginated subtemporal region, relatively elongate jugals, no pterygoid flange but a small ectopterygoidal boss, and an open basicranium, i.e., a basioccipital that is not underlapped by a posterior union of the pterygoids. It also has a short and nonspatulate mandibular symphysis, an elongate neck of about 1.5 times the body length (composed of approximately 40 cervical vertebrae), and vertebral centra whose edges are often markedly rugose and retains a prominent spatium interosseum while the pillar-like anterior epipodials of each limb (radius/tibia) reach distal to their crescentic posterior partners (ulna/fibula). In fully adult individuals, the forelimbs are slightly longer than the hindlimbs (although this relationship is reversed in young animals) and the coracoid possesses a significant posterolateral expansion. The anterior edge of the pectrum is characterized by a distinct notch within the clavicular arch. Many of these characters seem to be primitive for plesiosaurs and, indeed, the broad parietals, elongate jugals, curved humeral shaft, substantial spatium interosseum, and discordant epipodials are also variously found in stem-group sauropterygians (Storrs, 1991, 1993b). The Lower Liassic position of *P. dolichodeirus* is concordant with its largely plesiomorphic condition.

Plesiosaurs suffer from long, yet spotty, collection and description histories. Recognition of what truly constitutes *Plesiosaurus* is the first step in a fuller understanding of plesiosaur relationships. Future studies can only gain broader significance if placed in an evolutionary context, the basic phylogenetic and taxonomic framework of which does not yet exist, largely because of the prolonged "wastebasket taxon" status of *Plesiosaurus*. This is the fundamental lynchpin that has been missing from plesiosaur systematics for 170 years and is prerequisite for the progress of future investigations.

SUMMARY

Detailed examination of the broad range of English Liassic plesiosaur species indicates that only *P. dolichodeirus* and its junior synonyms can be assigned confidently to *Plesiosaurus*. Many named *Plesiosaurus* species are based on inadequate material and considered nomina dubia. Most others, even those contemporaneous with *P. dolichodeirus*, are certain or probable representatives of distinct genera exhibiting significant cranial autapomorphies. Elsewhere, only the German Lower Jurassic, notably the Holzmaden Lagerstätte, and the French upper Toarcian currently contain taxa that are potentially valid specific variants of *Plesiosaurus*. *Plesiosaurus dolichodeirus* has relatively short antorbital and temporal regions, subcircular temporal fenestrae, a broad but pointed rostrum, a large pineal opening, a nonspatulate mandibular symphysis, approximately 40 cervical vertebrae with distinctly doubled rib facets, slightly rugose articular edges to virtually all vertebrae, a sharp posterolateral expansion to the coracoid, slightly longer forelimbs than hindlimbs, a curved humerus with a shallow interepipodial groove on its ventral surface but no sharp anterodistal corner, and other identifying features.

To recapitulate, the genus *Plesiosaurus* currently contains only the English species *P. dolichodeirus* in the Lower Liassic, with the inclusion of one or perhaps two continental Toarcian species. Other valid Liassic species belong to other genera, many species are nomina dubia and no plesiosaurs dating from above the Early Jurassic are certainly known to represent *Plesiosaurus*, despite a long history of careless assignment to this genus.

ACKNOWLEDGMENTS

I thank H. P. Powell (OXFUM), N. Monahan (NMING), B. Battail (NMHN), D. Parris (NJSM), D. Harper (UG), D. Norman (CAMSM), W. Grange (DBYMU), J. C. van Veen (TM), M. A. Turner (YPM), T. Daeschler (ANSP), R. Wild (SMNS), W.-D. Heinrich (BM), A. Liebau (GIPT), P. R. Crowther and R. Clark (BRSMG), S. Tunnicliff (BGS), and A. C. Milner and S. Chapman (BMNH) for access to study material. Thanks are due to P. R. Crowther and R. F. Vaughn for the use of facilities at the Bristol City Museum and Art Gallery. Preparation aid was given by R. F. Vaughn, W. Lindsay, and A. Doyle. Portions of the figures are by P. Baldaro and B. Alexander. I am also grateful to D. S. Brown, A. R. I. Cruickshank, M. A. Taylor, and S. P. Welles for discussion. Two anonymous reviewers aided in the improvement of this chapter. This study was funded by the Natural Environment Research Council of the United Kingdom, the University of Bristol Department of Geology, and the Geier Collections and Research Center of the Cincinnati Museum of Natural History.

REFERENCES

Andrews, C. 1896. On the structure of the plesiosaurian skull. *Quarterly Journal of the Geological Society of London* 52:246-253.

Andrews, C. 1910. A descriptive catalogue of the marine reptiles of the Oxford Clay. Part I. *British Museum (Natural History)*, London, 205 pp.

Arkell, W. J. 1933. *The Jurassic System in Great Britain.* Oxford University Press, Oxford, 681 pp.

Bakker, R. 1993. Plesiosaur extinction cycles---events that mark the beginning, middle and end of the Cretaceous. IN W. G. E. Caldwell and E. G. Kauffman (Eds.), *Evolution of the Western Interior Basin*, pp. 641-664. *Geological Survey of Canada, Special Paper* 39.

Blainville, H. D. de. 1835. Description de quelques espèces de reptiles de la Californie, précédée de l'analyse d'un système general d'Erpetologie et d'Amphibiologie. *Nouvelles Annales du Muséum (National) d'Histoire Naturelle*, Paris 4:233-296.

Bogolubov, N. N. 1911. *On the History of Plesiosaurs in Russia* [In Russian]. Moscow Imperial University Press, Moscow, 412 pp.

Brown, B. 1913. A new plesiosaur, *Leurospondylus*, from the Edmonton Cretaceous of Alberta. *Bulletin of the American Museum of Natural History* 32:606-615.

Brown, D. S. 1981. The English Upper Jurassic Plesiosauroidea (Reptilia) and a review of the phylogeny and classification of the Plesiosauria. *Bulletin of the British Museum (Natural History)*, Geological Series 35:253-347.

Buckland, W. 1837. *Geology and Mineralogy, Considered With Reference to Natural Theology*. William Pickering, London, 605 pp.

Conybeare, W. D. 1824. On the discovery of an almost perfect skeleton of the *Plesiosaurus*. *Transactions of the Geological Society of London* 1:382-389.

Cope, J. C. W., T. A. Getty, M. K. Howarth, N. Morton, and H. S. Torrens. 1980. A correlation of Jurassic rocks in the British Isles. Part One: Introduction and Lower Jurassic. *Special Report of the Geological Society of London* 14:1-73.

Cruickshank, A. R. I. 1994a. Cranial anatomy of the Lower Jurassic pliosaur *Rhomaleosaurus megacephalus* (Stutchbury) (Reptilia: Plesiosauria). *Philosophical Transactions of the Royal Society of London* B 343:247- 260.

Cruickshank, A. R. I. 1994b. A juvenile plesiosaur (Plesiosauria: Reptilia) from the Lower Lias (Hettangian: Lower Jurassic) of Lyme Regis, England: a pliosauroid-plesiosauroid intermediate? *Zoological Journal of the Linnean Society* 112:151-178.

Cruickshank, A. R. I., P. G. Small, and M. A. Taylor. 1991. Dorsal nostrils and hydrodynamically driven underwater olfaction in plesiosaurs. *Nature* 352:62-64.

Cuvier, G. 1824. *Recherches sur les Ossemens Fossiles*, Volume 5, Part 2. G. Dufour and Ed. D'Ocagne, Paris, 547 pp.

Dames, W. 1895. Die Plesiosaurier der süddeutschen Liasformation. *Abhandlungen der Königliche Preussische Akademie der Wissenschaften zu Berlin* pp. 1-83.

De la Beche, H. T. and W. D. Conybeare. 1821. Notice of the discovery of a new fossil animal, forming a link between the *Ichthyosaurus* and the crocodile, together with general remarks on the osteology of *Icthyosaurus*. *Transactions of the Geological Society of London* 5:559-594.

Delair, J. B. 1959. The Mesozoic reptiles of Dorset. Part 2. *Proceedings of the Dorset Natural History and Archaeological Society* 80:52-90.

Edmonds, J. M. 1978. The fossil collection of the Misses Philpot of Lyme Regis. *Proceedings of the Dorset Natural History and Archaeological Society* 98:43-48.

Hawkins, T. 1834. *Memoirs on Ichthyosauri and Plesiosauri; Extinct Monsters of the Ancient Earth*. Relfe and Fletcher, London, 58 pp.

Hawkins, T. 1840. *The Book of the Great Sea-Dragons, Ichthyosauri and Plesiosauri, Gedolim Taninim of Moses. Extinct Monsters of the Ancient Earth.* W. Pickering, London, 27 pp.

Huene, F. R. F. von. 1923. Ein neuer Plesiosaurier aus dem oberen Lias Württembergs. *Jareshefte das Vereins für Vaterlandische Naturkunde in Württemberg* 79:1-21.

Hulke, J. W. 1883. The anniversary address of the president. *Quarterly Journal of the Geological Society of London* 39:38-65.

Lydekker, R. 1889. Catalogue of the fossil Reptilia and Amphibia in the British Museum (Natural History). Part II. Containing the orders Ichthyopterygia and Sauropterygia. *British Museum (Natural History)*, London, 307 pp.

Lydekker, R. 1890. Catalogue of the fossil Reptilia and Amphibia in the British Museum (Natural History). Part IV. The orders Anomodontia, Ecaudata, Caudata, and Labyrinthodontia; and supplement. *British Museum (Natural History)*, London, 295 pp.

Macfayen, W. A. 1970. *Geological Highlights of the West Country. A Nature Conservancy Handbook.* Butterworths, London, 296 pp.

Mansel-Pleydell, J. C. 1888. Fossil reptiles of Dorset. *Proceedings of the Dorset Natural History and Antiquarian Field Club* 9:1-40.

Novozhilov, N. 1964. Superfamily Plesiosauroidea [In Russian]. IN A. K. Rozhdestvensky and L. P. Tatarinov (Eds.), *Principles of Paleontology: Amphibians, Reptiles and Birds*, pp. 318-327. Nauka, Moscow.

Osborn, H. F. 1903. The reptilian subclasses Diapsida and Synapsida and the early history of the Diaptosauria. *Memoir of the American Museum of Natural History* 1:449-507.

Owen, R. 1840a. Report on British fossil reptiles. Part I. *Report of the British Association for the Advancement of Science* 1839, pp. 43-126.

Owen, R. 1840b. A description of a specimen of the *Plesiosaurus macrocephalus*, Conybeare, in the collection of Viscount Cole, MP, DCL, FGS, &c. *Transactions of the Geological Society of London* 5:559-594.

Owen, R. 1860. On the orders of fossil and Recent Reptilia, and their distribution in time. *Report of the British Association for the Advancement of Science* 29:153-166.

Owen, R. 1865. A monograph on the fossil Reptilia of the Liassic formations. Part I, Sauropterygia. *Palaeontographical Society Monograph* 17:1-40.

Parkinson, J. 1822. *Outline of Oryctology. An Introduction to the Study of Fossil Organic Remains; Especially of Those Found in British Strata.* London, 350 pp.

Persson, P. O. 1963. A revision of the classification of the Plesiosauria with a synopsis of the stratigraphical and geographical distribution of the group. *Lunds Universitet Årsskrift* 59:1-60.

Powell, H. P. and J. M. Edmonds. 1978. List of type-fossils in the Philpot Collection, Oxford University Museum. *Proceedings of the Dorset Natural History and Archaeological Society* 98:43-53.

Sander, P. M. 1989. The pachypleurosaurids (Reptilia: Nothosauria) from the Middle Triassic of Monte San Giorgio (Switzerland) with the description of a new species. *Philosophical Transactions of the Royal Society of London* B 325:561-670.

Sauvage, H.-E. 1882. Recherches sur les reptiles trouvé dans le Gault du l'est du Bassin de Paris. *Memoire Societé Géologique de France* 2:24-28.

Sauvage, H.-E. 1898. Les reptiles et les poissons des terrains Mésozoïques du Portugal. *Bulletin Societé Géologique de France*, 26:442-446.

Sciau, J., J.-Y. Crochet and J. Mattei. 1990. Le premier squelette de plesiosaure de France sur le Causse du Larzac (Toarcien, Jurassique Inférieur). *Geobios* 23:111-116.

Seeley, H. G. 1865. On two new plesiosaurs from the Lias. *Annals and Magazine of Natural History* 16:352-359.

Seeley, H. G. 1874. Note on the generic modifications of the plesiosaurian pectoral arch. *Quarterly Journal of the Geological Society of London* 30:436-449.

Sollas, W. J. 1881. On a new species of *Plesiosaurus* (*P. conybeari*) from the Lower Lias of Charmouth, with observations on *P. megacephalus*. *Quarterly Journal of the Geological Society* 37:440-481.

Storrs, G. W. 1991. Anatomy and relationships of *Corosaurus alcovensis* (Diapsida: Sauropterygia) and the Triassic Alcova Limestone of Wyoming. *Bulletin of the Peabody Museum of Natural History* 44:1-151.

Storrs, G. W. 1993a. Function and phylogeny in sauropterygian (Diapsida) evolution. *American Journal of Science* 293-A:63-90.

Storrs, G. W. 1993b. The systematic position of *Silvestrosaurus* and a classification of Triassic sauropterygians (Neodiapsida). *Paläontologische Zeitschrift* 67:177-191.

Storrs, G. W. 1994. Fossil vertebrate faunas of the British Rhaetian (latest Triassic). *Zoological Journal of the Linnean Society* 112:217-259.

Storrs, G. W. and M. A. Taylor. 1996. Cranial anatomy of a new plesiosaur genus from the lowermost Lias (Rhaetian/Hettangian) of Street, Somerset, England. *Journal of Vertebrate Palaeontology* 16:403-420..

Storrs, G. W. In press.. A juvenile specimen of ?*Plesiosaurus* sp. from the Lias (Lower Jurassic, Pliensbachian) near Charmouth, Dorset, England. *Proceedings of the Dorset Natural History and Archaeological Society*.

Stukely, W. 1719. An account of the impression of the almost entire skeleton of a large animal in a very hard stone from Nottinghamshire. *Philosophical Transactions of the Royal Society of London* 30:963-968.

Stutchbury, S. 1846. Description of a new species of *Plesiosaurus*, in the Museum of the Bristol Institution. *Quarterly Journal of the Geological Society of London* 2:411-417.

Taylor, M. A. 1992. Functional anatomy of the head of the large aquatic predator *Rhomaleosaurus zetlandicus* (Plesiosauria, Reptilia) from the Toarcian (Lower Jurassic) of Yorkshire, England. *Philosophical Transactions of the Royal Society of London* B 335:247-280.

Taylor, M. A., and A. R. I. Cruickshank. 1993. Cranial anatomy and functional morphology of *Pliosaurus brachyspondylus* (Reptilia: Plesiosauria) from the Upper Jurassic of Westbury, Wiltshire. *Philosophical Transactions of the Royal Society of London* B 341:399-418.

Torrens, H. S. 1995. Mary Anning (1799-1847) of Lyme: "the greatest fossilist the world ever knew." *British Journal for the History of Science* 28:257-284.

Tschanz, K. 1989. *Lariosaurus buzzii* n. sp. from the Middle Triassic of Monte San Giorgio (Switzerland) with comments on the classification of nothosaurs. *Palaeontographica* A 208:153-179.

Watson, D. M. S. 1909. A preliminary note on two new genera of Upper Liassic plesiosaurs. *Memoirs of the Manchester Museum* 54:1-26.

Watson, D. M. S. 1924. The elasmosaurid shoulder-girdle and forelimb. *Proceedings of the*

Zoological Society of London pp. 885-917.

Welles, S. P. 1943. Elasmosaurid plesiosaurs with description of new material from California and Colorado. *Memoir of the University of California* 13:125-215.

Welles, S. P. 1952. A review of North American Cretaceous elasmosaurs. *University of California Publications in the Geological Sciences* 29:47-144.

Welles, S. P. 1962. A new species of elasmosaur from the Aptian of Columbia and a review of the Cretaceous plesiosaurs. *University of California Publications in the Geological Sciences* 44:1-96.

White, T. E. 1940. Holotype of *Plesiosaurus longirostris* Blake and classification of the plesiosaurs. *Journal of Paleontology* 14:451-467.

Winkler, T. C. 1873. Le *Plesiosaurus dolichodeirus* Conyb. du Musée Teyler. *Achives du Musée Teyler* 3: 219-233.

Woodward, H. B. 1893. The Jurassic rocks of Britain. Volume III. The Lias of England and Wales (Yorkshire excepted). *Memoir of the Geological Survey of the United Kingdom*, London, 399 pp.

APPENDIX 1

Comparable and contrasting dimensions of holotype and referred specimens of *Plesiosaurus dolichodeirus*:

Measurement	Approximate Dimension (Cm)
BMNH 22656 (holotype)	
total length	290
length of cervical series	120
length of caudal series	68
distance between limb girdles	35
width of gastral basket	42
skull length (broken rostrum---to supraoccipital)	13
skull length (broken rostrum---to occipital condyle)	16
maximum skull breadth as preserved	11*
mandible length	21
length of pineal foramen	1
estimated orbital length	4*
supratemporal fenestra length	4
supratemporal fenestra width	3
length of exoccipital facets on basioccipital	1
length of basiociptal/basisphenoid complex	2
length of atlas/axis complex	3
length of third cervical centrum	2
height of tenth cervical centrum	2

APPENDIX 1 (continued)

breadth of tenth cervical centrum	3
length of posterior cervical centrum	4
height of posterior cervical centrum	4
breadth of posterior cervical centrum	5
height of dorsal neural spines	6
height of midseries dorsal centrum	4
breadth of midseries dorsal centrum	5
length of sacral ribs	4
height of midseries caudal centrum	3
breadth of midseries caudal centrum	4
length of scapular dorsal blade	10
length of scapular glenoid ramus	6
maximum breadth of pectoral girdle	27
length of forelimb	60
length of humerus	19
maximum breadth of humerus	9
length of ulna	8
breadth of ulna	5
length of radius	8
proximal breadth of radius	5
distal breadth of radius	4
length of radiale	2
breadth of radiale	3
length of carpal intermedium	3
breadth of carpal intermedium	3
diameter of ulnare	4
diameter of anterior distal carpal	2
diameter of middle distal carpal	2
diameter of posterior distal carpal	3
length of pelvis	22
maximum breadth across pubes	26
length of pubis	11
breadth of pubis	13
breadth of puboischiatic fenestra	5
length of puboischiatic fenestra	3
length of ischium	11
maximum breadth of ischium	10
length of ilium	11
length of hindlimb	59
length of femur	18
maximum breadth of femur	9
length of tibia	7

APPENDIX 1 (continued)

proximal breadth of tibia	6
distal breadth of tibia	4
length of fibula	7
breadth of fibula	5
diameter of tibiale	2
diameter of tarsal intermedium	4
diameter of fibulare	4
diameter of anterior distal tarsal	2
diameter of middle distal tarsal	2
diameter of posterior distal tarsal	3

OXFUM J.10304 (topotype)

length of caudal series	88
skull length (rostrum---to occipital condyle)	18
skull length (rostrum---to articular end of quadrate)	21
mandible length	22
maximum skull breadth as preserved (at fenestrae)	10*
length of forelimb	69
length of humerus	21
length of radius	8
length of ulna	10
length of pubis	12
breadth of pubis	15
length of femur	20
length of tibia	7
length of fibula	8
diameter of tarsal intermedium	4

BMNH 39490

skull length (rostrum---to occipital condyle)	19
skull length (rostrum---to articular end of quadrate)	21
antorbital length	8
length of temporal region (orbit to squamosal midline)	5
mandible length	24
maximum skull breadth (posterior end of maxilla/jugal)	12*
length of temporal fenestra after crushing	4
estimated length of temporal fenestra before crushing	5
width of temporal fenestra	5
diameter of orbit	3
minimum interorbital distance (across frontals)	3
length of pineal foramen	2
anteroposterior breadth of postorbital bar as preserved	3*

APPENDIX 1 (continued)

breadth of articular end of quadrate	2
length of paroccipital process	3
diameter of occipital condyle	2
length of longest tooth crown (fourth dentary)	2
length of third cervical centrum	2
length of seventh cervical centrum	2

OXFUM J.13809

skull length (rostrum---to articular end of quadrate)	17
maximum skull breadth	8

BMNH 41101

skull length (rostrum---to occipital condyle)	18
skull length (rostrum---to articular end of quadrate)	21
maximum skull breadth (posterior end of maxillae)	10*
length of internal nares	1
length of posterior interpterygoid vacuity	2
diameter of occipital condyle	2
breadth of articular end of quadrate	2

BMNH R.1313

total length	350
length of cervical series	145
length of caudal series	80
skull length (rostrum---to occipital condyle)	18
skull length (rostrum---to articular end of quadrate)	19
length of posterior interpterygoid vacuity	2
width of posterior interpterygoid vacuity	1
mandible length	22
length of mandibular symphysis	3
length of longest tooth crown	1

BMNH 36183

skull length (rostrum---to articular end of quadrate)	20
mandible length	22
height of skull from supraoccipital to quadrate	9*
supratemporal fenestra length	4
diameter of orbit	3
length of humerus	21
length of radius	9
proximal breadth of radius	7
distal breadth of radius	4

APPENDIX 1 (continued)

BMNH 39491

length of dentary	16
length of symphysis	2
anterior width of dentary	2
posterior width of dentary	1
median height of dentary	1
posterior height of dentary	2

YPM-PU 3352

total length	225
length of caudal series	59
length of pectrum	22
length of humerus	14
length of pelvis	15
length of femur	14
length of pubis	8
breadth of pubis	9

NMHN A.-C. 8592

length of humerus	22
length of femur	19

* slightly increased by crushing

APPENDIX 2

Anatomical Abbreviations: **aa** = atlas/axis complex, **aiv** = anterior interpterygoid vacuity, **an** = angular, **art** = articular, **bo** = basioccipital, **bs** = basisphenoid, **cl** = clavicular arch, **co** = coracoid, **cp** = coronoid process, **d** = dentary, **ect** = ectopterygoid, **eo** = exoccipital, **eof** = exoccipital facet, **f** = frontal, **h** = humerus, **in** = internal naris, **j** = jugal, **m** = maxilla, **o** = orbit, **p** = parietal, **pf** = postfrontal, **pif** = pineal foramen, **piv** = posterior interpterygoid vacuity, **pl** = palatine, **pm** = premaxilla, **po** = postorbital, **pop** = paroccipital process, **prf** = prefrontal, **ps** = parasphenoid, **pt** = pterygoid, **q** = quadrate, **s** = sclerotic plate, **san** = surangular, **sc** = scapula, **so** = supraoccipital, **sq** = squamosal, **st** = stapes, **v** = vomer.

Chapter 7

COMPARATIVE CRANIAL ANATOMY OF TWO NORTH AMERICAN CRETACEOUS PLESIOSAURS

KENNETH CARPENTER

INTRODUCTION

The taxonomy of Western Interior Cretaceous plesiosaurs is in disarray despite the efforts of Williston (1903, 1906, 1908) and Welles (1943, 1952, 1962). The problems are similar to those faced with English Jurassic plesiosaurs, including the inadequacy of type material, differential crushing, and skull-less skeletons. Nevertheless, revision of the English plesiosaurs has, in recent years, outpaced similar work in North America owing to the work of Storrs (Chapter 6), Taylor and Cruickshank (1993), Taylor (1992a), Brown (1981), and Tarlo (1960). Their revision of the plesiosaurs was made possible by the discovery of new specimens, advances in preparation, and new analytical techniques.

Perhaps the most fundamental change in the study of plesiosaurs, however, is philosophical, with the recognition that differences among specimens may not reflect taxonomic differences so much as individual variation, ontogeny, and sexual dimorphism. The significance of ontogenic change in plesiosaurs was best documented by Andrews (1910), who showed that the medial bar formed by the scapula and coracoid splits the coracoid foramen only in adult size. Thus, the presence or absence of this bar should not be used for taxonomic purposes. Even variation or dimorphism may be considerable among plesiosaurs, as briefly noted by Welles (1962) in his study of two specimens of *Aldzadasaurus columbiensis*. However, I differ from Welles in that I recognize the larger, more robust individual as a female because many female reptiles are two thirds to over twice longer than males, with one quarter to 15 times greater mass (Fitch, 1981).

Other advances in plesiosaur studies have occurred because of new preparation techniques (e.g., Taylor and Cruickshank, 1993; Cruickshank et al., 1991). These techniques were applied in the study of the uncrushed skull of "*Elasmosaurus*" *morgani* from the Cretaceous of North America. To avoid problems associated with differences between matrix and bone, the skull was first acid prepared before being CAT scanned and X-rayed. The skull was then compared with that of the short-necked plesiosaur *Dolichorhynchops osborni*, which also occurs in the marine Cretaceous of North America. The two skulls show similarities that may have evolutionary implications as noted below.

Welles (1949) referred "*E.*" *morgani* to *Elasmosaurus* primarily on the presence of a median pectoral bar, a condition first reported by Cope (1869). A historical account of the discovery of Cope's *Elasmosaurus platyurus* is presented by Almy (1987). Welles (1962) later used this median bar to define the subfamily Elasmosaurinae but, as noted above, this character may be ontogenetic. Indeed, the short and deep atlas-axis in "*E.*" *morgani* contrasts with the long and low one of *E. platyurus* (Figure 1). Welles (1952) noted that Cope's illustration of the pectoral girdles is probably not correct, but I consider the two versions given by Welles (1943, 1949) as unsubstantiated. Unfortunately, the girdles of *Elasmosaurus*

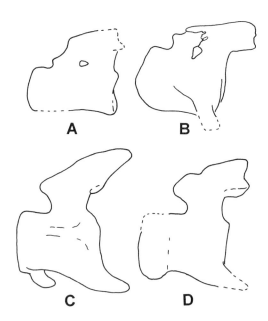

Figure 1. Comparison of the atlas-axis complex of **A)** *Elasmosaurus platyurus*, ANSP 10081; **B)** *Libonectes morgani*, SMUMP 69120; **C)** *Thalassomedon hanningtoni*, DMNH 1588; **D)** *Tuarangisaurus keysi* (after Wiffen and Moisley, 1986; reversed). Not to scale.

platyurus are missing, as first reported by Williston (1906), and recent attempts to relocate them have failed (Arnold Lewis, personal communications). Until the holotype elements are found and their exact morphology is reexamined, I restrict the name, *Elasmosaurus platyurus* to the holotype specimen, ANSP 10081 (not ANSP 18001 as reported by Welles, 1952, 1962). This restriction, and the differences between atlas-axis complexes (Figure 1), leave the holotype "*Elasmosaurus*" *morgani* (SMUSMP 69120) in need of a generic name, for which *Libonectes* is proposed (Appendix 1).

The other plesiosaur, *Dolichorhynchops osborni*, was named by Williston in 1903 based on a nearly complete skeleton from the Smoky Hill Chalk of Kansas. Later, he (Williston, 1906, 1908) concluded that the genus was synonymous with *Trinacromerum* Cragin, 1888. However, differences between the two genera make this synonymy implausible (Carpenter, 1989; in preparation). A revised diagnosis for *Dolichorhynchops osborni* is presented in Appendix 1 based on additional specimens not available to Williston.

Cretaceous plesiosaurs from North America have been divided into three families, Pliosauridae, Polycotylidae, and Elasmosauridae (Welles, 1962; Williston, 1907, 1908); Brown (1981), however, does not recognize Polycotylidae. As will be shown elsewhere (Carpenter, in preparation), two families of short-necked plesiosaurs should be recognized in North America, as first noted by Williston (1907, 1908). Here, I elaborate on the skulls of the elasmosaurid *Libonectes morgani* (SMUSMP 69120) and the polycotylid *Dolichorhynchops osborni* (holotype KUVP 1300, AMNH 5834, FHSM VP404, UCM 35059) as representatives of their two families.

Repository Abbreviations

The abbreviations used for the institutions referred to in the text are as follows: AMNH, American Museum of Natural History, New York; ANSP, Academy of Natural Sciences, Philadelphia; BMNH, British Museum of Natural History, London; DMNH, Denver Museum of Natural History, Denver; FHSM, Fort Hays Sternberg Museum, Hays, Kansas; HMG, Hunterian Museum, Glasgow University, Glasgow; KUVP, Kansas University, Vertebrate Paleontology, Museum of Natural History, Lawrence; SDSM, South Dakota School of Mines, Rapid City; SMUSMP, Southern Methodist University, Shuler Museum of Paleontology, Dallas; and UCM, University of Colorado Museum, Boulder.

COMPARISON OF *LIBONECTES* AND *DOLICHORHYNCHOPS*

The skull of *Libonectes* is considerably shorter relative to its width than is the case in *Dolichorhynchops*. This difference is due to the proportionally shorter preorbital segment of the skull of *Libonectes* (compare Figure 2A and C and Figure 3A and C). As noted by Williston (1903, 1906), this difference correlates well with neck length and has been used to separate the two families (e.g., Persson, 1963). Comparative measurements of *Libonectes* and *Dolichorhynchops* are given in Table 1 and Figure 4.

The paired premaxillaries form the dorsal rim of the snout, including the anterior and dorsal margins of the external nares of both plesiosaurs (Figures 2C and 3B). They are large, wide bones in *Libonectes*, whereas they are long and slender in *Dolichorhynchops*. *Dolichorhynchops* has six premaxillary teeth and *Libonectes* has five. The first tooth is procumbent and large in both taxa. The rest of the tooth-bearing portion of the premaxilla is proportionally shorter relative to the total length of the premaxilla in *Dolichorhynchops* than in *Libonectes*. The dorsal process separates the frontals to contact the parietals above the orbits in both *Dolichorhynchops* and *Libonectes*. In the primitive elasmosaur *Brancasaurus*, the frontals are apparently not separated (Wegner, 1914). However, this region is damaged, so the frontal premaxillary and frontal parietal contacts are equivocal. The suture between the orbits of *Libonectes* is unusual in that it is partially open, leaving a groove that extends posteriorly into the anteriormost portion of the parietals. Welles (1949) identified this groove as the pineal foramen, but it does not extend through the premaxilla. In addition, the only expansion in the dorsal part of the braincase that could have housed the pineal is located just in front of the supraoccipital (Figure 5D; see also below). In both *Libonectes* and *Dolichorhynchops*, the pineal foramen is closed dorsally, unlike *Brancasaurus* Jurassic plesiosaurs (Brown, 1981; Wegner, 1914; Linder, 1913). Its loss is considered synapomorphic for the Upper Cretaceous elasmosaurids and polycotylids. Ventrally, the premaxillaries in both animals are separated by the vomers, at the anterior end of which is a V-shaped slit here called the vomeronasal fenestra (Figures 2D and 3B), which is discussed further below.

The maxillaries are large bones and extend posterior of the orbits in both taxa. Laterally, the maxilla forms a portion of the external nares rim. The maxilla forms most of the lower rim of the orbit in *Dolichorhynchops* but only a small, anterior portion in *Libonectes*. Within the anterior edge of the orbit, the maxilla projects medially as a small, buttressed wall (Figures 5D and 6B). This wall marks the posterior limits of the nasal chamber formed by the premaxilla dorsally, the maxilla laterally and lateroventrally, and the vomer ventromedially. The lateral surface of the maxilla is pierced by numerous foramina, most likely for blood vessels and nerves. The maxilla forms part of the rim of the internal nares in *Libonectes*.

The preorbital portion of the maxilla in *Dolichorhynchops* is considerably longer than in *Libonectes*. It holds about 29 teeth, to 14 in *Libonectes*. In both taxa, the teeth are large anteriorly, but become progressively smaller posteriorly. A row of irregular pits containing the tips of replacement teeth occur in a shallow groove medial to the teeth in *Libonectes* (Figure 2D).

The external nares are just anterior to the orbits in both *Libonectes* and *Dolichorhynchops* and are separated from them by the prefrontals (Figures 2C and 3A). In *Brancasaurus* they are situated as they are in *Libonectes*; however, the external nares are proportionally smaller. The prefrontals are damaged in *Libonectes*, but form a small portion of the anterior edge of the orbits. In *Dolichorhynchops*, the prefrontals form most of the anterior edge.

The lachrymal is absent in both *Dolichorhynchops* and *Libonectes*. The presence of a lachrymal in plesiosaurs is problematical because the area in front of and above the orbits is usually damaged. Nevertheless, it was apparently absent in plesiosauroids, according to Andrews (1910) and Brown (1981), and alleged to be present in the pliosauroids *Peloneustes* and *Liopleurodon* (Taylor and Cruickshank, 1993; Andrews, 1913). Nicholls (personal communications) believes that the presence of the lachrymal in these two pliosaurids needs to be substantiated and questions their homology.

The nasals in plesiosaurs are also problematical because of the damage in front of the orbits. They were identified in "*Plesiosaurus*" *conybeari* by Sollas (1881), but it is more probable that they are prefrontals as in *Libonectes*. The element that Welles (1949) identified as the nasal is here reinterpreted to be the prefrontal. Storrs (1991) unites all plesiosaurs, except *Pistosaurus*, with the loss of this element; I accept it as a synapomorphy for all plesiosaurs above *Pistosaurus*.

The frontals in *Libonectes* and *Dolichorhynchops* are separated by the dorsal process of the premaxillaries (Figures 2C and 3A). Although incomplete in *Libonectes*, the frontals evidently formed the dorsal rim of the orbit. Medially, the frontals extend ventrally on each side of the premaxilla, forming a pair of partial walls inside the orbit (Figure 5D). In life, the cartilaginous internasal and interorbital septum extended from this wall to the top of the palate, as indicated by the scars on the bone surfaces. A gap between the two walls, the olfactory sulcus, probably housed the olfactory (I) nerve.

The frontals in *Dolichorhynchops* have a wing-like process that extends dorsally over the orbit. An oval fenestra, or frontal fenestra, is present in the frontal at its suture with the premaxilla (Figure 6A). The frontals extend posteriorly to contact the parietals. Lateral to this posterior portion of the frontal is a narrow strip of bone that may be a supraorbital. It is present in *Trinacromerum*, but apparently not in elasmosaurs.

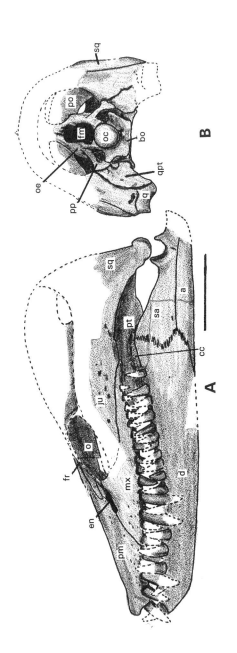

Figure 2. Skull of *Libonectes morgani*, SMUSMP 69120. **A)** Left lateral view. **B)** Posterior aspect. Abbreviations in Appendix 2. Scale bar = 10 cm.

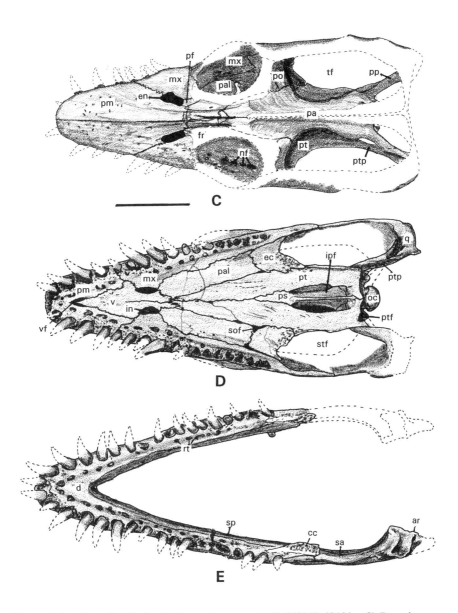

Figure 2 (continued). Skull of *Libonectes morgani*. SMUSMP 69120. **C**) Dorsal aspect. **D**) Ventral aspect. **E**) Dorsal view of mandible. Abbreviations in Appendix 2. Scale bar = 10 cm.

In both *Libonectes* and *Dolichorhynchops*, the postorbital contacts the parietal dorsally and the jugal ventrally (Figures 5B and 6A). In addition, the postorbital contacts the supraorbital in *Dolichorhynchops*. The postorbital forms the rear wall

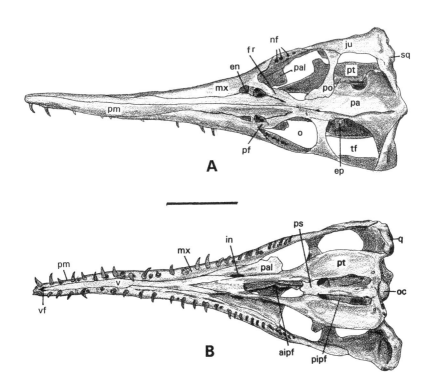

Figure 3. Skull of *Dolichorynchops osborni*, FHSM VP404. **A**) Dorsal aspect. **B**) Ventral aspect. Abbreviations in Appendix 2. Scale bar = 10 cm.

of the orbit to protect the eyeball from the pseudotemporalis muscle. The thinness of the bone makes them easily damaged, as has happened in *Libonectes* since additional preparation was done.

The jugals are not complete in the holotype of *Libonectes* but they apparently formed most of the lower edge of the orbits as in *Styxosaurus* and *Thalassomedon* (Carpenter, in preparation). In *Dolichorhynchops*, they formed only a small portion of the rear edge (Figure 6A). The jugals in *Libonectes* and *Dolichorhynchops* probably excluded contact between the postorbital and squamosal as in *Styxosaurus* and *Thalassomedon* (Carpenter, in preparation; Figure 6). The sutural contact of the jugal with the squamosal is unknown in *Libonectes*, but in *Styxosaurus* it almost divides the temporal bar obliquely in half (Williston, 1903).

The quadratojugal is not present in either taxon, contrary to Welles (1949). Its presence in other plesiosaurs as advocated by Welles (1943, 1949, 1952, 1962) is unsubstantiated. As early as 1881, the existence of the quadratojugal was questioned by Sollas, but Andrews (1896) suggested that it was fused to the

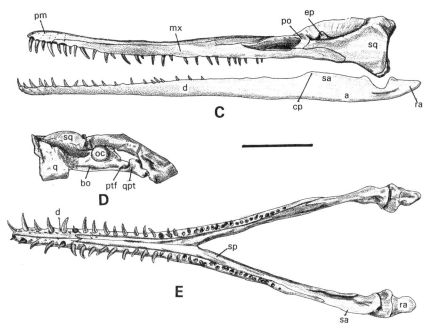

Figure 3 (continued). Skull of *Dolichorynchops osborni*, FHSM VP404. **C)** Left lateral view. **D)** Posterior aspect. **E)** Dorsal view of mandible. Abbreviations in Appendix 2. Scale bar = 10 cm.

Table 1. Cranial comparative measurements (in cm) of *Dolichorhynchops* and *Libonectes* (see Figure 4 for location of measurements).

	a	b	c	d	e
Libonectes (SMUSMP 69120)	18	45.9	46.6	-	7
Dolichorhynchops (FHSM VP404)	32	51.3	-	56.2	4

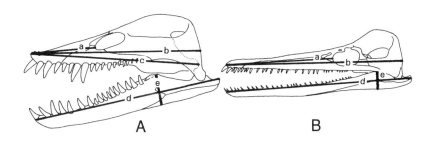

Figure 4. Locations for measurements in Table 1. **A)** *Libonectes*. **B)** *Dolichorhynchops*.

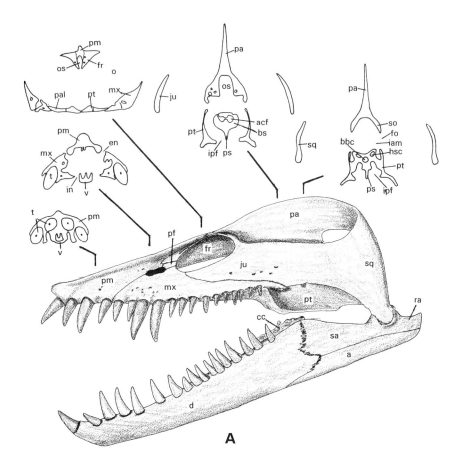

Figure 5. Skull of *Libonectes morgani*, SMUMP 69120. **A)** Lateral reconstruction with cross sections from CAT scans. Abbreviations in Appendix 2.

quadrate. More recently, Storrs (1991) considered its loss linked to the elimination of the diapsid lower temporal bar, a character uniting the Sauropterygia. This is the view supported here.

The squamosal is a large element in both *Libonectes* and *Dolichorhynchops* (Figures 5A and 6A), where it forms the posterior wall and part of the lateral wall of the temporal fenestra. The squamosal extends posteriorly beyond the occipital condyle in *Libonectes*; thus, the temporal fenestra is elongated compared to that in *Dolichorhynchops*.

The quadrate of *Libonectes* and *Dolichorhynchops* is short and wide (Figures 2B and 3D). It articulates laterally and dorsally to the squamosal, and medially to the quadrate process of the pterygoid. In *Dolichorhynchops* the points of

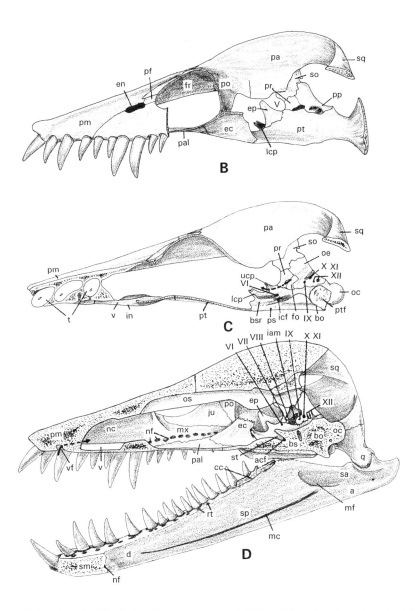

Figure 5 (continued). Skull of *Libonectes morgani*, SMUMP 69120. **B**) With postorbital section of skull removed. **C**) With side of skull removed exposing braincase. **D**) Sagittal section showing interior of braincase and internal view of right side of skull. Abbreviations in Appendix 2.

articulation with the squamosal and pterygoid processes are shallow fossae, or sockets in the quadrate.

The highest point on the skulls of *Libonectes* and *Dolichorhynchops* is the

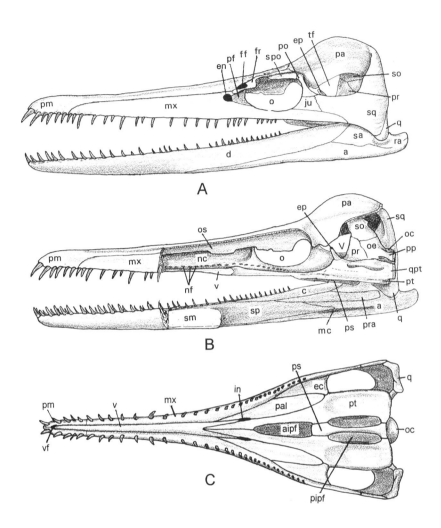

Figure 6. Composite reconstructed skull of *Dolichorhynchops osborni*. **A)** Left lateral view. **B)** With part of skull removed showing braincase. **C)** Palatal view. Abbreviations in Appendix 2.

parietal crest (Figures 5A and 6A), which rises more sharply relative to the face in *Dolichorhynchops* than in *Libonectes*; the crest is apparently not as tall in *Brancasaurus*. Laterally, the parietals expand dorsally over the braincase and ventrolaterally down its sides (Figures 5B and 6B). In *Libonectes*, the parietals contact the sides of the supraoccipital, but they are separated a little in *Dolichorhynchops*.

On the palate, the vomers partially separate the left and right premaxillaries and maxillaries in both taxa (Figures 2D and 3B). At the anterior end of the vomer is the vomeronasal fenestra, which opens dorsally into the nasal chamber. In life, a cartilaginous pocket in the anterior portion of the nasal chamber may have housed the Jacobson or vomeronasal organ. The presence of the vomeronasal organ for underwater olfaction in plesiosaurs was suspected by Cruickshank et al. (1991). The vomeronasal organ is present primitively in many tetrapods, including the most primitive plesiosaur, *Pistosaurus* (Meyer, 1855), but apparently was lost in the earliest plesiosaurs because it is absent in Jurassic plesiosaurs such as *Muraenosaurus, Pelonusteus, Liopleurodon ferox,* and *P. brachyspondylus* (Taylor and Cruickshank, 1993; Linder, 1913; Andrews, 1910, 1913; also *Plesiosaurus dolichodeirus,* Storrs, personal communications). Its "reappearance" in the advanced elasmosaurs and polycotylids is believed to be synapomorphic, as is discussed below.

Dorsally on the vomer of *Libonectes*, a thin sheet of bone extends from the premaxilla, dividing the anteriormost portion of the nasal chamber (Figure 5D). In life, this sheet of bone would have separated the two vomeronasal organs. Where the vomer forms the medial edge of the internal nares, three low ridges are present on the dorsal surface. Posteriorly, the vomers contact the palatines and anterior tips of the pterygoids.

The palatines are large, subrectangular sheets of bone in *Libonectes*, but triangular sheets in *Dolichorhynchops* (Figures 2D and 3B). They barely contact the internal nares in *Libonectes*, but form the lateral margins in *Dolichorhynchops*. In *Libonectes* a small fenestra, believed to be the remnant of the suborbital fenestra, is present at the conjuncture of the palatines, ectopterygoids, and pterygoids. The size of the two fenestra is not the same, indicating that the structure is variable in elasmosaurs. No such fenestra is present in *Dolichorhynchops* (Carpenter, in preparation).

Posterior to the palatines are the ectopterygoids (Figure 2C). These are somewhat C-shaped in *Libonectes*, but triangular in *Dolichorhynchops* (missing in the illustrated specimen, but known from the holotype KUVP 1300 and UCM 35059). In *Libonectes*, a small, triangular scar extends onto the pterygoids adjacent to the subtemporal fenestra, and it may have been formed by pterygoideus muscle. Dorsally, the ectopterygoids brace the postorbital-temporal fenestra bar, transferring stresses generated against the palate upward around the sides of the skull.

Medial to the ectopterygoids, long, slender pterygoids are present in both taxa (Figures 2D and 6B). Posteriorly, the pterygoids are plate-like structures that cover most of the posterior ventral part of the skull. They are partially separated from one another by the interpterygoid fenestra, which in turn is divided by a keel on the parasphenoid. Below the basioccipital, the pterygoids meet one another posterior to the parasphenoid keel. The pterygoids taper anteriorly into two wedges in *Libonectes*. In *Dolichorhynchops*, the pterygoids taper into two slender bars on e?

side of the anterior interpterygoid fenestra. Anterior to this fenestra they expand and meet, forming the medial edge of the internal nares.

In *Libonectes*, a vertical process of the pterygoids extends up the sides of the basisphenoid and basipterygoid, but is separated from them by a gap for the carotid artery and capitis vein. This process possibly prevented the temporalis muscle from impeding blood flow when the jaw closed. Posteriorly in *Dolichorhynchops* and *Libonectes*, this vertical process becomes the pterygoid process, articulating with the quadrate by a peg and socket (shallow fossa). The peg, developed in the pterygoid, is elongated dorsoventrally.

The epipterygoid in *Libonectes* extends dorsally from the anterior dorsal edge of the vertical pterygoid process (Figure 5B). It is triangular and projects somewhat anteriorly, leaving an ovoid foramen posterodorsally for the trigemial (V) cranial nerve. In *Dolichorhynchops*, the epipterygoid is a laterally compressed rod of bone that is situated lateral to the sella turcica (Figure 6B).

Anterior to the median keel that divides the interpterygoid fenestra, the parasphenoid expands to form a wedge between the pterygoids in *Libonectes*, and the posterior margin of the anterior interpterygoid fenestra in *Dolichorhynchops* (Figures 2D and 3B). Dorsally, the parasphenoid expands along the sides of the basioccipital and basisphenoid, forming the lower rim of a groove for the internal carotid.

The basioccipital forms the ventroposterior portion of the braincase in both taxa. The occipital condyle is spherical in *Dolichorhynchops*, and somewhat heart shaped in *Libonectes* (i.e., it is wider near the top than at the bottom; compare Figures 2B and 3D). A shallow groove rings the occipital condyle in *Libonectes*, forming a very small neck. The exoccipital-opisthotic in both taxa does not form part of the occipital condyle. This feature is shared with *Plesiosaurus dolichodeirus* (Owen, 1865:Plate 2), "*Plesiosaurus*" *macrocephalus* (Andrews, 1896), and the cryptoclidids *Muraenosaurus* and *Tricleidus* (Brown, 1981; Andrews, 1910). They do, however, contribute to the condyle in the pliosaurids (Andrews, 1913) and just barely in the cryptoclidid *Cryptoclidus* (Brown, 1981). Ventrally, the basitubera of *Libonectes* and *Dolichorhynchops* have been modified to form a facet to brace the pterygoid plates dorsally. A small anteroposteriorly ovoid fossa is present on the floor of the foramen magnum in *Libonectes*.

The basisphenoid is fused to the anterior of the basioccipital in both taxa. In *Libonectes*, however, fusion is not complete, leaving a cavity seen on the CAT scans (Figure 5A). The basisphenoid in both taxa is pierced laterally by the internal carotid foramen, which opens into the back of the sella turcica housing the pituitary (Figure 5D). The palatine artery and palatine nerve probably continued along the lateral surface of the basisphenoid beneath the pterygoid process.

The sella turcica opens anteroventrally to the dorsal surface of the basicranium.

The abducens (VI) foramen pierces the dorsal surface of the basisphenoid (floor of the braincase). In *Libonectes*, the lower cylindrical process is anterior and lateral to the sella turcica. It projects anterodorsally, ending in a small, grooved process that is aligned with a fenestra for the lateral head vein at the base of the upper cylindrical process. The fenestra opens dorsally in the prootic, just anterior to the fenestra ovalis. Most likely, the lateral head vein was supported by the grooved process of the lower cylindrical process. No such processes are known for *Dolichorhynchops*.

The braincase is a tube open at both ends in both taxa (Figure 7C and 7D). It is composed of the fused exoccipital-opisthotic posteriorly, the prootic anteriorly, and the supraoccipital dorsally (Figures 5C and 7A and B). As noted by Williston (1903), there is no trace of an alisphenoid or orbitosphenoid. The foramen magnum is taller than wide and slightly constricted at the supraoccipital-exoccipital-opisthotic suture (Figures 2B and 6B). The lateral surface of the braincase is pierced by the keyhole-shaped fenestra ovalis in *Libonectes* but this fenestra is partially blocked laterally by the vertical process of the pterygoid (compare Figure 5B and 5C). No stapes was found in *Libonectes* despite the exquisite preservation, including at least one hyoid. It is probable that none was present, in marked contrast to *Brachauchenius* (Williston, 1907) and *Rhomaleosaurus* (Taylor, 1992b). No stapes

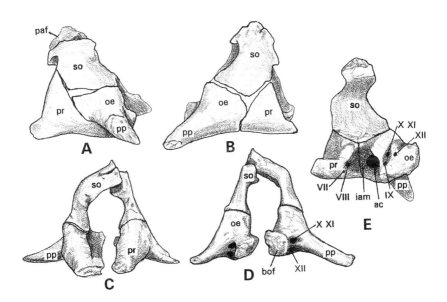

Figure 7. Braincase of *Dolichorhynchops osborni*, FHSM VP404. **A)** Left lateral view. **B)** Right lateral view. **C)** Anterior view. **D)** Posterior view. **E)** Medial view of right side. Abbreviations in Appendix 2.

was found in *Libonectes* despite the exquisite preservation, including at least one hyoid. It is probable that none was present, in marked contrast to *Brachauchenius* (Williston, 1907) and *Rhomaleosaurus* (Taylor, 1992b). No stapes is known for any specimen of *Dolichorhynchops* where the fenestra ovalis is closed (Figure 7A and B). Medially, the internal auditory meatus is a chamber or cavity developed primarily in the prootic and exoccipital-opisthotic in both taxa (Figures 5 and 7).

The exoccipital-opisthotic in *Libonectes* and *Dolichorhynchops* is pierced at the base of the paroccipital process by foramina for the vagus (X), accessory (XI), and hypoglossal (XII) nerves. These foramina open into the braincase posterior to the internal auditory meatus. Another foramen, for the glossopharyngeal (IX) nerve, pierces the exoccipital-opisthotic immediately ventral to the anterior edge of the paroccipital process. The posterior semicircular canal is contained mostly within the exoccipital-opisthotic. It opens into the internal auditory meatus in *Libonectes*, and indirectly through the acoustic chamber in *Dolichorhynchops*.

The paroccipital process extends laterally and braces the dorsal part of the quadrate process of the pterygoid in both *Libonectes* and *Dolichorhynchops* (Figure 2B). In life, the internal carotid and lateral head vein passed anteriorly, and the vagus and glossopharyngeal nerves passed posteriorly through the gap between the paroccipital process and the quadrate process of the pterygoid.

The prootic occurs above the sella turcica and anterior to the basisphenoid in *Libonectes* and *Dolichorhynchops*. In lateral view, it is a rectangular element in *Libonectes* (Figure 5D) and a triangular element in *Dolichorhynchops* (Figure 7). A foramen believed to be for the lateral head vein opens just anterior to the fenestra ovalis (see above). Medially, the facial (VII) foramen is present just anterior to the internal auditory meatus (Figure 5D). The anterior semicircular canal occupies the internal part of the prootic. In life, the trigeminal (V) nerve exited from the front of the prootic.

The supraoccipital roofs the braincase (Figures 5C and 7A and B). A short, rough process extends posteriorly over the foramen magnum for the nuchal ligament. Dorsally, the supraoccipital contacts the ventral surface of the parietal. This contact is more extensive in *Libonectes* than in *Dolichorhynchops*.

The only endocranial cast of a plesiosaur described is a partial one of the elasmosaur *Brancasaurus brancai* (Hopson, 1979, figured by Edinger, 1928;Figure 8A). Owing to the uncrushed state of the *Libonectes morgani* braincase, a latex peel was taken (Figure 8B). The endocast is short and deep, with a slight flexure. The olfactory sulcus is not ossified ventrally, so the maximum thickness of the olfactory tract is unknown. It angles downward posterior to the level of the orbits. Dorsally, an enlargement beneath the parietal may house the remnants of the pineal as well as dura to suspend the brain. The cerebrum is large, although its anterior limit is not known because the front of the prootic is open. The pituitary is small and is

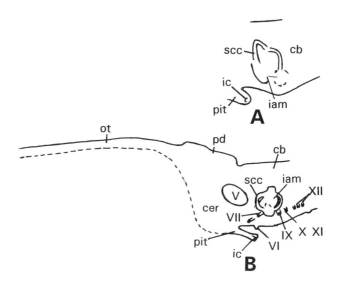

Figure 8. Comparison of the endocasts of **A)** *Brancasaurus brancai* (from Edinger, 1928) and **B)** *Libonectes morgani*, SMUMP 69120. Abbreviations in Appendix 2.

slightly taller than wide. The two internal carotids are present at the end of the pituitary cast.

SIGNIFICANCE OF THE SIMILARITIES BETWEEN *LIBONECTES* AND *DOLICHORHYNCHOPS*

Although the differences between the elasmosaur *Libonectes* and the polycotylid *Dolichorhynchops* seem most obvious and important, they are united by several synapomorphies that separate them from their Jurassic ancestors. In the past, phyletic relationships of the plesiosaurs were based primarily on postcranial features because skulls were typically missing or damaged (e.g., Welles, 1962; White, 1940). However, the postcrania are shaped by ecological needs, making their indiscriminate use suspect. For example, enlarged, plate-like pectoral and pelvic girdles have independently developed in aquatic reptiles, such as tangasaurid eosuchians (Currie, 1981). Others, such as the thalattosaurs, tanystrophids, and dolichosaurs (Carroll, 1988), have elongate necks, and still others have reduced metapodials relative to the propodial (e.g., metriorhynchid crocodiles, mosasaurs, and ichthyosaurs; Carroll, 1988). The result is that the postcrania within each group, including plesiosaurs, are conservative throughout their evolution.

The skull is thought to be less affected by ecological parameters, other than the elongation of the snout that has occurred independently among several reptile groups (e.g., mesosaurs, thalattosaurs, champsosaurs, pleurosaurs, aigialosaurs, mosasaurs, ichthyosaurs, phtyosaurs, and various crocodiles; Carroll, 1988). The utility of cranial features was demonstrated by Russell (1967), who used cranial features in his discussion of mosasaur-aigialosaur relations. Carroll (1988) also used cranial features to link the aquatic pleurosaurs with the sphenodontids.

In *Libonectes* and *Dolichorhynchops*, the synapomorphic cranial features include the presence of a vomeronasal fenestra, expansion of the pterygoids into plates beneath the braincase, and loss of both the pineal foramen and stapes.

An oval opening at the suture between the premaxilla and vomer in *Pistosaurus* is probably the vomeronasal fenestra (Meyer, 1855:Plate 21, Figure 3). It is apparently absent in Jurassic plesiosaurs, although this region of the palate is not documented in all taxa. The vomeronasal fenestra apparently reappeared in the Lower Cretaceous elasmosaurid *Brancasaurus* (Wegner, 1914:Figure 1). This reappearance may be due to uncovering of the vomeronasal organ in one lineage of plesiosaurs that gave rise to elasmosaurids and polycotylids.

The expansion of the pterygoids beneath the braincase has apparently occurred independently twice and these may be designated as types A and B. The type A pterygoids are folded under to meet on the midline below the basicranium. An incipient version is seen in *Pistosaurus* (Figure 9A); it is seen in the pliosaurs *Liopleurodon* (Figure 9B), *Peloneustes*, *Pliosaurus*, *Simolestes*, and *Brachauchenius* (Andrews, 1913; Linder, 1913; Williston, 1903). The type B pterygoids form a horizontal plate beneath the basicranium (Figure 9C and D) occurring in all elasamosaurids (e.g., *Libonectes*) and polycotylids (e.g., *Dolichorhynchops*). Its presence in the pliosaur *Rhomaleosaurus* (White, 1940) and the plesiosauroid *Plesiosaurus hawkensii* (Owen, 1865) suggests a possible closer relationship than previously recognized. Incipient type B pterygoids are seen in *Plesiosaurus dolichodeirus* and *Plesiosaurus macrocephalus* (Andrews, 1896).

Another feature uniting *Libonectes* and *Dolichorhynchops* is the closure of the pineal foramen. The presence of this foramen is considered plesiomorphic for reptiles, including sauropterygians, so its absence is a synapomorphy for elasmosaurids and polycotylids.

The stapes are apparently absent in both *Libonectes* and *Dolichorhynchops* (Storrs, personal communication, is skeptical), which, if true, may be a synapomorphy. Certainly, the presence of one is plesiomorphic for vertebrates.

CONCLUSIONS

Traditionally, the short-necked polycotylids were thought to have descended

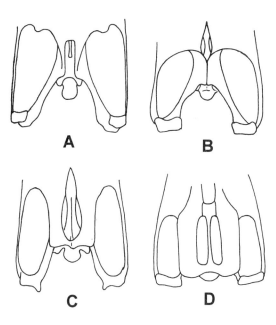

Figure 9. Comparison of the two pterygoid types. **A**) Incipient type A, *Pistosaurus* (after Meyer, 1855). **B**) Type A, *Liopleurodon* (after Andrews, 1913). **C**) Type B, *Libonectes*. **D**) Type B, *Dolichorhynchops*.

from a short-necked pliosaurid, while the elasmosaurids were thought to have descended from a long-necked cryptoclidid (Figure 10; Welles, 1943:Figure 37). While it is true that there are differences that separate *Libonectes* and *Dolichorhynchops*, there are a few synapomorphies in the skull which may indicate a closer relationship than hitherto realized. On the basis of these synapomorphies, I venture as a hypothesis that *Libonectes* and *Dolichorhynchops* share a common ancestor, and that both are more closely related to the plesiosaurids than either is to the pliosaurids. Williston (1906:226, 1907:485) independently reached a similar conclusion, stating that the short neck was not a primitive character and that it had been acquired independently in more than one phylum. This conclusion is in marked contrast to the more traditional approach (e.g., Brown, 1981) which places *Dolichorhynchops* with the pliosaurids and *Libonectes* with the elasmosaurids.

One major objection to this hypothesis is that it requires a reduction in the number of cervical vertebrae independent from that in pliosaurs. Brown (1981) estimated that the primitive number of cervicals in plesiosauroids is 28-32, whereas *Dolichorhynchops* has 19. However, considering that *Pistosaurus* probably had about 24 cervicals (Sues, 1991), Brown's estimation may be too high. In addition,

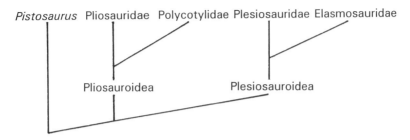
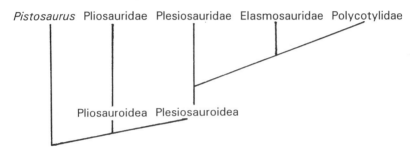

Figure 10. Contrasting phylogenetic schemes (see text for details).

Polycotylus has 26 cervicals (Williston, 1906), about the same as the primitive elasmosaurid *Tricleidus* (Brown, 1981). A reduction in the number of cervical vertebrae independent of the pliosaurs, therefore, is not unreasonable, and if true, then it is unlikely that Bakker's (1993) scheme of deriving *Libonectes* and *Dolichorhynchops* from an Upper Jurassic pliosaur is possible. Bakker based his conclusion on the sharing of a specialized palate (i.e., covering of the basicranium by the pterygoids) by Cretaceous pliosaurids, elasmosaurids, and the Jurassic *Pliosaurus*. However, as shown above, this condition in *Libonectes* and *Dolichorhynchops* is more similar to that in Jurassic plesiosaurids than in Jurassic pliosaurids.

SUMMARY

The skulls of "*Elasmosaurus*" *morgani* and *Dolichorhynchops osborni* are compared as representatives of the Cretaceous plesiosaur families Elasmosauridae and Polycotylidae, respectively. Cranial features and the atlas-axis complex appear to be more stable evolutionarily than postcranial features. Thus, similarities indicate that the short-necked Cretaceous polycotylids are the sister-group to long-necked elasmosaurids. This implies that the short-necked polycotylids of the Cretaceous are not descended from the short-necked pliosaurs of the Jurassic. The short neck has appeared independently at least twice in the Plesiosauria and the term *pliosaur* to refer to any short-necked plesiosaur should be abandoned to avoid any phyletic implications. Differences between "*Elasmosaurus morgani*" and *Elasmosaurus platyurus* demonstrate that the two species belong to different genera and a new name is proposed for "*E.*" *morgani.*

ACKNOWLEDGMENTS

The generic name *Libonectes* was suggested by Ben Cressler and I am pleased to acknowledge his expertise in etymologies of taxonomic names. CAT scans and X-rays of *Libonectes* were made possible by Janeth Alband, RT, and Robert Meals, DO (Chairperson), of the Department of Radiology, Osteopathic Medical Center of Philadelphia, and Luther Brady, MD (Chairperson), Joanne Bethold, RTT, and Mary Lansu, BS, of the Radiation Therapy Department, Hahnemann University, Philadelphia. Many individuals have been gracious in allowing me to study the plesiosaurs in their care: Phil Bjork (South Dakota School of Mines), Richard Zakrzewski (Sternberg Memorial Museum, Fort Hays State University), Charlotte Holten and Gene Gaffney (American Museum of Natural History), Robert Purdy and Nicholas Hotton (National Museum of Natural History), Larry Martin (University of Kansas Museum of Natural History), and Peter Robinson (University of Colorado Museum). Review comments by Elizabeth Nicholls and Glenn Storrs are appreciated.

REFERENCES

Almy, K. 1987. Thof's dragon and the letters of Capt. Theophilus H. Turner, M.D., U.S. Army. *Kansas History* 10:170-200.

Andrews, C. 1896. On the structure of the plesiosaurian skull. *Quarterly Journal of the Geological Society* 52:246-253.

Andrews, C. 1910. A descriptive catalogue of the marine reptiles of the Oxford Clay. Part I. *British Museum (Natural History)*, London, 205 pp.

Andrews, C. 1913. A descriptive catalogue of the marine reptiles of the Oxford Clay. Part II. *British Museum (Natural History)*, London, 206 pp.

Bakker, R. 1993. Plesiosaur extinction cycles---events that mark the beginning, middle and end of the Cretaceous. IN W. G. E. Caldwell and E. G. Kauffman (Eds.), *Evolution of the Western Interior Basin*, pp. 641-644. *Geological Association of Canada, Special*

Paper 39.
Brown, D. 1981. The English Upper Jurassic Plesiosauroidea (Reptilia) and a review of the phylogeny and classification of the Plesiosauria. *British Museum (Natural History)*, Geology 35:253-347.
Carpenter, K. 1989. *Dolichorhynchops* ≠ *Trinacromerum*. *Journal of Vertebrate Paleontology* 9, supplement 3:15A.
Carroll, R. 1988. *Vertebrate Paleontology and Evolution*. W. H. Freeman and Co., New York, 698 pp.
Cope, E. D. 1869. Extinct Batrachia, Reptilia and Aves of North America. *Transactions of the American Philosophical Society* 14:1-252.
Cragin, F. 1888. Preliminary description of a new or little known saurian from the Benton of Kansas. *American Geologist* 2:404-407.
Cruickshank, A., P. Small, and M. Taylor. 1991. Dorsal nostrils and hydrodynamically driven underwater olfaction in plesiosaurs. *Nature* 352:62-64.
Currie, P. 1981. *Hovasaurus boulei*, an aquatic eosuchian from the Upper Permian of Madagascar. *Palaeontographica Africana* 24:99-168.
Edinger, T. 1928. Über einige fossile Gehirne. *Paleontologische Zeitschrift* 9:379-402.
Fitch, H. 1981. Sexual size differences in reptiles. *University of Kansas Museum of Natural History, Miscellaneous Publication* 70:1-72.
Hampe, O. 1992. Ein großwüchsiger Pliosauride (Reptilia: Plesiosauria) aus der Unterkreide (oberes Aptium) von Kolumbien. *Courier Forschungsinstitut Senkenberg* 145:1-32.
Hopson, J. 1979. Paleoneurology. IN C. Gans (Ed.), *Biology of the Reptilia*, Volume 9A, pp. 39-146. Academic Press, New York.
Kuhn, O. 1964. *Fossilium Catalogus. I: Animalia: Sauropterygia*. Ysel Press, Nederland, 72 pp.
Linder, H. 1913. Beiträge zur Kenntniss der Plesiosaurier-Gattungen *Peloneustes* und *Pliosaurus*; nebst Anhang: Über die beiden ersten Halswirbel der Plesiosaurier. *Geologische und Palaeontologische Abhandlungen*, (N. S.) 11:339-409.
Meyer, H. 1855. *Zur Fauna der Vorwelt. Die Saurier der Muschelkalkes mit Rücksicht auf sie Saurier aus Buntem Sandstein un Keuper*. Heinrich Keller, Frankfurt am Main, 167 pp.
Owen, R. 1865. A monograph on the fossil Reptilia of the Liassic Formations. Part I, Sauropterygia. *Palaeontographical Society Monograph* 17:1-40.
Persson, O. 1963. A revision of the classification of the Plesiosauria with a synopsis of the stratigraphical and geographical distribution of the group. *Lunds Universitets Arsskrift* 59:1-59.
Riggs, E. 1944. A new polycotylid plesiosaur. *University of Kansas Science Bulletin* 30:77-87.
Russell, D. 1967. Systematics and morphology of American mosasaurs. *Peabody Museum of Natural History Bulletin* 23:1-230.
Russell, L. 1935. A plesiosaur from the Upper Cretaceous of Manitoba. *Journal of Paleontology* 9:385-389.
Sollas, W. 1881. On a new species of *Plesiosaurus* (*P. conybeari*) from the Lower Lias of

Charmouth; with observations on *P. megacephalus*, Stutchbury, and *P. brachycephalus*, Owen. *Quarterly Journal of the Geological Society of London* 7:440-480.

Storrs, G. 1991. Anatomy and relationships of *Corosaurus alcovensis* (Diapsida: Sauropterygia) and the Triassic Alcova Limestone of Wyoming. *Peabody Museum of Natural History Bulletin* 44:1-151.

Sues, H. 1991. Postcranial skeleton of *Pistosaurus* and interrelationships of the Sauropterygia (Diapsida). *Zoological Journal of the Linnean Society* 90:109-131.

Tarlo, B. 1960. A review of the Upper Jurassic pliosaurs. *British Museum (Natural History)*, Geology 4:145-189.

Taylor, M. 1992a. Taxonomy and taphonomy of *Rhomaleosaurus zetlandicus* (Plesiosauria, Reptilia) from the Toarcian (Lower Jurassic) of the Yorkshire coast. *Yorkshire Geological Society Proceedings* 49:49-55.

Taylor, M. 1992b. Functional anatomy of the head of the large aquatic predator *Rhomaleosaurus zetlandicus* (Plesiosauria: Reptilia) from the Toarcian (Lower Jurassic) of Yorkshire, England. Philosophical Transactions of the Royal Society of London, B 335:247-280.

Taylor, S., and A. Cruickshank. 1993. Cranial anatomy and functional morphology of *Pliosaurus brachyspondylus* (Reptilia: Plesiosauria) from the Upper Jurassic of Westbury, Whitshire. *Philosophical Transactions of the Royal Society of London* B 341:399-418.

Thurmond, J. 1968. A new polycotylid plesiosaur from the Lake Waco Formation (Cenomanian) of Texas. *Journal of Paleontology* 42:1289-1296.

Wegner, T. 1914. *Brancasaurus brancai* n.g. n. sp., ein Elasmosauridae aus dem Wealden Westfalens. IN F. Schoendorf (Ed.), *Branca Festschrift*, pp. 235-305. Berlin..

Welles, S. P. 1943. Elasmosaurid plesiosaurs with a description of new material from California and Colorado. *University of California Memoirs* 13:125-254.

Welles, S. P. 1949. A new elasmosaur from the Eagle Ford Shale of Texas. *Fondren Science Series, Southern Methodist University* 1:1-28.

Welles, S. P. 1952. A review of the North American Cretaceous elasmosaurs. *University of California Publications in Geological Sciences* 29:47-144.

Welles, S. P. 1962. A new species of elasmosaur from the Aptian of Columbia, and a review of the Cretaceous plesiosaurs. *University of California Publications in Geological Sciences* 46:1-96.

White, T. 1940. Holotype of *Plesiosaurus longirostris* Blake and classification of the plesiosaurs. *Journal of Paleontology* 14:451-467.

Wiffen, J. and W. Moisley. 1986. Late Cretaceous reptiles (families Elasmosauridae and Pliosauridae) from the Mangahouanga Stream, North Island, New Zealand. *New Zealand Journal of Geology and Geophysics* 29:205-252.

Williston, S. W. 1903. North American plesiosaurs. Part I. *Field Columbian Museum Publication* 73, Geological Series 2:1-77.

Williston, S. W. 1906. North American plesiosaurs: *Elasmosaurus, Cimoliasaurus*, and *Polycotylus*. *American Journal of Science, 4th Series* 21:221-236.

Williston, S. W. 1907. The skull of *Brachauchenius*, with special observations on the relationships of the plesiosaurs. *United States National Museum Proceedings* 32:477-489.

Williston, S. W. 1908. North American plesiosaurs: *Trinacromerum. Journal of Geology* 16:715-735.

APPENDIX 1

Diagnosis of *Libonectes morgani* and *Dolichorhynchops osborni*:

Class REPTILIA
Subclass SAUROPTERYGIA Owen, 1860
Order PLESIOSAURIA Blainville, 1835
Superfamily PLESIOSAUROIDEA
Family ELASMOSAURIDAE

Genus *LIBONECTES* n. g.

Etymology: *Libonectes* for "southwest diver," Greek masculine gender from *libo* "southwest wind" + *nektes* "swimmer," alluding to the American Southwest, where the holotype was found.

Diagnosis: Cretaceous elasmosaurid with 62 (?) cervical vertebrae as in *Thalassomedon* and *Styxosaurus*, compared to 37 in *Brancasaurus*, 46 in *Morenosaurus*, 56 in *Alzadasaurus*, 57 in *Aphrosaurus*, 60 in *Hydrotherosaurus*, 63 (?) in *Hydralmosaurus*, and 74 in *Elasmosaurus* (*sensu stricto*). Preorbital length/skull length (of limited taxonomic utility) 38.6, whereas ratio is 42.3 in *Alzadasaurus*, 39.1 in *Thalassomedon*, 38.6-41.6 in *Styxosaurus*, and 34.3 in *Tuarangisaurus*. Atlas-axis centrum short and deep like in *Tuarangisaurus* (Figure 1), whereas it is long and low in *Elasmosaurus platyurus*; neural spine low, with postzygapophyses of axis extending well beyond posterior face of centrum, whereas neural spine tall and postzygapophyses of axis do not extend beyond centrum in *Thalassomedon hanningtoni*; *Tuarangisaurus* atlas-axis intermediate between *Libonectes* and *Thalassomedon* in shape and size of neural spine.

LIBONECTES MORGANI (Welles, 1949)

Elasmosaurus morgani Welles, 1949

Holotype: SMUSMP 69120: skull, most of the cervicals, gastralia, and gastroliths. The pectoral girdle and forelimb were apparently discarded long ago

(Storrs, 1991).

Type locality: Britton Formation (Turonian), Eagle Ford Group, near Cedar Hill, Dallas County, Texas.

Diagnosis: As for the genus.

Family POLYCOTYLIDAE

DOLICHORHYNCHOPS OSBORNI Williston, 1903

Dolichorhynchops osborni Williston, 1903
Trinacromerum osborni Williston, 1908
Trinacromerum osborni Russell, 1935
Trinacromerum osborni Riggs, 1944
Dolichorhynchops osborni Persson, 1963
Dolichorhynchops osborni Kuhn, 1964
Dolichorhynchops osborni Thurmond, 1968
Dolichorhynchops osborni Hampe, 1992

Holotype: KUVP 1300 nearly complete skeleton (Williston, 1903).

Type locality: *Hesperornis* biozone, Smoky Hill Chalk Member, Niobrara Formation, Logan County, Kansas.

Referred specimens: FHSM VP404, a nearly complete skeleton from the Smoky Hill Chalk, near Russell Springs, Logan County, Kansas. MCZ 1064, partial skeleton and skull of a young individual from the Smoky Hill Chalk, in Logan County, Kansas. UCM 35059, a partial skeleton and skull from the Sharon Springs Member of the Pierre Shale, Mule Creek drainage, Niobrara County, Wyoming. AMNH 5834, skull from the Sharon Springs Member of the Pierre Shale, Mule Creek drainage, Niobrara County, Wyoming. UNSM 50133, skull from the Sharon Springs Member of the Pierre Shale, Hat Creek drainage, Fall River County, South Dakota. KUVP 40001, skull from the Sharon Springs Member of the Pierre Shale, Hat Creek drainage, Fall River County, South Dakota.

Diagnosis: More derived than *Trinacromerum* in having vertical suspensorium, supratemporal fenestra as wide as long, shorter pterygoids, interpterygoid fenestra, parasphenoid, jugals, horizontal process of the squamosal, and parietals; teeth covered with finer striae and more slender than in *Trinacromerum* and *Polycotylus*; 19 cervicals, 26 in *Polycotylus* and 20 in *Trinacromerum*; anterior chevrons borne almost exclusively on single caudal, shared equally by adjacent caudals in *Trinacromerum*.

APPENDIX 2

Anatomical Abbreviations: **a** = angular, **acf** = anterior carotid foramen, **ac** = acoustic chamber, **aipf** = anterior interpterygoid fenestra, **ar** = articular, **bbc** = basioccipital-basisphenoid cavity, **bo** = basioccipital, **bof** = basioccipital facet of opisthotic-exoccipital, **bs** = basisphenoid, **bsr** = basisphenoid rostrum, **c** = coronoid, **cb** = cerebellum, **cc** = coronoid cartilage (ossified), **cer** = cerebrum, **cp** = coronoid process, **d** = dentary, **ec** = ectopterygoid, **en** = external nares, **ep** = epipterygoid, **ff** = frontal fenestra (supraorbital fenestra), **fm** = foramen magnum, **fo** = fenestra ovalis, **fr** = frontal, **hsc** = horizontal semi-circular canal, **iam** = internal auditory meatus, **ic** = internal carotid, **icf** = internal carotid foramen, **in** = internal nares, **ipf** = interpterygoid fenestra, **ju** = jugal, **lcp** = lower cylindrical process, **mc** = Meckelian canal, **mf** = mandible fossa, **mx** = maxilla, **nc** = nasal chamber, **nf** = nutrient foramen, **o** = orbit, **oc** = occipital condyle, **oe** = opisthotic-exoccipital, **os** = olfactory sulcus, **ot** = olfactory tract, **pa** = parietal, **paf** = parietal facet of supraoccipital, **pal** = palatine, **pd** = pineal and dura, **pf** = prefrontal, **pipf** = posterior interpterygoid fenestra, **pit** = pituitary, **pm** = premaxilla, **po** = postorbital, **pp** = paroccipital process, **pr** = prootic, **pra** = prearticular, **ps** = parasphenoid, **pt** = pterygoid, **ptf** = pterygoid facet of basioccipital, **ptp** = pterygoid process, **q** = quadrate, **qpt** = quadrate ramus of pterygoid, **ra** = retroarticular, **rt** = replacement tooth, **sa** = surangular, **scc** = semicircular canal, **sm** = symphysis, **so** = supraoccipital, **sof** = suborbital fenestra, **sp** = splenial, **spo** = supraorbital, **sq** = squamosal, **st** = sella turcica, **stf** = subtemporal fenestra, **t** = tooth (in cross section), **tf** = temporal fenestra, **ucp** = upper cylindrical process, **v** = vomer, **vf** = vomeronasal fenestra, **V** = trigeminal foramen, **VI** = abducens foramen, **VII** = facial foramen, **VIII** = acoustic foramen, **IX** = glossopharyngeal foramen, **X** = vagus foramen, **XI** = accessory foramen, **XII** = hypoglossal foramen.

PART III
Testudines

Part III: Testudines

INTRODUCTION

ELIZABETH L. NICHOLLS

Turtles are one of the great success stories of marine reptiles. They are the only living reptiles fully adapted to the marine environment, returning to shore only to lay their eggs. The five living genera are found throughout the tropical and subtropical oceans of the world. Both the leatherback, *Dermochelys*, and the green turtle, *Chelonia*, extend well up into temporal waters (Ernst and Barbour, 1989; Goff and Lein, 1988), and *Chelonia* and *Caretta* are known to hibernate in cool waters (Mrosovsky, 1980).

Morphologically, sea turtles are a rather uniform group, and the differences between the living species appear to be minor. Behaviorally, however, modern sea turtles are more diverse than their rather uniform morphology suggests. They utilize a wide variety of food sources. The green turtle, *Chelonia mydas*, is a herbivore feeding mainly on sea grasses, although the Hawaiian and Galápagos populations are known to feed on algae (Hendrickson, 1980). Other genera are more omnivorous, feeding mainly on invertebrates, such as crustaceans (*Lepidochelys*), molluscs (*Caretta*), and jellyfish (*Dermochelys*) (Ernst and Barbour, 1989; Hendrickson, 1980). Unlike some of their freshwater cousins, they are not known to be piscivorous.

The migrations of the sea turtle are legendary, but most forms are neritic and remain coastally oriented. Only *Dermochelys* has become truely pelagic. It inhabits open seas and can dive to depths of up to 1000 m (McAuliffe, 1995).

Turtles were one of the first marine reptile fossils to be recognized as such by the scientific community. Cretaceous marine turtles from Maastricht, the Netherlands, were mentioned by Camper (1786) and figured by Cuvier (1812:Vol. 4). Subsequently, Cuvier (1824:Plate 15, Figure 7) figured the skull of a Jurassic sea turtle. This was later identified as *Plesiochelys etalloni* and is the earliest figure of a fossil turtle skull (Gaffney, 1975a).

These early workers identified the fossil sea turtles as living genera. Owen

(1842, 1851) put all fossil sea turtles in the genus *Chelone*, family Cheloniidae. Rütimeyer (1873) recognized the distinctiveness of some of the Jurassic forms, placing them in the genera *Plesiochelys* and *Thalassemys*. He referred *Plesiochelys* to the Pleurodira and *Thalassemys* to the Cryptodira. However, with the discovery of the huge protostegids from the Upper Cretaceous of North America (Wieland, 1896; Cope, 1871), it was soon apparent that there were many extinct lineages of sea turtles.

Lydekker (1889) referred the extinct Jurassic "thalassemyd" sea turtles to his new order, Amphichelydia, which was diagnosed as a primitive group, intermediate between pleurodires and cryptodires. This concept became entrenched in the literature with the work of Romer (1956), Williston (1914), and Hay (1908), although Romer acknowledged that it was a diverse assemblage, united only by primitive characters. The paraphyletic nature of the Amphychelydia was clearly demonstrated by Gaffney (1969, 1975b), who redefined the Cryptodira and Pleurodira on cranial characters. Since then, the primitive Jurassic forms have been recognized as cryptodires.

There appear to have been three separate radiations of marine turtles. The earliest marine radiation is seen in the Plesiochelyidae of the Jurassic of Europe. This primitive group of marine cryptodirans lacked central articulations (Gaffney and Meylan, 1988; Rieppel, 1980). Fore- and hindlimbs were about equally developed, and they lacked the limb modifications seen in the Chelonioidea (Młynarski, 1976; Zangerl, 1953). They probably stayed in shallow, coastal waters.

A second marine radiation is seen in the pleurodiran family, Pelomedusidae. All living members of this family inhabit freshwater, but during the Cretaceous and Tertiary, some genera invaded the sea. They are reported in the Upper Cretaceous and Tertiary deposits of North and South America, Europe, and Africa (Moody, 1993 and Chapter 10; De Broin, 1977; Gaffney, 1975c; Gaffney and Zangerl, 1968; Wood, 1972). These lack the obvious marine adaptations of the Chelonioidea, and the carapace and plastron are unreduced. However, they are always found in marine formations, associated with marine faunas (Gaffney, 1975c; Wood, 1972, 1976). They appear to have occupied the shallow inland seas and coastal areas on both sides of the opening Atlantic, and in the Mediterranean area of the Tethys. *Stupendemys*, from the Pliocene of Venezuela, may have been marine (Wood, 1976). This is the largest chelonian known, with a carapace length of 230 cm.

The most successful of the marine turtles is the Chelonioidea, which first appears in the fossil record in the late Early Cretaceous (Hirayama, Chapter 8). This taxon includes both living families of sea turtles, the Dermochelyidae and the Cheloniidae, as well as the protostegid turtles of the Late Cretaceous. This group is distinguished from other sea turtles by the elongation and modification of the forelimb into a wing for underwater flight (Zangerl and Sloan, 1960; Zangerl,

1953), and by having procoelous central articulations (Gaffney and Meylan, 1988).

The Dermochelyidae range from the Santonian to the Recent (Hirayama, Chapter 8). The only living species, *Dermochelys coriacea,* is the largest living turtle. The usual carapacial elements have been lost and replaced by a mosaic of small, epithecial bony plates embedded in the leathery skin. Once thought to be very primitive, and ancestral to all other turtles (Hay, 1908; Cope, 1872), it has been more extensively studied than any other species of turtle.

The Cheloniidae includes the other four genera of living sea turtle. The family ranges from the Aptian to the Recent, reaching a peak in diversity in the Late Cretaceous (Hirayama, Chapter 8).

The Protostegidae were restricted to the Cretaceous, first appearing in the Aptian and become extinct at the end of the Maastrichtian (Hirayama, Chapter 8). *Archelon* and *Protostega* were giants of the turtle world, and their large size was accompanied by a reduction in the bony elements of the carapace and plastron.

Relationships within the Chelonioidea, however, are still not clearly understood. Primitive members of the group lack the locomotor adaptations to a varying degree. Zangerl (1971, 1980) relied on the development of a secondary palate, but this may be more closely related to feeding habits than to phylogeny (Moody, 1984). Many fossil forms are still known only from the carapace and plastron, and shell morphology often reflects adaptions to the marine environment. Fontanelle development is known to vary with ontogeny (Moody, 1974, 1984). In this volume, for example, Elliott et al. (Chapter 9) and Moody (Chapter 10) both recognize the Desmatochelyidae as a separate family, while Hirayama (Chapter 8) puts it in the Protostegidae. The Osteopygidae and Toxochelyidae are recognized as separate families by Moody, while Hirayama refers them to the Cheloniidae. Hirayama refers *Corsochelys* to the Dermochelyidae, while previous authors (Gaffney and Meylan, 1988; Zangerl, 1960) consider it to be a cheloniid. The systematic relationships of *Allopleuron* are still controversial.

The Chelonioidea are one of the few groups of marine reptiles to survive the extinctions at the end of the Cretaceous. The only other marine reptile known to extend into the Paleocene is the crocodilian genus *Hyposaurus* (Denton et al., Chapter 13). As the sea turtle is the only living reptile totally adapted to the marine environment, studies of its locomotion, behavior, and physiology are pertinent to the understanding of marine reptiles of the past.

REFERENCES

Camper, P. 1786. Conjectures relative to the petrifications found in St. Peter's Mountain near Maestricht. *Philosophical Transactions of the Royal Society of London* 76:443-456.

Cope, E. D. 1872. A description of the genus *Protostega*, a form of extinct Testudinata. *Proceedings of the American Philosophical Society* 12:422-433.

Cuvier, G. 1812. *Recherches sur les ossements fossiles, où l'on rétablit les caractères de plusiers animaux dont les révolutions du globe ont détruit les espèces.* First edition, 4 volumes, Paris.

Cuvier, G. 1824. *Recherches sur les ossements fossiles, où l'on rétablit les caractères de plusiers animaux dont les révolutions du globe ont détruit les espèces.* Second edition, 5 volumes, Paris.

De Broin, F. 1977. Contribution à l'etude des chéloniens. Chéloniens continentaux du Crétacé et du Tertiaire de France. *Mémoires du Muséum National d'Histoire Naturelle. Nouvelle Série. Série C, Science de la Terre* 38:1-366.

Ernst, C. H. and R. W. Barbour. 1989. *Turtles of the World.* Smithsonian Institution Press, Washington, D. C., 313 pp.

Gaffney, E. S. 1969. *The North American Baenoid Turtles and the Cryptodire-Pleurodire Dichotomy.* Unpublished Ph. D. dissertation, Columbia University, 338 pp.

Gaffney, E. S. 1975a. A taxonomic revision of the Jurassic turtles *Portlandemys* and *Plesiochelys. American Museum of Natural History, Novitates* 2574:1-19.

Gaffney, E. S. 1975b. A phylogeny and classification of the higher catagories of turtles. *Bulletin of the American Museum of Natural History* 155(5):391-436.

Gaffney, E. S. 1975c. A revision of the side-necked turtle *Taphrosphys sulcatus* (Leidy) from the Cretaceous of New Jersey. *American Museum of Natural History, Novitates* 2571:1-24.

Gaffney, E. S. and P. A. Meylan. 1988. A phylogeny of turtles. IN M. J. Benton (Ed.), *The Phylogeny and Classification of the Tetrapods, Volume 1: Amphibians, Reptiles, Birds,* pp. 157-219. Systematics Association Special Volume, 35A. Clarendon Press, Oxford.

Gaffney, E. S. and R. Zangerl. 1968. A revision of the chelonian genus *Bothremys* (Pleurodira:Pelomedusidae). *Fieldiana, Geology Memoirs* 16:193-239.

Goff, G. P. and J. Lein. 1988. Atlantic leatherback turtles, *Dermochelys coriacea*, in cold water off Newfoundland and Labrador. *Canadian Field Naturalist* 102:1-5.

Hay, O. P. 1908. The Fossil Turtles of North America. *Carnegie Institute of Washington* 75, 568 pp.

Hendrickson, J. R. 1980. The ecological strategies of sea turtles. *American Zoologist* 20:597-609.

Lydekker, R. 1889. *Catalogue of the Fossil Reptilia and Amphibia in the British Museum (Natural History).* Longmans and Co., London, 239 p.

McAuliffe, K. 1995. Elephant seals, the champion divers of the deep. *Smithsonian,* September:45-56.

Młynarski, M. 1976. Testudines. IN O. Kuhn (Ed.), *Handbuch der Paläoherpetologie,* Gustav Fischer Verlag, Stuttgart, 128 pp.

Moody, R. T. J. 1974. The taxonomy and morphology of *Puppigerus camperi* (Gray), an Eocene sea turtle from northern Europe. *Bulletin of the British Museum (Natural History),* Geology 25:153-186.

Moody, R. T. J. 1984. The relative importance of cranial/post cranial characters in the classification of sea turtles. *Studia Palaeocheloniologica I, Studia Geologica Salmanticensia* 1:205-213.

Moody, R. T. J. 1993. Cretaceous-Tertiary marine turtles of North West Europe. *Revue de Paléobiologie*, Volume Spécial 7:151-160.

Mrosovsky, N. 1980. Thermal biology of sea turtles. *American Zoologist* 20:531-548.

Owen, R. 1842. Report on British fossil reptiles. Part II. *Report of the British Association of the Advancement of Science* 11:60-204.

Owen, R. 1851. *Monograph on the Fossil Reptilia of the Cretaceous Formations.* Palaeontographical Society, London, 118 pp.

Rieppel, O. 1980. The skull of the Upper Jurassic cryptodire turtle *Thalassemys*, with a reconsideration of the chelonian braincase. *Palaeontographica*, Abteilung A, Palaeozoologie-Stratigraphie 171(4-6):105-140.

Romer, A. S. 1956, *The Osteology of the Reptiles.* University of Chicago Press, Chicago, 772 pp.

Rütimeyer, L. 1873. Die fossilen Schildkröten von Solthurn und der übrigen Jura-formation. *Denkschriftender Allgemeinenschweizerischen Gesellschaft für die Gesammten Naturwissenschaften* 25:1-185.

Wieland, G. R. 1896. *Archelon ischyros*: a new gigantic cryptodire testudinate from the Fort Pierre Cretaceous of South Dakota. *American Journal of Science* 4(2):399-415.

Williston, S. W. 1914. *Water Reptiles of the Past and Present.* University of Chicago Press, Chicago, 251 pp.

Wood, R. C. 1972. A fossil pelomedusid turtle from Puerto Rico. *Breviora* 392:1-13.

Wood, R. C. 1976. *Stupendemys geographicus*, the world's largest turtle. *Breviora* 436:1-31.

Zangerl, R. 1953. The vertebrate fauna of the Selma Formation of Alabama. Part 4. The turtles of the family Toxochelyidae. *Fieldiana, Geology Memoirs* 3:136-277.

Zangerl, R. 1960. The vertebrate fauna of the Selma Formation of Alabama. Part 5. An advanced cheloniid sea turtle. *Fieldiana, Geology Memoirs* 3:279-312.

Zangerl, R. 1971. Two toxochelyid sea turtles from the Landenian Sands of Erquelinnes (Hain Aut) of Belgium. *Institut Royal des Sciences Naturelles de Belgique, Mémoires* 169:1-32.

Zangerl, R. 1980. Patterns of phylogenetic differentiation in the toxochelyid and cheloniid sea turtles. *American Zoologist* 20:585-596.

Zangerl, R. and R. E. Sloan. 1960. A new description of *Desmatochelys lowi* (Williston). A primitive cheloniid sea turtle from the Cretaceous of South Dakota. *Fieldiana, Geology Memoirs* 14:7-40.

Chapter 8

DISTRIBUTION AND DIVERSITY OF CRETACEOUS CHELONIOIDS

REN HIRAYAMA

INTRODUCTION

The superfamily Chelonioidea, sea turtles in the strict sense, is the only chelonian group highly adapted to the marine environment. Its limbs are modified into large paddles as propulsive organs, reflecting the unique swimming mode of "flying under water" (Walker, 1973; Zangerl, 1953). The fossil record of chelonioids can be traced into the Early Cretaceous, about 110 Ma (Hirayama, in press). Modern chelonioids are a rather small group, allocated to six genera and seven species of two families, Cheloniidae and Dermochelyidae. In the past, however, morphological diversity of the sea turtles appears to be much greater, particularly during the Cretaceous (Zangerl, 1953, 1980). This chapter deals with the distribution and diversification of the Cretaceous chelonioids, incorporating current investigation. The classification of chelonioids adopted here follows Nicholls (1988) and Hirayama (1992).

CLASSIFICATION OF THE CRETACEOUS CHELONIOIDS

The classification of Cretaceous chelonioids with synonymies is as follows (the maximum length of carapace, MCL, is shown in cm):

a. Family Cheloniidae Gray, 1825

Diagnosis: Incipient secondary palate involving palatine; apertura narium interna entirely formed by palatines and vomer; basisphenoid with ventral V-shaped crest; cervicals relatively elongate, with central articulation wider than tall, involving double central articulation between 7th and 8th cervicals; first thoracic vertebra with anterior articulation facing anteroventrally; ulna and radius in tight contact at distal end and with prominent rugositites; pelvic girdle with a large, confluent thyroid fenestra. *Toxochelys latiremis* **Cope, 1873** (= *T. serrifer* Cope, 1875 [part]; *T. brachyrhinus* Case, 1898; *Porthochelys browni* Hay, 1905; *Phyllemys barberi* Schmidt, 1944; *T. weeksi* Collins, 1951; *Lophochelys niobrarae* Zangerl, 1953). Coniacian to Lower Maastrichtian, USA (Kansas, South Dakota, Wyoming, Arkansas, and Tennessee) and Canada (Manitoba and Alberta). MCL = 80. Figures 1A (top), 7A. *Toxochelys moorevillensis* **Zangerl, 1953**. Lower Campanian, USA (Alabama). MCL = 59. Figure 1A (bottom). *Porthochelys laticeps* **Williston, 1901**. Coniacian or Santonian, USA (Kansas). MCL = 74. *Thinochelys lapisossea* **Zangerl, 1953**. Lower Campanian, USA (Alabama). MCL = 69. *Catapleura repanda* **(Cope, 1868)** (= *Toxochelys atlantica* Zangerl, 1953). Upper Maastrichtian, USA (New Jersey). MCL = 58. *Ctenochelys stenoporus* **(Hay, 1905)** (= *Toxochelys serrifer* Cope, 1875 [part]; *T. procax* Hay, 1905; *T. elkader*

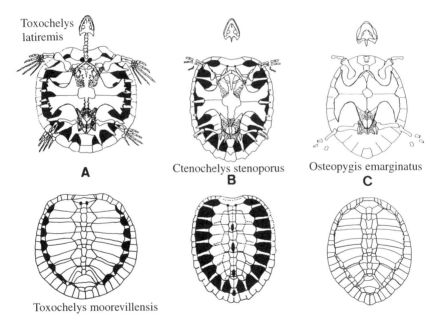

Figure 1. Primitive Cretaceous cheloniids. Ventral views of reconstructed skeletons (after Hirayama, in press) and dorsal views of carapaces. The stippled area shows the coracoid and pelvic girdle. **A)** Carapace after Zangerl (1953). **B)** Carapace after Hay, 1908; black area on neurals shows the epithecal region. **C)** Carapace after Hay (1908).

Hay, 1908; *Lophochelys natarix* Zangerl, 1953; *L. venatrix* Zangerl, 1953; *C. tenuitesta* Zangerl, 1953; *C. acris* Zangerl, 1953). Coniacian to Middle Campanian, USA (Kansas and Alabama) and United Kingdom. MCL = 80. Figures 1B, 7B. ***Prionochelys nauta* Zangerl, 1953** (= ?*P.matuina* Zangerl, 1953; *P. galeotergum* Zangerl, 1953). Campanian, USA (Arkansas, Alabama, and Kansas). MCL = 120. ***Peritresius ornatus* (Leidy, 1856)** (= *Taphrosophys nodosus* Cope, 1870). Maastrichtian, USA (New Jersey). MCL = 58. ***Osteopygis emarginatus* Cope, 1868** (= *O. sopita* Cope, 1868; *O. chelydrinus* Cope, 1868; *O. platylomus* Cope, 1869; *Lytoloma angusta* Cope, 1870; *O. erosus* Cope, 1875; *Catapleura chelydrina* Cope, 1875; *O. gibbi* Wieland, 1904; *Propleura borealis* Wieland, 1904; *O. borealis* Hay, 1908; *O. robustus* Hay, 1908; *Lytoloma wielandi* Hay, 1908; *Erquelinnesia molaria* Hay, 1908). Maastrichtian to Early Paleocene, USA (New Jersey, Maryland, and California). MCL = 69. Figures 1C, 7D, 8A. ***Lytoloma cantabrigiense* Lydekker, 1889**. Middle or Upper Albian to Cenomanian, United Kingdom. MCL = 70? ***Allopleuron hoffmanni* (Gray, 1831)**. Upper Maastrichtian, the Netherlands and Belgium. MCL = 136. Figures 2C, D, 7C, 8B.

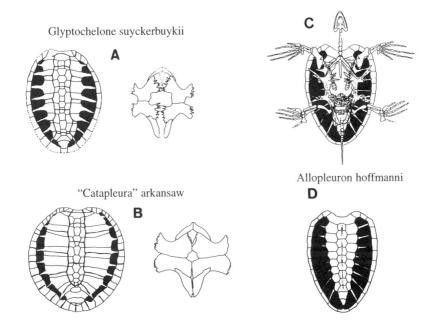

Figure 2. Advanced Cretaceous cheloniids. **A)** Based on the holotype (IRScNB Ht.R2). **B)** After Schmidt (1944). **C** and **D)** Reconstructed skeleton (ventral view; after Hirayama, in press) and carapace (based on IRScNB Reg.3668). The stippled area shows the coracoid and pelvic girdle.

"Catapleura" arcansaw Schmidt, 1944. Upper Campanian to ?Upper Maastrichtian, USA (Arkansas and ?New Jersey). MCL = 75. Figure 2B. *Glyptochelone suyckerbuykii* (Ubaghs, 1879). Upper Maastrichtian, Belgium, and the Netherlands. MCL = 105. Figure 2A.

b. Family Protostegidae Cope, 1873

Diagnosis: Large jugal with nearly straight ventral border, reaching to quadrate; posterior portion of vomer reduced, with palatines meeting medially; pterygoid thickened, reaching to mandibular articulating surface of quadrate; foramen posterius canalis carotici interni lying between basisphenoid and pterygoid; lateral process of humerus restricted to anterior portion of shaft, not visible in ventral view; middle portion of radius bent anteriorly; large hyo-hypoplastra star shaped. **Undescribed new genus and species (Hirayama, in preparation).** Upper Aptian or Lower Albian, Brazil (Ceara). MCL = 14.5. Figures 3, 7F. ***Rhinochelys pulchriceps* (Owen, 1842)** (= *R. cantabrigiensis* Lydekker, 1889; *R. elegans* Lydekker, 1889; *R. macrorhina* Lydekker, 1889; *R. jessoni* Lydekker, 1889; *Lytoloma cantabrigiensis* Lydekker, 1889 [part]; *R. amaberti* Moret, 1935; ?*Cimochelys benstedi* [Mantell,

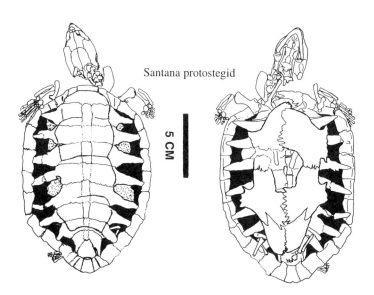

Figure 3. Skeleton of undescribed new genus and species of the Family Protostegidae (THUg1386) from the upper Aptian or lower Albian of Brazil (Santana Formation). Dorsal and ventral views after acid preparation.

1841]). Middle or Upper Albian to Turonian, United Kingdom and France. MCL = 30? Figure 7G. *Notochelone costata* (Owen, 1882). Upper Albian, Australia (Queensland). MCL = 60? Figure 4B. *"Protostega" anglica* Lydekker, 1889 (= ?*Chelone jessoni* Lydekker, 1889). Middle or Upper Albian to Cenomanian, United Kingdom. MCL = 50? *Desmatochelys lowi* Williston, 1894. Turonian, USA (Nebraska, South Dakota, and Kansas) and Japan (Hokkaido). MCL = 120. Figures 4C, D, 7H, 8C. *Chelosphargis advena* (Hay, 1908). Coniacian to Upper? Campanian, USA (Kansas and Alabama), ?Canada (Saskatchewan), ?New Zealand and ?Japan (Hokkaido). MCL = 50. Figure 4A. *Calcarhichelys gemma* Zangerl, 1953. Lower Campanian, USA (Alabama and ?Kansas). MCL = 21. *"Protostega" eaglefordensis* Zangerl, 1953. Turonian, USA (Texas). MCL = 135. *"Protostega" copei* Wieland, 1909. Coniacian, Santonian, or Lower Campanian, USA (Kansas). MCL = 80. Figure 5C. *Protostega gigas* Cope, 1871 (= *P. potens* Hay, 1908; *P. dixie* Zangerl, 1953; *Archelon copei* [Wieland, 1909; part]). ?Santonian to Lower Campanian, USA (Kansas, Alabama, and Arkansas), Canada (Manitoba) and Japan (Hokkaido). MCL = 150. Figures 5A, D, 7I. *Archelon ischyros* Wieland, 1896 (= *A. marshi* Wieland, 1900). Upper Campanian, USA

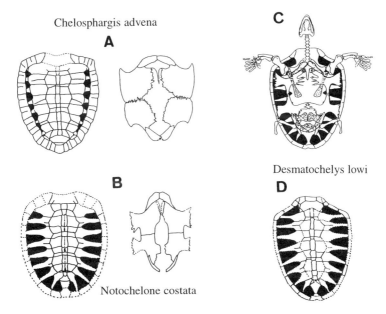

Figure 4. Primitive protostegids with carapace scutes retained. **A)** After Zangerl (1953), with additions from AMNH 1778 and YPM 40866. **B)** After Gaffney (1981), with additions from BMNH R77, R9705 and QM F uncatalogued specimen. **C** and **D)** Reconstructed skeleton (ventral view; after Hirayama and Chitoku, 1994) and carapace (FMNH PR385; slightly modified from Zangerl and Sloan, 1960). The stippled area shows the coracoid and pelvic girdle.

(South Dakota and Colorado). MCL = 220 (E. S. Gaffney, personal communication). Figures 5B, E, 8D. ***Atlantochelys mortoni* Agassiz, 1849** (= *?Pneumatoarthrus peloreus* Cope, 1870). Maastrichtian, USA (New Jersey).

c. Family Dermochelyidae Gray, 1825

Diagnosis: Ossification of rostrum basisphenoidale much reduced; lateral process of humerus anteroposteriorly elongate, with strong anterior projection; shell scute nearly entirely lost. ***Corsochelys haliniches* Zangerl, 1960**. Lower Campanian, USA (Alabama). MCL = 130. Figure 6A. **Undescribed new genus and species (Hirayama and Chitoku, in preparation)**. Santonian to Lower Maastrichtian, Japan (Hokkaido and Hyogo). MCL = 150. Figures 6B, 8E.

d. Chelonioidea, family indeterminate

Diagnosis: lateral process of humerus with U-shaped structure. **Genus and**

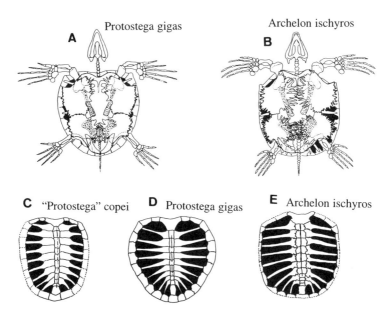

Figure 5. Advanced protostegids with carapace scutes lost. **A** and **B**) Reconstructed skeletons (ventral views; after Hirayama, in press). The stippled area shows the coracoid and pelvic girdle. **C** and **E**) After Wieland (1909). **D**) After Zangerl (1953).

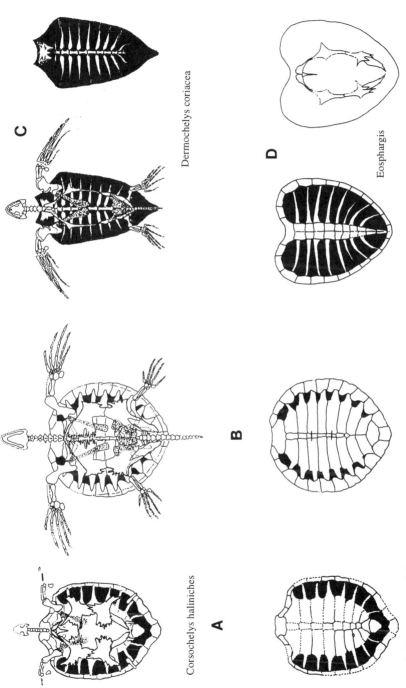

Figure 6. Reconstructed skeletons (ventral views) and shells of the dermochelyids. The stippled area shows the coracoid and pelvic girdle. **A)** After Zangerl (1960), with additions from observation of the holotype (FMNH PR249). **B)** Undescribed new genus and species from Japanese Maastrichtian (skeleton after Hirayama and Chitoku, 1994; carapace based on HMG5, 342, 369 and 1053). **C)** Modified from Volker (1913). **D)** Lower Eocene of Europe; based on BMNH R8312 and IRScNB Reg.1736, with additions from Nielsen (1964).

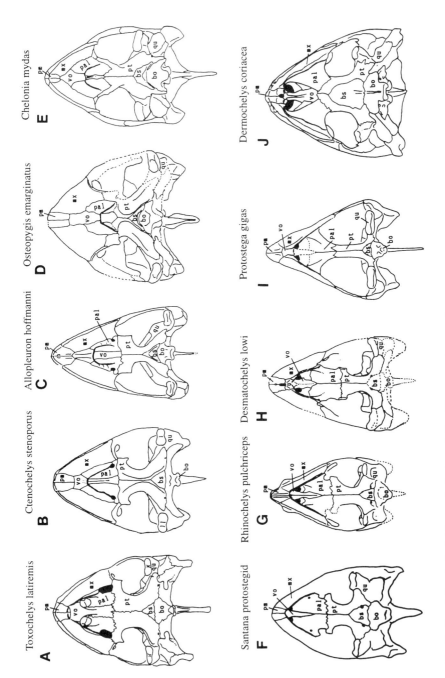

Figure 7. Skulls of the chelonioids, ventral views (after Hirayama, in press).

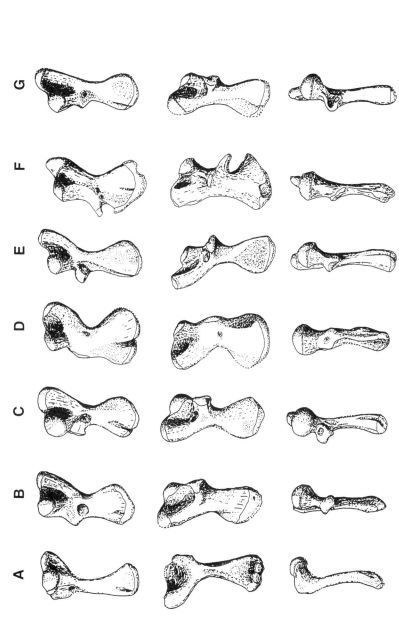

Figure 8. Left humeri of the chelonioids (after Hirayama, 1992). Dorsal, ventral, and anterior views. **A)** *Osteopygis emarginatus*. **B)** *Allopleuron hoffmanni*. **C)** *Desmatochelys lowi*. **D)** *Archelon ischyros*. **E)** Undescribed new dermochelyid from Japanese Maastrichtian. **F)** *Dermochelys coriacea*. **G)** An aberrant Albian chelonioid of family indeterminate.

species indet. Upper Albian, United Kingdom. MCL = 40? (known only from isolated humeri). Figure 8G.

DISCUSSION

The chelonioids can be traced back into the late Aptian; more than 700 specimens of the Cretaceous chelonioids, showing extensive morphological diversity, are known in the literature and from personal observation. Nonetheless, their fossil record is far from complete as shown by Figure 9. First, the origin and relationships of the chelonioids among other Mesozoic cryptodiran turtles, such as family Macrobaenidae, is still a subject of debate (Brinkman and Peng, 1993). The Aptian-Albian record of chelonioids demonstrates that the origin and basic radiation of the chelonioid sea turtles must have happened much earlier, possibly prior to the Aptian. One humeral type (Figure 8G) suggests a quite curious early chelonioid.

The family Cheloniidae is the largest group of chelonioids. It includes the turtles formerly placed in the family Toxochelyidae and most Cenozoic sea turtles as currently defined (Hirayama, in press). The toxochelyid sea turtles, including *Toxochelys*, *Ctenochelys*, and *Osteopygis*, were extensively described by Zangerl (1953). More recent studies, however, show that this family appears to be paraphyletic, not monophyletic (Gaffney and Meylan, 1988; Hirayama, 1992). The small plastron, with a narrow bridge, seems primitive for chelonioids, very possibly associated with the poor development of the pectoral and pelvic girdles of these turtles (Figure 1; Hirayama, in press). The movable articulations of the short first and second digits are primitively retained, and the orbit faces dorsally. These characters strongly suggest that so-called toxochelyids were bottom dwellers in the very shallow sea rather than good swimmers, perhaps like living chelydrids or trionychids. *Porthochelys* and *Thinochelys* may form a monophyletic group together with *Toxochelys* based on the possession of a very wide cervical scute on the nuchal plate, as broad as the first vertebral. *Ctenochelys* and *Osteopygis* are more advanced than *Toxochelys* in the possession of rod-like rostrum basisphenoidale and the more extensive secondary palate. *Prionochelys* and *Peritresius* appear to constitute a monophyletic group with *Ctenochelys*, defined by the additional epithecal ossifications on the neural area. The *Lophochelys* species of Zangerl (1953) appear to be based on poorly preserved juveniles of *Ctenochelys* or *Toxochelys* (personal observation), and the name of Lophochelyinae should be abandoned. *Allopleuron* is more advanced and closely related to Cenozoic cheloniids in the development of large paddles, whereas the foramen palatinum posterius is primitively retained (Figures 2C and D, 7C, and 8B). In this aberrant turtle, the scutes are completely lost, and the shell ossification is much reduced, being the most similar to

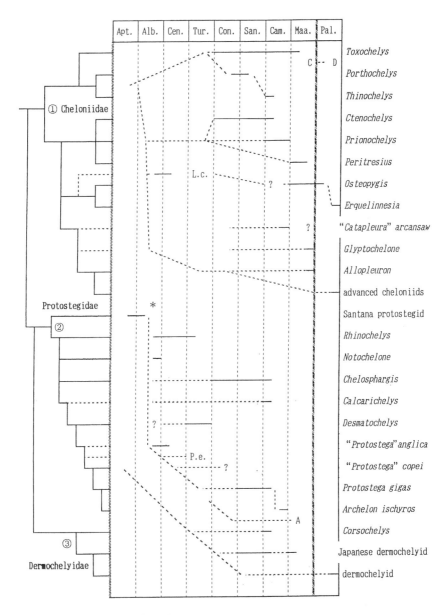

Figure 9. Cladogram showing the interrelationships and geological distribution of the Cretaceous (Aptian to Maastrichtian) and Paleocene chelonioids. Chelonioid relationships are adopted and modified from Hirayama (1992). Dotted line shows doubtful relationships or speculative geological distribution. * = aberrant chelonioid known only from humeri. Abbreviations: **A** = *Atlantochelys mortoni*, **C** = *Catapleura repanda*, **D** = *Dollochelys*, **L.c.** = *Lytoloma cantabrigiense*, **P.e.** = "*Protostega*" *eaglefordensis*.

Eosphargis, an Eocene dermochelyid (Figure 6D), as mentioned by Weems (1988). However, the structure of its skull, cervical vertebrae, humerus and pelvic girdle is essentially cheloniid. The shell similarity between *Allopleuron* and *Eosphargis* is considered a result of extreme convergence, possibly because of a pelagic mode of life. A few cheloniids known from the shell such as *Glyptochelone* and "*Catapleura*" *arcansaw* might be additional advanced cheloniids of the Cretaceous based upon the relatively longer plastron (Figure 2A and B). Though the better materials of the Cretaceous cheloniids are post-Coniacian, the isolated dentaries (and isolated humeri) named as *Lytoloma cantabrigiense* of the English Albian and Cenomanian might represent an early cheloniid with an extensively secondary palate, possibly related with *Osteopygis*.

The remaining two families, Protostegidae and Dermochelyidae, are considered to be more closely related to each other than either is to the cheloniids. They possess a very elongate coracoid, a pelvic girdle with a huge lateral process of the pubis, and a very long plastron. The neural plates of protostegids and dermochelyids are rectangular rather than hexagonal as in cheloniids, which may be another synapomorphy of these sea turtles. The cervical vertebrae of the protostegids and dermochelyids are very short, with nearly circular articular surface. Furthermore, there are no double central articulations like those in cheloniids, and the almost vertical central articular surface of the first thoracic is primitively retained in this group .

The skull of protostegids may be modified to feed on hard-shelled animals, judging from the extensive contact between pterygoid and quadrate, and the strong processus trochlearis oticum as seen in the living batagurid *Malayemys*, which is a typical molluscivorous freshwater turtle. The anterior portion of the triturating surfaces of the protostegids could have been covered and reinforced by the thick rhamphotheca in life, though the palate itself seems not well developed. Because the limbs were well-developed paddles in the protostegids, their main food might have been planktonic animals such as ammonites rather than benthonic forms. The distribution of the lateral process of the protostegid humerus is limited to the anterior shaft, extending little onto the ventral surface, so that movement in the horizontal direction might be more prominent during forelimb motion in protostegids than in the dermochelyids or advanced cheloniids. This kind of swimming mode suggests that the protostegids were surface dwellers, not good divers in the ocean.

A nearly complete skeleton of a protostegid (THUg1386; Figure 3) from the Santana Formation (late Aptian or early Albian) of Brazil, the oldest known chelonioid, retains a number of primitive characters of chelonioid and protostegid turtles: movable articulations of the first and second digits, elongate first thoracic rib, lateral process of humerus without median concavity seen in later protostegids, and large triangular xiphiplastra. Nonetheless, its skull is essentially protostegid. The protostegids were well diversified by the end of the Albian as shown by *Rhinochelys, Notochelone,* and *"Protostega" anglica,* and they are the most dominant chelonioids during Albian to Turonian time. A few protostegids, such as *Desmatochelys* and *Protostega gigas,* hitherto recorded only from North America, were probably cosmopolitan, as referred specimens are reported from Japan (Hirayama and Chitoku, 1994; Hirayama, 1994). In protostegid turtles, the evolutionary trends are development of a huge and massive head and appendicular skeleton, loss of scutes, reduced ossification of carapace, and gigantism which reached its peak in *Archelon* of the Late Campanian (Figures 5B and E and 8D). The protostegids seem to have drastically declined, and probably became extinct, during the Maastrichtian. *Atlantochelys mortoni,* based on a huge isolated humerus from New Jersey (ANSP 9234; Leidy, 1865), is the only Maastrichtian protostegid known with certainty.

The fossil record of dermochelyid turtles is poor in both quality and quantity even during the Cenozoic, possibly due to the poor ossification of the skeleton and an extremely pelagic mode of life. Recently, however, some dozen specimens referred to the Dermochelyidae were unearthed from the Upper Cretaceous post-Santonian of Japan (Hirayama and Chitoku, 1994; Hirayama, in press). In fact, the dermochelyids are the most abundant turtle among the Cretaceous chelonioids in Japan, whereas *Corsochelys haliniches,* known from the holotype only, is the only Mesozoic dermochelyid known with certainty outside Japan. Shell ossification of these Mesozoic dermochelyids was primitively well developed, whereas the scute sulci were almost entirely lost (Figure 6A and B). Cranial morphology is not completely known, but limited information, such as the narrow lower jaw and the weak processus trochlearis oticum, suggests that these turtles had a primary palate. The extreme specialization toward feeding mainly on jellyfish, as seen in the living leatherback turtle, seems to have developed during the Cenozoic as implied by reduced shell ossification (Figure 6C and D).

The temporal and geographic distribution of Cretaceous chelonioids provides only limited information about their paleobiogeography. Among living sea turtles, those with cosmopolitan distribution are common (*Caretta, Chelonia, Eretmochelys, Lepidochelys,* and *Dermochelys*). This is also the case in some Tertiary chelonioids such as *Syllomus* (Miocene-Pliocene, Cheloniidae) and *Psephophorus* (Oligocene-Pliocene, Dermochelyidae). During the Cretaceous, however, the geographic distribution of cheloniid and dermochelyid genera seems to have been very limited, whereas some protostegids, such as *Desmatochelys* and *Protostega gigas,* might be cosmopolitan. The various cheloniids were quite abundant from the Coniacian to Maastrichtian of North America. Dermochelyids were dominant sea turtles from the Santonian to Maastrichtian in Japan, where no Cretaceous cheloniid has yet been discovered. The Maastrichtian chelonioids are quite different even between the eastern coast of North America (New Jersey; represented by cheloniids *Osteopygis* and *Peritresius* and protostegid *Atlantochelys*) and western Europe (the Netherlands and Belgium; represented by cheloniids *Allopleuron* and *Glyptochelone*). This strongly suggests that restricted distribution of sea turtle species was rather common in the Cretaceous. The much higher chelonimid diversity in the Cretaceous than today may be a result of this endemism. This presumed endemic speciation among Cretaceous sea turtles is not a reflection of their functional primitiveness, because many Mesozoic chelonioids had well-developed "modern" paddles. The long-distance migration patterns, seen in living sea turtles, might be absent in many of the Cretaceous sea turtles.

Little can be said about how the chelonioid lineages relate to the so-called mass extinction at the Cretaceous-Tertiary boundary because of very limited information. The decrease and extinction of the protostegids, however, seems to have been a rather gradual process during the Late Cretaceous, possibly being concordant with the fate of the ammonites. A primitive member of the cheloniid (*Osteopygis*) survived into the Paleocene, together with more advanced cheloniids, whereas the aberrant genus *Allopleuron* left no descendants. In the case of dermochelyids, there is a rather large geological and morphological gap around the K-T boundary. The referred specimen of *Lembonax insularis*, allocated into *Allopleuron* by Weems (1988), seems to be the earliest occurrence of genus *Eosphargis*.

CONCLUSION

The Cretaceous chelonioid sea turtles show much higher morphological and presumed ecological diversity than today. The family Cheloniidae, well documented since the Coniacian, have their own derived characters, such as the secondary palate involving the palatine, the apertura narium interna entirely formed by palatines and vomer, the basisphenoid with ventral V-shaped crest, and the pelvic girdle with a large, confluent thyroid fenestra. The various cheloniids, such as primitive *Toxochelys* and aberrant *Allopleuron,* were quite abundant in the Late Cretaceous of North America and western Europe. The extinct family Protostegidae, including the largest known turtle (*Archelon*), is characterized by a number of autapomorphies such as the large jugal reaching the quadrate, the pronounced pterygoid-quadrate contact, the lateral process of the humerus restricted to the anterior portion of the shaft, and a large star-shaped hyo-hypoplastra. The protostegids were presumed to be specialized for feeding on hard shelled planktonic animals such as ammonites. The protostegids were once cosmopolitan and the most dominant chelonioids from the Albian to the Turonian, whereas they seem to have drastically declined and terminated during the Maastrichtian. The family Dermochelyidae is represented by the holotype of *Corsochelys* of North America (Campanian) and some dozen specimens from Japan (Santonian to Maastrichtian). These Cretaceous dermochelyids share some derived characters such as the short rostrum basisphenoidale, the elongate lateral process of the humerus, and the loss of shell scutes, with Cenozoic relatives, in which the shell ossification was primitively prominent. It is highly probable that cosmopolitan chelonioid species were quite rare during the Cretaceous unlike today. Such endemic geographical distribution of Mesozoic sea turtles emphasize their morphological diversity, and might imply their habit, possibly lacking the grand migrations of today.

SUMMARY

The Cretaceous Chelonioidea, which includes at least twenty-two valid genera, is divided into the three monophyletic families, Cheloniidae, Protostegidae, and Dermochelyidae. The cheloniids were well diversified and abundant in the Late Cretaceous of North America and western Europe, whereas they were absent from the Japanese Cretaceous. The protostegids were cosmopolitan, as were the dominant chelonioids from the Aptian to the Turonian; however, they drastically declined and became extinct during the Maastrichtian. The Mesozoic dermochelyids were dominant only in the Late Cretaceous of Japan. Most Mesozoic sea turtles were endemic, and, unlike today, cosmopolitan chelonioid species were quite rare.

ACKNOWLEDGMENTS

This work was partially supported by the GRANT-IN-AID from Teikyo Heisei University. I thank the curators of paleontology of the following institutions for access to material in their care: American Museum of Natural History, AMNH; British Museum (Natural History), BMNH; Field Museum of Natural History, FMNH; Hobetsu Museum, HMG; Institut Royal des Sciences Naturelles de Belgiques, IRScNB; Queensland Museum, QM; Teikyo Heisei University, THUg; and Yale Peabody Museum, YPM. I give my special thanks to Dr. Eugene S. Gaffney (AMNH) for his assistance and thought-provoking comments. Reviews of the manuscripts by Dr. J. Howard Hutchison (Museum of Paleontology, University of California, Berkeley) and Dr. Eugene S. Gaffney greatly improved this contribution.

REFERENCES

Brinkman, D. B. and J. Peng. 1993. *Ordosemys leios*, n. gen., n. sp., a new turtle from the Early Cretaceous of the Ordos Basin, Inner Mongolia. *Canadian Journal of Earth Sciences* 30:2128-2138.

Gaffney, E. S. 1981. A review of the fossil turtles of Australia. *American Museum Novitates* 2720:1-38.

Gaffney, E. S. and P. A. Meylan. 1988. A phylogeny of turtles. IN M. J. Benton (Ed.), *The Phylogeny and Classification of Tetrapods*. Vol. 1, *Amphibians, Reptiles, Birds*, pp. 157-219. Clarendon Press, Oxford.

Hay, O. P. 1908. The Fossil Turtles of North America. *Carnegie Institute of Washington* 75, 568 pp.

Hirayama, R. 1992. Humeral morphology of chelonioid sea-turtles; its functional analysis and phylogenetic implications. *Bulletin of the Hobetsu Museum* 8:17-57 (Japanese with English Abstract).

Hirayama, R. 1994. Fossil marine turtles from the Upper Cretaceous of Hokkaido, North Japan. *Abstracts, 101th Annual Meeting of the Geological Society of Japan*, p. 112 (Japanese).

Hirayama, R. In press. Phylogenetic systematics of chelonioid sea turtles. IN L. G. Barnes and Y. Hasegawa (Eds.), *Evolution and Migration of Fossil Marine Vertebrates in the Pacific Realm*. Island Arc, 3.

Hirayama, R. and T. Chitoku. 1994. Fossil turtles of Japanese marine strata. IN I. Kobayashi (Ed.), *Evolution and Adaptation of Marine Vertebrates* 2, pp. 17-24. The Association for the Geological Collaboration in Japan, Monograph 43 (Japanese with English Abstract).

Leidy, J. 1865. Memoir on the extinct reptiles of the Cretaceous formations of the United States. *Smithsonian Contributions to Knowledge* 14(6):1-165.

Nicholls, E. L. 1988. New material of *Toxochelys latiremis* Cope, and a revision of the genus *Toxochelys* (Testudines, Chelonioidea). *Journal of Vertebrate Paleontology* 8(2):181-187.

Nielsen, E. 1964. On the post-cranial skeleton of *Eosphargis breineri* Nielsen. *Bulletin of the Geological Society of Denmark* 15(3):281-328.

Schmidt, K. P. 1944. Two new thalassemyd turtles from the Cretaceous of Arkansas. *Field Museum of Natural History, Geological Series* 8:63-74.

Walker, W. E., Jr. 1973. The locomotor apparatus of Testudines. IN C. Gans and T. S. Parsons (Eds.), *Biology of Reptilia*, Volume 4, pp. 1-100. Academic Press, New York.

Weems, R. E. 1988. Paleocene turtles from the Aquia and Brightseat formations, with a discussion of their bearing on sea turtle evolution and phylogeny. *Proceedings of the Biological Society of Washington* 101:109-145.

Wieland, G. R. 1909. Revision of the Protostegidae. *American Journal of Science* 27(4):101-130.

Zangerl, R. 1953. The vertebrate fauna of the Selma Formation of Alabama. Part 3. The turtles of the family Protostegidae. Part 4. The turtles of the family Toxochelyidae.

Fieldiana, Geology Memoirs 3:61-277.

Zangerl, R. 1960. The vertebrate fauna of the Selma Formation of Alabama. Part 5. An advanced cheloniid sea turtle. *Fieldiana, Geology Memoirs* 3:281-312.

Zangerl, R. 1980. Patterns of phylogenetic differentiation in the toxochelyid and cheloniid sea turtles. *American Zoologist* 20:585-596.

Zangerl, R. and R. E. Sloan. 1960. A new specimen of *Desmatochelys lowi* Williston. A primitive cheloniid sea turtle from the Cretaceous of South Dakota. *Fieldiana, Geology* 14(2):7-40.

Chapter 9

DESMATOCHELYS LOWI, A MARINE TURTLE FROM THE UPPER CRETACEOUS

DAVID K. ELLIOTT, GRACE V. IRBY, and J. HOWARD HUTCHISON

INTRODUCTION

In 1983 the senior author (D. K. E.) found a large chelonioid turtle eroding from the Upper Cretaceous Mancos Shale at Blue Point on Black Mesa, northern Arizona (Figure 1). The remains were excavated during 1983 and 1984 (Elliott, 1984) and deposited in the collections of the Museum of Northern Arizona (MNA), whose collection number they now bear. This is the first reported occurrence of an identifiable turtle from the Mancos Shale, although fragmentary remains have been collected in the past and are held in the collections of the Museum of Northern Arizona.

The Mancos Shale is a 145-m-thick open marine deposit at Black Mesa representing the transgressive portion of the Greenhorn Cyclothem. The Mancos Shale consists of dark gray to brown silts and clays with minor fine-grained sandstones and limestones and occasional bentonites (Kirkland, 1991). The interval in which the turtle was found contains oysters, the inoceramid *Mytiloides opalensis*, and *Watinoceras* sp. However, the low abundance or total lack of infaunal elements may indicate low-oxygen conditions within the sediments (Elder, 1991).

The turtle was completely disarticulated, and based on the orientation of the skull and carapace elements may have arrived on the bottom in an upside-down position. A plesiosaur tooth found in association with the skeleton suggests a possible scavenger of the carcass; however, none of the bones show any evidence of predation. As this horizon is part of the latest Cenomanian and earliest Turonian Oceanic Anoxic Event (Elder, 1991; Eicher and Diner, 1985), which is marked by

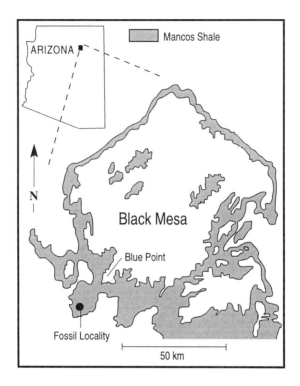

Figure 1. Locality map showing outcrop of the Mancos Shale at Black Mesa (based on Kirkland, 1991).

the extinction of numerous microfaunal and macrofaunal taxa, it may be that the animal was undisturbed by scavengers and was disassociated by currents.

Although the entire skeleton was probably present initially, part had been lost to erosion prior to collecting, and the difficulties of removing overburden as the excavation progressed probably resulted in some parts being left.

Despite this the skull and anterior part of the lower jaw were recovered, together with most of the cervical vertebrae, substantial portions of the carapace, part of the plastron, one humerus, one femur, and some distal elements of the flippers. Notably lacking are the limb girdles, although there are some fragments that may represent parts of the scapula and coracoid. All the material has been crushed to some degree. In the limb elements this has resulted in some flattening and changing of proportions but in the carapace and plastron, elements which are relatively thin, the result has been cracking and fragmentation. The skull has been flattened dorsoventrally, and material has been lost to cracking and weathering in the posterior part. In addition, the development of selenite crusts between skull elements has forced a number of them apart.

Repository Abbreviations

The abbreviations used for the institutions referred to in the text are as follows: CNHM, Field Museum of Natural History, Chicago; CVM, Comox Valley Museum, Courtenay, British Columbia; KUVP, University of Kansas, Museum of Natural History, Lawrence; MNA, Museum of Northern Arizona, Flagstaff; and UCMP, Museum of Paleontology, University of California, Berkeley.

AGE OF THE MATERIAL

The Mancos Shale ranges in age from upper Cenomanian to middle Turonian in the Black Mesa area (Kirkland, 1991). The turtle was collected about 12 m above the base of the Mancos Shale in the lower shale member. It occurs between the bentonite marker horizons BM 13 and BM 15 (Kirkland, 1991) which interval spans the Cenomanian-Turonian boundary. However, the presence of *Watinoceras* sp. and *Mytiloides opalensis* together with the turtle suggests that the horizon is in the lowermost Turonian.

PREVIOUS WORK

Williston (1894) reported the collection of a turtle (KUVP 1200) from the Upper Cretaceous Benton Formation of Fairbury, Nebraska. This included a well-preserved skull and lower jaw together with portions of the carapace, plastron, limbs, and limb girdles. He described this as a new genus and species of marine turtle, *Desmatochelys lowi*, and placed it in a new family, Desmatochelyidae. No further material was discovered until 60 years later when parts of the carapace, limbs, and limb girdles of a second specimen (CNHM PR 385) were discovered in Cretaceous sediments deposited on Precambrian granites in a quarry on the South Dakota-western Minnesota border (Zangerl and Sloan, 1960). They did not accept the validity of the family Desmatochelyidae, recognizing the species instead as a primitive cheloniid and placing it in the Cheloniidae.

Most recently, Nicholls (1992) described some material from the Upper Cretaceous of Vancouver Island (CVM 003) that was attributed to *Desmatochelys* cf. *D. lowi*. This included some postcranial elements and part of the mandible, but not the skull. Although no other material has been reported as belonging to this species, Smith (1989) has provided a preliminary description of a skull from the Early Cretaceous of Colombia (UCMP 38346) that he considers belonging to the genus *Desmatochelys*. This specimen has yet to be prepared and fully described.

SYSTEMATIC PALEONTOLOGY

Class REPTILIA
Order CHELONIA
Family DESMATOCHELYIDAE Williston, 1894
Genus *DESMATOCHELYS* Williston, 1894
DESMATOCHELYS LOWI Williston, 1894
Figures 2-6

Material: MNA V4516; skull, lower jaw, hyoid, cervical centra 3-8, one neurapophysis, two shell vertebrae, carapace and plastron plates, right humerus, right femur, phalanges, fragments of (?) coracoid and scapula.

Locality and horizon: MNA Locality 262, 12 m above the base of the Mancos Shale at Blue Point on Black Mesa, near Oraibi, northern Arizona.

DESCRIPTION

Skull

The skull is 28.8 cm in length with an elongated and cordiform appearance (Figures 2A and D and 3A and C). It is compressed, which has resulted in some fracturing and distortion, and in addition the posterior margin is crushed. Dorsally the large dermal bones show well-developed sutures that allow them to be clearly distinguished. Anteriorly the posterior margin of the external nares are formed by a pair of nasal bones which contact the frontals posteriorly and are bounded laterally by the maxilla. The frontals are large and irregularly shaped with an anterior process that meets the nasals and a lateral process that forms part of the dorsal margin of the orbits. The prefrontals are small bones that form dorsal lappets in the anterodorsal corner of the orbits and are separated from each other in the midline by the anterior processes of the frontals. Large postorbital bones contact the frontals anteriorly, form the posterior margin of the orbit, and extend to the posterior margin of the skull. These separate the squamosals (which cannot be distinguished due to crushing) from the parietals, which are elongated with rounded anterior and lateral margins. In the midline and close to the anterior margin of the parietals a parietal foramen may be present. This area is somewhat crushed, but thinning of the bone suggests the presence of the foramen. Posteriorly the supraoccipital forms an extensive projection 11 cm in length.

In palatal view (Figures 2D and 3C) the specimen is seen to be badly crushed, which has brought some of the more dorsal parts of the skull into close proximity to ventral elements. In addition, small postcranial elements drifted into skull apertures prior to crushing, contributing to further distortion and concealment of detail. The internal nares are deeply recessed, but the external openings are not visible in this view due to the distortion of the anterior part of the skull and the presence of small postcranial elements in this area. The vomer extends only a short distance into this region and hence the palatines meet in the midline. The maxilla and premaxilla form a high and robust lingual ridge that constitutes the lateral and anterior margins of this region. No secondary palate is developed. A small foramen palatinum posterius is present. The pterygoids form the posterior part of the palate and occupy less area than the palatines. Posteriorly they are in contact with the basisphenoid, which is a narrow triangular bone in ventral view. Toward the anterior part of the pterygoid-basisphenoid suture on each side there is a small foramen, which may be the posterior opening of the canalis caroticus internus. The basioccipital region extends prominently from the posterior margin of the skull and is flanked by prominent paroccipital processes.

Mandible

The anterior part of the lower jaw is preserved, as is the posterior part of the left ramus (Figure 2C). The jaw is long and narrow and the dentary shows a strong symphysis and a single, simple labial ridge. The posterior part of the left ramus shows the articulating surface and the posterior part of the fossa meckelii. Posteriorly the foramen posterius chorda tympani is visible.

Hyoid

Part of the hyoid apparatus is preserved and is reported here for the first time in *D. lowi* (Figure 2B). This is the right cornu branchiale I, which is the largest of the three hyoid horns and consists of a slightly curved, flattened, bony rod, 11 cm in length, that in life would have run posterolaterally from the intermediate lateral process of the hyoid.

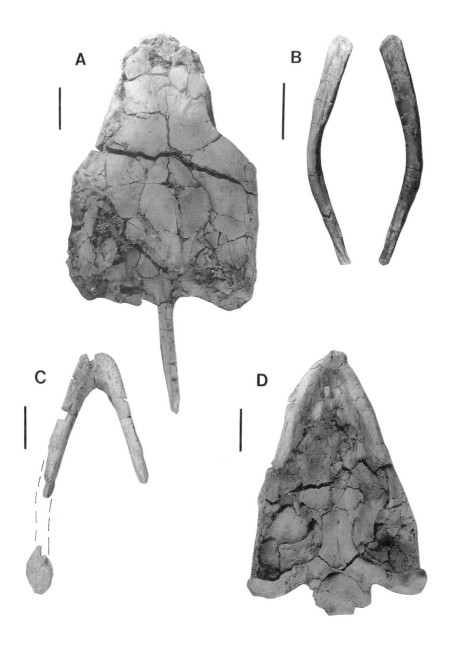

Figure 2. *Desmatochelys lowi* Williston, MNA V4516. **A**) Skull in dorsal view. **B**) Right cornu branchiale I in ventral and dorsal aspects. **C**) Mandible in dorsal aspect. **D**) Skull in palatal view. Scale bars = 3 cm.

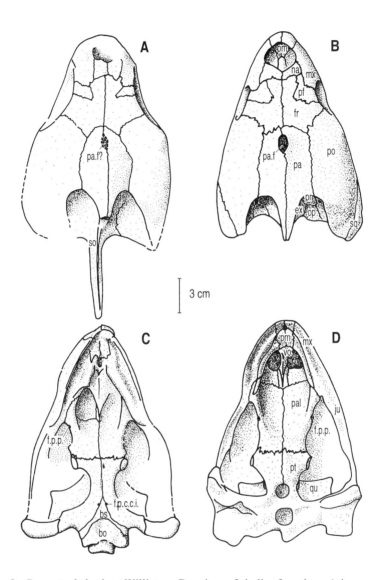

Figure 3. *Desmatochelys lowi* Williston. Drawings of skulls of northern Arizona and type specimens. **A)** Skull of northern Arizona specimen, MNA V4516, in dorsal view. **B)** Skull of holotype, KUVP 1200, in dorsal view. **C)** Skull of MNA V4516 in palatal view. **D)** Skull of KUVP 1200 in palatal view. Abbreviations: **bo** = basioccipital, **bs** = basisphenoid, **ex** = exoccipital, **f.p.c.c.i.** = foramen posterius canalis carotici interni, **f.p.p.** = foramen palatinum posterius, **fr** = frontal, **ju** = jugal, **mx** = maxilla, **na** = nasal, **op** = opisthotic, **pa** = parietal, **pa.f** = parietal foramen, **pal** = palatine, **pf** = prefrontal, **pm** = premaxilla, **po** = postorbital, **pr** = prootic, **pt** = pterygoid, **qu** = quadrate, **so** = supraoccipital, **sq** = squamosal, **vo** = vomer.

Axial Skeleton

Of the cervical vertebrae six centra and one neurapophysis are complete (Figure 4) and there are fragments of several other neurapophyses. The centra appear to represent C3 through C8, of which C4 through C8 are dorsoventrally compressed while C3 is compressed laterally. All the vertebrae are short and show the probably primitive ball-and-socket articulation rather than the transversely expanded flat-jointed situation found in modern sea turtles. The C4 is biconvex, as appears to be normal for *Desmatochelys* (Zangerl and Sloan, 1960). The C3 has not been described before and one notable feature is that the hypapophysis consists of a single sheet of bone divided ventrally by a longitudinal groove that is deep and narrow.

Two carapace centra are present, of which only one is complete. This is long and therefore not the most anterior, and is typically chelonian in construction with a rounded ventral face.

Carapace and Plastron

All the plates of the carapace were disarticulated and fragmented, and many are missing (Figures 5I-K and 6). The nuchal is present along with portions of seven costals, four peripherals, and the suprapygal. The nuchal has a somewhat concave anterior margin and together with the anterior segments of the marginal plates produces a shape that is typical for sea turtles with slight excavations in the shield above the forelimbs. The carapace narrows from this point posteriorly to produce a generally cordiform outline. The dorsal surface of the carapace is patterned by faint ridges that indicate the junctions of the scutes of the epidermal shield cover, in contrast to the normal arrangement in which sulci mark the scute boundaries. The carapace is estimated to be 100 cm long and 75 cm wide. The plastron is represented by the right hypoplastron only.

Appendicular Skeleton

The right humerus, right femur, and 14 carpal and tarsal elements are preserved (Figure 5A-H) together with fragments of what may be the coracoid and scapula. The right humerus is complete, 20.5 cm long, and is a massive, stout bone with a pronounced ulnar and an unusually large radial tuberosity.

The right femur is missing the distal end, the remaining portion is 12.8 cm long. The bone is slender and elongated and the head is compressed. The

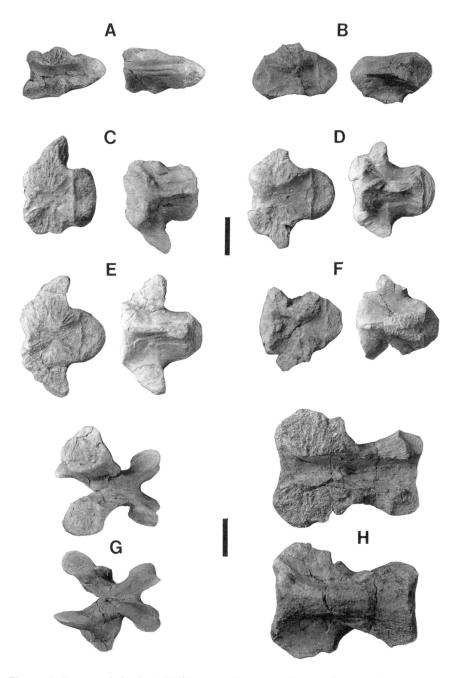

Figure 4. *Desmatochelys lowi* Williston. **A-F**, vertebral centra from northern Arizona specimen, MNA V4516, in dorsal and ventral views. **A**) C3. **B**) C4. **C**) C5. **D**) C6. **E**) C7. **F**) C8. **G**) Neurapophysis of C6 in dorsal and ventral views. **H**) Thoracic vertebra in dorsal and ventral views. Scale bars = 2 cm.

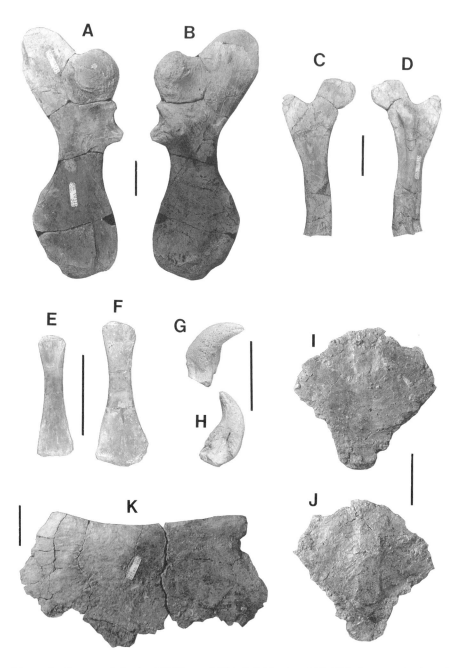

Figure 5. *Desmatochelys lowi* Williston, MNA V4516. **A** and **B**) Right humerus in dorsal and ventral views. **C** and **D**) Right femur in ventral and dorsal views. **E** and **F**) Metacarpal elements. **G** and **H**) Claw phalanx. **I** and **J**) Suprapygal in dorsal and ventral views. **K**) Nuchal in dorsal view. Scale bars = 3 cm.

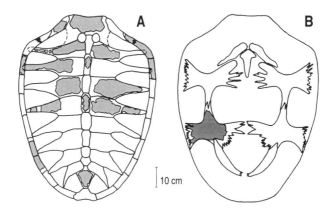

Figure 6. *Desmatochelys lowi* Williston, MNA V4516, carapace and plastron. Plates present in the northern Arizona specimen are denoted by stippling on an outline prepared from the reconstruction by Zangerl and Sloan (1960).

trochanters are separate but connected basally by a well-developed ridge.

DISCUSSION

The new skull accords well with the type although it is dorsoventrally crushed. Smith (1989) has noted three skull roof features of *Desmatochelys* that differ from those of modern sea turtles and that he considers to be diagnostic. All three features can be seen in the new specimen. First, the posterior margin of the external nares are formed by well-defined nasal bones and the prefrontals play no part in the structure of this area but are confined to forming dorsal lappets at the anterodorsal corners of the orbits (Figure 3B). As noted by Smith (1989), both *Glyptops* and *Rhinochelys* show a similar relationship. In the new specimen the nasals are relatively longer and narrower than those in the type and so are not contacted by the anterior projection of the prefrontals (Figure 3A). This may be an individual variation as marine turtles are known to show a high degree of variation in the relationship of skull bones (Gaffney, 1979). Second, Smith (1989) has noted that the prefrontals lack medial contact, being separated by the anterior projection of irregularly shaped frontal bones. Third, the large postorbitals separate the parietals from the squamosals and make a major contribution to the posterior margin of the skull.

The new skull does differ from the type in having what appear to be proportionally larger orbits. However, this may simply be due to the great degree of dorsoventral crushing experienced by the new specimen. The type specimen of

D. lowi has a parietal foramen (Figure 3B). It is unclear whether the new specimen has a similar structure as the area is damaged. However, thinning of the bone suggests that it was present. This is not considered to be a diagnostic feature, as it can vary individually; skulls of *Geochelone radiata* examined by Crumly (1982), for instance, showed a parietal foramen in only 59% of examples. One additional feature shown in the new skull is the posterior spine of the supraoccipital, which is a slender, tapering bone 11 cm in length. This bone is missing in both the type specimen from Kansas and the skull from Colombia.

Ventrally the new skull is very similar to the type (Figure 3C and D). Posteriorly it provides new information in the area of the basisphenoid and basioccipital, which have been destroyed by a drill-hole in the type specimen. The basisphenoid has been shown by Smith (1989) to contain a number of diagnostic characters. However, these are mostly on the dorsal surface, and in the new specimen crushing has obscured the more important areas. The anterior part of the basisphenoid-pterygoid suture shows a foramen on each side, which may indicate the presence of the foramen posterius canalis carotici interni. Similarly placed foramina are visible in the skull of *Desmatochelys* sp. from the Early Cretaceous of Colombia described by Smith (1989). This anterior position of the foramina is considered to be primitive (Smith, 1989) and is also shared by *Rhinochelys*.

Postcranially the new material compares closely with that previously described. One exception is the presence of a third cervical vertebra that has not previously been described in *Desmatochelys* (Figure 4A). This is similar to the same element in *Caretta caretta* but differs in having a hypapophysis that bears a narrow, longitudinal slit ventrally.

Recent authors (Moody, Chapter 10; Hirayama, Chapter 8; Smith, 1989; Gaffney and Meylan, 1988) agree that *Desmatochelys* is aligned with the protostegids and dermochelyids but differ in the placement within this broad grouping. Gaffney and Meylan (1988) leave *Desmatochelys* a sister-group of protostegids or in an unresolved multicotomy. Hirayama (Chapter 8) placed it within his expanded Protostegidae. Moody (Chapter 10) and Smith (1989) retained it as a separate family. We follow the latter position for the moment on the grounds that the Protostegidae *sensu stricto* are well defined on a number of derived features (e.g., lateral expansion of the entoplastron and probable loss of the epiplastra, loss of nasals, loss of pterygoid processes, frontals excluded from the orbits, etc.). Desmatochelyids are clearly an outgroup to the Protostegidae *sensu stricto*.

The Arizona specimen shares with protostegids and desmatochelyids such derived features as the position of the f.p.c.c.i. between the basisphenoid and pterygoid, mutual contact of the palatines, and the pterygoid extending to the articular surface of the quadrate. It differs from protostegids and agrees with desmatochelyids in the retention of nasals, retention of frontal exposure at the orbits, retention of pterygoid processes, and retention of postorbital separation of the parietal and squamosal. Within chelonioids, it agrees uniquely with desmatochelyids in the contact of the frontal with the nasals, thus separating the prefrontals.

AGE AND DISTRIBUTION

The type specimen of *Desmatochelys lowi* comes from the Hartland Shale Member of the upper Greenhorn Formation (Williston, 1894), which places it in the upper Cenomanian. The specimen from South Dakota was thought to have come from a horizon equivalent to the Carlisle Shale (Zangerl and Sloan, 1960), which is middle to upper Turonian in age. Additional isolated material noted by Nicholls (1992) comes from the Fairport Chalk of Ellis County, Kansas, which is the basal member of the Carlisle Shale and is middle Turonian in age. As the Arizona specimen comes from the base of the Turonian, all four occurrences are late Cenomanian to late Turonian in age (Figure 7). Although this suggests a narrow age distribution for the species, it should be noted that the recently described material from Vancouver Island (Nicholls, 1992) comes from the late Campanian to early Santonian Trent River Formation, extending the genus, if not the species, into the uppermost Cretaceous. The undescribed skull from Colombia (Smith, 1989) appears to extend the genus back into the Early Cretaceous, although precise stratigraphic information is not available for this specimen.

The North American specimens were found at the margins of the Cretaceous Interior seaway (Figure 8), with the exception of the Vancouver Island specimen, which is to the west of the Cordilleran highlands. The seaway formed in the Western Interior Basin, which itself developed in response to accelerated spreading, convergence, and subduction along the western margin of North America during Late Jurassic to Cretaceous time. The basin was flooded by shallow epicontinental seas from the north and south in Barremian-Aptian time; the arms joined in the late Albian and the seaway remained flooded until late middle Maastrichtian (Kauffman, 1985). Flooding reached a peak in early Turonian to middle Santonian, which induced a broad mixing of warm temperate to subtropical water masses from the south with mild to cool temperate water masses from the north (Kauffman, 1985). As the Arizona, South Dakota, and Nebraska specimens are placed around the margins of the seaway during peak transgression, they seem to indicate a basin-wide distribution at this time. The presence of a member of the genus in the Early Cretaceous of Colombia suggests an origin for the genus in that area with dispersion northward as the Western Interior Basin developed and the seaway was connected. The presence of *Desmatochelys* to the west of the Cordilleran highlands late in the Cretaceous probably represents a northward dispersal as the seaway began to close.

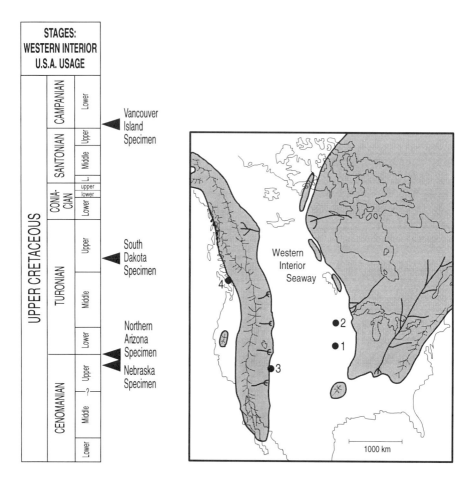

Figure 7 (left). Age distribution of principal specimens of *Desmatochelys lowi*.

Figure 8 (right). Map of the Western Interior seaway in the Late Cretaceous showing the locations of the principal specimens of *Desmatochelys lowi*. **1**) Holotype (Williston, 1894). **2**) South Dakota specimen (Zangerl and Sloan, 1960). **3**) Arizona specimen. **4**) Vancouver Island specimen (Nicholls, 1992).

SUMMARY

A new specimen of *Desmatochelys lowi* from the Mancos Shale of northern Arizona was described. It has a well-preserved skull and contributes to our knowledge of the cranial anatomy of the species and its geographic distribution.

ACKNOWLEDGMENTS

We would like to thank H. G. Honanie of the Office of the Vice-Chairman of the Hopi Tribe for assistance in obtaining permission to excavate this specimen, and W. Downs for valuable assistance during the excavation. This research was supported by a Northern Arizona University Organized Research Grant award to D. K. Elliott.

REFERENCES

Crumly, C. R. 1982. The "parietal" foramen in turtles. *Journal of Herpetology* 16:317-320.

Eicher, D. L. and R. Diner. 1985. Foraminifera as indicators of water mass in the Cretaceous Greenhorn Sea, Western Interior. IN L. M. Pratt, E. G. Kauffman, and F. B. Zelt (Eds.), *Fine-Grained Deposits and Biofacies of the Cretaceous Western Interior Seaway: Evidence of Cyclic Sedimentary Processes*, pp. 60-71. SEPM Field Trip Guidebook 4.

Elder, W. P. 1991. Molluscan paleoecology and sedimentation patterns of the Cenomanian-Turonian extinction interval in the southern Colorado Plateau region. IN J. D. Nations and J. G. Eaton (Eds.), *Stratigraphy, Depositional Environments, and Sedimentary Tectonics of the Western Margin, Cretaceous Western Interior Seaway*, pp. 113-137. Geological Society of America Special Paper 260.

Elliott, D. K. 1984. A large marine turtle from the Upper Cretaceous Mancos Shale. *Abstracts of the Symposium on Southwestern Geology and Paleontology, Museum of Northern Arizona*:7.

Gaffney, E. S. 1979. Comparative cranial morphology of recent and fossil turtles. *Bulletin of the American Museum of Natural History* 164:65-376.

Gaffney, E. S. and P. A. Meylan. 1988. A phylogeny of turtles. IN M. J. Benton (Ed.), *The Phylogeny and Classification of Tetrapods, Volume I: Amphibians, Reptiles, Birds*, pp. 157-219. Systematics Association Special Volume, 35A. Clarendon Press, Oxford.

Kauffman, E. G. 1985. Cretaceous evolution of the Western Interior Basin of the United States. IN L. M. Pratt, E. G. Kauffman, and F. B. Zelt (Eds.), *Fine-Grained Deposits and Biofacies of the Cretaceous Western Interio r Seaway: Evidence of Cyclic Sedimentary Processes*, pp. 4-13. SEPM Field Trip Guidebook 4.

Kirkland, J. I. 1991. Lithostratigraphic and biostratigraphic framework for the Mancos Shale (Late Cenomanian to Middle Turonian) at Black Mesa, northeastern Arizona. IN J. D. Nations and J. G. Eaton (Eds.), *Stratigraphy, Depositional Environments, and Sedimentary Tectonics of the Western Margin, Cretaceous Western Interior Seaway*, pp. 85-11 Geological Society of America Special Paper 260.

Nicholls, E. L. 1992. Note on the occurrence of the marine turtle *Desmatochelys* (Reptilia: Chelonioidea) from the Upper Cretaceous of Vancouver Island. *Canadian Journal of Earth Sciences* 29:377-380.

Smith, D. T. J. 1989. *The Cranial Morphology of Fossil and Living Sea Turtles (Cheloniidae, Dermochelyidae and Desmatochelyidae)*. Ph. D. thesis, Kingston Polytechnic, Surrey, England, 310 pp.

Williston, S. W. 1894. A new turtle from the Benton Cretaceous. *Kansas University Quarterly* 3:5-18.

Zangerl, R. and R. E. Sloan. 1960. A new specimen of *Desmatochelys lowi* Williston, a primitive cheloniid sea turtle from the Cretaceous of South Dakota. *Fieldiana, Geology* 14:7-43.

Chapter 10

THE PALEOGEOGRAPHY OF MARINE AND COASTAL TURTLES OF THE NORTH ATLANTIC AND TRANS-SAHARAN REGIONS

RICHARD T. J. MOODY

INTRODUCTION

Marine and coastal marine turtles are recorded from Cretaceous, Paleogene, and Neogene strata along the opening Atlantic coastline of west Africa, throughout the western interior of the Sahara, North Africa, and across western Europe. A diverse fauna of coastal, hemipelagic, and pelagic species is recorded, with representatives of extant families such as the Pelomedusidae, Cheloniidae, and Dermochelyidae present from the early Cretaceous onward. The fossil record of the cheloniids and dermochelyids along the evolving eastern Atlantic margin is somewhat limited in comparison with the numerous species recorded from North America or the Paleogene strata in northern Europe. The distribution and relative abundance of the two families are thought to be closely associated with the paleogeography of both the Cretaceous Chalk Basin and the London-Paris-Belgium Basin during the Tertiary. The presence of sea grass in the chalks of the Maastricht region of Holland may indicate a possible feeding site for *Allopleuron*, whereas a mixed assemblage of juveniles and adults suggests that the Isle of Sheppey was a nesting site.

The Pelomedusidae are common to abundant in the essentially shallow marine sediments of the Cretaceous-Tertiary Trans-Saharan Seaway and North Africa and from various sites in Europe. Extant pelomedusids live in fresh water but many workers, including Wood (1973), think that many fossil species were adapted to a marine environment.

Taphrosphys sp., well known from North (Gaffney, 1975) and South America

(Wood, 1973) is also recorded from Zaire and France. Other fossil pelomedusid genera are recorded from Niger, Morocco, Tunisia, Egypt, Portugal, Spain, France, and England. They are usually found in association with other marine organisms and appear to have migrated freely along the shoreline of the widening Atlantic and through the relatively narrow Trans-Saharan Seaway (see Moody and Sutcliffe, 1990; Walker, 1979).

Extinct families such as the Desmatochelyidae and Toxochelyidae are also recorded from North America (Irby et al., 1994; Nicholls, 1992; Zangerl and Sloan, 1960; Williston, 1898), northern Europe, and possibly North Africa. In Europe, the Desmatochelyidae are represented by the genus *Rhinochelys* (see Smith, 1989) and possibly *Allopleuron hoffmani* (see Gaffney and Meylan, 1988) and the Osteopygidae, by the subfamilies Osteopyginae and Lophochelyinae.

Stratigraphic occurrences of the various genera and species are documented by Smith (1989), Seago (1980), Russell (1982), Moody (1974, 1980, 1993), and De Broin (1977). The stratigraphic distribution of named taxa is set out in Figure 1.

MARINE TURTLES OF THE DESMATOCHELYIDAE, OSTEOPYGIDAE, CHELONIIDAE, AND DERMOCHELYIDAE

Marine turtles referred to the micro-order Chelonioidea by Gaffney and Meylan (1988), are characterized by the modification of the forelimb as a flipper. This involves the extension of radius and ulna, a straightening of the humerus, elongation of the third and fourth digits, and the development of flattened carpals resulting in the restricted movement of the wrist (Gaffney et al., 1991; Gaffney and Meylan, 1988). Well-developed flippers are diagnostic of the Cretaceous and Paleogene marine turtles of Europe and Africa reviewed by Smith (1989), Seago (1980), and Moody (1970, 1974, 1980, 1993). Brief characterizations of the Desmatochelyidae, Osteopygidae, Cheloniidae, and Dermochelyidae are given below.

Desmatochelyidae

The skull of *Desmatochelys* is characterized by the presence of palatine foraminae, a descending process of the parietal and nasal bones in the skull mosaic (Williston, 1898). Gaffney and Meylan (1988) noted that a parietal-squamosal contact is absent in both *Desmatochelys* and *Allopleuron*. The postcranial skeleton of the Desmatochelyidae is typical of open marine turtles with reduced carapace and plastron and an advanced humeral morphology, with a straightened shaft and a distal lateral process. In contrast to *Desmatochelys*, the carapace of *Allopleuron* bears no

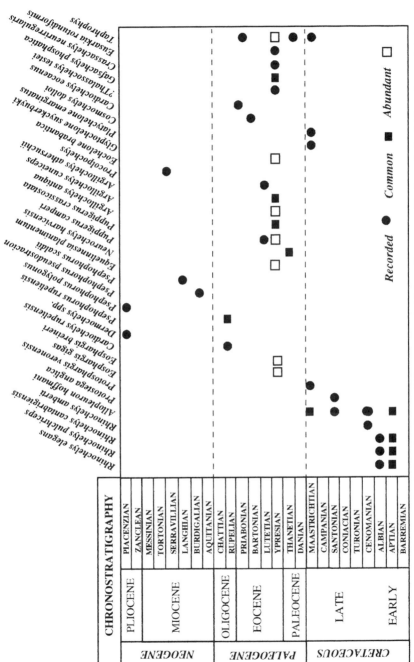

Figure 1. Stratigraphic distribution and relative abundance of European and African Cretaceous-Tertiary chelonioids and associated marine pelomedusids.

sign of epidermal scute furrows. This may indicate the absence of an epidermal armor as a more advanced feature or simply that the epidermal scutes were less intimately bound to the underlying bony plates.

The remains of *Allopleuron hoffmani* and other sea turtles are frequently found in the Cretaceous of Limburg (southern Holland-northern Belgium). Felder (1980a, b) noted that 257 sea turtle fossils were found in the region between 1766 and 1978. The vast majority is postcranial material belonging to *Allopleuron hoffmani*, with *Glyptochelone suyckerbuyki* and *Platychelone emarginata* also represented.

The statuses of the latter two species are not fully resolved, but Gaffney and Meylan (1988) claimed that *Allopleuron hoffmani* was more closely related to *Desmatochelys* than other chelonioids. Gaffney and Meylan (1988) claimed that *Desmatochelys* is very closely related to another North American species, *Notochelone costata*. Smith (1989) thought that these genera are not members of the Protostegidae. As in the more primitive cheloniids, such as *Puppigerus*, the anterior foramen for the carotid canal emerges beneath the wall of the dorsum sellae in *Notochelone, Desmatochelys,* and *Rhinochelys*. The fusion and elongation of the basisphenoid rostrum in these genera are more characteristic of the cheloniids than either the protostegids or dermochelyids, in which the basisphenoid remains cartilaginous.

Smith (1989) confirmed the presence of well-defined nasal bones in *Desmatochelys* and listed ten other characteristics that suggested a close affinity with the European genus *Rhinochelys*. Moody (1993) noted that *Rhinochelys* is known from over 100 cranial specimens and that it is extremely small in comparison to *Desmatochelys lowi*, having a typical skull length of 4-4.5 cm. In contrast, skull length of *Desmatochelys lowi* (Ku 1200, University of Kansas, Lawrence) is recorded as 22.5 cm. Collins (1970) referred the British material associated with *Rhinochelys* to the species *R. cantabrigiensis, R. pulchiceps,* and *R. elegans,* thus reducing the number of species from 21 to 3. Collins (1970) regarded the species *R. amaberti* (see Moret, 1935), from France, as distinct and valid.

Rhinochelys is known only from skulls, while the species *Cimiochelys benstedi*, from the Albian to Turonian of southeast England, is represented only by shell material. By reasoning that both were chelosphargid members of the Protostegidae, Collins claimed that the skulls and shells were probably from the same genus.

Smith (1989) cast doubt on several of the conclusions reached by Collins, including the existence of three species and the referral of *Rhinochelys* to the Protostegidae, subfamily Chelospharginae. Moody (1993) reported that *Rhinochelys* and *Desmatochelys* are chelonioids and closely related to *Allopleuron hoffmani*. It should be noted, however, that Hirayama (Chapter 8) argues that *Allopleuron* is a cheloniid turtle.

Geographic and Stratigraphic Distribution of the Desmatochelyidae (Figures 1 and 2)

Remains of *Rhinochelys* are recorded from the Cambridge Greensand, Gault Clay and lower Chalk of England, and from Vraconian sediments of France. Material collected from the Cambridge Greensand, near Cambridge, England, accounts for the vast majority of known samples, with the type material of the various species being housed either in The Natural History Museum (formerly the British Museum [Natural History]) or the Sedgwick Museum, Cambridge. The Cambridge Greensand is of lower Cenomanian age; the Gault Clay is Albian. The species *Rhinochelys amaberti*, described by Moret (1935), is known from southeast France (La Fauge Valley, near Grenoble).

New material associated with the genus *Allopleuron* has been uncovered regularly over the last decade. *Allopleuron* is one of the largest Cretaceous turtles, and is specifically adapted to a pelagic lifestyle.

Allopleuron hoffmani is well known from numerous specimens recovered from the Maastrichtian strata of Holland. No well-defined material is recorded from the English Chalk, but various bones have been recently identified as belonging to *Allopleuron* from the Costatus subzone of the middle Cenomanian of the Petreval district, Seine-Maritime, Normandy (E. Buffetaut, personal communication) and from the Turonian of Montrichard, Loire et Cher, Loire Valley, France (N. Bardet, personal communication). Additional material, including elements of the girdle, have been discovered from the Santonian of the Spanish Basque country, near La Cuenca Vasco, Alava Province (Bardet et al., 1993). The indications are that *Allopleuron* was one of the most long-ranging and widespread of European sea turtles.

Osteopygidae

The Osteopygidae are characterized by low, rounded skulls with short, pointed snouts, dorsally directed orbits, extensive secondary palates with choanal passages separated by a vomerine pillar, and a very long symphysis on the lower jaws (Gaffney and Meylan, 1988). These features draw together the American genera *Osteopygis* and *Erquelinnesia* from the Eocene of northern Europe. Gaffney and Meylan (1988) thought them to be distinctive synapomorphies and although Moody (1974) and Zangerl (1971) had recognized these characteristics earlier, they had failed to separate *Erquelinnesia* from the Toxochelyidae. Based on characters analyzed by Fastovsky (1985), Baird (1986) referred the genus and subfamily Osteopyginae to the Cheloniidae. Close affinities of the osteopygines to the

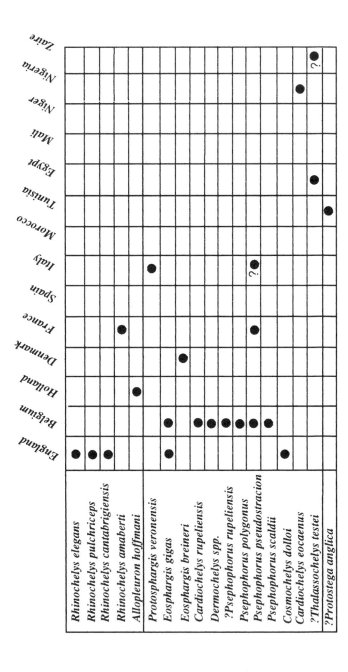

Figure 2. Geographic distribution of European and African Cretaceous-Tertiary species of the Desmatochelyidae, Dermochelyidae, and Protostegidae.

cheloniids were noted by Moody (1974) and Zangerl (1971). However, Gaffney and Meylan (1988) stated that the Toxochelyidae, as defined by Zangerl (1953b), is not a monophyletic group "since some of the included taxa are more closely related to cheloniids." These taxa include *Osteopygis* and *Erquelinnesia*. Gaffney and Meylan (1988) regarded *Toxochelys* as a sister taxon to all other chelonioids.

Moody (1970), in a review of Eocene turtles, referred various material attributed to *Lytoloma planimentum*, *L. crassicostatum*, *L. gosseleti*, and *Chelone harvicensis* to the Toxochelyidae, as defined by Zangerl (1953b), and later (Moody, 1980) consolidated the various material within the two species *Neurochelys harvicensis* and *Erquelinnesia planimentum*. Moody referred *N. harvicensis* to the subfamily Lophochelyinae (see Zangerl, 1953b) and *E. planimentum* to a new subfamily named the Erquelinnesinae.

Neurochelys is known only from postcranial material which bears comparison with that of *Ctenochelys* and *Lophochelys*. As with *Toxochelys*, *Ctenochelys* is seen by Gaffney and Meylan (1988) to be a sister taxon to all other chelonioids. It is possible that *Neurochelys* occupies the same status. *Toxochelys* sp. has recently been recorded from the Upper Cretaceous chalks of the Loire region of France by Bardet et al. (personal communication). Moody (1972), in a review of the turtle fauna of the Eocene phosphates of south central Tunisia, synonymized *Lytoloma elegans* and *L. crassa* of Bergounioux (1952). The Tunisian material is well preserved but the species are known only from skulls.

Geographic and Stratigraphic Distribution of the Osteopygidae (Figures 1 and 3)

The genus *Osteopygis* is known from the Upper Cretaceous of North America. It therefore predates related European species such as *Erquelinnesia planimentum*, recorded from the Ypresian of England and Belgium, and *Neurochelys harvicensis*, from the lower Ypresian of England. It is probable that *Erquelinnesia* also existed in the Paris Basin during the Ypresian. Zangerl (1953b) referred the type specimen of *Erquelinnesia molaria* to the species *Osteopygis emarginatus*. However, it appears that the American and European osteopygids were stratigraphically isolated.

Moody (1980) noted the similarity between the European species *Neurochelys harvicensis* and *Lophochelys* from the Upper Cretaceous of North America but thought that the two were separate in time and space, but with the American species as precursor stocks.

Specimens referred to the species *Erquelinnesia planimentum* are recorded from the London Clay Formation (lower Eocene) of both Harwich and the Isle of Sheppey in England and from lower Landenian deposits of Belgium.

	England	Belgium	Holland	Denmark	France	Spain	Italy	Morocco	Tunisia	Egypt	Mali	Niger	Nigeria	Zaire
Equelinnesia planimentum	●													
Neurochelys harvicensis	●													
?Lytoloma														?
Puppigerus camperi	●	●												
Puppigerus crassicostata	●	●							●					
Argillochelys antiqua	●	●												
Argillochelys cuneiceps	●	●												
Argillochelys athersuchii	●													
Procolpochelys spp.														
Eochelone brabantica	●	●												
Syllomus aegypticus										●				
Glyptochelone suyckerbuyki			●											
Platychelone emarginatus			●											
Gafsachelys phosphatica									●					
Crassachelys neurirregularis									●					
Eusarkia rotundiformis					●				●					
Taphrosphys				●					? ●	●	●		●	●

Figure 3. Geographic distribution of European and African Cretaceous-Tertiary species of the Osteopygidae, Cheloniidae, and marine Pelomedusidae.

Evidence of the widespread distribution of osteopygid turtles is presented by Wood (1973), who described lophochelyine material from the Paleocene of Landana in the Republic of Zaire.

Cheloniidae

The Cheloniidae are characterized by the presence of a platycoelus articulation between the sixth and seventh cervical vertebrae (Williams, 1950). Living cheloniids are divided into the tribes Chelonini and Carettini. The division is based on both cranial and postcranial characters. Gaffney and Meylan (1988) noted that the chelonines have an unusually large coracoid, scapular processes that form a relatively wide angle, and a relatively broad bridge. The carettines possess relatively low triturating surfaces without strong ridges, five or more pairs of pleural scales, and twelve pairs of peripheral bones.

Among fossil cheloniids from the Cretaceous and Tertiary, Moody (1974) recognized the presence of a platycoelus articulation in *Puppigerus*, but noted a convexo-concave (procoelus) relationship in *Argillochelys*. Zangerl and Sloan (1960) and Zangerl (1953b) recorded a procoelus articulation in *Toxochelys* and *Desmatochelys,* respectively.

Moody (1968, 1970, 1974, 1980, 1993) persisted in the recognition of *Puppigerus, Argillochelys* and *Eochelone* as early cheloniid turtles. Of the various species described by Moody (1974, 1980), Owen and Bell (1849), Dollo (1903), Gray (1831), and Konig (1825), six survive. They are: *Puppigerus camperi, Puppigerus crassicostata, Argillochelys cuneiceps, Argillochelys antiqua, Argillochelys athersuchii,* and *Eochelone brabantica.*

All six species share characteristics with living cheloniids, with the flipper-like forelimb, composed of a straight humerus, flattened carpals, and elongate third and fourth digits. The skull roof is only slightly emarginated and the orbits are directed outward and slightly forward. The carapace is frequently cordiform, and the plastron consists of nine bones as in all other chelonioids, with the possible exception of the dermochelyids.

Gaffney and Meylan (1988) cast doubt on the familiar status of *Eochelone* and stated that, while "there is enough material to argue that it is a cheloniid," it is possible that sufficient characteristics exist to relate *Eochelone* with the dermochelyids.

The skull of *Eochelone* is superficially similar to that of *Puppigerus* but lacks a primary palate and has broad triturating surfaces and "apparently has a smaller canalis caroticus lateralis in comparison with the canalis caroticus internus" (Gaffney and Meylan 1988). The shell is small, not reduced, and similar in most respects to those of *Puppigerus* and *Argillochelys*. Zangerl (1971) referred *Eochelone* to the family Toxochelyidae but Gaffney and Meylan (1988) prefer it to be an unnamed taxon.

Moody (1980) thought that *Puppigerus*, *Argillochelys*, and *Eochelone* represented a conservative root stock for recent cheloniids. However, one could argue for a review of this grouping, particularly as *Argillochelys* shares certain characteristics with *Toxochelys* and *Eochelone* does so with the dermochelyids.

Procolpochelys melii from the Miocene of Lecce in southern Italy is thought to be an early representative of the Carettini (see Seago, 1980).

Geographic and Stratigraphic Distribution of the Cheloniidae (Figures 1 and 3)

Of the three genera referred to from this family, both *Puppigerus* and *Argillochelys* are known from Ypresian-Lutetian sediments. *Puppigerus camperi* is recorded from the London Clay (lower Ypresian) and Bracklesham Beds (Lutetian) of England, and from the Sables de Bruxelles and Sables de Wemmel of Belgium. The last two formations are of middle to late Eocene (Bruxellian-Wemmelian) age. *Puppigerus crassicostata* is known only from the London Clay (lower Ypresian) of Sheppey and Harwich.

Of the species referred to the genus *Argillochelys*, *A. cuneiceps* is known only from the lower Eocene (London Clay) of Sheppey and Harwich, whereas *A. antiqua* is also known from the Landenian of Belgium and possibly from the Eocene of Holland. *Argillochelys athersuchii* appears to be restricted to the upper Eocene of the Hampshire Basin.

Eochelone brabantica is common in the Bruxellian (Sables de Bruxelles) of Belgium. Only two skulls, previously referred to the species *Argillochelys antiqua*, from the London Clay of Sheppey can be assigned to this species. Isolated and fragmentary remains of *Eochelone brabantica* are known to exist, in private collections, from the upper Eocene of the Hampshire Basin.

Dermochelyidae

Characteristically, these turtles have a comparatively large, subtriangular skull, with no emargination of the parietal area, have, a primary palate, and lack a truly ossified rostrum basisphenoid. The external nares are directed forward, and the orbit outward and slightly forward. The lower jaw has a distinct "tooth-like" projection, which slots into a socket or notch in the premaxilla. The humerus is short, typically flattened, with a large radial process near the middle of the shaft. The shell is greatly reduced and *Dermochelys* lacks an epidermal armor.

Seago (1980) undertook a comprehensive review of the fossil turtles of the Dermochelyidae, and noted that the taxonomic status of *Protosphargis veronensis* from northern Italy is uncertain. Seago (1980) suggested that the shell in particular was closer to that of the dermochelyids than those of its Cretaceous contemporaries, i.e., *Desmatochelys* and *Corsochelys*, and noted that although the centra of the dorsal vertebrae were like those of most marine chelonians their arrangement and the form of the ribs had dermochelyid affinities. Only the eighth cervical vertebra is known from this species, and the lack of a double anterior articulation surface on the centrum is a rather primitive condition.

Eosphargis gigas and *Eosphargis breineri*, however, are undisputed dermochelyids (Seago, 1980), although the size of the parasphenoid is smaller than that of *Dermochelys* in both species.

Bergounioux (1952) and Moody (1972), writing on the turtles of the Ypresian phosphates of south central Tunisia, noted that the skull of *Thalassochelys testei* bore chelonioid affinities and referred it to the Cheloniidae. Bergounioux (1952) thought the species was conspecific with *Lytoloma elegans*, whereas Moody (1972) noted that the generic name is a junior synonym of *Caretta*. Subsequent preparation of the lower jaw and palate reveals that the skull is that of a small dermochelyid with direct affinities to *Eosphargis*.

Seago (1980) figured various material referred to the species *Psephophorus scaldii* and *Psephophorus polygonus*. Both have a well-developed epithecal carapace, composed of a mosaic of closely spaced tesserae. The humerus is typical of the seagoing dermochelyids. Gaffney and Meylan (1988) linked the dermochelyids and protostegids within the epifamily Dermochelyoidae, noting that the femur possessed a very large *trochanter major* in relation to the *trochanter minor*, with the former projecting proximal to the femoral head. It should be noted that the femur of *Eosphargis* is more elongate than those of either *Dermochelys* or *Psephophorus*, which are broader, with slightly reduced radial processes. De Broin and Pironon (1980), however, thought it was closer to the protostegids.

The species *Cardiochelys rupeliensis* is also recorded from the lower Oligocene of Belgium. Seago (1980) figured the skull of *Cardiochelys rupeliensis* and noted that the basisphenoid area appeared to be more ossified than in *Dermochelys*. Smith (1989) noted that *Cardiochelys rupeliensis* was characterized by the lack of a truly ossified rostrum basisphenoid and the apparent passage of the abducens nerve through a groove running laterally to the area of the clinoid processes. Smith (1989) suggested that the deossification of the dermochelyid endocranial region may therefore be a recent phenomenon. It is possible that this character state may distinguish *Dermochelys* from all other dermochelyids.

Gaffney and Meylan (1988) discussed the close relationship of the protostegids and dermochelyids. They regarded the chelosphargids as less specialized protostegids, with *Archelon* and *Protostega* as a monophyletic group. They were less convinced about the grouping of *Dermochelys*, *Eosphargis,* and *Psephophorus*, regarding the latter as something like a wastebasket taxon. Seago (1980) considered *Cardiochelys rupeliensis* to be a dermochelyid. Moody (1993) stressed that a full review of the dermochelyid-protostegid plexus was required, but that these various species noted above would probably remain within the Dermochelyidae.

Geographic and Stratigraphic Distribution of the Dermochelyidae (Figures 1 and 2)

Protosphargis veronensis is known from the Upper Cretaceous of northern Italy.

Eosphargis gigas and *E. breineri* appear to be confined to the Ypresian stage, but Seago (1980) suggests that *E. breineri* was probably confined geographically to the area of the North Sea as known today. He indicated that *Eosphargis gigas* migrated into this area during the Ypresian, but evidence for an earlier presence outside this area is restricted to dubious data from southern Africa and Tunisia. There is little doubt that *Eosphargis* is a large, seagoing turtle of dermochelyid affinity, but the presence of a thecal dermal shell, and the primitive cheloniid nature of the flipper, led Seago (1980) to refer the species concerned to the subfamily Eospharginae.

According to Lydekker (1889), the remains of the dermochelyid genus *Psephophorus* are known from the middle Eocene of Bracklesham, Sussex; however, the most abundant and best preserved material referred to this genus in Europe is found in the Oligocene of Belgium. *Cardiochelys rupeliensis* is recorded from the Argile de Boom (Rupelian), middle Oligocene, of Belgium. *P. scaldii* is known from the upper Miocene Antwerp Sands (Anversian) of Belgium and *P. polygonus* is known from the lower Miocene Molasse de Vendargues of Herault, the Vindabonian of Neudorfl (near Bratislava, Slovakia), and the upper Miocene Antwerp Sands (Anversian) of Belgium. *Psephophorus pseudostracion* is recorded from the Pliocene of Vendarques, France, and *Dermochelys* sp. from the Pliocene of Belgium.

Of these species, *P. polygonus* is also known from the middle Miocene of Calvert County, Maryland. Another North American species of dermochelyid turtle is *Miochelys fermini* of the Altamira Shale of Point Fermin, California.

De Broin and Pironon (1980) reported the discovery of material belonging to the genus *Psephophorus* from the Matse Mountains, Benevento Province, Italy, and described it as the first dermochelyid from that country.

African dermochelyids such as *Cosmochelys dolloi* from the middle Eocene of Ameki, southern Nigeria, and *Cardiochelys eocaenus* from the upper Eocene of Fayum, Egypt, testify to the cosmopolitan distribution of the dermochelyids and their ancestors.

MARINE TURTLES OF THE PELOMEDUSIDAE

Pelomedusid turtles belong to the gigaorder Casichelyidia, Eupleurodira as defined by Gaffney and Meylan (1988). They lack nasals and the splenial bone is absent. The prefrontals meet in the midline. Williams (1950) noted that the second cervical vertebra was biconvex, and numbers 3 to 8 procoelus. The carapace oval is subcordiform in outline, moderately arched with 5 to 7 neurals, behind which the pleurals meet in the midline.

Moody (1972) reviewed the earlier work of Bergounioux (1952) on the turtle fauna of the Metlaoui region of south central Tunisia. He noted the relative abundance of pelomedusid turtles, confirming the existence of three species: *Gafsachelys phosphatica, Crassachelys neurirregularis,* and *Eusarkia rotundiformis.*

De Broin (1977) referred the Tunisian material to the genus *Taphrosphys*, which Gaffney and Meylan (1988) subsequently referred to an unnamed taxon. De Broin (1977) provides the most detailed listing of African and European species and notes that the family ranges from Aptian to the present. *Taphrosphys* ranges from Campanian through to Montian, with the earliest known material found in the United States (Gaffney, 1975). It is well known from South America (Wood, 1973).

Geographic and Stratigraphic Distribution of the Marine Pelomedusidae (Figures 1 and 3)

The African pelomedusids noted herein are best known, in terms of numbers, from the Mosasaurus Shales (Upper Cretaceous) of east central Niger and the Chouabine Formation phosphates of south central Tunisia. Both may be regarded effectively as "inland" seas at the time of deposition. The turtle material is mostly that of large to medium-size adults, with the Niger material from one locality consistently found in an upside-down position. The distribution of *Taphrophys* spp. and other large seagoing pelomedusids appears to be influenced by the opening of the Atlantic during the Cretaceous-Tertiary and the presence of the Trans-Saharan Seaway during the Upper Cretaceous and Paleogene. The distribution of Cretaceous-Tertiary European pelomedusids, particularly those of France and Spain, is detailed by De Broin (1977).

CONCLUSIONS

The Cretaceous and Tertiary deposits of northern Europe and northern and western Africa yield a rich and diverse fauna of marine chelonioids and sublittoral pelomedusid turtles. Figure 1 records the stratigraphic distribution and abundance of various species from the Apto-Albian to the Pliocene. The cladogram, Figure 4, (see listed synapomorphies, Table 1) is essentially a modified version of cladogram D of Gaffney and Meylan (1988), and follows their hypothesis that *Toxochelys* is a sister-group to all other chelonioids (see also Hirayama, Chapter 8).

The records of *Allopleuron* material from northern France, western France, and northern Spain (Moody, 1993; Bardet et al., 1993) is evidence of the longevity and widespread distribution of this taxon. Gaffney and Meylan (1988) follow Zangerl (1953a) in the definition of the family Protostegidae, and place members of this family closer to Dermochelyidae than to any other group of chelonioids. Zangerl (1953a) referred only to North American species, but the imperfectly known remains of *Protostega anglica* (Lydekker, 1889), from the Chalk and Cambridge Greensand of England, indicate a more cosmopolitan distribution of this taxon.

The relative abundance of some species at specific localities raises the question of whether the localities were feeding sites or breeding sites. The most prolific faunas are of *Allopleuron hoffmani* from the Chalk of Holland and northern Belgium, *Puppigerus camperi* from the London Clay of Sheppey and the Sables de Bruxelles of Belgium, and *Psephophorus* from the Argile de Boom of Belgium. The majority of *Allopleuron* and *Psephophorus* material is in the form of large, presumably adult specimens. Evidence of sea grass exists in the collections at Maastricht and one would assume that the shallower chalk seas of the region offered an ideal feeding site, assuming that *Allopleuron* occupied a niche similar to the extant *Chelone mydas*. Obvious comparisons can be made between the abundance of *Rhinochelys* and *Allopleuron* material in northern Europe and the fossiliferous Coniacian-Maastrichtian deposits of the Cretaceous Sea of the United States and Canada (Baird and Case, 1966; Zangerl, 1953a, b).

Of the smaller species such as *Puppigerus camperi, Argillochelys crassicostata*, and *Eochelone brabantica*, the range of specimen size is greater with obvious juveniles present in United Kingdom and Belgium collections. It is possible that these were restricted in terms of area and that both nesting and feeding sites were proximal to each other.

It is inevitable that the paleogeographic distribution of certain fossil chelonioids will change regularly as a result of new discoveries. This has been true

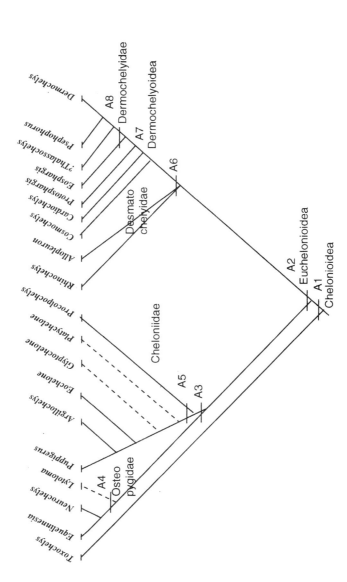

Figure 4. Cladogram of Cretaceous-Tertiary marine turtles (Chelonioidea) of Europe and Africa. Clades are based on criteria erected by Gaffney and Meylan (1988) and Moody (1993).

Table 1. Synapomorphies used for the cladogram of chelonioid groups in Figure 4.

A1	A1.1	Specialized paddle.
	A1.2	Nuchal bone with knob or facet on ventral surface.
	A1.3	Ossified trabeculae of basisphenoid close together.
	A1.4	Paired foramina canalis carotici interni lie close together.
A2	A2.1	Lateral trabeculae of rostrum basisphenoidale fused.
	A2.2	Canalis caroticus lateralis larger in diameter than canalis caroticus interius.
	A2.3	Parietal-squamosal contact.
	A2.4	Lateral process of humerus lies more distal to head of humerus than in *Toxochelys*.
	A2.5	Foramen palatinum posterius absent.
A3	A3.1	Secondary palate formed by broad maxilla with contribution from palatines.
	A3.2	Foramen caroticum laterale and canalis caroticus lateralis much larger in diameter than foramen anterius canalis carotici interni and medial branch of canalis caroticus internus.
A4	A4.1	Secondary palate extensively developed.
	A4.2	Long symphysis to lower jaw.
	A4.3	Ridged neurals in *Neurochelys* as in Lophochelyinae (see Zangerl, 1953b).
A5	A5.1	Cervical vertebrae with platycoelus articulation between sixth and seventh vertebrae (except in *Argillochelys*).
	A5.2	Unusually large coracoid in tribe Chelonini.
	A5.3	Triturating surfaces relatively low without strong ridging in Carettini.
A6	A6.1	Median process of jugal absent.
	A6.2	Scapular angle wide in comparison with other chelonioids.
	A6.3	Coracoid narrow and longer than scapula.
	A6.4	Plastron with large fontanelles and frequently with stellate arrangement of hyo- and hypoplastra.
A7	A7.1	Skull broad with downturned premaxillary beak.
	A7.2	Tooth-like projection of lower jaw.
	A7.3	Shell greatly reduced, often with epidermal armor.
A8	A8.1	Deossification of endocranial region.
	A8.2	Tooth and notched arrangement of upper and lower jaws.
	A8.3	Humerus short, wide, and flattened with large radial process.

Based on Gaffney and Meylan, 1988.

for both *Allopleuron* and *Toxochelys* over the last decade and would appear to be true for the rhinochelyids, as Hirayama (personal communication) has recently prepared a perfect specimen from the Santana deposits of South America.

SUMMARY

Representative species of the suborder Chelonioidea are commonplace in Cretaceous-Neogene strata of Africa, Europe, and North America. They are frequently found in association with large pelomedusid turtles which appear to have occupied coastal and shallow marine niches during the Cretaceous and early Tertiary. Among the chelonioid species the small eochelyines and osteopygids exhibit a strong provincialism, whereas much larger forms such as *Allopleuron* and the leathery turtle *Eosphargis gigas* have a cosmopolitan distribution similar to that of the extant *Dermochelys coriacea*. The abundance of several genera and species, such as *Rhinochelys*, *Allopleuron hoffmani*, *Puppigerus camperi*, and *Psephophorus*, is probably associated with location of paleo-feeding and paleo-nesting sites. This may be simply an artifact of preservation and discovery but it may also reflect a greater diversity and abundance of chelonioids during the Cretaceous-Tertiary Periods. Provincialism is also evident among the large pelomedusids of this period, although *Taphrophys* appears to be more widespread, occupying coastal niches on both sides of the opening Atlantic Ocean and in the Mediterranean area of Tethys. The taxonomic relationships and distribution of both chelonioids and pelomedusids were discussed in this chapter.

ACKNOWLEDGMENTS

The author wishes to acknowledge the efforts of Simon Cope, Linda Parry, and Pat Fawcett in the preparation of the manuscript. Acknowledgment is also given to the thorough research carried out by Dr. A. Seago and Dr. D. Smith on marine chelonioids during their time at Kingston Polytechnic.

REFERENCES

Baird, D. 1986. A skull fragment of the Cretaceous cheloniid turtle *Osteopygis* from the Atlantic highlands, New Jersey. *The Mosasaur, Delaware Paleontological Society* 3:47-52.

Baird, D. and R. C. Case. 1966. Rare marine turtles from the Cretaceous of New Jersey. *Journal of Paleontology* 40(5):1211-1215.

Bardet, N., J. C. Corral, and J. Pereda. 1993. Primeros restos de reptiles marinos en del Cretacio superior de la Cuenca Vasco-Cantabrica. *Estudios del Museo de Ciencias Naturales de Alvava, Vitoria* 8:27-35.

Bergounioux, F. M. 1952. Les chéloniens fossiles de Gafsa. IN C. Arambourg and J. Signeux (Eds.), Les vertebrés fossiles des gisement de phosphates (Maroc-Algérie-

Tunisie). *Service Géologique Maroc, Notes, Mémoires* (Appendix) 92:377-396.
Collins, J. I. 1970. The chelonian *Rhinochelys* (Seeley) from the Cretaceous of England and France. *Palaeontology* 13:355-378.
De Broin, F. 1977. Contribution à l'etude des chéloniens. Chéloniens continentaux du Crétacé et du Tertiaire de France. *Mémoires du Muséum National d'Histoire Naturelle. Nouvelle Série. Série C, Science de la Terre* 38:1-366.
De Broin, F. and B. Pironon. 1980. Découverte d'une tortue dermochelyidée dans le Miocène d'Italie centro-méridionale (Matese oriental), province de Benevento. *Rivista Italiana di Paleontologia e Stratigrafia, Universita degli Studi di Milano* 86(3):589-604.
Dollo, M. L. 1903. *Eochelone brabantica*. Tortue marine nouvelle du Bruxellien (Éocène Moyen) de la Belgique. *Bulletin Académie Royale des Sciences Belgique, Classe des Sciences* 8:792-801.
Fastovsky, D. E. 1985. A skull of the Cretaceous chelonioid turtle *Osteopygis* and the classification of the Osteopyginae. *New Jersey State Museum, Trenton* 3:1-28.
Felder, P. J. 1980a. Resten van Fossiele Zeeschildpadden gevonden in het Krijt van Limburg. *Natuurhistorisch Maanblad* 69(5):100-104.
Felder, P. J. 1980b. Resten van Fossiele Zeeschildpadden gevonden in het Krijt van Limburg. *Natuurhistorisch Maanblad* 69(6/7):117-124.
Gaffney, E. S. 1975. A revision of the side-necked turtle *Taphrosphys sulcatus* (Leidy) from the Cretaceous New Jersey. *American Museum of Natural History, Novitates* 2571:1-24.
Gaffney, E. S. and P. A. Meylan. 1988. A phylogeny of turtles. IN M. J. Benton (Ed.), *The Phylogeny and Classification of the Tetrapods*, Volume 1: *Amphibians, Reptiles, Birds*, pp. 157-219. Systematics Association Special Volume 35A. Clarendon Press, Oxford.
Gaffney, E. S., P. A. Meylan, and A. Wyss. 1991. A computer-assisted analysis of the relationships of the higher categories of turtles. *Cladistics* 7:313-335.
Gray, J. E. 1831. Synopsis Reptilium: or short descriptions of the species of Reptiles. *Synopsis Reptilium Part 1; Tortoises, Crocodiles and Enaliosaurians, London* III:85 pp.
Irby, G. V., D. K. Elliot, and J. H. Hutchinson. 1994. New material of the Cretaceous turtle *Desmatochelys lowi*. *Journal of Vertebrate Paleontology* (abstract) 14, 3 (Supplement):31a.
Konig, C. D. E. 1825. *Icones Fossilium Sectiles*. Centuria Prima, London, 4 pp., 100 figures.
Lydekker, R. 1889. *Catalogue of the Fossil Reptilia and Amphibia in the British Museum (Natural History)*. Longmans and Co., London, 239 pp.
Moody, R. T. J. 1968. A turtle, *Eochelys crassicostata* (Owen) from the London Clay of the Isle of Sheppey. *Proceedings of the Geologists Association of London* 79(2):129-140.
Moody, R. T. J. 1970. *A Revision of the Taxonomy and Morphology of Certain Eocene Cheloniidae*. Unpublished Ph. D. thesis, University of London, 2 volumes.
Moody, R. T. J. 1972. The turtle fauna of the Eocene Phosphates of Metlaoui, Tunisia. *Proceedings of the Geologists Association of London* 83(3):327-336.

Moody, R. T. J. 1974. The taxonomy and morphology of *Puppigerus camperi* (Gray), an Eocene sea turtle from northern Europe. *Bulletin of the British Museum (Natural History)*, Geology 25:153-186.

Moody, R. T. J. 1980. Notes on some European Palaeogene turtles. *Tertiary Research* 2(4):161-168.

Moody, R. T. J. 1993. Cretaceous-Tertiary marine turtles of North West Europe. *Revue de Paléobiologie,* Volume Spécial 7:151-160.

Moody, R. T. J. and P. J. Sutcliffe. 1990. Cretaceous-Tertiary crossroads of migration in the Sahel. *Geology Today* 1990:19-23.

Moret, L. 1935. *Rhinochelys amaberti,* nouvelle espèce de tortue marine du Vraconien de la Fauge pres du Villard-de-Lans (Isère). *Bulletin de la Société Géologique de France* 1:605-619.

Nicholls, E. L. 1992. Note on the occurrence of the marine turtle *Desmatochelys* (Reptilia: Chelonioidea) from the Upper Cretaceous of Vancouver Island. *Canadian Journal of Earth Sciences* 29:377-380.

Owen, R. and A. Bell. 1849. *Monograph of the Fossil Reptilia of the London Clay. Reptilia of the Tertiary Beds.* Part 1. *Chelonia,* pp. 1-76. Palaeontographical Society, London.

Russell, D. E. 1982. Tetrapods of the North-West Europe Tertiary Basin. *International Geological Correlation Programme, Project 124, Geologisches Jahrbuch. Reihe A* 60:1-77.

Seago, A. K. J. 1980. *A Review of the Fossil Turtles of the Family Dermochelyidae.* Unpublished Ph. D. Thesis, University of London, 2 volumes.

Smith, D. T. J. 1989. *The cranial morphology of fossil and living sea turtles (Cheloniidae, Dermochelyidae and Desmatochelyidae).* Unpublished Ph. D. Thesis, Kingston Polytechnic University, The Council for National Academic Awards, London.

Walker, C. A. 1979. New turtles from the Cretaceous of Sokoto. IN L. B. Halstead (Ed.), *International palaeontological expedition to Sokoto State 1977-1978,* pp. 42-48. Nigerian Field Monograph 1 (Volume 44, Supplement).

Williams, E. E. 1950. Variation and selection in the cervical central articulations of living turtles. *Bulletin of the American Museum of Natural History* 94:505-562.

Williston, S. W. 1898. *Desmatochelys lowi. University Geological Survey, Kansas* 4:353-368.

Wood, R. C. 1973. Fossil marine turtle remains from the Paleocene of the Congo. *Annales Musée Royal de l'Afrique Centrale. Tervuren, Belgique. Serie IN-8. Sciences Géologiques* 75:2-35.

Zangerl, R. 1953a. The vertebrate fauna of the Selma Formation of Alabama. Part 3. The turtles of the family Protostegidae. *Fieldiana, Geology Memoirs* 3:63-133.

Zangerl, R. 1953b. The vertebrate fauna of the Selma Formation of Alabama. Part 4. The turtles of the family Toxochelyidae. *Fieldiana, Geology Memoirs* 3:136-277.

Zangerl, R. 1971. Two toxochelyid sea turtles from the Landenian Sands of Erquelinnes (Hain Aut) of Belgium. *Institut Royal des Sciences Naturelles de Belgique, Mémoires* 169:1-32.

Zangerl, R. and R. E. Sloan. 1960. A new description of *Desmatochelys lowi* (Williston). A primitive cheloniid sea turtle from the Cretaceous of South Dakota. *Fieldiana, Geology* 14:7-40.

PART IV
Mosasauridae

Part IV: Mosasauridae

INTRODUCTION

GORDEN L. BELL, JR.

The mosasaurs were a large group of squamates that invaded Late Cretaceous seas following the demise of the last of the ichthyosaurs. Their invasion was so successful that they rapidly diversified to explore and exploit numerous ecological niches throughout the world's oceans and epeiric seas. Coincidently, world sea level was at its highest position during the Mesozoic, a situation that created tremendous areas of habitat and provided superior conditions for preservation of the remains of individuals whose carcasses sank to the bottoms. The sediment-fossil record of the Late Cretaceous, especially in North America, is punctuated by layers of altered volcanic ash, or bentonite, which provide minerals capable of yielding unusually precise radiometric dates. These factors have combined to create a 27 million-year procession of vertebrate evolution so complete that it may very well rival the example provided by the fossil record of horses.

The original mosasaur discovery was made in 1780 in Maastricht, the Netherlands. The fossil itself rapidly became famous and highly coveted. The story, as related by Faujas Saint Fond (quoted by Williston, 1898:84-85, and Leidy, 1865:31) and Baron Cuvier (1808), descibes the efforts of a surgeon, Dr. C. K. Hoffman, to recover the specimen. Found at a depth of 90 ft below the surface and 500 ft from the entrance of a limestone mine, Hoffman paid the workers who had discovered the alarming petrification to quarry a large block of rock containing the specimen and lift it to the surface. The wonder and excitement that followed the discovery attracted the attention of a clergyman who owned the surface land above where the specimen was found. Canon Goddin demanded that the fossil monster was his property and soon sued Dr. Hoffman for its possession. When other clergymen from the area supported Goddin, Hoffman lost the creature---and was forced to pay court costs! Goddin then built a chapel near his estate home so that interested persons could come view his miraculous monster. But by then the fossil had caught the attention of a very influential man far away.

In 1795, the invading army of the French Republic drove off the defending army of Austrians and turned their attack toward Maastricht and Fort St. Peter. Either by ultimate imperial authority or widespread reputation of the fossil, the French cannoniers were ordered to avoid hitting Goddin's chapel. Goddin interpreted this seeming act of kindness with some measure of trepidation and had the fossil hidden in a safe place in the town. But the French army soon took control of the area and, upon finding the fossil missing, offered a reward of 600 bottles of wine for its return. Needless to say, a small fortune in wine can easily harvest vagrant loyalty and the next day several grenadiers came bearing the prize to the victors.

From there the fossil was sent directly to Paris and in 1808, Baron Cuvier issued the first comprehensive discussion of "le grand animal fossile de Maastricht." Cuvier agreed with Adrian Camper that the fossil's relationships lay somewhere between the iguanas and varanids, but also used it to support his radical idea of extinction.

Not until 1822 did the specimen receive a Linnean nominal, when W. D. Conybeare dubbed it *Mosasaurus*, *Mosa-*, the Latin name for the Meuse River near the town of Maastricht and, of course, *-saurus*, for lizard (Parkinson, 1822:298). Gideon Mantell (1829:207) used *Mosasaurus hoffmani* as the first binomial applied to the famous Maastricht specimen.

Richard Harlan (1834) described the first North American mosasaur (as *Ichthyosaurus missouriensis*) from a snout fragment found by a fur trader on the upper Missouri River.

Sir Richard Owen was first to generate a name for inclusion of multiple mosasauroid forms when he erected the tribe Natantia (Owen, 1849-1884) to include *Mosasaurus* and his newly described genus from the English Chalk, *Liodon* (Owen, 1840-1845). Owen did not include characters which would diagnose the new taxon, however, and the name was not used subsequently for mosasaurs. The major portion of mosasaurian morphology first became known when Hrn. Dr. August Goldfuss (1845) described *Mosasaurus maximiliani* based on a relatively complete specimen recovered from the Big Bend of the Missouri River in present-day South Dakota. The family name, Mosasauridae, was generated by Paul Gervais (1853), who associated a diverse array of non-mosasaurian fossils with the two European genera and an American genus, *Macrosaurus* (Owen, 1849:380). Unfortunately, Gervais must have been unaware of several new North American forms described by Gibbes (1850, 1851), mainly from unassociated teeth and vertebrae.

The next great leaps in the study of mosasaurs came at the hands of E. D. Cope and O. C. Marsh. Both were seemingly inexhaustable and, as such, were responsible for greatly increasing the knowledge of mosasaurs with specimens from New Jersey, the Gulf Coastal Plain, and also Kansas, to which they both traveled (separately) in 1871. Cope described several species each of *Clidastes, Platecarpus, Mosasaurus,* and *Liodon* (= *Tylosaurus* and *Leiodon*) in several major articles (1869a, 1869-1870, 1872a, b, 1874, 1875, 1881). Early on, Cope (1869b) proposed a new order, Pythonomorpha, to include the mosasaurs. He enumerated the diagnostic characters to imply a systematic position between the order Lacertilia and the order Ophidia but clearly closer to snakes than lizards (see Figure 1A). Numerous authors took issue with this suggestion, notably Williston (1898), Merriam (1894), Dollo (1894), Baur (1890, 1892, 1895), Marsh (1880), and Owen (1877). Regardless of Cope's views of mosasaur relationships, there is no doubt as to his abilities in describing the morphology of various specimens and taxa, and he provided the first element-by-element comparison of the various genera (Cope, 1875).

Marsh (1869, 1871, 1872) also described many mosasaurs, all but one being on the heels of Cope. The only taxa Marsh named that are presently recognized as valid are *Halisaurus platyspondylus* and *Tylosaurus*. Even though most of Marsh's taxa are now considered junior synonyms, he was uncannily good at finding and reporting various portions of anatomy or orientations of elements that were unknown to Cope. It is important to note that several of these were Marsh's independent rediscovery of findings originally made by Goldfuss (1845).

During this same interval, Leidy (1865, 1870, 1873) reported several specimens from New Jersey, Alabama, Mississippi, and the upper Missouri River, most of which were based on fragmentary material.

Discoveries of mosasaur fossils began to be reported from areas more distant from the earlier known occurrences. James Hector (1874) described *Leiodon haumuriensis* and *Taniwhasaurus oweni* from what were thought to be late Cenomanian rocks (later found to be Campanian-Maastrichtian) of New Zealand. This erroneous information led Dollo (1904) to propose New Zealand as the center of mosasaur radiation.

Meanwhile in Europe, Louis Dollo (1882, 1888, 1889a, b, 1890, 1894) described many new taxa from Belgium, including several species of *Hainosaurus, Plioplatecarpus, Prognathodon,* and *Phosphorosaurus*. Albert Gaudry (1892) described two new specimens of *Leiodon* and Armand Thevinin (1896) described *Platecarpus somenensis* from France. N. N. Iakovlev (1901) described *Dollosaurus lutugini* as the first mosasaur from European Russia.

Near the turn of the century, many of the ideas about the taxonomy and evolution of mosasauroids began to take on a modern shape. Merriam's dissertation (1894) described several new forms from the Niobrara chalks in Kansas, including a second species of *Halisaurus, H. onchognathus,* two new species of *Platecarpus*, and a new species of *Clidastes*. Samuel W. Williston (1898) gave the best review of the known mosasauroids of that time and revised the internal classification of the

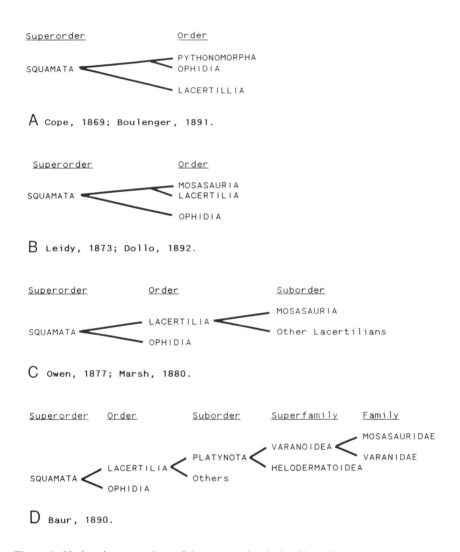

Figure 1. Various interpretations of the systematic relationships of mosasaurs used by early authors. Figure continues on following page.

family to its modern form. He also gave a comprehensive and detailed osteological comparison of the elements from each North American genus.

To summarize the views of mosasaur relationships adhered to until the turn of the century (Figure 1), Leidy (1873) felt that the systematic relationships of mosasaurs were closer to lizards than snakes but equal in rank, a view shared by Dollo (1894). Most other authors agreed that mosasaurs constituted some subunit

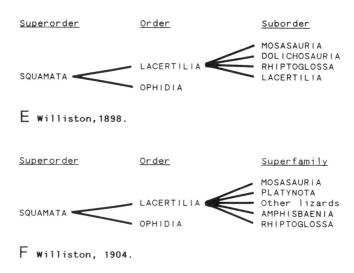

Figure 1 (continued). Various interpretations of the systematic relationships of mosasaurs used by early authors.

of "Lacertilia" (Williston, 1898; Merriam, 1894; Baur, 1890; Marsh, 1880; Owen, 1877; Goldfuss, 1845). Of these authors, Goldfuss and Baur supported close relationships with varanids. In fact, Baur (1890), Cope's literary nemesis, was the first to use the formal name, Platynota (presumably at a suborder ranking), to include varanids, helodermatids, and mosasaurs.

About this same time, Kornhuber (1893, 1901) and Kramberger (1892) described new taxa, *Carsosaurus marchesetti, Opetiosaurus bucchichi*, and *Aigialosaurus dalmaticus* from rocks of late Cenomanian to early Turonian age near Komen in present-day Slovenia, and Lesina in present-day Croatia. Kramberger thought these newly discovered aigialosaurs were ancestral to both lizards and mosasaurs, while Kornhuber contended these specimens were intermediate between varanids and mosasaurs. Several authors quickly adopted a similar view (Williston, 1904 [see Figure 1F]; Nopcsa, 1903; Dollo, 1894), believing that the root relationship was within more primitive varanoid lizards.

Then, in 1912, C. W. Gilmore described the most bizarre mosasaur yet. Based on fragments of a robust skull containing almost spherical teeth, he named *Globidens alabamaensis*, the first distinctively shell-crushing mosasaur known to science. The next year, Dollo (1913) described *Globidens fraasi* (which he would later rename *Compressidens*) from Belgium, another shell-crusher, but with laterally compressed teeth.

Dr. Carl Wiman (1920) described additional morphology of several Kansas mosasaurs and an almost complete specimen of a new species, *Halisaurus sternbergi*.

Dr. Charles L. Camp (1923), in a landmark systematic treatment of lizards, supported the close relationships between varanids and mosasaurs. He maintained that aigialosaurs were derived from close relatives of varanids and gave rise to both dolichosaurs and mosasaurs. In 1942, Camp again stated this conviction in a major work which described the geologically youngest and distinctly most derived mosasauroids, consisting of three taxa from California. He named two species of *Kolposaurus* (= *Plotosaurus*) and *Plesiotylosaurus*. The illustrations of *K. bennisoni* indicate a mosasaur with hyperphalangy and skull modifications strikingly convergent with moderately derived ichthyosaurs.

S. B. McDowell and C. M. Bogert (1954) reviewed the morphology of anguinomorphan lizards and concluded that mosasaurs, aigialosaurs, dolichosaurs, and *Lanthanotus*, although still platynotans, formed their own distinct group and should be separated from a group containing varanids and helodermatids.

Dale A. Russell's classic work of 1967 described mosasauroid morphology in exquisite detail and also synonymized many taxa, as Williston (1898:170) had hinted might be necessary. The book was organized in much the same way as Williston's 1898 work, with sections comparing each individual skeletal element according to genus and a section systematically describing each taxon. In addition, he included reconstructions of the musculature and circulation based on a dissection of *Varanus* and provided the first interpretations of the functional morphology of the entire body. His was the first detailed comparison between derived mosasauroids, the generalized aigialosaurs and platynotan lizards, and he presented the first tree of mosasauroid phylogeny (Russell, 1967:Figure 99). He further added to our knowledge of mosasaurs by describing several taxa from Alabama (Russell, 1970) and a new species of *Globidens* from South Dakota (Russell, 1975). Later, Kenneth R. Wright and Samuel W. Shannon (1988) described an unusual plioplatecarpine, *Selmasaurus russelli*, from Alabama.

M. Telles-Antunes (1964) and A. Azzaroli et al. (1972) described previously unknown mosasaurs from Africa. Antunes described the only two well-represented taxa from late Turonian rocks, *Angolosaurus bocagei* and *Mosasaurus iembeensis* from Angola. Azzaroli et al. described *Goronyosaurus* from Nigeria and erected a new subfamily, Goronyosaurinae, to contain it.

Samuel P. Welles and D. R. Gregg (1971) reevaluated Hector's (1874) work and reassigned *Leiodon haumuriensis* to *Tylosaurus*, separated some of the type material from *Taniwhasaurus* and referred some of that material to a new species, *Mosasaurus mokoroa*, meanwhile referring *Taniwhasaurus* to Tylosaurinae. In the same work they also described *Prognathodon waiparaensis* from newly discovered materials. Joan Wiffen also described new mosasaurs, *Mosasaurus mangahouangae*

(1980) and *Mosasaurus flemingi* and *Rikisaurus tehoensis* (1990), from New Zealand.

Theagarten Lingham-Soliar recently redescribed several of the northern European taxa and revised the nomenclature in numerous papers (Lingham-Soliar, 1992, 1993, 1994a; Lingham-Soliar and Nolf, 1989). He also redescribed and revised the nomenclature of all of the well-known African mosasaurs (Lingham-Soliar, 1988, 1991, 1992, 1994b) and described a new type, *Igdamanosaurus*, from Niger (Lingham-Soliar, 1991).

Mosasaur fossils are apparently rare in Asia, the only significant specimens reported from Japan. Those include the Early Maastrichtian *Mosasaurus hobetsuensis* (Suzuki, 1985) and Campanian material referred to *Tylosaurus* (Chitoku, 1994).

Antarctica has recently produced very fragmentary mosasaur material from Seymour Island (Gasparini and del Valle, 1984).

Robert Carroll and Michael deBraga (1992) published improved descriptions of the aigialosaurs and supported their sister-group relationship with mosasaurs. They maintained the concept of close relationships with varanids, but without testing this relationship. They also (Carroll and deBraga, 1992:Figure 14) employed a cladogram of relationships which includes only terminal taxa previously referred to Varanoidea. Shortly thereafter, deBraga and Carroll (1993) presented results of a cladistic analysis incorporating Necrosauridae, Helodermatidae, Varanidae, Aigialosauridae, and most of the well-known genera of North American mosasaurs as terminal taxa. Again, they assumed close mosasauroid (= mosasaurs + aigialosaurs) relationships with varanids. The relationships among mosasaurs in their study (deBraga and Carroll, 1993:Figure 4) closely paralleled those identified by Russell (1967), with the exception that *Halisaurus* became the sister taxon to all other mosasaurs.

Also in 1993, I completed a phylogenetic study of mosasauroids incorporating alpha-level taxa as terminal taxa. The relationships among mosasauroids identified by cladistic analysis are the subject of Chapter 11. However, it was necessary to the analysis to consider the extant sister-group relationships of mosasauroids. A PAUP analysis was performed using the same characters that Estes et al. (1988) used for extant squamates, to which were added the mosasauroid data. The results placed mosasauroids within Anguimorpha, but could not further resolve their phylogenetic relationships. Recently, an analysis by Michael Caldwell et al. (1995), using a slightly modified version of the Estes et al. (1988) character matrix, removed mosasauroids from Anguimorpha.

After more than 200 years of scientific inquiry into the relationships of these fascinatingly savage sea beasts, what can we say is acceptable as well-tested knowledge? In summary, although many authors have favored varanid-mosasauroid relationships, such a hypothesis has not been rigorously tested using modern phylogenetic methods. The only two such attempts have produced results which either do not support or falsify that hypothesis. At this point, what remains to be done is to search carefully among both fossil and extant taxa for other morphological characters which may be used in phylogenetic analysis.

As for relationships among mosasauroids, the picture is much clearer. Many of the relationships proposed by Russell (1967) have been supported by two fairly rigorous phylogenetic analyses using many characters. The "aigialosaurs" still present problems in that there is no support for the monophyly of Aigialosauridae and the systematic position of all but one basic taxon is in doubt. This, in part, is a result of the manner in which these fossils are preserved---as relatively complete, articulated specimens embedded on slabs of rock. Because only a single specimen from each taxon is known, many aspects of the morphology are hidden and characters cannot be assessed. The only known North American "aigialosaur" (Bell, 1993 and Chapter 11), the Dallas specimen, preserves very little skull morphology. Otherwise, the only controversy to be further addressed is the placement of the clade of *Prognathodon* + *Plesiotylosaurus* among derived mosasaurs. But more Turonian mosasaurs have been recovered in Texas and they, no doubt, will have a bearing on future phylogenetic models.

REFERENCES

Azzoroli, A., C. De Guili, and D. Torre. 1972. An aberrant mosasaur from the Upper Cretaceous of North Western Nigeria. *Accademia Nazionale dei Lincei, Rendiconti, Classe di Scienze Fisiche Matematiche e Naturali* 52:53-56.

Baur, G. H. 1890. On the characters and systematic position of the large sea-lizards, Mosasauridae. *Science* 16(405):262.

Baur, G. H. 1892. On the morphology of the skull in Mosasauridae. *Journal of Morphology* 7:1-22.

Baur, G. H. 1895. The paroccippital of the Squamata and the affinities of the Mosasauridae once more: a rejoinder to Prof. E. D. Cope. *American Naturalist* 30:143-147.

Bell, G. L., Jr. 1993. *A Phylogenetic Revision of Mosasauridae (Squamata)*. Ph. D. dissertation, University of Texas at Austin, Austin, Texas, 293 pp.

Boulenger, G. A. 1891. Notes on the osteology of *Heloderma horridum* and *H. suspectum*, with remarks on the systematic position of the Helodermatidae and on the vertebrae of the Lacertilia. *Zoological Society of London Proceedings* pp. 109-118.

Caldwell, M. W., R. L. Carroll, and H. Kaiser. 1995. The pectoral girdle and forelimb of *Carsosaurus marchesetti* (Aigialosauridae), with a preliminary phylogenetic analysis of mosasauroids and varanoids. *Journal of Vertebrate Paleontology* 15:516-531.

Camp, C. L. 1923. Classification of the lizards. *Bulletin of the American Museum of Natural History* 49:289-481.

Camp, C. L. 1942. California mosasaurs. *University of California Memoir* 13:1-68.
Carroll, R. L. and M. deBraga. 1992. Aigialosaurs: mid-Cretaceous varanoid lizards. *Journal of Vertebrate Paleontology* 12:66-86.
Chitoku, T. 1994. *Tylosaurus* sp. indet. (Reptilia, Mosasauridae) from the Upper Cretaceous of the Hobetsu District, Hokkaido, Japan. *Bulletin of the Hobetsu Museum* 10:39-54.
Cope, E. D. 1869a. On some Cretaceous reptilia. *Proceedings of the Academy of Natural Sciences of Philadelphia* 20:233-234.
Cope, E. D. 1869b. On the reptilian orders Pythonomorpha and Streptosauria. *Boston Society of Natural History Proceedings* 12:250-266.
Cope, E. D. 1869-1870. Synopsis of the extinct Batrachia, Reptilia, and Aves of North America. *Transactions of the American Philosophical Society* 2:106-235.
Cope, E. D. 1872a. Catalogue of the Pythonomorpha found in the Cretaceous strata of Kansas. *American Philosophical Society Proceedings* 12:264-287.
Cope, E. D. 1872b. On the geology and palaeontology of the Cretaceous strata of Kansas: Pythomomorpha. IN F. V. Hayden (Ed.), *United States Geological Survey of Montana and Portions of Adjacent Territories*, pp. 400-414. 5th Annual Report, Washington, D. C.
Cope, E. D. 1874. Review of the vertebrata of the Cretaceous period found west of the Mississippi River. United States Geological Survey of the Territories Bulletin 1:3-48.
Cope, E. D. 1875. The Vertebrata of the Cretaceous formations of the West. *United States Geological Survey of the Territories Report* 2:1-303.
Cope, E. D. 1881. A new *Clidastes* from New Jersey. *American Naturalist* 15:587-588.
Cuvier, G. C. 1808. Sur le grand animal fossile des carrieres de Maastricht. *Musée du Historie Naturale de Paris Annales* 12:145-176.
deBraga, M. and R. L. Carroll. 1993. The origin of mosasaurs as a model of macroevolutionary patterns and processes. *Evolutionary Biology* 27:245-322.
Dollo, L. 1882. Note sur l'osteologie des Mosasauridae. *Bulletin Musée Royale de la Naturale Historie de Belgique* 1:55-80.
Dollo, L. 1888. Sur le crane des mosasauriens. *Scientifique France Belgique Bulletin* 19:1-11.
Dollo, L. 1889a. Nouvelle note sur les vertebres fossiles recemment offerts au Musée de Bruxelles par M. Alfred Lemmonier. *Société Belge de Geologie Proces-Verbaux* 3:214-215.
Dollo, L. 1889b. Premiere note sur les mosasauriens de Mesvin. *Bulletin de la Société Belge de Geologie, de Paleontolgie et d'Hydrologie* 3:271-304.
Dollo, L. 1890. Premiere note sur les mosasauriens du Maestricht. *Bulletin de la Société Belge de Geologie, de Paleontolgie et d'Hydrologie* 4:151-169.
Dollo, L. 1894. Nouvelle note sur l'osteologie des mosasauriens. *Bulletin de la Société Belge de Geologie, de Paleontolgie et d'Hydrologie* 6:219-259.
Dollo, L. 1904. L'origine des mosasauriens. *Société Belge de Geologie, de Paleontolgie et d'Hydrologie* 18:217-222.
Dollo, L. 1913. *Globidens fraasi*, mosasaurien mylodonte nouveau du Maestrichtien (cretace superior) du Limbourg, et l'ethologie de la nutrition chez les mosasauriens. *Archives*

de Biologie 28:609-626.

Estes, R., K. deQueiroz, and J. Gauthier. 1988. Phylogenetic relationships within Squamata. IN R. Estes and G. Pregill (Eds.), *Phylogenetic Relationships of the Lizard Families*, pp. 119-281. Stanford University Press, Stanford, California.

Gasparini, Z. and R. del Valle. 1984. Mosasaurios (Reptilia, Sauria) Cretacicos, en el continente Antartico. *Noveno Congreso Geologico Argentino, S. C. de Bariloche, Actas* 4:423-431.

Gaudry, A. 1892. Le pythonomorphes de France. *Société Geologie de France (Paleontologie) Memoir* 10:1-13.

Gervais, P. 1853. Observations relative aux reptiles fossiles de France. *Academie Scientifique de Paris Comptu Rendus* 36:374-377, 470-474.

Gibbes, R. W. 1850. On *Mosasaurus* and other allied genera in the United States. *American Association for the Advancement of Science, Proceedings of 2nd Meeting.* Cambridge, 1849, p. 77.

Gibbes, R. W. 1851. A memoir on *Mosasaurus* and three allied genera, *Holcodus*, *Conosaurus*, and *Amphorosteus*. *Smithsonian Institution Contributions to Knowledge* 2:1-13.

Gilmore, C. W. 1912. A new mosasauroid reptile from the Cretaceous of Alabama. *Proceedings of the United States National Museum* 41:479-484.

Goldfuss, A. 1845. Der Schädelbau des *Mosasaurus*, durch Beschreibung einer neuen Art dieser Gattung erläutert. *Nova Acta Academica Caesar Leopoldino-Carolinae Germanicae Natura Curiosorum* 21:1-28, pl. VI-IX.

Harlan, R. 1834. On some new species of fossil saurians found in America. *British Association for the Advancement of Science, Report of the 3rd Meeting*, Cambridge, 1833, p. 440.

Hector, J. 1874. On the fossil reptilia of New Zealand. *New Zealand Institute, Transactions and Proceedings* 6:333-358.

Iakovlev, N. N. 1901. Restes d'un mosasaurien trouve dans le Cretace superieur de sud de la Russie. *Izvestiya Geologicheskago Komiteta* 20:507-520.

Kornhuber, A. 1893. *Carsosaurus marchesetti*, ein neuer fossiler Lacertilier aus den Kreideschichten des Karstes bei Komen. *Abhandlungen der Kaiserlich-Koeniglich Geologischen Reichsanstalt Wein* 17:1-15.

Kornhuber, A. 1901. *Opetiosaurus bucchichi*, eine neue fossile Eidechse aus der unteren Kreide von Lesina in Dalmatien. *Abhandlungen der Kaiserlich-Koeniglich Geologischen Reichsanstalt Wein* 17:1-24.

Kramberger, K. G. 1892. *Aigialosaurus*, eine neue Eidechse us den Kreideschiefern der Insel Lesina, mit Rücksicht auf die bereits beschreibenen Lacertiden von Comen und Lesina. *Glasnik Huvatskoga Naravoslovnoga Drustva (Societas Historico-Natulis Croatica) u Zagrebu* 7:74-106.

Leidy, J. 1865. Memoir on the extinct reptiles of the Cretaceous formations of the United States. *Smithsonian Institution Contributions to Knowledge* 14:1-165.

Leidy, J. 1870. (Remarks on *Clidastes intermedius* and *Liodon proriger*). *Proceedings of the Academy of Natural Sciences of Philadelphia* 22:3-5.

Leidy, J. 1873. Contributions to the extinct vertebrate fauna of the western territories. *United States Geological Survey of the Territories Report* 1:14-385.
Lingham-Soliar, T. 1988. The mosasaur *Goronyosaurus* from the Upper Cretaceous of Sokoto State, Nigeria. *Palaeontology* 31:747-762.
Lingham-Soliar, T. 1991. Mosasaurs from the Upper Cretaceous of the Republic of Niger. *Palaeontology* 34:653-670.
Lingham-Soliar, T. 1992. The tylosaurine mosasaurs (Reptilia, Mosasauridae) from the Upper Cretaceous of Europe and Africa. *Bulletin de l'Institut Royal des Sciences Naturelles de Belgique* 62:171-194.
Lingham-Soliar, T. 1993. The mosasaur *Leiodon* bares its teeth. *Modern Geology* 18:443-458.
Lingham-Soliar, T. 1994a. The mosasaur *Plioplatecarpus* (Reptilia, Mosasauridae) from the Upper Cretaceous of Europe. *Bulletin de l'Institut Royal des Sciences Naturelles de Belgique* 64:177-211.
Lingham-Soliar, T. 1994b. The mosasaur "*Angolosaurus*" *bocagei* (Reptilia, Mosasauridae) from the Turonian of Angola reinterpreted as the earliest member of the genus *Platecarpus*. *Palaeontologie Zietschrift* 68:267-282.
Lingham-Soliar, T. and D. Nolf. 1989. The mosasaur *Prognathodon* (Reptilia, Mosasauridae) from the Upper Cretaceous of Belgium. *Bulletin de Institute Royal de Sciences Naturelles de Belgique, Sciences de la Terre* 59:137-190.
Mantell, G. A. 1829. A tabular arrangement of the organic remains of the county of Sussex. *Geological Society of London Transactions, 2nd series* 3:201-216.
Marsh, O. C. 1869. Notice of some new mosasauroid reptiles from the greensand of New Jersey. *American Journal of Science (2nd series)* 48:392-397.
Marsh, O. C. 1871. Notice of some new fossil reptiles from the Cretaceous and Tertiary formations. *American Journal of Science (3rd series)* 1:447-459.
Marsh, O. C. 1872. On the structure of the skull and limbs in mosasauroid reptiles, with descriptions of new genera and species. *American Journal of Science (3rd series)* 3:448-464.
Marsh, O. C. 1880. New characters of mosasauroid reptiles. *American Journal of Science (3rd series)* 19:83-87.
McDowell, S. B., Jr. and C. M. Bogert. 1954. The systematic position of *Lanthanotus* and the affinities of the anguinomorph lizards. *Bulletin of the American Museum of Natural History* 105:1-142.
Merriam, J. C. 1894. Über die Pythonomorphen der Kansas Kreide. *Palaeontographica* 41:1-39.
Nopcsa, F. 1903. Über die Varanus-artigen Lacerten Istriens. *Beitrag Paleontologie Geologie Osterreich-Ungarns* 15:30-42.
Owen, R. 1840-1845. *Odontography*. London, 655 pp.
Owen, R. 1849. Notes on remains of fossil reptiles discovered by Prof. Henry Rogers of Pennsylvania, U. S., in greensand formations of New Jersey. *Geological Society of London Quarterly Journal* 5:380.
Owen, R. 1849-1884. *A History of British Fossil Reptiles*. London, 4 volumes.

Owen, R. 1877. On the rank and affinities of the reptilian class of the Mosasauridae. *Geological Society of London, Quarterly Journal* 33:682-719.

Parkinson, J. 1822. *Outlines of Orycytology. An Introduction to the Study of Fossil Organic Remains.* Sherwood, Neely and Jones, London, 345 pp.

Russell, D. A. 1967. Systematics and morphology of American mosasaurs. *Bulletin of the Peabody Museum of Natural History* 23:1-241.

Russell, D. A. 1970. The vertebrate fauna of the Selma Formation of Alabama. The mosasaurs. *Fieldiana, Geology Memoirs* 3:365-380.

Russell, D. A. 1975. A new species of *Globidens* from South Dakota, and a review of globidentine mosasaurs. *Fieldiana, Geology Memoirs* 33:235-256.

Suzuki, S. 1985. A new species of *Mosasaurus* (Reptilia, Squamata) from the Upper Cretaceous Hakobuchi Group in central Hokkaido, Japan. IN M. Goto et al. (Eds.), *Evolution and Adaptation of Marine Vertebrates*, pp. 45-66. Association for Geological Collaboration in Japan, Monograph 30.

Telles-Antunes, M. T. 1964. O Neocretacico e o Cenozoico do litoral de Angola. *Junta de Investigacoes do Ultramar*, Lisbon, 255 pp.

Thevinin, A. 1896. Mosasauriens de la Craie Grise de Vaux-Eclusier pres Peronne (Somme). *Société Geologique de la France Bulletin (3rd series)* 24:900-916.

Welles, S. P. and D. R. Gregg. 1971. Late Cretaceous marine reptiles of New Zealand. *Records of the Canterbury Museum* 9:1-111.

Wiffen, J. 1980. *Moanasaurus*, a new genus of marine reptile (Family Mosasauridae) from the Upper Cretaceous of North Island, New Zealand. *New Zealand Journal of Geology and Geophysics* 23:507-528.

Wiffen, J. 1990. New mosasaurs (Reptilia; Family Mosasauridae) from the Upper Cretaceous of North Island, New Zealand. *New Zealand Journal of Geology and Geophysics* 33:67-85.

Williston, S. W. 1898. Mosasaurs. *University Geological Survey of Kansas* 4:83-221.

Williston, S. W. 1904. The relationships and habits of the mosasaurs. *Journal of Geology* 12:43-51.

Wiman, C. 1920. Some reptiles from the Niobrara Group in Kansas. *Bulletin of the Geological Institute of Uppsala* 18:9-18.

Wright, K. R. and S. W. Shannon. 1988. *Selmasaurus russelli*, a new plioplatecarpine mosasaur (Squamata, Mosasauridae) from Alabama. *Journal of Vertebrate Paleontology* 8:102-107.

Chapter 11

A PHYLOGENETIC REVISION OF NORTH AMERICAN AND ADRIATIC MOSASAUROIDEA

GORDEN L. BELL, JR.

INTRODUCTION

The last comprehensive revision of mosasauroids was published nearly 30 years ago (Russell, 1967), just 1 year after the first English edition of the pivotal work *Phylogenetic Systematics* (Hennig, 1966). Since then phylogenetic systematics, or cladistics, has become the premier methodology utilized in biological systematics. The major objective of this chapter is to place most mosasauroids into a phylogenetic systematic framework, identify current taxonomic names that are valid within that hierarchical framework, and diagnose those taxa with synapomorphic characters.

Fossils of mosasauroid lizards have been found on every continent and their great diversity is a matter of record, but a modern hypothesis of phylogenetic relationships among mosasauroids is necessary in order to test ideas about evolutionary concepts relating to the taxon. The varanid-aigialosaur-mosasaur transition, for example, has been cited as an excellent model of macroevolution (deBraga and Carroll, 1993), without any unambiguous evidence that such a transition ever took place.

New fossils recovered since 1967 must also be added to the analysis. Two basal taxa, the Trieste and Dallas specimens, have been found in Slovenia (Calligaris, 1988:Figure 14) and Texas (Bell, 1993), respectively. New basic taxa have been described and previously unknown taxa were discovered in museum drawers. There are also better specimens of taxa known previously only from fragments, and invalidated taxa, now known to be distinct and recognizable, that should be resurrected.

This study focuses on only those mosasauroid taxa known from North America and three basal taxa (heretofore referred to family Aigialosauridae) from Slovenia and Croatia. Herein I test the more recent hypotheses of mosasauroid relationships (deBraga and Carroll, 1993; Russell, 1967) using modern phylogenetic methods and an enlarged group of basic taxa and characters. Because only common ancestry (sister-group) relationships are testable (Schoch, 1986), no attempt shall be made to identify direct ancestors.

Mosasauroidea as used here includes all basic taxa from the families "Mosasauridae" and "Aigialosauridae," as they were used by Russell (1967), Romer (1956), and many others in a traditional ranked classification. However, no assumption is made at the outset of this study concerning the monophyly of either taxon, and the names will not be formally recognized until their separate monophyletic status is demonstrated. For purposes of comparison, Russell's tree is converted to a cladogram and shown in Figure 1.

POLARITY DECISIONS

Preliminary condideration of outgroup relationships was crucial to this analysis and an attempt was made to identify the closest extant sister taxa. Mosasauroids were scored using the same characters that Estes et al. (1988) used for extant squamates. The mosasauroid data (Appendix 2) were added to their matrix and analyzed using PAUP. Strict and Adam's consensuses of the six most equally parsimonious trees placed mosasauroids within Anguimorpha, but provided no better lower-level resolution of their phylogenetic relationships.

Final determination of character states at the outgroup node were based on the algorithm described by Maddison et al. (1984). Each of these character states was then used to determine polarity of the ingroup character states. Lack of resolution of the relationships of mosasauroids among the outgroup taxa (Bell, 1993:10-16) contributed to uncertain polarity in 26 of 142 characters, or 18% of the total number. In the ingroup analysis, these were treated as nonpolarized and are identified within the data matrix with a question mark as the outgroup state. Also, 10 characters are based on structures or modifications of structures not represented in the outgroups and are also identified in the same manner. For the ingroup taxa within the data matrix, question marks are used for both nonrepresentation of the attribute due to extreme modification, and missing information due to incomplete specimens. The data matrix used in this analysis is contained in Appendix 2.

Phylogenetic Revision of Mosasauroidea

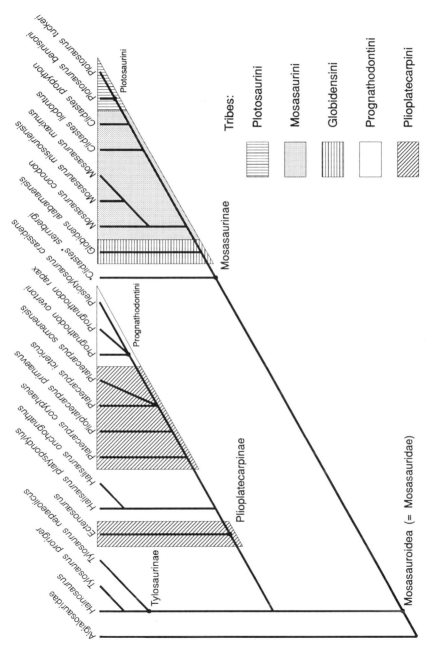

Figure 1. Phylogeny of Russell (1967), conservatively converted to standard cladogram format for purposes of comparison.

PHYLOGENETIC ANALYSIS

The data matrix was entered into the PAUP 3.0 program (Swofford, 1990) and run on an Apple/Macintosh Powerbook 145B computer. The DELTRAN option of character optimization was selected because it permits assignment of synapomorphies at the most specific taxonomic level. All multistate characters were unordered and all trees were rooted in the outgroup. Analysis was performed using the heuristic mode due to the large number of characters and taxa. Heuristic search options included: simple addition sequence with one tree held at each step, TBR branch swapping performed, the MULPARS option in effect, and steepest descent option not in effect.

The analysis produced 99 equally parsimonious trees with a total length of 351 steps. The consistency index for any representative tree is 0.477 with all uninformative characters removed and the retention index is 0.778.

Three types of tests were performed to evaluate the quality of the data or the stability of portions of the tree. Such tests have limitations of the number of characters and taxa that can be accommodated, but these problems were circumvented by dividing the tree into various units or component groups. The $g1$ skewness test statistic (Hillis and Huelsenbeck, 1992:189) measures the amount of signal (phylogenetic information) in a data set and results indicate significant values (95-99% confidence interval) for all of the major clades and most of the subsets of those clades. A set including the more labile taxa, indicated by lack of resolution of their phylogenetic relationships, had very low skewness values and were thus not significant. Bootstrapping tests (Felsenstein, 1985) were also run for various parts of the tree. Results identified good to strong support (70-100%) for each of the following taxa: Mosasauridae, Mosasaurinae, Globidensini, Plotosaurini, and five clades containing the basic taxa of *Halisaurus*, *Tylosaurus*, *Ectenosaurus*, *Platecarpus* + *Plioplatecarpus*, and *Prognathodon* + *Plesiotylosaurus*. Poor support (50-60%) was indicated for "Russellosaurinae," Plioplatecarpinae, and a clade containing the basic taxa of *Globidens*. Several bootsrapping tests involving a variety of different taxa within Natantia produced results ranging from less than 50% to 75%. Results for the "aigialosaurs," taxon novum, and YPM 40383 were even lower. Finally, node collapsing tests were performed on the matrix. Adam's consensus trees were calculated using groups of trees with successively larger numbers of minimum steps. A general evaluation of the results indicates poor support for the positions of any of the conservative "aigialosaurs," as well as taxon novum and YPM 40383. The entire constituency of "Russellosaurinae" has poor to moderate support. The clade containing the basic taxa of *Halisaurus* is fairly well supported as the sister-group to Natantia, as is the clade containing *Platecarpus* + *Plioplatecarpus*. The best supported relationships are within and among

Mosasaurinae.

CONSENSUS TREES AND PREFERRED HYPOTHESIS OF RELATIONSHIPS

There is very little difference in the strict and Adam's consensus trees calculated from 99 trees generated by the phylogenetic analysis. These are illustrated in Figure 2 and Figure 3, respectively. The strict consensus tree was chosen to represent the preferred hypothesis of relationships among Mosasauroidea (Figure 4) because it provides a great deal of resolution, but does not overstep the limitations of the data that are available at this time. It provides but one major surprise and only a moderate amount of incongruency with Russell's tree (see Figure 1) and even less with that of deBraga and Carroll (1993:Figure 4). Unfortunately, the analysis did little to resolve the relationships of basal taxa with derived mosasauroids, largely due to incompleteness of the data.

CHARACTER ANALYSIS

Skull Characters

1. Premaxilla predental rostrum I: total lack of a bony rostrum (0), or presence of any predental rostrum (1) (Figure 5). In lateral profile, the anterior end of the premaxilla either exhibits some bony anterior projection above the dental margin (Figure 5B-D), or the bone recedes posterodorsally from the dental margin (Figure 5A). The derived condition produces a relatively taller lateral profile with an obvious "bow" or "prow."

2. Premaxilla predental rostrum II: rostrum very short and obtuse (0) (Figure 5D), or distinctly protruding (1) (Figure 5B), or very large and inflated (2) (Figure 5C). In *Clidastes* a short, acute, protruding rostrum (state 1) produces a "V-shaped" dorsal profile (Russell, 1967:128, 130) and, as far as is known, is peculiar to that genus. An alternative condition, described as U-shaped, include those taxa whose rostral conditions span the whole range of states of characters 1 and 2. Hence, the descriptive character is abandoned in favor of a more informative structure-based series.

3. Premaxilla shape: bone broadly arcuate anteriorly (0), or relatively narrowly arcuate or acute anteriorly (1). In virtually all lizards the premaxilla is a very widely arcuate and lightly constructed element, and the base of the internarial process is quite narrow as in *Opetiosaurus*. All other mosasaurids have the derived

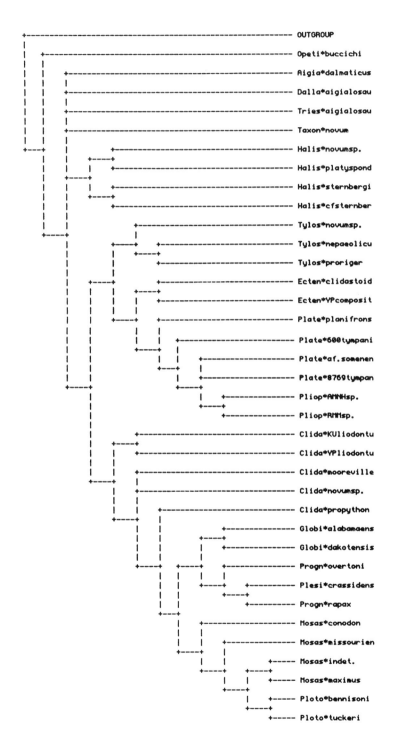

Figure 2. Strict consensus of 99 trees generated by PAUP analysis.

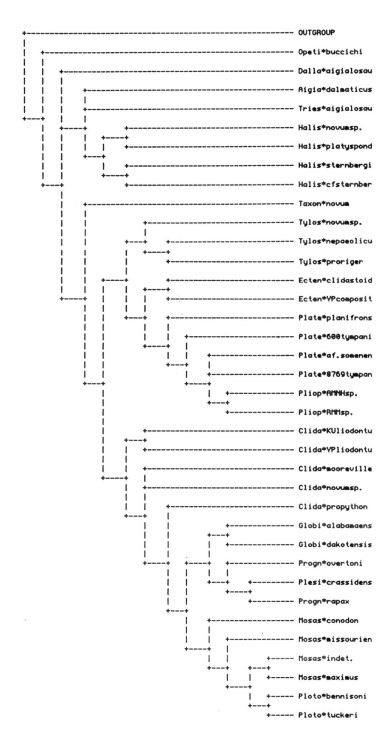

Figure 3. Adams consenus of 99 trees generated by PAUP analysis.

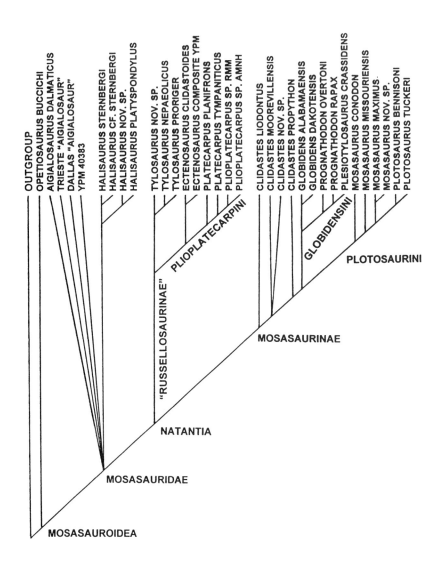

Figure 4. Preferred hypothesis of relationships among Mosasauroidea. Unresolved relationships among more basal taxa are reserved pending a more focused analysis.

condition of a very narrowed premaxilla with the teeth forming a tight curve and the internarial process being proportionally wider.

4. Premaxilla internarial bar width: narrow, distinctly less than half of the maximum width of the rostrum in dorsal view (0), or wide, being barely narrower than the rostrum (1).

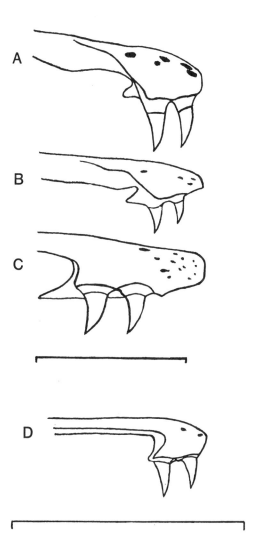

Figure 5. Comparison of mosasauroid premaxillae. **A)** *Platecarpus tympaniticus* (AMNH 1820). **B)** *Clidastes propython* (YPM 1319). **C)** *Tylosaurus proriger* (AMNH 1592). **D)** *Halisaurus* nov. sp. (AUMP 408). **A, B**, and **C** modified from Russell (1967:Figure 2); **D** modified from Shannon (1975:Figure 12). Scale bars = 10 cm.

5. Premaxilla internarial bar base shape: triangular (0), or rectangular (1). A vertical cross section through the junction of the internarial bar and the dentigerous rostrum produces an inverted triangle in most taxa. But in the derived state, this cross section is transversely rectangular because the broad ventral surface of the bar is planar.

6. Premaxilla internarial bar dorsal keel: absent (0), or present (1). In the derived condition a ridge rises above the level of a normally smoothly continuous transverse arch formed by the bones of the anterior muzzle. Because of the distinct difference of form, these structures may not be homologous in *Tylosaurus* and *Mosasaurus-Plotosaurus*.

7. Premaxilla internarial bar venter: with entrance for the fifth cranial (facial) nerve close to rostrum (0), or far removed from rostrum (1). The conduit that marks the path of the fifth cranial nerve from the maxilla into the premaxilla is expressed as a ventrolateral foramen within the premaxillo-maxillary sutural surface at the junction of the internarial bar and the dentigerous rostrum. The derived condition includes a long shallow groove on the ventral surface of the bar. Anteriorly the groove becomes a tunnel entering the bone at an extremely shallow angle, but disappearing below the surface at least 1 cm behind the rostrum.

8. Nasals: present (0) (Figure 6D), or absent or fused to other elements (1) (Figure 6A-C, E-I). In most mosasauroids, nasals are not present even in well-preserved skulls and must be either *always* lost as a result of taphonomic processes, embryonically fused to the frontal or premaxilla, or lost evolutionarily.

9. Frontal shape: sides sinusoidal (0) (Figure 6A-D, H), or bone nearly triangular and sides relatively straight (1) (Figure 6E-G, I). In the derived condition, the area above the orbits is expanded and an isosceles triangle is formed by the rectilinear sides. In certain taxa a slight concavity is seen above the orbits, but anterior and posterior to this there is no indication of a sinusoidal or recurved edge.

10. Frontal width: element broad and short (0) (Figure 6B, C, E-G, I), or long and narrow (1) (Figure 6A, D, H). Mosasauroid frontals can be separated into a group that generally has a maximum length to maximum width ratio of about 2:1 and a group which has a ratio being generally equal to or less than 1.5:1.

11. Frontal narial emargination: frontal not invaded by posterior end of nares (0) (Figure 6A, E, H, I), or distinct embayment present (1) (Figure 6B-D, F, G). In some mosasauriods, the posterior ends of the nares are concomitant with the anterior terminus of the frontal-prefrontal suture and, therefore, there is no marginal invasion of the frontal by the opening. However, in other mosasauroids this suture begins anterior and lateral to the posterior ends of the nares, causing a short emargination into the frontal.

12. Frontal midline dorsal keel: absent (0), or low, fairly inconspicuous (1), or high, thin, and well developed (2).

13. Frontal ala shape: sharply accuminate (0) (Figure 6D), or more broadly pointed or rounded (1) (Figure 6A-C, E-I). In state 0, the anterolateral edge of the ala is smoothly concave, thus helping to form the sharply pointed and laterally oriented posterior corners. In some Natantia, the anterolateral edge of the ala may

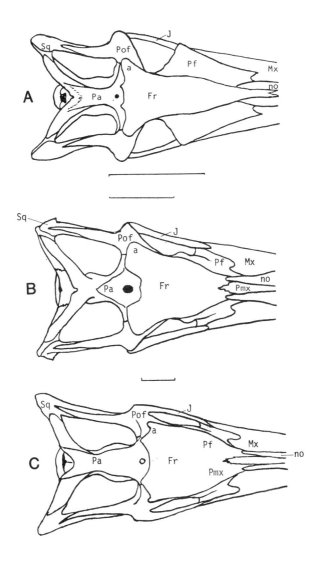

Figure 6. Comparison of mosasauroid skulls, dorsal view. **A)** *Clidastes liodontus*. **B)** *Platecarpus tympaniticus*. **C)** *Tylosaurus proriger*. **D)** *Halisaurus sternbergi*. **E)** *Prognathodon overtoni*. **F)** *Globidens dakotensis*. **G)** *Plotosaurus bennisoni*. **H)** *Ectenosaurus clidastoides*. **I)** *Mosasaurus missouriensis*. A, C, and I modified from Williston (1898). B, E, F, and H modified from Russell (1967). G modified from Camp (1942). D based on UPI R163. Abbreviations: **a** = ala of frontal, **Fr** = frontal, **J** = jugal, **Mx** = maxilla, **N** = nasal, **no** = narial opening, **Pa** = parietal, **Pf** = prefrontal, **Pmx** = premaxilla, **Pof** = postorbitofrontal, **Sq** = squamosal. Scale bars = 10 cm. Figure continues on next two pages.

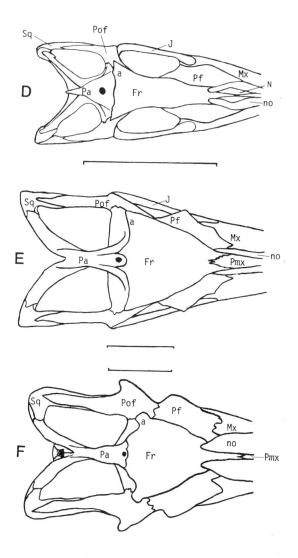

Figure 6 (continued). Comparison of mosasauroid skulls, dorsal view.

be concave, but the tip is not sharp and directed laterally.

14. Frontal olfactory canal embrasure: canal not embraced ventrally by descending processes (0), or canal almost or completely enclosed below (1). In state 1, very short descending processes from sides of the olfactory canal surround and almost, or totally, enclose the olfactory nerve.

15. Frontal-prefrontal overlap: Suture consists of laterally abutting surfaces (0), or suture largely a horizontally oriented, broadly overlapping flange (1).

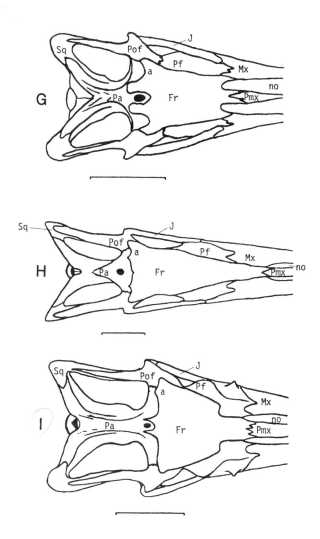

Figure 6 (continued). Comparison of mosasauroid skulls, dorsal view.

16. Frontal posteroventral midline: tabular boss immediately anterior to the frontal-parietal suture absent (0), or present (1). A triangular boss with a flattened ventral surface at the posterior end of the olfactory canal is the derived condition.

17. Frontal-parietal suture: apposing surfaces with low interlocking ridges (0), or with overlapping flanges (1). Plesiomorphically, an oblique ridge on the anterior sutural surface of the parietal intercalates between a single median posterior and a single lateral posterior ridge from the frontal. In the derived condition, these ridges

are protracted into strongly overlapping flanges. The dorsal trace of the suture can be quite complex with a portion of the parietal embraced by the posterior extension of these frontal flanges.

18. Frontal-parietal suture overlap orientation: suture with oblique median frontal and parietal ridges contributing to overlap (0), or with all three ridges almost horizontal (1). In state 0, the median ridge from the frontal and the single parietal ridge are oriented at a distinct angle to the upper skull surface while the outer, or lateral, frontal ridge appears to be nearly horizontal. In *Tylosaurus nepaeolicus* and *T. proriger* (state 1), the obliquity of the intercalating ridges is reclined almost to the horizontal, greatly extending the amount of lateral overlap.

19. Frontal invasion of parietal: lateral sutural flange of frontal posteriorly extended (0) (Figure 6B), or median frontal sutural flange posteriorly extended (1) (Figure 6A, C, E-G, I), or both extended (2) (Figure 6H). In all mosasaurines and in *Tylosaurus* nov. sp., the oblique median frontal sutural ridge extends onto the dorsal surface of the parietal table and embraces a portion of the anterior table within a tightly crescentic midline embayment. In *Plioplatecarpus*, and *Platecarpus*, the lateral oblique sutural ridge from the frontal is greatly protracted posteriorly to cause a large, anteriorly convex embayment in the dorsal frontal-parietal suture. In this case the entire posterolateral corners of the frontal are extended backward to embrace the anterolateral portions of the parietal table. Consequently, the parietal foramen is very widely embraced laterally and the oblique anterior sutural ridge of the parietal occupies a position inside the embayment within the frontal.

20. Frontal medial invasion of parietal II: if present, posteriorly extended median sutural ridge short (0) (Figure 6A, C, F), or long (1) (Figure 6E, G, I). The median oblique sutural ridge discussed in character 19 is either short, not reaching back to the parietal foramen (state 0), or tightly embraces the foramen while extending backward to a position even with or beyond its posterior edge (state 1).

21. Parietal length: dorsal surface relatively short with epaxial musculature insertion posterior between suspensorial rami only (0) (Figure 6D), or dorsal surface elongate with epaxial musculature insertion dorsal as well as posterior (1) (Figure 6A-C, E-I).

22. Parietal table shape: generally rectangular to trapezoidal with sides converging but not meeting (0) (Figure 6A, C, E-I), or triangular with straight sides contacting in front of suspensorial rami (1) (Figure 6B, D).

23. Parietal foramen size: relatively small (0) (Figure 6A, C-F, I), or large (1) (Figure 6B, G, H). If the foramen is smaller than or equal to the area of the stapedial pit, it is considered small. If the foramen is significantly larger or if the distance across the foramen is more than half the distance between it and the nearest edge of the parietal table, the derived state is achieved.

24. Parietal foramen position I: foramen generally nearer to center of parietal table, well away from frontal-parietal suture (0) (Figure 6D, H), or close to or barely touching suture (1) (Figure 6A-C, E-G, I), or huge foramen straddling suture and deeply invading frontal (2) (not figured). Generally in state 1, the distance from the foramen to the suture is about equal to or less than one foramen length.

25. Parietal foramen ventral opening: opening is level with main ventral surface (0), or opening surrounded by a rounded, elongate ridge (1).

26. Parietal posterior shelf: presence of a distinct horizontal shelf projecting posteriorly from between the suspensorial rami (0), or shelf absent (1). In some mosasauroids, a somewhat crescent-shaped shelf (in dorsal view) lies at the posterior end of the bone medial to and below the origination of the suspensorial rami.

27. Parietal suspensorial ramus compression: greatest width vertical or oblique (0), or greatest width horizontal (1). In *Tylosaurus*, the anterior edge of the ramus begins very low on the lateral wall of the descending process, leading to formation of a proximoventral sulcus, but the straps are horizontal distally.

28. Parietal union with supratemporal: suspensorial ramus from parietal overlaps supratemporal without interdigitation (0), or forked distal ramus sandwiches end of supratemporal (1).

29. Prefrontal supraorbital process: process absent, or present as a very small rounded knob (0) (Figure 6B-D, H), or a distinct to large, triangular or rounded, overhanging wing (1) (Figure 6A, E-G, I).

30. Prefrontal contact with postorbitofrontal: no contact at edge of frontal (0) (Figure 6A, D, H), or elements in contact there (1) (Figure 6B, C, E-G, I). State 1 is usually described as the frontal being emarginate above the orbits. Often this character can be evaluated by examining the ventral surface of the frontal where depressions outline the limits of the sutures for the two ventral elements.

31. Prefrontal-postorbitofrontal overlap: prefrontal overlapped ventrally by postorbitofrontal (0), or prefrontal overlapped laterally (1). Postorbitofrontal ventral overlap of the prefrontal is extreme in *Platecarpus tympaniticus,* and *Plioplatecarpus*, such that there is even a thin flange of the frontal interjected between the prefrontal above and the postorbitofrontal below. In *T. proriger*, the postorbitofrontal sends a long narrow process forward to fit into a lateral groove on the prefrontal. In *Plesiotylosaurus*, the overlap is relatively short and more oblique, and there is no groove on the prefrontal.

32. Postorbitofrontal shape: narrow (0) (Figure 6B-E, G-I), or wide (1) (Figure 6A, F). In *Clidastes* and Globidensini, the lateral extent of the element is almost equal to half the width of the frontal and the outline of the bone is basically squared, while in all other ingroup and outgroup taxa, it is a fairly narrow hourglass shape.

33. Postorbitofrontal transverse dorsal ridge: absent (0), or present (1). In state

1, an inconspicuous, low and narrowly rounded ridge traces from the anterolateral corner of the parietal suture across the top of the element to disappear behind the origin of the jugal process.

34. Postorbitofrontal squamosal ramus: does (0) (Figure 6B, C, E-I), or does not (1) (Figure 6A), reach end of supratemporal fenestra.

35. Maxilla tooth number: 20-24 (0), or 17-19 (1), or 15-16 (2), or 14 (3), or 13 (4), or 12 (5).

36. Maxillo-premaxillary suture posterior terminus: suture ends above a point that is anterior to or even with the midline of the fourth maxillary tooth (0), or between the fourth and ninth teeth (1), or even with or posterior to the ninth tooth (2). These somewhat arbitrary divisions of the character states are meant to describe in more concrete terms those sutures that terminate far anteriorly, those that terminate less anteriorly, and those that terminate near the midlength of the maxilla, respectively.

37. Maxilla posterodorsal process: recurved wing of maxilla dorsolaterally overlaps a portion of the anterior end of prefrontal (0) (Figure 6A-C, E-I), or process absent (1) (Figure 6D).

38. Maxilla posterodorsal extent: recurved wing of maxilla prevents emargination of prefrontal on dorsolateral edge of external naris (0) (Figure 6C, G, H), or does not (1) (Figure 6A, B, D-F, I).

39. Jugal posteroventral angle: angle very obtuse or curvilinear (0), or slightly obtuse, near 120 degrees (1), or 90 degrees (2).

40. Jugal posteroventral process: absent (0), or present (1).

41. Ectopterygoid contact with maxilla: present (0), or absent (1).

42. Pterygoid tooth row elevation: teeth arise from robust, transversely flattened, main shaft of pterygoid (0), or teeth arise from thin pronounced vertical ridge (1). Plesiomorphically, the teeth emanate from the relatively planar surface of the thick, slightly dorsoventrally compressed main shaft of the pterygoid. In the derived state a tall, thin dentigerous ridge emanates ventrally from a horizontal flange that forms the base of the quadratic ramus and the ectopterygoid process, thus causing the main shaft to be trough-shaped.

43. Pterygoid tooth size: anterior teeth significantly smaller than marginal teeth (0), or anterior teeth large, approaching size of marginal teeth (1).

44. Quadrate suprastapedial process length: process short, ends at a level well above midheight (0) (Figure 7D), or of moderate length, ending very near midheight (1) (Figure 7B), or long, ending distinctly below midheight (2) (Figure 7A).

45. Quadrate suprastapedial process constriction: distinct dorsal constriction (0), or virtually no dorsal constriction (1). Lack of constriction results in an essentially parallel-sided process in posterodorsal view, but can also include the tapering form characteristic of some *Tylosaurus*. Russell (1967:148) used parallel-sided

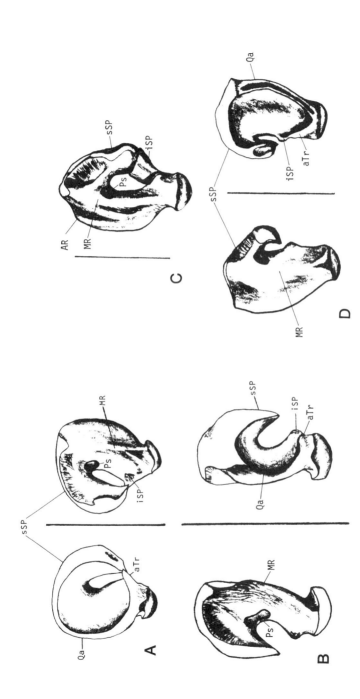

Figure 7. Comparison of mosasauroid quadrates. **A)** *Platecarpus tympaniticus*, left quadrate, lateral view (left) and medial view (right). **B)** *Tylosaurus nepaeolicus*, left quadrate, medial view (left) and lateral view (right). **C)** *Prognathodon rapax*, medial view. **D)** *Plotosaurus bennisoni*, right quadrate, medial view (left) and lateral view (right). **A** and **B** modified from Russell (1967). **C** based on NJGS 9827. **D** modified from Camp (1942). Abbreviations: **AR** = anterodorsal ridge of ala, **aTr** = posteroventral ascending tympanic rim, **iSP** = infrastapedial process, **MR** = median ridge, **Ps** = stapedial pit, **Qa** = quadratic ala, **sSP** = suprastapedial process. Scale bars = 10 cm.

suprastapedial processes as diagnostic for members of the Plioplatecarpinae and distal expansion of the process as a character of the Mosasaurinae (1967:124). Distribution on the preferred tree indicates separate derivations for state 1, supporting the possibility that the tapered "*Tylosaurus* form" is distinct from the parallel-sided "*Platecarpus* form."

46. Quadrate suprastapedial ridge: if present, ridge on ventromedial edge of suprastapedial process indistinct, straight and/or narrow (0), or ridge wide, broadly rounded, and curving downward, especially above stapedial pit (1).

47. Quadrate suprastapedial process fusion: no fusion present (0) (Figure 7A, B, D), or process fused to elaborated process from below (1) (Figure 7C). A posterior rugose area may be inflated and broadened mediolaterally to partially enclose the ventral end of a broad and elongate suprastapedial process as in *Halisaurus*. In *Globidens, Prognathodon,* and *Plesiotylosaurus*, the process is fused ventrally to a narrow pedunculate medial extension of the tympanic rim. A similar condition is present in *Ectenosaurus*, except that the tympanic rim is not medially extended and has a short projection that overlaps a portion of the suprastapedial process posteriorly.

48. Quadrate stapedial pit shape: pit broadly oval to almost circular (0) (Figure 7C, D), or relatively narrowly oval (1) (Figure 7A), or extremely elongate with a constricted middle (2) (Figure 7B). In state 0, the length to width ratio is less than 1.8:1; in state 1, it ranges from 1.8:1 to 2.4:1; and in state 2, it is greater than 2.4:1.

49. Quadrate infrastapedial process: absent (0) (not figured), or present (1) (Figure 7A-D). The infrastapedial process can be expressed as an elaborated or swollen rugose area on the posteroventral face of the main quadrate shaft, as an extension of the ascending posteroventral portion of the tympanic rim, or as a small protuberance emanating from the ventral end of the anterior meatal wall. These various stuctures are probably not homologous.

50. Quadrate posteroventral ascending tympanic rim condition: small, low ridge present (0) (Figure 7B), or a high, elongate crest (1) (Figure 7A, D), or crest extremely produced laterally (2) (not figured). In state 1, this extended rim causes a fairly deep sulcus in the ventral portion of the intratympanic cavity. In *Plioplatecarpus* (not figured), the entire lower tympanic rim and ala are expanded into a large conch (state 2) which tremendously increases the depth of the intratympanic cavity.

51. Quadrate ala thickness: ala thin (0), or thick (1). In state 0, the bone in the central area of the ala is only about 1 m thick in medium-sized specimens and that area is usually crushed or completely destroyed. Alternatively, the ala extends from the main shaft with only minor thinning, providing a great deal of strength to the entire bone.

52. Quadrate conch: ala and main shaft encompassing a deeply bowled area

(0), or alar concavity shallow (1). A relatively deeper sulcus in the anterior part of the intratympanic cavity and more definition between the ala and the main shaft are features of the plesiomorphic state.

53. Quadrate ala shape I: anterodorsal segment of tympanic rim more tightly curved than rest of rim (0), or rim with a uniformly circular curve throughout (1). A slight expansion of the anterodorsal segment of the alar wing produces a lateral profile that very much resembles an ear or a question mark in the plesiomorphic condition.

54. Quadrate ala shape II: angular protuberance on anterodorsal edge of ala absent (0) (Figure 7A, B), or angular protuberance present (1) (Figure 7D).

55. Quadrate ala ridge: no vertical ridge present dorsolaterally on anterior face of ala (0) (Figure 7A, B, D), or strong obtuse ridge present in that position (1) (Figure 7C). *Prognathodon overtoni* and *P. rapax* exhibit a broad vertical ridge which rises near the center of the anterior face of the ala and becomes more pronounced as it approaches the anterodorsal edge.

56. Quadrate ala groove: absent (0) (Figure 7A, B), or long, distinct, and deep groove present in anterolateral edge of ala (1) (Figure 7D).

57. Quadrate tympanic rim size: large, almost as high as quadrate (0) (Figure 7A, D), or smaller, about 50-65% of the height (1) (Figure 7B). In the derived state, a large portion of the dorsal articular surface and the ventral end of the main shaft is exposed in lateral view.

58. Quadrate dorsal median ridge: ridge is a relatively thin and high crest (0) (Figure 7B-D), or low, broadly inflated dome present (1) (Figure 7A). In state 1, the dorsal median ridge barely rises above the rest of the bone surface.

59. Quadrate central median ridge: relatively thin and distinct (0) (Figure 7A-C), or in form of smooth broadly inflated dome around stapedial pit (1) (Figure 7D). In state 1, the sharp median ridge loses definition in the area of the stapedial pit.

60. Quadrate ventral median ridge: a single thin ridge present (0) (Figure 7B, C), or thin ridge diverging ventrally (1) (Figure 7A, D). In the derived state, the anterior ray continues to the ventromedial corner of the bone, while the posterior ray gradually curves posteriorly and merges into the posteromedial face of the bone near the infrastapedial process.

61. Quadrate ventral condyle: condyle saddle-shaped, concave in anteroposterior view (0), or gently domed, convex in any view (1).

62. Quadrate ventral condyle shape: articular surface mediolaterally elongate (0), or very narrow and subtriangular or teardrop-shaped (1).

63. Quadrate anterior ventral condyle modification: no upward deflection of anterior edge of condyle (0), or distinct deflection present (1). A relatively narrow bump in the otherwise horizontal trace of the anterior articular edge is also

supertended by a sulcus on the anteroventral face of the bone.

64. Basisphenoid pterygoid process shape: process relatively narrow with articular surface facing mostly anterolateraly (0), or somewhat thinner, more fan-shaped with a posterior extension of the articular surface causing a more lateral orientation (1).

65. Basioccipital tubera size: short (0), or long (1). Long tubera are typically parallel-sided in posterior profile and protrude ventrolaterally at exactly 45 degrees from horizontal. Short tubera have relatively large bases that taper distally, and emanate more horizontally.

66. Basioccipital tubera shape: tubera not anteroposteriorly elongate (0), or anteroposteriorly elongate with rugose ventrolateral surfaces (1).

67. Basioccipital canal: absent (0), or canal through basioccipital and basisphenoid present (1). In *Ectenosaurus, Platecarpus*, and *Plioplatecarpus* a bilobate tunnel enters the basioccipital dorsally, passes forward into the basisphenoid, and exits that bone dorsally and laterally to the braincase. Russell (1967:148) virtually equates this tunnel with a similarly positioned groove described in *Prognathodon*. However, more complete preparation of both the type and the only referred specimen of the latter has revealed only small, rounded, blind vestibules not quite symmetrical with the midline and, therefore, of dubious homology.

68. Dentary tooth number I: 20-24 (0), or 17-19 (1), or 15-16 (2), or 14 (3), or 13 (4), or 12 (5), or <12 (6). It is easy to assume this character is correlated with the number of maxillary teeth, except that is not the case in *Ectenosaurus clidastoides*, which has 16 or 17 maxillary teeth and only 13 dentary teeth. Therefore, this character is included as different from character 35.

69. Dentary anterior projection: projection of bone anterior to first tooth present (0), or absent (1). Although Russell (1967:149) used "dentary terminates abruptly in front of first dentary tooth" as a character diagnosing *Platecarpus*, *P. planifrons* clearly shares this character. In fact, the dentary he used to illustrate a character that distinguished *P. coryphaeus* from *P. ictericus* (1967:Figure 85) does not represent *P. coryphaeus*, but is referable to *P. planifrons*.

70. Dentary anterior projection length: short (0), or long (1). In the derived state, the projection of bone anterior to the first tooth is at least the length of a complete tooth space.

71. Dentary medial parapet: parapet positioned at base of tooth roots (0), or elevated and straplike, enclosing about half of height of tooth attachment in shallow channel (1), or strap equal in height to lateral wall of bone (2). The two derived states are possibly sequential stages of modification from a classically pleurodont dentition to the typical mosasaur "subthecodont" dentition (Romer, 1956:442).

72. Splenial-angular articulation shape: splenial articulation in posterior view

almost circular (0), or laterally compressed (1), or intermediate (2).

73. Splenial-angular articular surface: essentially smooth concavo-convex surfaces (0), or distinct horizontal tongues and grooves present (1). Angulars of many *P. tympaniticus* have three or four arcuate ridges that fit into grooves in the contacting surface of the splenial. Thus, the splenial rotated horizontally on the angular about a vertical line. In *Plioplatecarpus*, the tongues and grooves are longer and deeper, thinner, and more numerous, a condition that absolutely restricts the anterior mandible to mediolateral movement in a horizontal plane. This supports conclusions of Camp (1942:35), Antunes (1964:157), and Callison (1967).

74. Coronoid shape: coronoid with slight dorsal curvature, posterior wing not widely fan-shaped (0), or very concave above, posterior wing greatly expanded (1).

75. Coronoid posteromedial process: small but present (0), or absent (1).

76. Coronoid medial wing: does not reach angular (0), or contacts angular (1).

77. Coronoid posterior wing: without medial crescentic pit (0), or with distinct excavation (1). In state 1, there is a posteriorly open, C-shaped excavation in the medial side of the posterior wing of this element.

78. Surangular coronoid buttress: low, thick, about parallel to lower edge of mandible (0), or high, thin, rapidly rising anteriorly (1). A rounded dorsal edge of the surangular remains almost parallel to the ventral edge as it approaches the posterior end of the coronoid, meeting the latter element near its posteroventral edge in state 0. In the derived condition, the dorsal edge rises and thins anteriorly until meeting the posterior edge of the coronoid near its apex, producing a triangular posterior mandible in lateral aspect.

79. Surangular-articular suture position: behind condyle in lateral view (0), or at middle of glenoid on lateral edge (1). In the derived condition, there is usually an interdigitation in the dorsal part of the suture.

80. Surangular-articular lateral suture trace: suture descends and angles or curves anteriorly (0), or is virtually straight throughout its length (1). In state 1, the suture trails from the glenoid posteriorly about halfway along the dorsolateral margin of the retroarticular process, then abruptly turns anteriorly off the edge and strikes in a straight line for the posterior end of the angular.

81. Articular retroarticular process inflection: moderate inflection, less than 60 degrees (0), or extreme inflection, almost 90 degrees (1). In *Mosasaurus*, *Plotosaurus*, and *Prognathodon overtoni* the posterior terminus of the lower jaw lies almost horizontal, probably allowing for more muscle attachment.

82. Articular retroarticular process innervation foramina: no large foramina on lateral face of retroarticular process (0), or one to three large foramina present (1).

83. Tooth surface I: teeth finely striate medially (0), or not medially striate (1). In "Russellosaurinae" medial tooth striations are very fine and groups of tightly spaced striae are usually set apart by facets, leading to a fasciculate appearance.

84. Tooth surface II: teeth not coarsely textured (0), or very coarsely ornamented with bumps and ridges (1). In both species of *Globidens* and in *Prognathodon overtoni*, the coarse surface texture is extreme, consisting of thick pustules, and vermiform or anastomosing ridges. Although Russell (1967) described the teeth of *P. overtoni* as smooth, a single-well preserved tooth and replacement teeth in crypts of the type show this observation to be in error. Teeth in *P. rapax* are smooth over the majority of their surface, but usually a few widely scattered large, very long, sharp-crested, vermiform ridges are present.

85. Tooth facets: absent (0), or present (1). *Halisaurus* teeth are smoothly rounded except for the inconspicuous carinae. *Clidastes* is described in numerous places as having smooth unfaceted teeth, but many immature individuals and some larger specimens have teeth with three distinct facets on the medial faces. Adult *Tylosaurus proriger* has indistinct facets. *Mosasaurus* has taken this characteristic to the extreme (see Russell, 1967:138).

86. Tooth fluting: absent (0), or present (1). In *Ectenosaurus*, some *Platecarpus planifrons*, and taxon novum, several broadly rounded vertical ridges alternate with shallow, round-bottomed grooves completely around the teeth.

87. Tooth inflation: crowns of posterior marginal teeth conical, tapering throughout (0), or crowns of posterior marginal teeth swollen near the tip or above the base (1). The rear teeth of *Globidens* and *Prognathodon overtoni* are distinctly fatter than other mosasauroid teeth, but those of *P. rapax* are also swollen immediately distal to the base.

88. Tooth carinae I: absent (0), or present but extremely weak (1), or strong and elevated (2). *Halisaurus* exhibits the minimal expression of this character (state 1) in that its marginal teeth are almost perfectly round in cross section; the carinae are extremely thin and barely stand above the suface of the teeth. *Globidens* is convergent in the strict sense of the character, but this is probably a result of obliteration of the carinae by extreme inflation.

89. Tooth carinae serration: absent (0), or present (1).

90. Tooth replacement mode: replacement teeth form in shallow excavations (0), or in subdental crypts (1). All mosasauroids that can be evaluated have an "anguimorph" type of tooth replacement (Estes et al., 1988:221), which is to have interdental positioning of replacement teeth and resorption pits associated with each. In some taxa, the resorption pits remained shallow, but in others, the pits deeply invaded the bony bases of the functional teeth and replacement teeth developed mostly in chambers underneath them. Such a change may be a necessary consequence of the "subthecodont" type of tooth implantation.

Postcranial Axial Skeleton

91. Atlas neural arch: notch in anterior border (0), or no notch in anterior border (1). Although Russell (1967:71) reported that *Tylosaurus* has no notch, it is distinctly present in all basic taxa of that clade.

92. Atlas synapophysis: extremely reduced (0), or large and elongate (1). In state 1, a robust synapophysis extends well posteroventral to the medial articular surface for the atlas centrum, and it may be pedunculate (*Clidastes*) or with a ventral "skirt" that gives it a triangular shape (*Mosasaurus*). A very small triangular synapophysis barely, if at all, extends posterior to the medial articular edge in state 0.

93. Zygosphenes and zygantra: absent (0), or present (1). This character assesses only the presence of zygosphenes and zygantra, not their relative development. Nonfunctional and functional are considered as present.

94. Zygosphene and zygantra number: present on many vertebrae (0), or present on only a few (1).

95. Hypapophyses: last hypapophysis occurs on or anterior to seventh vertebra (0), or on ninth or tenth vertebra (1). State 1 is reported in *Mosasaurus missouriensis* (Goldfuss, 1845:191), but this author noted that the last hypapophysis occurred on the seventh cervical.

96. Synapophysis height: facets for rib articulations tall and narrow on posterior cervicals and anterior trunk vertebrae (0), or facets ovoid, shorter than the centrum height on those vertebrae (1).

97. Synapophysis length: synapophyses of middle trunk vertebrae not laterally elongate (0), or distinctly laterally elongate (1). The lateral extension of the synapophyses from the middle of the trunk is as much as 70-80% of the length of the same vertebra in the derived state.

98. Synapophysis ventral extension: synapophyses extend barely or not at all below ventral margin of cervical centra (0), or some extend far below ventral margin of centrum (1). In the derived state, two or more anterior cervical vertebrae have rib articulations that dip well below the centrum, causing a very deeply concave ventral margin in anterior profile.

99. Zygopophysis development: zygopophyses present far posteriorly on trunk vertebrae (0), or zygopophyses confined to anterior trunk series (1). Plesiomorphically, zygopophyses extend at least to the sacral area.

100. Vertebral condyle inclination: condyles of trunk vertebrae inclined (0), or condyles vertical (1).

101. Vertebral condyle shape I: condyles of anteriormost trunk vertebrae extremely dorsoventrally depressed (0), or slightly depressed (1), or essentially equidimensional (2). In state 0, posterior height:width ratios of anterior trunk

vertebrae are close to 2:1. In state 1, they are close to 4:3, but posterior to this the ratio decreases as the vertebrae become proportionally higher.

102. Vertebral condyle shape II: condyles of posterior trunk vertebrae not higher than wide (0), or slightly compressed (1). In the derived condition, the posterior condylar aspect reveals outlines that appear to be higher than wide and even perhaps slightly subrectangular, due to the slight emargination for the dorsal nerve cord.

103. Vertebral synapophysis dorsal ridge: sharp ridge absent on posterior trunk synapophyses (0), or with a sharp-edged and anteriorly precipitous ridge connecting distal synapophysis with prezygopophysis (1). In the plesiomorphic condition, the ridge in question, if present, may be incomplete or it may be rounded across the crest with the anterior and posterior sides about equally sloping.

104. Vertebral length proportions: cervical vertebrae distinctly shorter than longest vertebrae (0), or almost equal or are the longest (1).

105. Presacral vertebrae number I: relatively few, 32 or less (0), or numerous, 39 or more (1). Here presacral vertebrae are considered to be all those anterior to the first bearing an elongate transverse process, following Williston (1898:139), Osborn (1899:177), and Russell (1967:78).

106. Presacral vertebrae number II: if few, then 28 or 29 (0), or 30 or 31 (1). *Tylosaurus* is reported to have 29 or 30 presacral vertebrae (Russell, 1967:171), but in every specimen that seems to be trustworthy (all *Tylosaurus proriger*), I have counted 31.

107. Sacral vertebrae number: two (0), or less than two (1). Numerous well-preserved specimens of derived mosasauroids have failed to show any direct contact of the pelvic girdle with vertebrae in the sacral area. Certainly, no transverse processes bear any type of concave facet for the ilium, and so it is generally assumed that a ligamentous contact was established with only one transverse process (Camp, 1942; Huene, 1911; Osborn, 1899; Merriam, 1894; Williston and Case, 1892). Depending on one's perspective, it could be said that derived mosasauroids have either no or one sacral vertebra.

108. Caudal dorsal expansion: neural spines of tail all uniformly shortened posteriorly (0), or several spines dorsally elongated behind middle of tail (1).

109. Hemal arch length: hemal arches about equal in length to neural arch of the same vertebra (0), or length about 1.5 times greater than neural arch length (1). This ratio may be as great as 1.2:1 in state 0. Comparison is most accurate in the middle of the tail and is consistent even on those vertebrae in which the neural spines are also elongated.

110. Hemal arch articulation: arches articulating (0), or arches fused to centra (1). Among the outgroups, the hemal arches fuse to the centra generally only in the latest life stages. These are considered to have state 0 of this character. All

Phylogenetic Revision of Mosasauroidea

mosasaurines have fused hemals and all "russellosaurines" have fused hemals. No mosasaurian is known to change this condition in any ontogenetic stage represented by fossil material, but the possibility that some very old "russellosaurines" fused the hemals awaits discovery.

111. Tail curvature: no structural downturn of tail (0), or tail with decurved posterior portion (1). See Bell (1993:125) for a discussion of this character.

112. Body proportions: head and trunk shorter than or about equal to tail length (0), or head and trunk longer than tail (1).

Appendicular Skeleton

113. Scapula/coracoid size: both bones about equal (0), or scapula about half the size of coracoid (1).

114. Scapula width: no anteroposterior widening (0), or distinct fan-shaped widening (1), or extreme widening (2). In the plesiomorphic condition, the anterior and posterior edges of the scapula encompass less than one quarter of the arc of a circle, but in state 1, the arc is increased to approximately one third. In state 2, the distal margin encompasses almost a half-circle and the anterior and posterior borders are of almost equal length.

115. Scapula dorsal convexity: if scapula widened, dorsal margin very convex (0), or broadly convex (1). In state 0, the anteroposterior dimension is almost the same as the proximodistal dimension. In state 1, the anteroposterior dimension is much larger.

116. Scapula posterior emargination: posterior border of bone gently concave (0), or deeply concave (1). In the derived condition, there is a deeply arcuate emargination on the posterior scapular border just dorsal to the glenoid. It is immediately bounded dorsally by a corner which begins a straight-edged segment that continues to the dorsal margin.

117. Scapula-coracoid fusion: ontogenetic fusion occurs (0), or no fusion at any life stage (1). Fully grown representatives of the outgroups and the only known specimen of *Opetiosaurus* have fused the scapula and coracoid. This occurs in no other mosasauroid specimen, regardless of size.

118. Scapula-coracoid suture: unfused scapula-coracoid contact has interdigitate suture anteriorly (0), or apposing surfaces without interdigitation (1). Although all outgroup taxa fuse this suture ontogenetically, an interdigitate suture is present early in life; therefore, I assigned the interdigitate suture as the plesiomorphic state.

119. Coracoid neck elongation: neck rapidly tapering from medial corners to a relatively broad base (0), or neck gradually tapering to a relatively narrow base (1). The derived state of this character gives an outline of the bone which is nearly

symmetrically and gracefully fan-shaped with gently concave, nearly equidistant sides.

120. Coracoid anterior emargination: present (0), or absent (1). See Bell (1993:132) for a discussion of this character.

121. Humerus length: humerus distinctly elongate, about three or more times longer than distal width (0) (Figure 8E), or greatly shortened, about 1.5 to 2 times longer than distal width (1) (Figure 8B, F), or length and distal width virtually equal (2) (Figure 8A, C), or distal width slightly greater than length (3) (Figure 8D).

122. Humerus postglenoid process: absent or very small (0) (Figure 8B, C, E, F), or distinctly enlarged (1) (Figure 8A, D).

123. Humerus glenoid condyle: if present, condyle gently domed and elongate ovoid in proximal view (0) (Figure 8A, E), or condyle saddle-shaped, subtriangular in proximal view, and depressed (1) (Figure 8D), or condyle highly domed or protuberant and short ovoid to almost round in proximal view (2) (not figured). In some taxa, the condylar surfaces of the limbs were finished in thick cartilage and there was no bony surface of the condyle to be preserved. This condition is scored as not represented. In some taxa, the glenoid condyle extends more proximally than does the postglenoid process (state 2) and it is not as ovoid as in the plesiomorphic state.

124. Humerus deltopectoral crest: crest undivided (0), or split into two separate insertional areas (1). In the derived state, the deltoid crest occupies an anterolateral or anterior position confluent with the glenoid condyle, while the pectoral crest occupies a medial or anteromedial area that may or may not be confluent with the glenoid condyle. The deltoid crest is often quite short, broad, and indistinct, being easily erased by degradational taphonomic processes.

125. Humerus pectoral crest: located anteriorly (0) (Figure 8A, D), or medially (1) (Figure 8B, C). In the derived condition, the pectoral crest is located near the middle of the flexor (or medial) side on the proximal end of the bone.

126. Humerus ectepicondylar groove: groove or foramen present on distolateral edge (0), or absent (1).

127. Humerus ectepicondyle: absent (0) (Figure 8B, C, E, F), or present as a prominence (1) (Figure 8A, D). A radial tuberosity is reduced in size in *Prognathodon*, but very elongated in *Plesiotylosaurus*.

128. Humerus entepicondyle: absent (0) (Figure 8B, E, F), or present as a prominence (1) (Figure 8A, C, D). The ulnar tuberosity protrudes posteriorly and medially from the posterodistal corner of the bone immediately proximal to the ulnar facet, causing a substantial dilation of the posterodistal corner of the humerus.

129. Radius shape I: radius not expanded anterodistally (0) (Figure 8E), or slightly expanded (1) (Figure 8B, F), or broadly expanded (2) (Figure 8A, C, D).

130. Ulna contact with centrale: broad ulnare prevents contact (0) (Figure 8B,

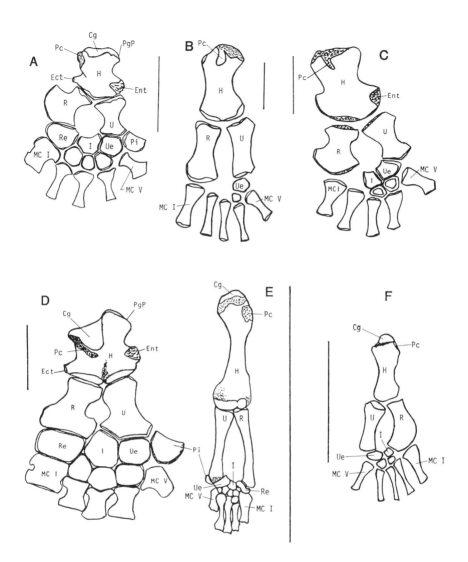

Figure 8. Comparison of mosasauroid proximal anterior limbs. A) *Clidastes liodontus*, flexor aspect of right limb. B) *Tylosaurus proriger*, flexor aspect of right limb. C) *Platecarpus tympaniticus*, flexor aspect of right limb. D) *Plotosaurus bennisoni*, flexor aspect of right limb. E) *Opetiosaurus bucchichi*, flexor aspect of left limb. F) *Halisaurus sternbergi*, extensor aspect of right limb. A, B, and C modified from Russell (1967). D based on LACM (CIT)2750. E modified from Carroll and deBraga (1992). F based on UPI R163. Abbreviations: **Cg** = glenoid condyle, **Ect** = ectepicondyle, **Ent** = entepicondyle, **H** = humerus, **I** = intermedium (centrale), **MC I** = metacarpal of digit I, **MC V** = metacarpal of digit V, **Pc** = pectoral crest, **PgP** = postglenoid process, **Pi** = pisiform, **R** = radius, **Re** = radiale, **U** = ulna, **Ue** = ulnare. Scale bars = 10 cm.

C, E, F), or ulna contacts centrale (1) (Figure 8A, D). The derived condition causes the ulnare to be omitted from the border of the antebrachial foramen. There is usually a well-developed faceted articulation between the ulna and the centrale (or intermedium, as used by Russell, 1967:93).

131. Radiale size: large and broad (0) (Figure 8A, D), or small to absent (1) (Figure 8B, C, E, F).

132. Carpal reduction: carpals number six or more (0) (Figure 8A, D, E), or five or less (1) (Figure 8B, C, F). One limb of *Ectenosaurus* was reconstructed to have seven carpals (Russell, 1967:Figure 54), although the other apparently intact manus of the same specimen bears only six.

133. Pisiform: present (0) (Figure 8A, D, E), or absent (1) (Figure 8B, C, F).

134. Metacarpal I expansion: spindle-shaped, elongate (0) (Figure 8B, C, E, F), or broadly expanded (1) (Figure 8A, D). The broad expansion is also associated with an anteroproximal overhanging crest in every case observed.

135. Phalanx shape: phalanges elongate, spindle-shaped (0), or blocky, hourglass-shaped (1). All the basic taxa of *Mosasaurus* and *Plotosaurus* have phalanges that are slightly compressed and anteroposteriorly expanded on both ends.

136. Ilium crest: crest blade-like, points posterodorsally (0), or elongate, cylindrical (1).

137. Ilium acetabular area: arcuate ridge supertending acetabulum (0), or acetabulum set into broad, short V-shaped notch (1). The primitive ilium has the acetabulum impressed into the lateral wall of the bone, with a low narrow crest anterodorsally as the only surrounding topographic feature. In the derived condition, the acetabular area is set into a short, broadly V-shaped depression that tapers dorsally. The lateral walls of the ilium are therefore distinctly higher than the rim of the acetabulum.

138. Pubic tubercle condition: tubercle an elongate protuberance located closer to midlength of shaft (0), or a thin semicircular crest-like blade located close to acetabulum (1). In *Tylosaurus proriger*, this was scored as missing data because the structure is apparently lost.

139. Ischiadic tubercle size: elongate (0), or short (1). In the ancestral condition the tubercle is as long as the shaft of the ischium is wide, but it is only a short narrow spur in the derived state.

140. Astragalus: notched emargination for crural foramen, without pedunculate fibular articulation (0), or without notch, pedunculate fibular articulation present (1). For state 0, the tibia and fibula are of equal length about the crural foramen and the astragalus contacts both to about the same degree. The form of the latter element is symmetrical and subcircular with a sharp proximal notch. In state 1, the outline of the element is basically reniform and the tibial articulation is on the same line as the crural emargination. The fibula is also shortened and its contact with the

astragalus is narrow.

141. Appendicular epiphyses: formed from ossified cartilage (0), or from thick unossified cartilage (1), or epiphyses missing or extremely thin (2). Ends of limb bones show distinct vascularization and rugose surfaces indicating an apparently thick nonvascularized, unossified cartilage cap. Extremely smooth articular surfaces suggest the epiphyses were excessively thin or pehaps even lost.

142. Hyperphalangy: absent (0), or present (1).

CONCLUSIONS

Although some of the evidence is equivocal, the analysis does not support a monophyletic "Aigialosauridae" as maintained by deBraga and Carroll (1993:Figure 4), because *Opetiosaurus* is identified as the sister-group to all other mosasauroids. There are no known synapomorphies that diagnose a monophyletic grouping and among the 99 trees produced by this analysis none linked these basal mosasauroids into a single clade. Rather, two or three "aigialosaurs" were linked to basal positions along the *Halisaurus* clade.

A direct comparison with Russell's tree (1967:Figure 99; or Figure 1 herein) leads to several observations. Most of his subgroups are coherent sets supported in this phylogeny. One of those subgroups, Prognathodontini, was reassigned wholesale to a different major clade, yet maintained its internal relationships. *Halisaurus* was removed from Mosasaurinae and placed in a lineage that diverged before any other major clades. Globidensini and *Mosasaurus* have a common ancestor within a paraphyletic *Clidastes* on the preferred tree, rather than each having a common ancestor with *Clidastes* as opposed to Russell's tree. *Plotosaurus* is derived from within a paraphyletic *Mosasaurus* rather than evolving from *Clidastes* as Russell believed. Both trees have the same relationships among *Tylosaurus, Ectenosaurus*, and the *Platecarpus-Plioplatecarpus* lineage.

All trees demonstrate that some previously recognized taxa (genera) are paraphyletic. These include *Mosasaurus, Prognathodon*, and *Platecarpus*. Although *Hainosaurus* was not included in this analysis, characters recently published by Nicholls (1988) and Lingham-Soliar (1992) support a common ancestry with *Tylosaurus proriger*. This would also require that *Tylosaurus* as a taxon be paraphyletic. Of the tribes recognized by Russell, only Mosasaurini is paraphyletic.

REVISED PHYLOGENY

Where possible, I have tried to use names already available from the literature, rather than erecting a series of new and unfamiliar designators. The one exception to this is the informal use of a new name, "Russellosaurinae," in anticipation of formally designating the taxon and describing a new taxon, *Russellosaurus*, from new Turonian material from Texas. Another, Natantia (Owen, 1849-1884), is resurrected from mid-nineteenth-century literature. Plotosaurini (Russell, 1967:122) is expanded to include basic taxa previously referred to *Mosasaurus*, and Mosasaurini (Russell, 1967:121) is abandoned because of paraphyly of *Mosasaurus*. Prognathodontini (Russell, 1967:123) is removed from Plioplatecarpinae and placed within Globidensini (Russell, 1967:122). *Halisaurus* is removed from Plioplatecarpinae and arranged as the sister taxon to all other derived mosasaurs. Finally, Aigialosauridae is abandoned as a general taxon due to paraphyly and the constituent basic taxa are included within Mosasauroidea.

The monophyly of Mosasauroidea (Camp, 1923:297) is supported by the following nine characters, the first four from Carroll and deBraga (1992), and the last five from Bell (1993). They may be taken as diagnostic for the taxon:

1) Circular configuration of the quadrate.
2) Well-developed hinge joint between angular and splenial.
3) Articulating surface of lower jaw formed equally by surangular and articular.
4) Reduction of transverse processes and zygopophyses of caudal vertebrae.
5) Frontal invaded by posterior end of nares (assigned by analysis, but reversed in Mosasaurinae and some basic taxa).
6) Teeth with expanded bony base of attachment.
7) Extreme forward extension of the prearticular process.
8) Extreme reduction of posteromedial process of coronoid.
9) Elongation of hemal and neural processes of caudal vertebrae.

Following are the sets of characters and states assigned to the various nodes and taxa on the preferred tree. Note that the large number of synapomorphies supporting *Halisaurus* is to some degree an artifact of collapsed nodes around the unresolved portion of the tree. Unequivocal character transformations are preceded by U. Equivocation is indicated by E. Numbers correspond to characters as numbered in the character list. These are followed by a dash and the character state as numbered in the character list.

Mosasauridae (Gervais, 1853) - U:3-1, 117-1; E:15-1, 83-1.
Halisaurus (Marsh, 1869) - U:47-1, 74-1, 79-1, 88-1, 93-0, 103-1, 107-1, 137-1; E:23-1, 25-1, 36-1, 37-1, 61-1, 69-1, 71-1, 92-1, 93-1, 98-1, 101-0, 110-1, 138-1, 139-1.
Node A - U:21-0, 36-2.

Node B - U:42-1; E:1-1.
Natantia (Owen, 1849-1884) - U:13-1, 17-1, 24-1, 27-1, 28-1, 44-1, 57-1, 63-1, 85-1, 90-1, 100-1, 107-1; E:1-1, 26-1, 53-1, 71-2, 93-1, 114-1, 121-2, 124-1, 129-2, 136-1.
"Russellosaurinae" - U:16-1, 34-1, 40-1, 48-2, 64-1, 83-0, 111-1, 115-0, 116-1, 118-1, 125-0, 131-1, 133-1; E:39-1, 68-1, 141-1.
Tylosaurus (Marsh, 1872) - U:2-2, 4-1, 6-1, 30-1, 70-1, 94-1, 113-1, 120-1, 121-1, 129-1; E:7-1, 35-4, 38-1.
Node C - U:5-1, 18-1, 66-1, 89-1, 93-0; E:132-1.
Plioplatecarpini (Russell, 1967) - U:18-1, 26-1, 27-1, 73-1, 90-1; E:37-1, 136-1.
Plioplatecarpus (Dollo, 1882) - U:24-2, 24-3, 50-2, 62-1.
Ectenosaurus (Russell, 1967) - U:47-1, 72-0, 74-1; E:19-2, 50-1, 52-1, 86-1.
Node D - U:44-2, 48-1, 57-0, 60-1, 68-5, 91-1; E:10-0, 19-0, 35-5, 45-1, 61-1, 75-1, 132-1.
Node E - U:12-1, 30-1, 46-1, 58-1, 69-1, 73-1; E:50-1, 139-1.
Node F - U:1-1, 24-1, 93-0; E:41-1.
Mosasaurinae (Williston, 1897) - U:2-1, 11-0, 29-1, 32-1, 42-1, 74-1, 78-1, 101-2, 103-1, 105-1, 108-1, 109-1, 110-1, 122-1, 126-1, 127-1, 130-1, 134-1; E:8-1, 25-1, 35-2, 36-1, 39-2, 65-1, 68-1, 92-1, 128-1, 140-1, 141-2.
Node G - U:51-1.
Node H - E:68-2.
Node I - U:9-1, 76-1.
Node J - U:10-0, 30-1, 34-1, 77-1, 119-1.
Globidensini (Russell, 1967) - U:12-1, 47-1, 50-1, 84-1, 85-0, 87-1; E:35-4.
Globidens (Gilmore, 1912) - U:88-1.
Node K - U:43-1, 55-1, 72-1, 123-2; E:20-1, 89-1.
Node L - U:34-0, 61-1; E:120-1.
Plotosaurini (Russell, 1967) - U:94-1, 104-1, 121-3, 123-1; E:89-1.
Node M - U:80-1, 81-1; E:6-1, 20-1, 32-0, 35-3, 49-0, 51-0, 56-1, 59-1, 60-1, 61-1, 63-0, 135-1.
Node N - U:53-0, 54-1, 114-2; E:97-1, 120-1, 142-1.
Node O - U:52-1, 57-0, 58-1.
Plotosaurus (Camp, 1951) - U:2-0, 11-1, 23-1, 35-1, 38-1, 70-1, 95-1, 99-1.

SUMMARY

This study is an examination of previously published mosasaur phylogenies using the technique of phylogenetic systematics (cladistics) and an enlarged data set. Taxa included are described and undescribed North American mosasaurs, and three specimens previously referred to Aigialosauridae. Ingroup phylogenetic analysis of 142 characters among 36 terminal taxa produced 99 equally parsimonious trees. Consensus trees demonstrate poor resolution of relationships among the more incomplete basal taxa. Monophyly of the traditional "Aigialosauridae" is falsified by lack of a natural grouping of those basic taxa. *Halisaurus* is placed as the sister-group to Natantia Owen, 1849-1884 (resurrected). Natantia is a large taxon consisting of the derived mosasauroids and is diagnosed by 12 unequivocal synapomorphies. Mosasaurinae Williston, 1897, and the soon-to-be-formalized taxon "Russellosaurinae" are sister-groups comprising Natantia and are diagnosed, respectively, by 18 and 13 unequivocal synapomorphies. "Russellosaurinae" consists of *Tylosaurus* and Plioplatecarpini. The latter includes *Ectenosaurus* and a clade consisting of "*Platecarpus*" (paraphyletic) + *Plioplatecarpus*. Mosasaurinae contains paraphyletic "*Clidastes*," Globidensini, and Plotosaurini. Within Globidensini, *Globidens* is the sister-group to "*Prognathodon*" (paraphyletic) + *Plesiotylosaurus*. Plotosaurini is comprised of "*Mosasaurus*" (paraphyletic) + *Plotosaurus*. Placement of *Halisaurus* as a sister-group to derived mosasauroids and *Prognathodon* + *Plesiotylosaurus* with Globidensini among Mosasaurinae are the major inconsistencies with Russell's (1967) phylogeny.

ACKNOWLEDGMENTS

There are numerous people who provided access to specimens in their care and who assisted me during my quest to see "every mosasauroid in North America and then some." In order of recollection only, those are: James P. Lamb, Jr., Susan Henson, Doug Jones, John C. Hall, Brown Hawkins, Andrew Rindsberg, Charlie Copeland, Chuck Finsley, Lloyd Hill, Mike Polcyn, Van Turner, Lou Jacobs, Dale Winkler, Ernest Lundelius, Melissa Winans, Larry Martin, Orville Bonner, John Chorn, Desui Miao, H. P. Schultze, Richard Zakrzewski, Greg Liggett, Jim Martin, Phillip Bjork, John Bolt, Bill Simpson, J. Howard Hutchison, Mike Greenwald, Arthur Staebler, J. D. Stewart, Mary Dawson, Earle Spamer, David Parris, Barbara Grandstaff, Bob Denton, Nicholas Hotton, Michael Brett-Surman, Bob Purdy, Bob Emry, Dan Chaney, Mark Norell, Lowell Dingus, Charlotte Holton, Willard Whitson, John Ostrom, Mary Ann Turner, Bob Allen, Chuck Schaff, A. W. Crompton, Drs. Sergio Dolce and Ruggero Calligaris of Trieste, Drs. Hermann Kollman and Karl Summesberger of Vienna, Dr. Peter Wellenhoffer of Munich, Drs. Wighart von Koenigswald and Martin Sander of Bonn, and Ms. Solweg Stuenes of Uppsala.

This research was funded in part by grants from the American Museum of Natural History, Austin Paleontological Society, Dee Fund of the Field Museum of

Natural History, Ernst Mayr Award of the Museum of Comparative Zoology at Harvard, Geological Foundation of the University of Texas at Austin, Geological Society of America Grant #4623-91, Paleontological Society/Margaret C. Wray Trust, Sigma Xi Grants-in-Aid of Research, the Haslem Postdoctoral Fund from the South Dakota School of Mines and Technology Foundation, and from my supervisor, Tim Rowe.

REFERENCES

Antunes, M. T. 1964. O Neocretacico e o Cenozoico do litoral de Angola. *Junta de Investigacoes do Ultramar*. Lisbon, 255 pp.

Bell, G. L., Jr. 1993. *A Phylogenetic Revision of Mosasauroidea (Squamata)*. Unpublished Ph. D. thesis, University of Texas at Austin, 293 pp.

Calligaris, R. 1988. I rettili rossili degli "Strati calcarei ittiolitici di Comen" e dell'Isola di Lesina. *Atti del Museo Civico di Storia Naturale Trieste* 41:85-125.

Callison, G. 1967. Intracranial mobility in Kansas mosasaurs. *University of Kansas Paleontological Contributions* 26:1-15.

Camp, C. L. 1923. Classification of the lizards. *Bulletin of the American Museum of Natural History* 49:289-481.

Camp, C. L. 1942. California mosasaurs. *University of California Memoir* 13:1-68.

Camp, C. L. 1951. *Plotosaurus*, a new generic name for *Kolposaurus*, preoccupied. *Journal of Paleontology* 25:822.

Carroll, R. L. and M. deBraga. 1992. Aigialosaurs: mid-Cretaceous varanoid lizards. *Journal of Vertebrate Paleontology* 12:66-86.

deBraga, M. and R. L. Carroll. 1993. The origin of mosasaurs as a model of macroevolutionary patterns and processes. *Evolutionary Biology* 27:245-322.

Dollo, L. 1882. Note sur l'osteologie des Mosasauridae. *Bulletin Musée Royale de la Naturale Historie de Belgique* 1:55-80.

Estes, R., K. deQueiroz, and J. Gauthier. 1988. Phylogenetic relationships within Squamata. IN R. Estes and G. Pregill (Eds.), *Phylogenetic Relationships of the Lizard Families*, pp. 119-281. Stanford University Press, Stanford, California.

Felsenstein, J. 1985. Confidence limits on phylogenies: An approach using the bootstrap. *Evolution* 39:783-791.

Gervais, P. 1853. Observations relative aux reptiles fossiles de France. *Academie Scientifique de Paris Comptu Rendus* 36:374-377, 470-474.

Gilmore, C. W. 1912. A new mosasauroid reptile from the Cretaceous of Alabama. *Proceedings of the United States National Museum* 41:479-484.

Goldfuss, A. 1845. Der Schädelbau des *Mosasaurus*, durch Beschreibung einer neuen Art dieser Gattung erläutert. *Nova Acta Academica Caesar Leopoldino-Carolinae Germanicae Natura Curiosorum* 21:1-28, Plates VI-IX.

Hennig, W. 1966. *Phylogenetic Systematics*. University of Illinois Press, Chicago.

Hillis, D. M. and J. P. Huelsenbeck. 1992. Signal, noise, and reliability in molecular phylogenetic analyses. *Journal of Heredity* 83:189-195.

Huene, F. von. 1911. Über einem *Platecarpus* in Tübingen. *Neues Jahrbuch für Mineralogie, Geologie, und Paläontologie* 2:48-50.

Lingham-Soliar, T. 1992. The tylosaurine mosasaurs (Reptilia, Mosasauridae) from the Upper Cretaceous of Europe and Africa. *Bulletin de l'Institut Royal des Sciences Naturelles de Belgique* 62:171-194.

Maddison, W. P., M. J. Donoghue, and D. R. Maddison. 1984. Outgroup analysis and parsimony. *Systematic Zoology* 33:83-103.

Marsh, O. C. 1869. Notice of some new mosasauroid reptiles from the greensand of New Jersey. *American Journal of Science* (*2nd series*) 48(144):392-397.

Marsh, O. C. 1872. Note on *Rhinosaurus*. *American Journal of Science* (*3rd series*) 4(20):47.

Merriam, J. C. 1894. Über die Pythonomorphen der Kansas Kreide. *Palaeontographica* 41:1-39.

Nicholls, E. L. 1988. The first record of the mosasaur *Hainosaurus* (Reptilia:Lacertilia) from North America. *Canadian Journal of Earth Sciences* 25:1564-1570.

Osborn, H. F. 1899. A complete mosasaur skeleton, osseous and cartilaginous. *Memoirs of the American Museum of Natural History* 1:165-188.

Owen, R. 1849-1884. *A History of British Fossil Reptiles*. London, 4 volumes.

Romer, A. S. 1956. *Osteology of the Reptiles*. University of Chicago Press, Chicago, 772 pp.

Russell, D. A. 1967. Systematics and morphology of American mosasaurs. *Bulletin of the Peabody Museum of Natural History* 23:1-241.

Schoch, R. M. 1986. *Phylogeny Reconstruction in Paleontology*. Van Nostrand-Reinhold Co., New York.

Swofford, D. 1990. PAUP: Phylogenetic Analysis Using Parsimony, version 3.0. Privately printed documentation. *Illinois Natural History Survey,* Champaign, Illinois.

Williston, S. W. 1897. *Brachysaurus*, a new genus of mosasaurs. *Kansas University Quarterly* 6:95-98.

Williston, S. W. 1898. Mosasaurs. *University Geological Survey of Kansas* 4:83-221.

Williston, S. W. and E. C. Case. 1892. Kansas mosasaurs. *Kansas University Quarterly* 1:15-32.

APPENDIX 1

Institutional abbreviations and specimens examined:

AMNH, American Museum of Natural History
ANSP, Academy of Natural Sciences, Philadelphia
AUMP, Auburn University Museum of Paleontology, Auburn, Alabama
BSPhG, Bayerische Staatssammlung für Paläontologie und Historische Geologie, Munich
CM, Carnegie Museum, Pittsburgh

APPENDIX 1 (continued)

DMNH, Dallas Museum of Natural History
FHM, Sternberg Museum, Fort Hays State University, Hays, Kansas
FMNH, Field Museum of Natural History, Chicago
FWU, Palaontologische Institut, Frederick Wilhelm Universitat, Bonn
GSA, Geological Survey of Alabama, Tuscaloosa
KU, Kansas University Museum of Natural History
LACM, Los Angeles County Museum of Natural History
MCSN, Museo Civico di Storia Naturale, Trieste
MCZ, Museum of Comparative Zoology, Harvard University
NJSM, New Jersey State Museum, Trenton
NMW, Naturhistorisches Museum Wien, Vienna
RMM, Red Mountain Museum, Birmingham, Alabama
SDSM, South Dakota School of Mines and Technology
TMM, Texas Memorial Museum, University of Texas at Austin
UAMNH, University of Alabama Museum of Natural History, Tuscaloosa
UCBMP, Museum of Paleontology, University of California, Berkeley
UNSM, University of Nebraska State Museum, Lincoln
UPI, Uppsala Palaeontological Institute
USNM, United States National Museum of Natural History
YPM, Yale Peabody Museum

Ingroup specimens:

Ingroup specimens are all fossil materials prepared and conserved in a variety of ways. Specimens directly contributing to character scores in the final data matrix are marked with an asterisk. Those without asterisks provided supporting information concerning character variation and distortion during fossilization.

Aigialosaurus dalmaticus - BSPhG 1902II501.*
Clidastes liodontus - KU 1022*; YPM 1335*; FHM VP2071; YPM 1333; UNSM 11647, 11719.
Clidastes novum sp. - TMM 43208-1.*
Clidastes moorevillensis - RMM 070*; GSA TC218; FMNH PR495.
Clidastes propython - ANSP 10193*; AMNH 1513; FMNH P27324; KU 1000; UCBMP 34535; USNM 3765; YPM 1105, 1310, 1368.
Dallas specimen - TMM 43209-1.*
Ectenosaurus clidastoides - FHM VP7937 (also 401).*
Ectenosaurus composite - YPM 4673,* 4674,* 4671, 4672; KU 1024.
Globidens alabamaensis - UAMNH 9850017*; USNM 6527.*
Globidens dakotensis - FMNH PR846.*
Halisaurus novum sp. - RMM 6890,* 3284*; AUMP 408.*

Halisaurus platyspondylus - NJSM 12146,* 12259*; USNM 442450.*
Halisaurus sternbergi - UPI R163.*
Halisaurus cf. *H. sternbergi* - USNM 3777.*
Mosasaurus conodon - AMNH 1380.*
Mosasaurus indeterminate - UNSM 77040* (28-16-39).
Mosasaurus maximus - NJSM 11052,* 11053*; TMM 313-1*; YPM 430, 1504.
Mosasaurus missouriensis - FWU Goldfuss 1327*; KU 1034.*
Mosasaurus sp. (misc.) - CM 8941, 6424; SDSM 452.
Opetiosaurus bucchichi - NMW (specimen unnumbered).*
Platecarpus planifrons - KU 14349,* 75037,* 84853; AMNH 1491, 1511; FHM 2116, 2181; MCZ 1610, 1614; YPM 3971, 40409, 40434, 40439, 40506.
Platecarpus af. *P. somenensis* - FMNH PR674,* PR465, PR466; YPM 40734.
Platecarpus tympaniticus - DMNH 8769*; FMNH UC600*; AMNH 1488, 1559, 1563, 1566; ANSP 10193; CM 1511; FHM VP322; KU 1001; LACM 128319; UCBMP 34538; YPM 1258, 1269, 4003, 24931, 40436, 40728.
Plesiotylosaurus crassidens - LACM (CIT)2759*; UCBMP 126716,* 137249; AMNH 1490 (in part).
Plioplatecarpus (indeterminate) - AMNH 2182*; RMM 2048*; USNM 18254; YPM 40735, 55673.
Plotosaurus bennisoni - UCBMP 32778,* 137247.
Plotosaurus tuckeri - LACM (CIT)2750,* (CIT)2804, (CIT)2945; UCBMP 126278, 33913.
Plotosaurus (indeterminate) - UCBMP 57582, 126278, 126283, (CNS 82-303), (CNS 85-103).
Prognathodon overtoni - SDSM 3393*; KU 950.*
Prognathodon rapax - NJSM (GSNJ)9827*; UCBMP 126280,* 126715*; AMNH 1490 (in part).
Taxon novum - YPM 40383.*
Trieste specimen - MCSN 11430, 11431, 11432, 1 pc. unnumbered (all one specimen).*
Tylosaurus nepaeolicus - AMNH 124,* 134,* FHM VP2209,* VP2292; YPM 3974, 40761.
Tylosaurus novum sp. - FHM VP2295,* VP78, VP2495; LACM 127815*; MCZ 1589; YPM 3392, 40796.
Tylosaurus proriger - AMNH 4909*; DMNH 8100*; FHM VP3(4)*; YPM 1268*; KU 1032, 1033, 1075; MCZ unnumbered type, 1030.

APPENDIX 2. Ingroup data matrix:

Character number

Taxon	5	10	15	20	25	30	35	40
OUTGROUP	0?000	0000?	00??0	00???	00000	??000	?000?	0?00?
Aigia*dalmaticus	???0?	0?101	100??	?0???	1?10?	10?00	?000?	1?000
Dalla*aigialosau	?????	???0?	?0??1	?????	?????	?????	?????	?????
Opeti*buccichi	0?00?	???01	100??	?0???	10?0?	?0??0	?0???	0??00
Tries*aigialosau	?????	?????	?????	?????	?????	?????	?????	?????
Clida*KUliodontu	11100	0?101	00101	?1010	10011	11?10	?1002	1002?
Clida*YPliodontu	11100	0?101	001?1	?1010	1001?	11?10	?1002	100??
Clida*mooreville	11100	00101	00101	01010	10011	11?10	?1002	1?020
Clida*novumsp.	11100	00101	00101	?1010	10011	11110	?10?2	10020
Clida*propython	11100	0??11	00101	01010	10011	1??10	?1002	1?0??
Ecten*clidastoid	1110?	0?101	001?1	?1020	1110?	01?00	?0111	10111
Ecten*YPcomposit	10100	01?01	?0?11	11020	11100	?1???	?0?1?	0??11
Globi*alabamaens	?????	???10	?1111	0101?	?????	???10	?10??	1????
Globi*dakotensis	1110?	0?110	111?1	?1010	1001?	11?11	?1014	10020
Halis*novumsp.	10100	10?01	00001	000??	?????	???00	????0	??0??
Halis*platyspond	10100	00?01	11001	000??	10101	?0?00	?????	11???
Halis*sternbergi	?????	??001	100??	?0???	0010?	00000	????0	210??
Halis*cfsternber	0?100	?0?0?	??00?	000??	00111	00??0	????0	2????
Mosas*conodon	?????	?????	?????	?????	?????	?????	?????	?????
Mosas*indet.	1110?	1?111	001?1	?1011	1001?	11?11	?0013	100??
Mosas*maximus	11100	1?100	011?1	?1011	10021	11?11	00013	1??21
Mosas*missourien	11100	1?110	001?1	?1011	1001?	11?11	?00?3	100??
Plate*planifrons	10100	00100	10111	?100?	1110?	11?00	??115	000?1
Plate*af.somenen	?????	???00	111?1	?100?	1112?	11?01	00??5	0??11
Plate*8769tympan	0?100	00?00	11111	1100?	11110	11?01	001?5	00011
Plate*600tympani	1010?	0?100	111?1	?100?	1110?	?1?01	00115	?0011
Plesi*crassidens	1110?	0?111	011?1	?1011	1001?	11?11	11004	11???
Pliop*AMNHsp.	?????	???00	111?1	?100?	11130	11?01	0001?	0??11
Pliop*RMMsp.	?????	???00	??101	0100?	11130	11???	?0???	?????
Ploto*bennisoni	10100	1?010	101?1	?1011	1011?	11111	?0011	10120
Ploto*tuckeri	1010?	1??10	101?1	?1011	1011?	11111	?0011	101?0
Progn*overtoni	10100	0??10	01101	01011	1001?	11111	?1015	00?20
Progn*rapax	10100	0??10	00101	01011	10011	11?11	?1?04	1?020
Tylos*nepaeolicu	12111	1?000	10101	111??	1000?	11101	?0014	0???1
Tylos*novumsp.	12110	11000	11101	11011	1001?	11?01	?0114	10121
Tylos*proriger	12111	11000	111?1	111??	1001?	11101	10?14	10111
Taxon*novum	1011?	11001	10101	?01??	0001?	00000	?0?00	000??

APPENDIX 2 (Continued)

	Character number							
Taxon	45	50	55	60	65	70	75	80
OUTGROUP	000??	?0???	000?0	00000	0?0?0	00?00	0??00	00000
Aigia*dalmaticus	???2?	?0???	1?00?	00???	??000	0????	1??0?	??010
Dalla*aigialosau	?0???	?????	?????	?????	?????	???00	020??	0?0??
Opeti*buccichi	???2?	?0???	??10?	00???	???1?	??000	0??00	0?0??
Tries*aigialosau	???20	?0?10	00000	00???	10???	?????	?????	??000
Clida*KUliodontu	?1010	?0000	00100	01000	00101	00200	22?10	00100
Clida*YPliodontu	?1010	?0000	00100	01000	00101	0?100	2??10	?01??
Clida*mooreville	?1010	?0010	10100	01000	00101	00200	22010	00100
Clida*novumsp.	01010	?0010	10100	01000	001?1	0?200	22010	00100
Clida*propython	?1010	?0010	10100	01000	00101	00100	2201?	10100
Ecten*clidastoid	?0010	01?11	0110?	11000	???10	0?3??	20?1?	00000
Ecten*YPcomposit	00010	01211	01100	?1000	???1?	?1400	2001?	?0000
Globi*alabamaens	?1010	?1111	?0?00	?1000	001??	?????	?2010	?1100
Globi*dakotensis	?1010	?1011	10100	01001	001?1	00???	?????	?????
Halis*novumsp.	?1?20	01010	0????	?0000	10??0	0001?	12011	00010
Halis*platyspond	?1?20	01010	00000	00000	100??	?????	?201?	00010
Halis*sternbergi	??020	?1??0	0000?	01???	10???	?????	1??1?	??010
Halis*cfsternber	?00??	?????	?????	?????	?????	?0?1?	120??	0?010
Mosas*conodon	?????	?????	?????	?????	?????	??1??	2201?	???00
Mosas*indet.	?1?00	?001?	01010	10111	10001	0??00	220??	????1
Mosas*maximus	?1000	?0010	01010	10111	100?1	00300	22010	?1101
Mosas*missourien	?1000	?0010	00100	11011	100?1	0?300	2?010	11101
Plate*planifrons	00021	00110	00100	00001	101?0	01500	22001	00000
Plate*af.somenen	10021	10?11	00100	00101	101?0	0?51?	2??01	00000
Plate*8769tympan	?0021	10111	00100	00101	101?0	0151?	22101	00000
Plate*600tympani	?0021	10111	00100	00101	10110	0151?	22101	00000
Plesi*crassidens	?1110	?1011	101??	01???	101??	??200	2??10	1?100
Pliop*AMNHsp.	00021	10??2	00100	00011	111?0	0?51?	22101	00000
Pliop*RMMsp.	?0021	10012	00100	?0100	11???	01???	?21??	??000
Ploto*bennisoni	01000	?0010	00010	11011	10001	0?101	2??1?	??101
Ploto*tuckeri	?1000	?0010	00010	11??1	100?1	0?101	2??11	1?101
Progn*overtoni	?1110	?1011	10?01	?1000	001?1	00300	21010	11100
Progn*rapax	?1110	?1011	10101	01000	10101	00?1?	21010	???00
Tylos*nepaeolicu	10011	?0210	11100	01000	00110	1?401	22000	00000
Tylos*novumsp.	10011	?0200	11100	01001	00110	00401	22000	00000
Tylos*proriger	00000	?0210	00100	01000	00110	10401	22000	10000
Taxon*novum	?0011	?0?10	00???	?0000	00001	0????	?????	?????

APPENDIX 2 (Continued)

Taxon	85	90	95	100	105	110	115	120
OUTGROUP	?0?00	00?00	0?000	?0000	?00?0	?00?0	0000?	?0000
Aigia*dalmaticus	0??00	0020?	?010?	?000?	???00	00???	??00?	01???
Dalla*aigialosau	??1??	????0	011??	00101	1110?	??0??	???0?	010??
Opeti*buccichi	0?000	00200	?????	000?0	???00	?0000	0000?	00?00
Tries*aigialosau	0????	?????	111??	?0?00	1??00	?0?00	?????	?????
Clida*KUliodontu	0?101	00201	01100	00001	20101	?1111	01011	01000
Clida*YPliodontu	??101	00201	0110?	00001	20101	?1???	?????	?????
Clida*mooreville	00101	00201	01100	00001	2?101	?1111	??011	01000
Clida*novumsp.	00101	00201	0110?	00?01	2110?	??1?1	??011	01000
Clida*propython	00101	00201	01100	00001	2?10?	??1?1	?????	??000
Ecten*clidastoid	?1001	10201	??100	10001	???0?	?????	??010	11100
Ecten*YPcomposit	01001	10201	001??	?????	?????	?????	?????	??1??
Globi*alabamaens	00110	01101	011??	0?0?1	2??0?	?????	?????	?10??
Globi*dakotensis	??110	01101	011?0	0?0?1	2?10?	?????	?????	?????
Halis*novumsp.	00100	00100	??0?0	00100	00100	?1??1	?????	?????
Halis*platyspond	00100	00101	??0??	????0	0????	????1	?????	?????
Halis*sternbergi	00???	?????	?10??	00100	???00	11101	01111	11100
Halis*cfsternber	0010?	00100	?????	?????	0????	?????	?????	?????
Mosas*conodon	00101	00211	01110	010?1	2011?	????1	???1?	01010
Mosas*indet.	??101	00211	010??	?10?1	2??1?	?????	????1	010??
Mosas*maximus	10101	00211	01110	11001	20?1?	?1111	??021	01011
Mosas*missourien	10101	002?1	011?0	000?1	20111	?1??1	??01?	?100?
Plate*planifrons	01001	10201	10100	000?1	1??0?	?????	??010	11100
Plate*af.somenen	01001	00201	100??	?????	1????	?????	?????	?????
Plate*8769tympan	?1001	00201	100?0	000?1	1??0?	????0	?????	?????
Plate*600tympani	0?001	00201	1011?	00001	10000	01000	10010	11100
Plesi*crassidens	???1?0	002?1	?????	00001	2?10?	?????	??011	01011
Pliop*AMNHsp.	01???	?0??1	?????	?????	?????	?????	??010	11100
Pliop*RMMsp.	11001	0020?	??0??	00001	10000	???0	?????	?????
Ploto*bennisoni	10000	012??	010?1	0?011	2??1?	?????	?????	?????
Ploto*tuckeri	?0???	????1	????1	01011	20111	???1?	??021	01011
Progn*overtoni	10110	01211	?????	?0??1	2????	????1	?????	?????
Progn*rapax	00110	01211	0110?	00001	2110?	????1	??01?	01011
Tylos*nepaeolicu	00001	00211	00010	000?1	100??	?????	?????	?????
Tylos*novumsp.	00001	00201	00110	0?0??	1??0?	?????	??110	11101
Tylos*proriger	00001	00211	100?0	00001	10000	11000	10110	11101
Taxon*novum	??101	00200	000?0	00100	1?000	01???	?????	?????

APPENDIX 2 (Continued)

Taxon	Character number				
	125	130	135	140	142
OUTGROUP	00000	00000	00000	0000?	00
Aigia*dalmaticus	000??	00000	00000	001??	00
Dalla*aigialosau	01?00	000??	????0	00???	0?
Opeti*buccichi	000?0	?0000	00000	?????	00
Tries*aigialosau	?????	?0000	?????	0011?	00
Clida*KUliodontu	21010	11121	00010	10001	20
Clida*YPliodontu	21010	11121	00010	?????	2?
Clida*mooreville	21010	11121	00010	10001	20
Clida*novumsp.	21010	11121	00010	?????	2?
Clida*propython	21010	11121	0?0??	?????	2?
Ecten*clidastoid	21???	00120	10100	?????	10
Ecten*YPcomposit	?????	?????	?????	?????	??
Globi*alabamaens	21010	111??	?????	?????	2?
Globi*dakotensis	?????	?????	?????	?????	??
Halis*novumsp.	?????	?????	?????	?????	??
Halis*platyspond	?????	?????	?????	?1???	??
Halis*sternbergi	10010	00010	11100	1111?	00
Halis*cfsternber	?????	?????	?????	?????	??
Mosas*conodon	31110	111?1	?????	?????	2?
Mosas*indet.	31110	11121	00011	?????	2?
Mosas*maximus	31110	11121	00011	10001	21
Mosas*missourien	?????	???21	00?11	?????	2?
Plate*planifrons	20?11	00?20	11100	?????	1?
Plate*af.somenen	?????	?????	?????	???1?	??
Plate*8769tympan	?????	?????	?????	?????	??
Plate*600tympani	20??1	?0120	11100	10110	10
Plesi*crassidens	31210	11121	00010	?????	2?
Pliop*AMNHsp.	20?11	00120	11100	?????	1?
Pliop*RMMsp.	?????	?????	?????	?????	??
Ploto*bennisoni	?????	?????	?????	?????	??
Ploto*tuckeri	31110	11121	00011	?????	21
Progn*overtoni	21210	1?1??	?????	?????	??
Progn*rapax	21210	11121	00010	?????	2?
Tylos*nepaeolicu	10?1?	0001?	11100	?????	??
Tylos*novumsp.	10?11	0001?	?????	?????	1?
Tylos*proriger	10?11	0001?	11100	10?00	11
Taxon*novum	?????	?????	?????	?????	??

Chapter 12

ECOLOGICAL IMPLICATIONS OF MOSASAUR BONE MICROSTRUCTURE

AMY SHELDON

INTRODUCTION

The pelagic habit of the mosasaurs has inhibited an analysis of possible partitioning of the water column. After death, individuals may have bloated and drifted considerable distances, even hundreds of miles, before they sank to the bottom. Additionally, mosasaurs may have entered shallow, nearshore waters where they became stranded and died, as some large marine animals do today. Interpretation of possible water column partitioning by mosasaurs using geographic distribution is constrained by both phenomena.

Russell (1967) stated that mosasaurs had a worldwide distribution and inhabited subtropical epicontinental seas of depths less than 600 ft. Based on the geographic position of recovered specimens, Russell (1967) indicates that *Clidastes* inhabited the nearshore environment. Further, he indicates that *Platecarpus* and *Tylosaurus* frequented deeper water areas, farther from shore. He does not indicate whether they frequented deeper parts of the water column on a regular basis, but he only suggests that they were farther from shore.

The largest number of specimens of *Clidastes, Tylosaurus*, and *Platecarpus* was recovered from the Smoky Hill Member of the Niobrara Formation of Kansas (Sheldon, 1990). Subsequent work, especially in the Smoky Hill, found that the rocks exposed in the eastern, nearshore, region of the Niobrara are generally the lower part of the stratigraphic column. The rocks exposed in the western area are generally at the top of the section (Bennett, 1990; Hattin, 1982). More recent collection data show that *Clidastes, Platecarpus*, and *Tylosaurus* are distributed throughout the column and geographic range of the Smoky Hill Member (Sheldon,

1990, 1995). Therefore, a new analysis of the previous model presented by Russell (1967) for partitioning the water by mosasaurs is appropriate.

Marine mammals and turtles often partition the water column. Their bone microstructure is very different from that of their terrestrial relatives, and among themselves (Buffrénil et al., 1990; Buffrénil and Mazin, 1989; Buffrénil and Schoevaert, 1988; Rhodin, 1981, 1985; Felts, 1966; Felts and Spurrell, 1965). Bone microstructure seems to correlate with ecology. In an investigation of buoyancy, Taylor (1994) indicated that dense bone results in neutral buoyancy at shallow depth. With increased bone density animals can lose buoyancy so quickly that they become negatively buoyant even in very shallow water. Increased bone density therefore requires an increase in lung volume to maintain neutral buoyancy. The increase in lung volume often results in a bigger chest cavity, which may result in an increase in drag during swimming. The penalty for increased drag seen in some marine mammals, like sirenians, is slower swimming speeds. Alternatively, Taylor (1994) and Wall (1983) noted that reduced density of bone was linked with an increase in lipid storage and allowed for reduction of gas retained in the body; i.e., neutral buoyancy can be achieved with smaller lung volume because negative buoyancy can be produced by compression of air in the lungs with depth (Kooyman and Ponganis, 1994; Ridgeway and Harrison, 1986; Matthews, 1978). This combination results in more energy-efficient dives. It also permits a wider range of depths for the maximum vertical force exerted, i.e., "the difference between upward buoyancy and the animals' weight in air" (Taylor, 1994; see also Wall, 1983), and it maximizes the efficiency of hydrodynamic buoyancy control (Taylor, 1994; Wall, 1983). Reduction of bone mass and an increase in lipids is useful for deep-diving animals. Many cetaceans, such as dolphins and some whales, ichthyosaurs, and some turtles, have very porous, light bone, and many swim swiftly (Taylor, 1994; Buffrénil and Mazin, 1989; Buffrénil et al., 1985; Ridgeway and Harrison, 1985, 1986; Rhodin, 1981, 1985; Nopcsa, 1923). Today those marine mammals that have reduced bone density frequent the deeper portion of the water column.

TERRESTRIAL VERSUS AQUATIC BONE MICROSTRUCTURE

Significant microstructural differences exist between the bones of marine and terrestrial tetrapods. These differences are the direct result of mechanical stresses on the bone as it matures. Bone maturation and modification in terrestrial vertebrates are consistent with the physical impact of living under the influence of gravity (Meulen et al., 1993; Beaupre et al., 1990; Malacinski, 1990; Carter et al., 1989, 1991; Carter and Wong, 1988; Globus et al., 1986; Wolff, 1892). In early neonatal development of all tetrapods, endochondral cartilage template (anlagen) formation and morphology are controlled by the organism's genetic code (Carter et al., 1991; Malacinski, 1990; Martin and Burr, 1989; Jee, 1983). Later in skeletal development, however, the bone responds to the influence of mechanical stresses imposed by the physical environment or biological need of the animal (Meulen et al., 1993; Beaupre et al., 1990; Malacinski, 1990; Leclair, 1990; Martin and Burr, 1989; Carter et al., 1989, 1991; Carter and Wong, 1988; Carter, 1987; Globus et al., 1986; Wolff, 1892). Adult terrestrial endochondral bone is characterized by a shaft composed of a dense tube of periosteally deposited cortical bone with an open medullary cavity (Castanet, 1985; Castanet and Naulleau, 1985; Figure 2A). The medullary cavity is constantly being internally remodeled by a complex combination of resorption and deposition (endosteal processes, Figure 2A). Layers of cortical bone are added peripherally to the tube by periosteal deposition. The distal and proximal ends of the bone have a dense layer of periosteally deposited bone, and the interior is a complex network of trabecular bone laid down in response to repeated alternating directional stresses (Jee, 1983; Jee et al., 1983; Koch, 1917; Wolff, 1892).

Postnatal change in bone density has long been known as a response to varying levels of stimulus to the bone. Modification of intermittent maximum load can cause differences in the rate and manner of ossification. Characteristic variations resulting from a reduction of maximum gravitational load include a delay in the calcification of secondary ossification nuclei and a reduction of periosteal lamellar bone. Additionally, endosteal deposition of trabeculae results in a reduction or loss of medullary cavities (Barreto et al., 1993; Meulen et al., 1993; Beaupre et al., 1990). The development of ossified epiphyses and dense cortical bone is dependent on multiple loading events along the longitudinal axis of the bone (Delgodo-Baeza et al., 1991; Leach and Gay, 1986; Kugler et al., 1979; Wilsman and van Sickle, 1970). If the loading events do not occur (e.g., in aquatic settings where gravity is not important), secondary ossification nuclei do not form secondary ossification centers, resulting in the epiphyses remaining hyaline cartilage (Carter et al., 1989, 1991; Martin and Burr, 1989; Carter and Wong, 1988; Carter, 1987; Lufti, 1970; Haines, 1969; Moody, 1908). Additionally, cortical bone is either rapidly resorbed or never deposited by periosteal processes. However, the same absence of stress results in a trabecular infilling of the medullary cavity. These features can be expected in aquatic tetrapods as a result of their maturation in water, where loading due to gravity is negligible (Wong and Carter, 1990a, b; Martin and Burr, 1989; Bikle et al., 1987; Eurell and Kazarian, 1983; Jee et al., 1983; Patterson-Buckendahl et al., 1983; Pitts et al., 1983; Simmons et al., 1983; Little, 1973; Wolff, 1892).

METHODS

Mosasaur bone samples (ribs) were obtained from the American Museum of Natural History, New York (AMNH); Red Mountain Museum (RMM), Birmingham, Alabama; University of California (UCB), Berkeley; University of Wisconsin, Madison (UW); and Museum of the Rockies, University of Montana, Bozeman (MOR). Caution was used in selecting ribs to analyze because ribs of adults and juveniles of different taxa may mimic each other in size. Compounding this difficulty is that ribs from a single specimen vary in size depending on their position along the vertebral column. Thus, diagnostic data derived from other bones were used to determine growth stage and taxon identity. Size of a rib alone was not used to determine either. Taxonomic placement was based on the most recent revision of the Mosasauridae (Bell, 1993). Ontogenetic placement was determined using the size-independent criteria outlined by Sheldon (1987, 1989, 1990, 1992, 1993; Bell and Sheldon, 1986).

Before sectioning, rib bone was embedded in either Silmar or Buehlar epoxy resin to prevent shattering during cutting and polishing. Embedded bone was cut with either an Isomet Buehlar water-cooled saw or an Ingram thin-section water-cooled saw. Sections were mounted on petrographic glass slides. Slides were ground to desired thickness beginning with Al_2O_3 grit the same size as the grit on the blade used, and stepped to finer grit, with a final polishing using 5 μm Al_2O_3 grit on a felt disk lap table. When complete ribs were available, cross sections were cut from a proximal end of the rib (metaphysis) and a medial position of the shaft (diaphysis).

Color Snap Frame Grabber and software (Computer Friends Inc., Portland, Oregon) was used on a Macintosh 2CI system to capture images from slides and analysis was carried out using Image Analyst software (Automatix, Billerica, Massachusetts). In addition, Image Pro Plus (Media Cybernetics, Silver Springs, Maryland) was used on a Zenith 386 system to complete the analysis. A thickness of 60 μm for bone samples was the most useful. Thicker sections did not allow transmission of light; therefore, observation of detail was impossible. Thinner sections resulted in hot spots that read as vacuities rather than bone under digital analysis. All analyses were carried out at 200X magnification. Porosity was calculated as a percent of vacuities with respect to the total cross-sectional area. Roundness was calculated using the formula $R = P/4\pi A$ (where R is roundness, P is perimeter, A is surface area). Perfectly round vacuities have a roundness of 1 (Figure 1). When vacuity shape is elliptical or is highly irregular, the roundness value becomes larger proportional to the change from round.

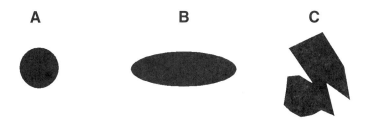

Figure 1. Diagrams represent vacuities which were measured for roundness using R = P/4πA. **A)** Roundness value of 1. **B)** Roundness value of 2.5. **C)** Roundness value of 12.6.

BONE MICROSTRUCTURE OF ADULT *CLIDASTES*

In proximal cross sections of adult *Clidastes*, primary bone has been totally remodeled into secondary bone, and primary periosteal lamellae have been remodeled into trabecular bone (Figure 2B). Convoluted, very thin trabeculae, deposited by endosteal processes (complex series of events occurring in the internal region of the bone that continually reshaped the inner part of the bone in response to stress and biological factors), form irregular and randomly distributed vacuities (Figure 2B). The trabeculae have numerous reversal lines indicating extensive remodeling by absorption and deposition by endosteal processes. Isolated, small masses of undifferentiated endochondral calcified cartilage are in scattered vacuities. Roundness values of the vacuities range from 1.0 to 5.8, with an average of 2.1 (Table 1). Volkmann's canals, revision lines between the periosteal deposition and endosteal deposition, and secondary osteons are absent. The mean porosity of the bone is 41.8% (Table 1), with a standard deviation of 11.4%.

In medial cross section of adult *Clidastes*, the anterior edge cortical bone is not completely remodeled (Figure 2C). Therefore, a revision line exists between the endosteal and the periosteal lamellae, but only in these locations. Periosteal lamellae are thus absent from the posterior, medial, and lateral edges of these ribs. The outermost periosteal lamellae are composed of primary bone, and it is vascularized by primary osteons. There are a very few, simple Volkmann's canals. The medullary cavity is filled by trabeculae of endosteal origin. In the most central area, the trabeculae are very thin but thicker trabeculae occur near the periphery. In most areas, the trabeculae are composed of secondary bone. *Clidastes* vacuities are subcircular and random in placement, with roundness values ranging from 1.2 to 8.2 (average of 2.4) across the diameter of the bone. Distal cross-sectional mean porosity is 11.3%, with a standard deviation of 11.4 (Table 1; Sheldon, 1995). High cortical bone porosity and the large vacuities (osteoporosis) of *Clidastes* suggest that significant quantities of lipids were deposited in these spaces, as seen in extant mammals and turtles (Taylor, 1994). The osteoporotic condition found in *Clidastes* is consistent with the low-gravity hypothesis (Meulen et al., 1993; Beaupre et al., 1990; Esteban, 1990; Wong and Carter, 1990a, b; Carter et al., 1989, 1991; Hunziker and Schenk, 1989; Carter and Wong, 1988; Morey, 1978; Carter, 1987; Globus et al., 1986; Jee et al., 1983; Cann and Adachi, 1983; Little, 1973). The proximal cross section showed no evidence of external gravitational stress, instead the architecture lacks dense cortical bone. However, in regions of muscle attachment in medial cross section, the torque applied to the bone by muscular action resulted in deposition of an anterior and posterior arch of lamellar cortical bone.

The loss of a medullary cavity, as seen in medial cross section of adult *Clidastes*, is a predicted effect on bone in a low-gravity environment (Carter et al., 1991; Wong and Carter, 1990a, b). It can also be seen in marine turtles, ichthyosaurs, dolphins, and some whales (Table 2; Buffrénil and Mazin, 1989; Buffrénil and Schoevaert, 1988; Rhodin, 1981, 1985; Felts, 1966; Felts and Spurrell, 1965). The occurrence of thin trabeculae in the inner medullary and thicker trabeculae on the periphery suggests that the medullary cavity in *Clidastes* closed from the periphery toward the central region with maturation (Sheldon, 1995; Wronski and Morey, 1987; Wronski et al., 1987).

BONE MICROSTRUCTURE OF ADULT *PLATECARPUS*

In proximal cross sections of adult *Platecarpus*, primary woven bone is totally remodeled to secondary trabecular bone, and no periosteal circumlamellae are present. Primary and secondary osteons result in a mean porosity of 9.4% for a single individual (Table 1; Sheldon, 1995). Analyses of these sections show that the vacuities have roundness values that range from 1.0 to 5.7, with an average of 1.5 (Table 1). Trabeculae are twice as thick as those seen in *Clidastes*. Secondary osteons penetrate many thick trabeculae. Random and isolated vacuities contain

---

Figure 2. A) Cross section of extant lizard limb bone, MOR uncatalogued comparative collection: arrowhead---reversal line showing extent of endosteal remodeling; solid arrow---periosteal lamellae. The photomicrograph is about 2.3 mm across.. B) Proximal rib cross section of *Clidastes propython*, RMM 5830 rib: arrowhead---very thin trabecular bone; solid arrow---totally remodeled, very thin cortical bone. Note the absence of growth lamellae in cortical bone. Greatest diameter is 0.8 cm. C) Distal rib cross section of *Clidastes propython*, RMM 1788 rib: arrowheads---anterior and posterior arches of periosteal lamellae; arrow---primary osteons. Trabeculae deposited by endosteal processes in medullary cavity region. Greatest diameter is 1.5 cm.

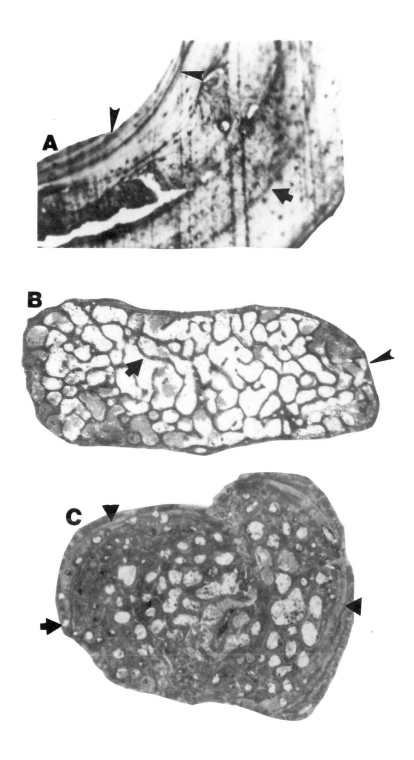

Table 1. Comparison of porosity and roundness values for adult *Clidastes*, *Platecarpus*, *Tylosaurus*, whale, manatee, and terrestrial lizard bone. Given are the mean; a range which is the standard deviation; and, in parentheses, the number of individuals in the sample.

	Clidastes	Tylosaurus	Platecarpus	whale	manatee	lizard limb
Porosity						
proximal	41.8+/-14.9(2)	62.3+/-1.7(2)	9.4(1)	88.5(1)	2(1)	1(1)
medial	11.3+/-11.4(2)	30.2(1)	12.8+/-3.7(2)			
Roundness values						
proximal	2.1(2)	2.3(2)	1.5(1)	2.3(1)	1.4(1)	1(1)
medial	2(2)	2.2(1)	1.6(2)			

Table 2. Bone characteristics of *Clidastes*, *Tylosaurus*, *Platecarpus*, whale, manatee, and a terrestrial lizard.

Characteristics	*Clidastes*	*Platecarpus*	*Tylosaurus*	whale	manatee	lizard limb
Cortical bone	limited	yes	no	no	yes	yes
Periosteal lamellae	no	yes	no	no	yes	yes
Secondary osteon	no	yes	yes	no	yes	no
Endochondral calcified cartilage mass	no	no	yes	no	no	no
Woven bone	no	no	yes	no	no	no
Vacuities arrangement	random	random	linear	random	random	random
Trabeculae	thin	very thick	thick	thin	thick	
Volkmann's canal	simple	simple and complex	simple and complex	simple and complex	simple	simple

endochondral calcified cartilage. Volkmann's canals are numerous and join adjacent primary and secondary osteons. Numerous resorption bays are present in many trabeculae, as are secondary osteons that have been completely in-filled by lamellae (Figure 3A).

In medial cross sections, the outermost periosteal primary bone contains no primary osteons. Interior to the outer lamella are some previously deposited lamellae penetrated by a few primary osteons. Further interior is an expanse of bone that has been vigorously remodeled and is penetrated by both primary and secondary osteons. The inner margin is marked by a revision line that separates the area of endosteal activities from the periosteally deposited bone (Figure 3B). *Platecarpus* microstructure is notable because fewer primary osteons are in the cortical bone than seen in other genera, *Clidastes* or *Tylosaurus*. Secondary osteons have invaded some areas of cortical bone and thick trabeculae. The medullary cavity region is crossed by a complex lattice of thick trabeculae, which were deposited by endosteal process. There is no open central cavity in the medullary region (Figure 3B). *Platecarpus* vacuities are subcircular and random in placement in the medial cross sections. Roundness values range from 1 to 2.5, with an average of 1.6 (Table 1). Mean porosity of the medial cross section is 12.8%, with a standard deviation of 3.7% (Table 2; Sheldon, 1995). The highest porosity is found in the most central area of the medullary cavity region. The least porous area is in the outermost periosteal lamellae. Volkmann's canals may be long and vermiform with bifurcation, or short and joining two closely adjacent osteons. Several Volkmann's canals leave or enter a single osteon and are large enough to cause difficulty with digital analysis (Figure 3B).

Adult *Platecarpus* bone is significantly more dense than that of *Clidastes* (Tables 1 and 2). The mean porosity for *Platecarpus* is 9.4% in proximal sections for one individual and 12.8% in medial sections, whereas *Clidastes* is 41.8% porosity in proximal sections and 11.3% porosity in medial sections (Table 1; Sheldon, 1995). These differences in porosity between the genera are not statistically significant, but they are based on very small sample sizes (one or two individuals per genus). These data seem to indicate that mechanical stress endured by *Platecarpus* was greater than that experienced by *Clidastes*. *Platecarpus* exhibit a condition known as pachyostosis, which is an increase in bone density across the entire bone. Pachyostosis is a condition correlated with an increase in lung capacity,

---

Figure 3. A) Adult proximal rib cross section of *Platecarpus* sp., RMM 1903: solid arrow---thick trabeculae; arrowhead---cortical bone. The photomicrograph is about 18 mm across. B) Adult medial rib cross section of *Platecarpus* sp., RMM 1501.35: solid arrow---medullary cavity partially obscured by endosteal deposition of trabeculae; open arrows---Volkmann's canals. Greatest diameter is 1.7 cm.

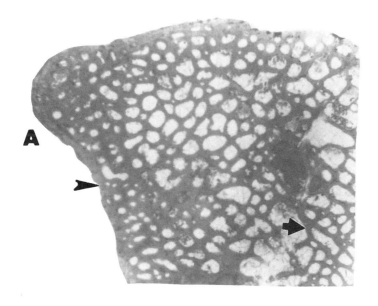

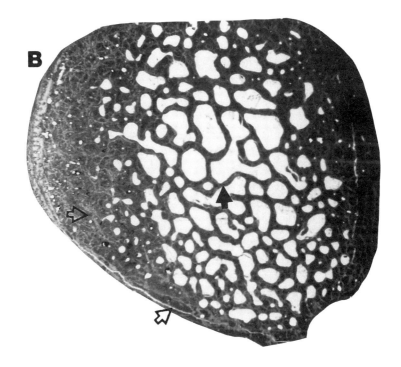

drag, increased work done during dives, increased rate of change of buoyancy with depth, and increased risk of decompression sickness (Taylor, 1994). Further, Taylor's (1994) analysis suggests that *Platecarpus* was a slower swimmer as penalty for the increased drag.

Volkmann's canals in adult *Platecarpus* are long, vermiform, and bifurcating, connecting several distant osteons. Both the pachyostosis and the complex morphology of the Volkmann's canal are found in manatees, some marine turtles, and beluga whales (Domning, 1991; Buffrénil and Schoevaert, 1988; Rhodin, 1981; Felts, 1966; Felts and Spurrell, 1965; Fawcett, 1942; Figure 6; Table 2).

BONE MICROSTRUCTURE OF ADULT *TYLOSAURUS*

The proximal cross sections of adult *Tylosaurus* ribs exhibit a very thin rim of periosteal primary bone, which encloses almost all of the circumference of the rib. However, masses of calcified cartilage and newly formed trabeculae occur in several places along the medial and lateral edges of the bone but lack a rim of primary bone. Volkmann's canals are absent; both primary bone and secondary bone are present (Figures 4A and 5A and B).

The proximal sections of *Tylosaurus* have a mean porosity of 62.3%, with a standard deviation of 1.7 (Table 1; Sheldon, 1995). Vacuities formed by the endosteal trabeculae are abundant. Large vacuities are oblong in shape and often linear in arrangement. Trabeculae have been extensively remodeled by secondary osteons and endosteal processes. Undifferentiated calcified cartilage is present in isolated vacuities. A large endochondral calcified cartilage mass is present in the central region of most proximal sections. Around the edges of the endochondral mass are transitional zones (Figure 5). In these zones, undifferentiated cartilage was in the process of transforming into bone (Figures 4A and 5A and B). A succession of transformation stages can be observed, starting with undifferentiated calcified cartilage, and then hypertrophied cartilage grading into primary bone (Barreto et al.,

---

Figure 4. Adult *Tylosaurus proriger*, RMM 1913, showing a greatly enlarged region of the area between the endochondral cartilage mass and primary bone. **A)** Proximal rib cross section: solid arrow---endosteal calcified cartilage; open arrow---region of transformation between endochondral cartilage and primary bone. Photomicrograph about 25 mm across. **B)** Medial rib cross section: arrowhead---remnant of endochondral cartilage mass; open arrow---periosteal calcified cartilage; solid arrow---primary osteon. Photomicrograph about 25 mm across. **C)** Periosteal cartilage: solid arrow---undifferentiated periosteal cartilage on outer edge of bone; arrowhead---openings for passage of connective tissue; open arrow---region of dense cartilage. Photomicrograph about 1 mm across.

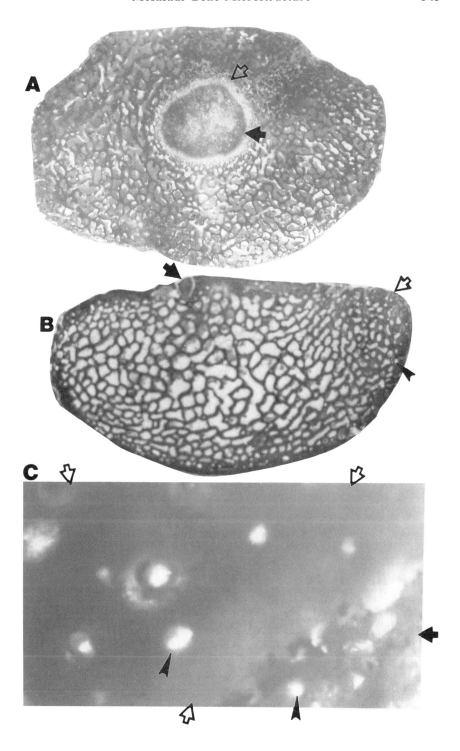

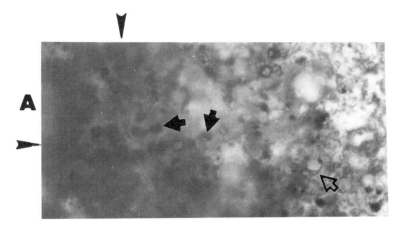

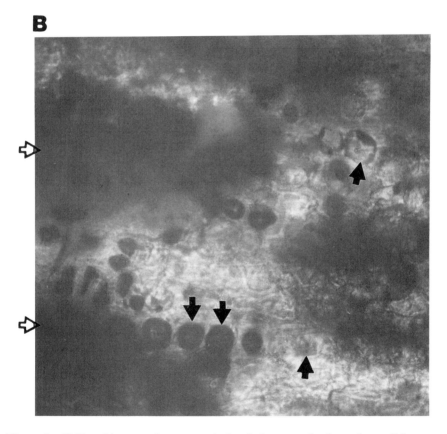

Figure 5. A) Transition zone between endochondral mass and primary bone, *Tylosaurus proriger*, RMM 1913: solid arrow---hypertrophied cartilage cells; arrowhead---primary bone; open arrow---proliferating cells. The photomicrograph is about 1 mm across. B) Enlarged transition zone: solid arrow---palisade cells; arrowhead---hypertrophied cells; open arrow---primary bone. The photomicrograph is about 1 mm across.

1993; Figures 4A and 5A and B). Hypertrophied cartilage forms a template for the trabeculae that will be ossified into primary bone. Vacuities in proximal sections have an oblong shape and are linear in orientation (Figure 4A). Roundness values range from 1.3 to 2.9, with an average of 2.2 (Table 1).

Periosteal cartilage surrounds the bone and has openings for passage of connective tissue (Figure 4C). In the anterior quadrant of the medial cross section, a rim of periosteal bone shows growth lamellae invaded by primary osteons. Just interior to the outer lamellae is secondarily remodeled bone with secondary osteons. These secondary osteons often have as many as five lamellae. In denser areas of the bone, Volkmann's canals are present. Globular osteocytes are found near the differentiating mass of endochondral calcified cartilage or near depositionally active regions of the periosteal calcified cartilage. On the medial edge of the rib there is a remnant of the endochondral mass, the edges of which were transforming into primary bone (Figure 4B).

The medullary cavity is filled by a lattice of trabeculae. These trabeculae have been remodeled by endosteal processes but there are remains of primary bone sandwiched between secondary endosteal lamellae.

The low bone density of adult *Tylosaurus* indicates osteoporosis. Osteoporosis is consistent with bone growth in very low gravity, and is found in deep-diving and swift-swimming whales, dolphins, and marine-adapted turtles (Taylor, 1994; Pitts et al., 1983; Figure 6, Tables 1 and 2). As in *Clidastes*, the microstructure of *Tylosaurus* bone suggests deposition of significant quantities of lipids in the bone vacuities. As Taylor (1994) proposed, this may result in a deeper but wide range of neutral buoyancy and energy-efficient diving.

Carter et al. (1991) found cartilage transformed into bone when the bone experiences impact of force (i.e., gravity). When the gravitational stress is low capillaries do not rapidly invade the cartilage and the cartilage is not quickly transformed into bone (Lufti, 1970; Haines, 1969; Moody, 1908). Finding active growth plates in an adult, indicated by the central endochondral mass, is contrary to the expected pattern of bone formation (Figure 4). A life history spent with a considerable portion of time within its neutral buoyancy zone, beyond the impact of gravity, may explain the presence of the growth plates in the proximal end of adult ribs (Delgodo-Baeza et al., 1991; Martin and Burr, 1989; Carter and Wong, 1988; Bikle et al., 1987; Jee, 1983; Jee et al., 1983; Cann and Adachi, 1983; Murry, 1936).

CONCLUSION

Ribs were the bones used in this analysis, as they are available for sectioning.

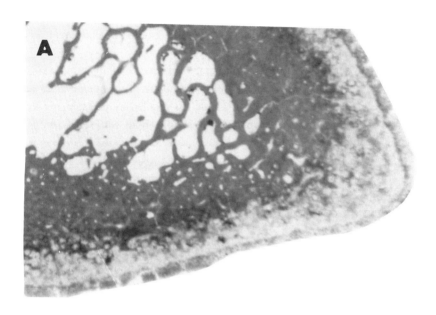

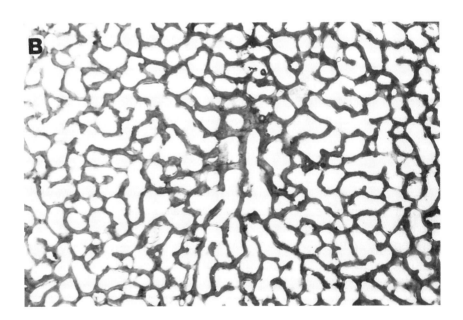

Figure 6. A) Cross section of deep-diving whale bone, MOR comparative collection. The photomicrograph is about 25 mm across. B) Cross section of manatee bone, RMM uncatalogued. The photomicrograph is about 15 mm across.

Observation of broken bones indicates that ribs can serve as a model for other adult skeletal elements.

Taylor (1994) suggested that buoyancy of an animal can be changed by altering its volume and density. Two strategies used by mosasaurs were to reduce body density and lung volume, and to increase bone density and lung volume. Reduced bone density and lung volume result in deeper and wider range of neutral buoyancy. Increased bone density and lung volume result in a shallow and narrow range of neutral buoyancy (Taylor, 1994). The increased porosity of *Clidastes* and *Tylosaurus* bone indicated that their neutral buoyancy was deep (Taylor, 1994; Martin and Burr, 1989). It also implies that their range of neutral buoyancy was larger than that of *Platecarpus*, with its denser bone and less potential lipid for deposition (Taylor, 1994). Reduced bone density is consistent with the reduced effect of gravity (Taylor, 1994; Carter et al., 1991; Martin and Burr, 1989). Evidently both *Tylosaurus* and *Clidastes* were frequenting the deeper portion of the water column.

Platecarpus shows pachyostosis, which requires an increase in lung volume. This suggests a shallower and narrower range of neutral buoyancy. A further result of increased bone density is that buoyancy is lost much more rapidly (Taylor, 1994; Domning, 1991). Increased lung volume, with its resultant increase in drag, is evidence for slower swimming speeds (Taylor, 1994). Taylor (1994) suggested that dense bone and slow-swimming animals would be expected to be more susceptible to decompression sickness, as Martin and Rothschild (1989) data confirm for *Platecarpus*. Martin and Rothschild (1989) noted that all of the *Platecarpus* specimens that they examined were afflicted with avascular necrosis, which they interpreted as the result of decompression syndrome.

Data for extant marine mammals show that if an animal's lung volume can be reduced during a deep dive as a response to increasing pressure, the animal is less likely to be afflicted with the bends (Kooyman and Ponganis, 1994; Ridgeway and Harrison, 1986; Matthews, 1978). It seems the lung membranes become thicker, when the lung can be collapsed with increased pressure. A lesser volume of gas can be diffused to the circulatory system (Ridgeway and Harrison, 1986; Matthews, 1978). However, the opposite is also true, and animals whose lung volume is not reduced are more likely to have the bends. Martin and Rothschild (1989) found that all the *Platecarpus* and *Tylosaurus* examined exhibited avascular necrosis. However, *Clidastes* was never found to exhibit avascular necrosis (Martin and Rothschild, 1989). These data may indicate that *Clidastes* thorax construction was conducive to lung compression. *Tylosaurus* does exhibit the confusing combination of osteoporosis and avascular necrosis. Ribs of *Tylosaurus* extend further ventrally than those of *Clidastes*. Additionally, *Tylosaurus* has a sternum and sternal ribs are calcified. A possible explanation is: (1) they exhibit osteoporosis and (2) they have a rigid rib cage that precludes lung collapse resulting in avascular necrosis. *Platecarpus* rib cage was more rigid because of pachyostosis in the ribs, which therefore did not allow for lung compression. Massare's (1988) analysis of swimming capabilities of Mesozoic marine reptiles found that, based on body shape and mode of swimming, mosasaurs were ambush predators. Belgula whales, which spend most of their time in shallow waters, have pachyostosis, and are subject to the bends (Taylor, 1994; Norris and Prescott, 1961). They are also known to rest on shallow-bottom sediments. It is possible that *Platecarpus* engaged in similar behavior to lie in wait for passing prey (Norris and Prescott, 1961).

SUMMARY

Marine mammals and turtles often partition the water column. They have bone microstructure very different from each other. Bone architecture can be correlated with life in specific portions of the water column in living marine mammals. A similar correlation is possible with mosasaurs.

Thin sections of ribs from adult *Clidastes, Platecarpus,* and *Tylosaurus* have been studied using digital imaging analysis. Microstructure of *Clidastes* and *Platecarpus* shows that each genus has unique architecture. Adult *Clidastes* exhibits osteoporosis, while adult *Platecarpus* microstructure shows pachyostosis. Comparison of mosasaur microstructure with extant marine vertebrates implies an ecological correlation. *Clidastes* and *Tylosaurus* microstructure correlates with life in deeper water, whereas *Platecarpus* microstructure correlates with life in the shallower part of the water column.

ACKNOWLEDGMENTS

I wish to thank Dr. Bruce Rothschild for his useful review. I also appreciate the constructive comments of an anonymous reviewer. Dr. Judy Massare's early reviews greatly improved this chapter.

I am grateful for the faith in this research demonstrated by the following collection managers: Susan Henson, Discovery 2000 (formerly Red Mountain Museum), Birmingham, Alabama; Charlotte Holton, American Museum of Natural History, New York; Dr. Westfal, University of Wisconsin Geology Museum, Madison; Dr. Melissa Winans, Texas Memorial Museum, University of Texas, Austin; and Mike Greenwald, Museum of Paleontology, University of California, Berkeley. They allowed the destructive analysis of their specimens' ribs. I appreciate the access to the Museum of the Rockies histological collection afforded by Dr. Jack Horner.

This project was funded in part by the Geological Society of America, Society of Paleontology, Sigma Xi, the American Museum of Natural History Visiting Scholar Grant and Theodore Roosevelt Grant, and Edwin A. Sheldon.

REFERENCES

Barreto, C., R. M. Albrecht, D. E. Bjorling, J. Horner, and N. J. Wilsman. 1993. Evidence of the growth plate and the growth of long bones in juvenile dinosaurs. *Science* 262:2020-2023.

Beaupre, G. S., T. E. Orr, and D. R. Carter. 1990. An approach for time-dependent bone. *Journal of Orthopaedic Research* 8:662-670.

Bell, G. L., Jr. 1993. *A Phylogenetic Revision of Mosasauroidea (Squamata)*. Unpublished Ph. D. dissertation, University of Texas, Austin, Texas, 293 pp.

Bell, G. L., Jr. and M. A. Sheldon. 1986. Description of a very young mosasaur from Greene County, Alabama. *Journal of the Alabama Academy of Science* 57(2):76-82.

Bennett, S. C. 1990. Inferring stratigraphic position of fossil vertebrates from the Smoky Hill Chalk Member locality data. IN S. C. Bennett (Ed.), *1990 Society of Vertebrate Paleontology Niobrara Chalk Excursion Guidebook*, pp. 43-72. Museum of Natural History and the Kansas Geological Survey, Lawrence, Kansas.

Bikle, D. D., B. P. Halloran, C. M. Cone, R. K. Globus, and E. Morey-Holton. 1987. The effects of simulated weightlessness on bone maturation. *Endocrinology* 120:678-683.

Buffrénil, V. de and J.-M. Mazin. 1989. Bone histology of *Claudiosaurus germaini* (Reptilia, Claudiosauridae) and the problem of pachyostosis in aquatic tetrapods. *Historical Biology* 2:311-322.

Buffrénil, V. de and D. Schoevaert. 1988. On how the periosteal bone of delphinid humerus becomes cancellous: Ontogeny of histological specialization. *Journal of Morphology* 198:149-164.

Buffrénil, V. de, A. Collet, and M. Pascal. 1985. Ontogenetic development of skeletal weight in the small delphinid, *Delphis delphis* (Cetacea, Odontoceti). *Zoomorphology* 105:336-334.

Buffrénil, V. de, A. de Ricqlès, and C. E. Ray. 1990. Bone histology of the ribs of the archaeocetes (Mammalia: Cetacea). *Journal of Vertebrate Paleontology* 10:455-466.

Cann, C. E. and R. R. Adachi. 1983. Bone resorption and mineral excretion in rats during spaceflight. *American Journal of Physiology* 244:R327-R331.

Carter, D. R. 1987. Mechanical loading and skeletal biology. *Journal of Biomechanics* 20(11,12):1095-1109.

Carter, D. R. and M. Wong. 1988. The role of mechanical loading histories in the development of diarthrodial joints. *Journal of Orthopaedic Research* 6:804-816.

Carter, D., T. E. Orr, and D. P. Fyhrie. 1989. Relationships between loading history and femoral cancellous bone architecture. *Journal of Biomechanics* 22:231-244.

Carter, D. R., M. Wong, and T. E. Orr. 1991. Musculoskeletal, ontogeny, phylogeny and

functional adaptation. *Journal of Biomechanics* 24:(suppl. 1):1-16.
Castanet, J. 1985. La squelettochronologie chez les reptiles I: Resultats experimentaux sur la signification des marques de croissance squelettiques chez les lezards et les tortues. *Annales des Sciences Naturelles, Zoologie*,13^e Serie 7:23-40.
Castanet, J. and G. Naulleau. 1985. La squelettochronologie chez les reptiles II. Resultats experimentaux sur la signification des marques de croissance squelettiques les serpents. Remarques cur la croissance et la longevete de la *Vipere aspic*. *Annales des Sciences Naturelles, Zoologie*, 13^e Serie 7:41-62.
Delgodo-Baeza, E., M. Gimenz-Robotta, C. Miralles-Flores, A. Nieto-Chaguaceda, and I. Santos-Alverez. 1991. Morphogenesis of cartilage canals: Experimental approach in rat tibia. *Acta Anatomica* 142:132-137.
Domning, D. P. 1991. Hydrostasis in the Sirenia, quantitative data and functional interpretations. *Marine Mammal Sciences* 7:331-368.
Esteban, M. 1990. Environmental influences on the skeletochronological record among Recent and fossil frogs. *Annales des Sciences Naturelles, Zoologie*, 13^e Serie 11:201-204.
Eurell, J.A. and L.E. Kazarian. 1983. Quantitative histochemistry of rat lumbar vertebrae following spaceflight. *American Journal of Physiology* 244:R315-R318.
Fawcett, D. W. 1942. The amedullary bones of the Florida manatee (*Trichechus latirostris*). *American Journal of Anatomy* 7:271-309.
Felts, W. J. L. 1966. Some structural and developmental characteristics of cetacean (Odontocete) radii. A study of adaptive osteogenesis. *American Journal of Anatomy* 118:103-134.
Felts, W. J. L. and F. A. Spurrell. 1965. Structural orientation and density in cetacean humeri. *American Journal of Anatomy* 116:171-204.
Globus, R. K., D. D. Bikle, and E. Morey-Holton. 1986. Temporal response of bone to unloading. *Endocrinology* 118:733-742.
Haines, W. 1969. Epiphysis and sesamoids. IN C. Gans, A. F. Bellairs, and T. Parsons (Eds.), *Biology of the Reptilia*, pp. 81-114. Academic Press, London and New York.
Hattin, D. E. 1982. Stratigraphy and depositional environment of Smoky Hill Chalk Member, Niobrara Chalk (Upper Cretaceous) of the type area, western Kansas. *Kansas Geological Survey Bulletin* 225:1-108.
Hunziker, E. B. and R. K. Schenk. 1989. Physiological mechanisms adopted by chondrocytes in the regulating of long bone growth in rats. *Journal of Physiology* 414:55-71.
Jee, W. S. S. 1983. The skeletal tissues. IN L. Weiss (Ed.), *Histology, Cell and Tissue Biology*, pp. 55-71. Elsevier Biomedical.
Jee, W. S. S., T. J. Wronski, E. R. Morey, and D. B. Kimmel. 1983. Effects of spaceflight on trabecular bone in rats. *American Journal of Physiology* 244:R310-R314.
Koch, J. C. 1917. The laws of bone architecture. *American Journal of Anatomy* 21:177-298.
Kooyman, G. L. and P. J. Ponganis. 1994. Emperor penguin oxygen consumption, heart rate, and plasma lactate levels during graded swimming. *Journal of Experimental*

Biology 195:199-209.

Kugler, J. H., A. Tomlinson, A. Wagstaff, and S. Ward. 1979. The role of cartilage canals in the formation of secondary centres of ossification. *Journal of Anatomy* 129:493-506.

Leach, R. and C. V. Gay. 1986. Role of epiphyseal cartilage in endochondral bone formation. *American Institute of Nutrition* 87:785-790.

Leclair, R., Jr. 1990. Relationships between relative mass of the skeleton, endosteal resorption, habitat and precision of age determination in ranid amphibians. *Annales des Sciences Naturelles, Zoologie*, 13ᵉ Serie 11:205-208.

Little, K. 1973. *Mechanical Influences, Bone Behavior*, Chapter 5, pp. 191-288. Academic Press, New York.

Lufti, A. M. 1970. Mode of growth, fate and functions of cartilage canals. *Journal of Anatomy* 106:135-145.

Malacinski, G. M. 1990. Reproduction and development of animals in space. IN M. Asahima and G. Malacinski (Eds.), *Fundamentals of Space Biology*, pp. 241-280. Springer-Verlag, New York.

Martin, L. D. and B. Rothschild. 1989. Paleopathology and diving mosasaurs. *American Scientist* 77:460-467.

Martin, R. B. and D. B. Burr. 1989. *Structure and Adaptation of Compact Bones*. Raven Press, New York, pp. 50-85.

Massare, J. 1988. Swimming capabilities of Mesozoic marine reptiles: implications for method of predation. *Paleobiology* 14:187-205.

Matthews, L. H. 1978. *Natural History of the Whale*. Columbia Press, New York, 321 pp.

Meulin, M. C. H. van der, G. S. Beaupre, and D. R. Carter. 1993. Mechanical influences in long bone cross-sectional growth. *Bone* 14:635-642.

Moody, R. 1908. Reptilian epiphyses. *American Journal of Anatomy* 7:447-469.

Morey, E. R. 1978. Inhibition of bone formation during space flight. *Science* 201:1138-1141.

Murry, P. D. F. 1936. *The Development of the Bony Skeleton*. Cambridge University Press, London, 135 pp.

Nopcsa, F. 1923. Volaufige Notiz die Pachyostose und Osteosklerose eineger marine Wirbeltier. *Anatomischer Anzeiger* 56:355-359.

Norris, K. S. and J. H. Prescott. 1961. Observations on Pacific cetaceans in Californian and Mexican waters. *University of California Publications in Zoology* 63:291-402.

Patterson-Buckendahl, P., S. Arnaud, G. L. Mechanic, R. B. Martin, R. E. Grindeland, and C. E. Cann. 1983. Fragility and composition of growing rat bone after one week in space flight. *American Journal of Physiology* 244:R240-R245.

Pitts, G. C., A. S. Ushakov, N. Pace, A. H. Smith, D. F. Rohlmann, and T. A. Smirnova. 1983. Effects of weightlessness on body composition in the rat. *American Journal of Physiology* 244:R332-R337.

Rhodin, A. G. J. 1981. Chondro-osseous morphology of *Dermochelys coriacea*, a marine reptile with mammalian skeletal feature. *Nature* 290:244-246.

Rhodin, A. G. J. 1985. Comparative chondro-osseous development and growth in marine turtles. *Copeia* 3:752-771.

Ridgeway, S. H. and R. Harrison. 1985. *Handbook of Marine Mammals*. Academic Press, New York, 312 pp.

Ridgeway, S. H. and R. J. Harrison. 1986. Diving Dolphins. IN M. M. Bryden and R. Harrison (Eds.), *Research on Dolphins*, pp. 33-58. Cambridge University Press, New York.

Russell, D. A. 1967. Systematics and morphology of American mosasaurs. *Bulletin of the Peabody Museum of Natural History* 23:1-240.

Sheldon, M. A. 1987. Juvenile mosasaurs from the Mooreville Chalk of Alabama. *Journal of Vertebrate Paleontology* 7(3):25A.

Sheldon, M. A. 1989. Implications of juvenile mosasaur recognition on taxonomy. *Journal of Vertebrate Paleontology* 9(3):38A.

Sheldon, M. A. 1990. Immature mosasaurs from the Niobrara: a sampling problem? *Journal of Vertebrate Paleontology* 10:181.

Sheldon, M. A. 1992. Ontogenetic changes in mosasaur bone microstructure. *Journal of Vertebrate Paleontology* 12(3):150.

Sheldon, M. A. 1993. *Ontogenetic Study of Selected Mosasaurs of North America*. Unpublished M. Sc. thesis, University of Texas at Austin, Austin, Texas, 184 pp.

Sheldon, M. A. 1995. *Ontogeny, Ecology and Evolution of North American Mosasaurids (Clidastes, Platecarpus and Tylosaurus): Evidence from Bone Microstructure*. Unpublished Ph. D. dissertation, University of Rochester, Rochester, New York..

Simmons, D. J., J. E. Russell, F. Winter, P. Tran Van, A. Vignery, R. Baron, G. D. Rosenberg, and W. V. Walker. 1983. Effect of spaceflight on non-weight bearing bones of rat skeleton. *American Journal of Physiology* 244:R319-R326.

Taylor, M. A. 1994. Stone, bone and blubber? Buoyancy control strategies in aquatic tetrapods. IN L. Maddock, Q. Bone, and J. M. V. Rayner (Eds.), *Mechanisms and Physiology of Animal Swimming*, pp. 151-161. Cambridge University Press, New York.

Wall, W. P. 1983. The correlation between high limb-bone density and aquatic habits in Recent mammals. *Society of Economic Paleontologists and Mineralogists* 57:197-208.

Wilsman, N. J. and D. C. van Sickle. 1970. The relationship of cartilage canals to the initial osteogenesis of secondary centers of ossification. *Anatomy Records* 168:381-392.

Wolff, J. 1892. *The Law of Bone Remodelling (Des Gesetz der Transformation der Knochen)*. Translated by P. Marquet, and R. Furlong, 1986. Springer, Berlin.

Wong, M. and D. Carter. 1990a. Mechanical stress and morphogenetic endochondral ossification of the sternum. *Journal of Bone and Joint Surgery* 70A(7):992-1000.

Wong, M. and D. Carter. 1990b. A theoretical model of endochondral ossification and bone architectural construction in long bone ontogeny. *Anatomy and Embryology* 181:523-532.

Wronski, T. J. and E. R. Morey. 1987. Effect of spaceflight on periosteal bone formation in rats. *American Journal of Physiology* 252:R252-R255.

Wronski, T. J., E. R. Morey-Holton, S. B. Doty, A. C. Maese, and C. C. Walsh. 1987. Histomorphometric analysis of rat skeleton following spaceflight. *American Journal of Physiology* 252:R252-R255.

PART V
Crocodylia

Part V: Crocodylia

INTRODUCTION

STÉPHANE HUA and ERIC BUFFETAUT

Although extinct marine crocodilians were among the first fossil vertebrates to be subjected to scientific studies (Buffetaut, 1987), their adaptations to marine life and their paleoecology have been largely neglected by authors. However, contrary to such other fossil marine reptiles as ichthyosaurs and sauropterygians, they have living relatives, including *Crocodylus porosus*, a species which frequents salt water and can, to some extent, be used as an actualistic model for reconstructions of their physiology and ecology.

Although some Tertiary eusuchians, including early gavialids (Buffetaut, 1982b), apparently also inhabited coastal waters and dispersed across oceanic barriers, this introduction concentrates first on extinct crocodilians traditionally placed in the suborder Mesosuchia (which may well correspond to a grade rather than a clade), whose adaptations to marine life were especially marked. The main families to be considered are the Teleosauridae, the Metriorhynchidae, the Pholidosauridae, and the Dyrosauridae. Our aim is not to discuss the largely unclear phylogenetic relationships between and within these groups, but rather to investigate their physiology and paleoecology, partly on the basis of what is known about the living "saltwater crocodile," *Crocodylus porosus*. In addition, the possible importance of transoceanic migrations for the biogeography of extant eusuchians is also discussed.

WHAT IS A MARINE REPTILE?

When one thinks of marine reptiles, ichthyosaurs or marine turtles, whose body and limbs are deeply modified in relation to their adaptation to life in the sea, come easily to mind. However, reptiles in which morphological adaptations are less obvious, such as the marine iguana, may also be considered as marine reptiles. We therefore consider as a marine reptile any reptile able to grow and feed in a saltwater environment, without frequent access to fresh water, and which inhabits the marine environment on an occasional or permanent basis. Reproduction is not considered within this definition, because egg-laying occurs on land, and various extinct marine reptiles (including plesiosaurs, whose remains are sometimes found in freshwater deposits) possibly returned to a nonmarine environment to reproduce.

According to our definition, the "Indo-Pacific," "estuarine," or "saltwater" crocodile, *Crocodylus porosus*, can be considered as a marine reptile, because it seems to be able to grow in salt water, without any influx of fresh water (Grigg et al., 1980)---although this conclusion may need to be confirmed by further experiments (Mazzotti and Dunson, 1989). The diet of *Crocodylus porosus* in a marine environment (Allen, 1974) shows that it can easily feed in salt water, and its ability to swim in the open sea is well established (Bustard and Choudhury, 1980).

THE PHYSIOLOGY OF MARINE CROCODILIANS

Heat loss is important to inhabitants of the marine environment. The living Crocodylidae, which are ectotherms, live only in tropical waters (the only extant crocodilians which are able to live in warm temperate zones are alligatorids). There, the thermal gradient between the organism and its environment is reduced, and the animal runs no risk of a lethal drop of its internal temperature. In living crocodilians, salinity regulation is effected by a salt gland located on the tongue, which is especially active in *Crocodylus porosus* (Taplin and Grigg, 1981). In crocodilians, cutaneous permeability is limited, which prevents lethal osmotic dehydration (see the synthesis by Mazzotti and Dunson, 1989).

Besides temperature, the only factor which seems to have limited the dispersal of *Crocodylus porosus* may have been the lack of suitably located islands, not too far from each other, on which the animals could have laid their eggs. This may explain why this species failed to colonize the central Pacific (although the part played by adverse oceanic currents deserves investigation). On the other hand, the main structure that enabled *Crocodylus porosus* to occupy its vast geographical range, from India to the northern coast of Australia, seems to be its salt gland---a structure which is undetectable on fossils. The only feature which could be observed on fossil specimens and can be interpreted as an increasing adaptation to marine life is the regression of post-occipital osteoderms (see Hua, 1994).

There is no evidence that endothermy ever developed in marine mesosuchians (Buffetaut et al., 1982; Hua and Buffrénil, in press), so that temperature probably acted as a limiting factor for these animals (including highly specialized forms such as the metriorhynchids), as it does for living forms. The following review of the main groups of marine mesosuchians, based on their paleontological record, also tries to integrate data from living eusuchians, in order to reconstruct mesosuchians' adaptations to life in the sea.

THE TELEOSAURIDAE

The Teleosauridae are known from the Early Jurassic to the Early Cretaceous (Figure 1; Buffetaut, 1982a). The earliest well-known teleosaurids, from the Toarcian (including, for instance, early species of *Steneosaurus* Geoffroy Saint-Hilaire, 1825), show few adaptations to marine life, and might have been mistaken for freshwater forms, had not their remains been found in abundance in marine deposits. *Peipehsuchus teleorhinus* Young, 1935, a teleosaurid from the Early Jurassic of China, comes from nonmarine deposits and is not markedly different from marine teleosaurids in skull anatomy (Li, 1993). Liassic teleosaurids probably spent much time in coastal waters and on shore (Westphal, 1962), and the juveniles may have been mainly freshwater animals, as suggested by the low number of young individuals found in marine deposits. A tendency to move downstream as the animals grow has been reported in such living crocodilians as *Crocodylus niloticus* (Hutton, 1989).

European paleogeography during the Jurassic consisted of island groups (Godefroit, 1994), and the distribution of teleosaurids within this geographical framework is reminiscent of that of *Crocodylus porosus* among the archipelagoes of the Indian Ocean. No material evidence is available about the reproduction of teleosaurids (the so-called teleosaurid "nest" reported by Buckman [1860] has turned out to be a concretion [Benton and Taylor, 1984]).

During the Bathonian, the genus *Teleosaurus* Geoffroy Saint-Hilaire, 1825, probably still lived mainly in protected littoral waters, to judge from both its strong dermal armor and the depositional environment of the rocks in which its remains are found (Buffetaut, 1982a). Conversely, during the Jurassic, the genus *Steneosaurus* (Figures 2B and 3) seems to have become better adapted to marine life, with an increasingly streamlined skull, a reduction of the bony armor, and a reduction and transformation of the forelimb (Buffetaut, 1980a). Studies on the sedimentary environment of *Steneosaurus* fossils (Rieppel, 1981; Buffetaut and Thierry, 1977) suggest a platform environment. Rieppel (1981) has suggested that in life teleosaurids avoided the contemporary metriorhynchids, which were better adapted to life in the open sea.

During the Callovian and Oxfordian, teleosaurids show three main trends in

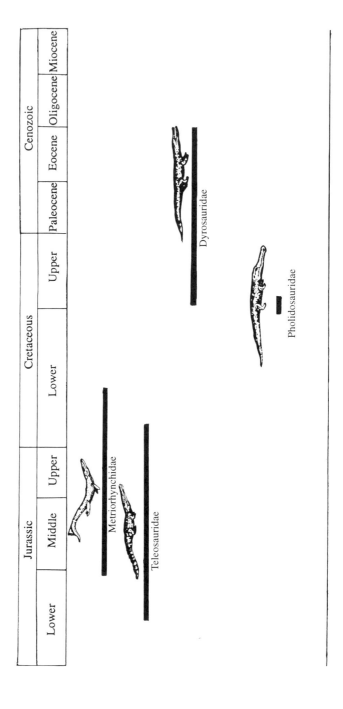

Figure 1. Stratigraphic ranges of the main families of marine mesosuchians discussed in the text.

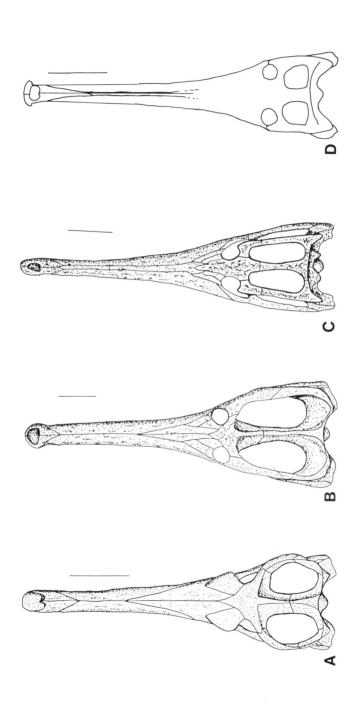

Figure 2. A comparison of the skull (in dorsal view) in several families of marine mesosuchians. **A)** *Metriorhynchus*, a metriorhynchid (after Wenz, 1968). **B)** *Steneosaurus*, a teleosaurid (after Andrews, 1913). **C)** *Dyrosaurus*, a dyrosaurid (after Buffetaut, 1982a). **D)** *Teleorhinus*, a pholidosaurid (after Buffetaut and Wellnhofer, 1980). The limits of the skull bones are poorly known in *Teleorhinus*. Scale bars = 10 cm.

skull and tooth shape. Longirostrine forms such as *Steneosaurus leedsi* Andrews, 1909, had long slender teeth and were undoubtedly piscivorous. Forms with much more robust jaws and blunt teeth, such as *Machimosaurus hugii* von Meyer, 1838, were apparently more durophagous. A third group, represented, for instance, by *Steneosaurus larteti* Eudes-Deslongchamps, 1866, exhibited a somewhat intermediate condition.

Teleosaurids must have been good swimmers. Increasing adaptation to marine life is shown notably by a reduction of the bony armor and of the forelimbs, which become flattened dorsoventrally (Buffetaut, 1980a, 1982a). The latter transformation may be interpreted as a means of better inserting the limbs in axillary depressions in order to reduce drag during fast swimming; the forelimbs were probably of little use for propulsion, except possibly at low speeds. The hindfeet were webbed (Berckhemer, 1929), as in living crocodilians. One of the most spectacular transformations is the "closure" of the angle between the zygapophysial facets of thoracal vertebrae (Krebs, 1962, 1967), which allowed the animal to practice a kind of imperfect oscillatory axial swimming (with the trunk held rigid, the limbs closely appressed against the body, and lateral oscillations of only the tip of the tail), preventing lateral movements which could have generated useless turbulences. The rather dorsal position of the orbits has recently been interpreted (Martill et al., 1994) as an adaptation for hunters lying in wait on the bottom and stalking prey outlined on the surface. A recent histological study by Hua and Buffrénil (in press) supports this hypothesis and suggests that teleosaurids were a kind of "marine gavials," which lay in wait on the bottom and caught their prey by sudden lateral movements of the jaws. This interpretation would explain why teleosaurids are usually found in relatively shallow-water deposits (Buffetaut, 1982a; Rieppel, 1981).

Callovian marine communities are well known from the Oxford Clay of England (Martill et al., 1994; Massare, 1987). Teleosaurids and metriorhynchids are found there in association with other marine reptiles and they seem to have occupied niches of ichthyophagous predators, *sensu lato*, with a diet including both fish and cephalopods. In the Kimmeridgian and Tithonian, only two of the three main Callovo-Oxfordian groups of teleosaurids persisted and they were highly specialized. One group is represented by very gracile representatives of *Steneosaurus*, such as the Tithonian species *S. priscus* Soemmering, 1814. This group may have suffered from the competition of longirostrine metriorhynchids, which probably had a similar

---

Figure 3. Reconstructions of the skeleton in several groups of marine mesosuchians, showing different adaptations to marine life. **Top**) The metriorhynchid *Metriorhynchus*. **Middle**) The teleosaurid *Steneosaurus*. **Bottom**) The dyrosaurid *Dyrosaurus*. For greater clarity the dermal armor has been deleted (except in *Metriorhynchus*, in which it was absent). Scale bar = 1 m.

Crocodylia

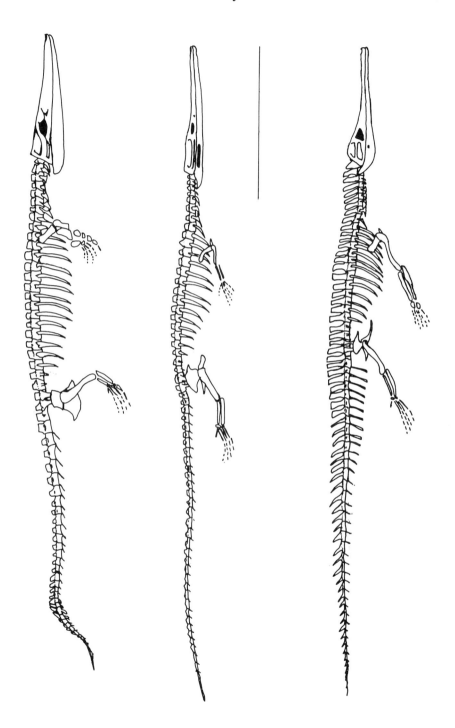

diet and occupied a similar ecological niche. This may have led to an early extinction of that group of slender-jawed teleosaurids, whereas the other, more robust, group survived longer. The latter group is represented mainly by *Machimosaurus mosae* Sauvage and Liénard, 1879, a massive form with thick dermal scutes, robust jaws and blunt teeth (a set of characters not encountered in metriorhynchids). The postcranial skeleton, with a stout trunk, a strongly developed ventral shield, and thick gastralia, suggests adaptations to a high-energy environment (Hua, in press). Teeth of *Machimosaurus* have been found in association with plates of marine turtles (Meyer, 1991), which suggests a possible diet for this animal. This specialized teleosaurid lineage apparently persisted until the Early Cretaceous (a Valanginian specimen is described by Cornée and Buffetaut, 1979), when the family Teleosauridae died out.

THE METRIORHYNCHIDAE

The Metriorhynchidae (Figure 1) are frequently grouped with the Teleosauridae in an infraorder Thalattosuchia (Buffetaut, 1982a), on the basis of morphological resemblances which are especially noticeable among early (Liassic) representatives of both families. As early as the Toarcian, the genus *Pelagosaurus* Bronn, 1841, a small form with lateral orbits, incipiently enlarged preorbital bones, and a still fairly extensive bony armor, can be viewed as a primitive representative of the family Metriorhynchidae (Buffetaut, 1980b).

During the Bajocian and Bathonian, the genus *Teleidosaurus* Eudes-Deslongchamps, 1869, exhibits more advanced metriorhynchid features (Hua and Atrops, in press), with its laterally facing orbits and protruding prefrontals. Two evolutionary trends seem to be recognizable within this rather poorly known genus, with longirostrine (*T. bathonicus* Mercier, 1933) and brevirostrine forms (*T. calvadosi* Eudes-Deslongchamps, 1869). In the Callovian, the same trends are recognizable within the genus *Metriorhynchus* von Meyer, 1830 (Figures 2A and 3), with both longirostrine (*M. superciliosus* Blainville, 1853) and brevirostrine (*M. brachyrhynchus* Eudes-Deslongchamps, 1868) forms. The differences in snout proportions probably reflect dietary differences. The brevirostrine forms likely could tackle larger prey, so competition between the two groups may have been limited. A few stomach contents from Oxford Clay specimens provide some evidence as to the diet of *Metriorhynchus* (Martill, 1986), which apparently included ammonites, belemnites, pterosaurs (*Rhamphorhynchus*), and even the giant fish *Leedsichthys*.

Advanced metriorhynchids such as *Metriorhynchus* exhibit numerous and far-reaching adaptations to marine life. The skull and body were highly streamlined, and the skull resembled that of mosasaurs (Langston, 1973), a convergence certainly linked to similar modes of life as pelagic predators. The enlarged prefrontals overhung the anterodorsal part of the orbits, which may have enhanced streamlining and protected the eyes (Buffetaut, 1982a). The cranial bones were extremely cancellous, which made the skull lighter, reduced inertia, and displaced the buoyancy center forward. This facilitated passive upward movements (Buffetaut, 1979) and made acceleration easier through a rearward displacement of the center of gravity (Hua, 1994). The external nares never showed the posterodorsal displacement often seen in sustained swimmers, which may have been linked to the lightness of the skull, thus making it easier to keep the nostrils out of the water. The laterally oriented orbits provided a wide field of vision, which was of advantage to animals hunting by sight in a nonturbid environment (Hua, 1994; Martill et al., 1994; Massare, 1988). The characteristic crocodilian external mandibular fenestra had been lost, and the shape of the lower jaw was convergent with that of pliosaurs. The closure of the fenestra is indicative of a regression of the *musculus intramandibularis,* which, in living crocodilians, helps to keep the mouth open when the animals are "basking" on the shore; this regression may thus be linked to an increasingly marine habitat. On the vertebrae, the zygapophysial facets showed the subvertical orientation already noted in teleosaurids, which conferred a certain lateral rigidity to the body when swimming. The posterior end of the tail was deflected downward. Although only a single impression of a cutaneous dorsal lobe is known (in a Tithonian specimen of *Geosaurus*), there is no doubt that a tail fluke was present in all advanced metriorhynchids. The tail was apparently carchariniform in shape, which made a wide range of speeds possible (Hua, 1994). The chevron bones of the distal caudal vertebrae were greatly elongated anteroposteriorly and may have provided insertion surfaces for strong propulsive muscles. The loss of bony armor may be interpreted in several ways. It may have reduced the body mass not used for acceleration, as in transient swimmers (Webb and Skadsen, 1979). It may also have increased streamlining through the loss of irregular turbulence-producing surface features (Hua, 1994; Gasparini, 1978). The loss of the osteoderms may also correspond to the disappearance of a sustaining structure which had become useless in fully marine animals (Frey, 1988). The forelimbs were very short and paddle-like, while the hindlimbs were longer but also paddle-like (although there was neither polydactyly nor hyperphalangy). The regression of the fourth trochanter of the femur apparently indicates a regression of the *musculus caudofemoralis,* which may be linked to the fact that propulsion was effected only by movements of the distal end of the tail, instead of undulations of the whole tail (Buffetaut, 1977). Some authors (Gasparini, 1978, 1980, 1981; Mateer, 1977; Neill, 1971) think that metriorhynchids were too vulnerable when out of the water to have laid eggs on the shore, and that they must have been ovoviviparous. According to others (Hua and Buffrénil, in press; Buffetaut, 1982a), oviparity is more likely.

The above-mentioned specializations suggest that metriorhynchids practiced a kind of carangiform swimming which was energetically more efficient than that of the Teleosauridae. According to Hua (1994), they were efficient hunters which stalked their prey on the surface, caught it by a sudden acceleration, and easily came back to the surface because of the good buoyancy of their skulls. The recent histological study by Hua and Buffrénil (in press) shows that metriorhynchids were ectotherms. However, thermoregulation probably presented few problems in a world without polar ice caps and limited latitudinal thermal gradients. The wide geographical distribution of metriorhynchids accords well with this hypothesis.

During the Kimmeridgian, *Metriorhynchus* was replaced by *Dakosaurus* Quenstedt, 1856, and *Geosaurus* Cuvier, 1824. While *Dakosaurus* was the largest known metriorhynchid, reaching a length of more than 4 m, *Geosaurus* was a smaller and more lightly built form. These differences in size and build probably reflect different dietary adaptations.

The Metriorhynchidae, notably *Dakosaurus*, survived into the Early Cretaceous and became extinct in the Hauterivian. No obvious explanation is available for this extinction.

THE PHOLIDOSAURIDAE

Following the Early Cretaceous, after the extinction of the Thalattosuchia, representatives of the family Pholidosauridae made a brief incursion into the marine realm during the middle part of the Cretaceous (Figure 1). Most pholidosaurids were freshwater longirostrine crocodilians of the Late Jurassic and Early Cretaceous, and only one genus, *Teleorhinus* Osborn, 1904 (Figure 2D), has been found exclusively in marine deposits. Species of *Teleorhinus* are known from the Albian and Cenomanian of North America (Erickson, 1969; Mook, 1934; Osborn, 1904) and the Cenomanian of Bavaria (Buffetaut and Wellnhofer, 1980). *Teleorhinus* was a very long-snouted, certainly piscivorous, form, which, apart from some reduction of the bony armor, resembled the other, nonmarine, pholidosaurids and apparently did not show many adaptations to marine life (although *Pholidosaurus* is still a rather poorly known genus). Its way of life may have been similar to some extent to that of the living *Crocodylus porosus*, although the latter is not a longirostrine form.

THE DYROSAURIDAE

The Dyrosauridae (Figure 1) represent the last "attempt" by mesosuchian crocodilians to colonize the marine environment. Fragmentary remains of what appear to be early dyrosaurids occur in probably Cenomanian nonmarine deposits in Sudan (Buffetaut et al., 1990), but they are best known from Maastrichtian to Eocene shallow marine deposits in Africa, North America, and South America. However, dyrosaurid remains are also known from freshwater beds of Eocene age in Pakistan (Buffetaut, 1978a). This suggests that, as in some modern crocodilians, the young escape the competition of adults by living farther upstream (Hutton, 1989; Webb and Messel, 1978), so juvenile dyrosaurids lived in a freshwater environment, whereas adults lived in the sea.

The adaptations of dyrosaurids to marine life are still poorly known, because, although very abundant remains have been found, especially in the Early Tertiary of Africa, few articulated skeletons are known. Nevertheless, on the basis mainly of good cranial material from the phosphates of North Africa, two primary groups corresponding to different dietary adaptations can be recognized on the criterion of skull morphology (Buffetaut, 1982a). The Phosphatosaurinae, represented by *Phosphatosaurus* Bergounioux, 1955, had robust jaws with stout, blunt teeth reminiscent of those of the Jurassic teleosaurid *Machimosaurus*; they are known from the Maastrichtian to the Ypresian, mainly in Africa, and may have fed on turtles and nautiloids, the remains of which are frequently found in the same deposits. The Hyposaurinae, known from the Maastrichtian to the middle Eocene (with a number of genera, including *Dyrosaurus* (Figures 2C and 3), *Hyposaurus*, *Rhabdognathus*, and *Tilemsisuchus*), had much more lightly built skulls, with long slender jaws and pointed teeth; they were probably ichthyophagous and enjoyed their greatest development during the Paleocene (Buffetaut, 1980c, 1982a).

Three more or less complete articulated dyrosaurid skeletons were known before specimens described by Denton et al. (Chapter 13). One comes from the Maastrichtian of New Jersey and was described by Troxell (1925) as *Hyposaurus natator* (considered by Buffetaut [982a] and Denton et al. as a synonym of *Hyposaurus rogersii* Owen, 1849, but see Norell and Storrs [1989] for a different opinion). A skeleton of *Hyposaurus bequaerti* (Dollo, 1914) from the Paleocene of Cabinda has been described by Dollo (1914) and Swinton (1950), and a headless specimen of an indeterminate dyrosaurid has been described by Storrs (1986) from the Paleocene of Pakistan. A reexamination by one of us (S. H.) of the skeleton from Cabinda has revealed various peculiarities which provide useful data concerning the mode of life of these crocodilians. The osteodermal armor is reduced both ventrally and dorsally, a feature typical of crocodilians living in a marine habitat. The long-snouted skull was lightly constructed (Buffetaut et al.,

1982) and its inertia was thus reduced. The cervical muscles were very powerful, as shown by their insertion areas on the elevated occipital surface of the skull and by the tall neural spines and hypapophyses of the cervical vertebrae. This suggests that very rapid lateral movements of the head were possible, as in many longirostrine crocodilians. All along the vertebral column, the neural spines are tall, which indicates a strong musculature favoring powerful undulations of the trunk and the tail. The lateral area of the tail was increased by very long chevron bones. Lateral bending of the vertebral column was not hindered by the reduced scutes. Contrary to the reduction observed in thalattosuchians, the appendicular skeleton of dyrosaurids was well developed. The forelimbs were as long as the hindlimbs, with well-marked reliefs for muscle insertions. The terminal phalanges were claw-shaped. The strong development of the limbs is not associated with polydactyly or hyperphalangy, as in some other marine reptiles, and dyrosaurids certainly had an axial, not paraxial, type of locomotion (as suggested by the above-mentioned characters of the vertebral column). The strong limbs suggest that terrestrial locomotion was less limited than in thalattosuchians. However, the reduction of the external mandibular fenestra (also encountered in metriorhynchids) limited the action of the *musculus intramandibularis*, which is used for keeping the mouth open when basking, and this may be seen as adaptation to a more aquatic life. The long limbs may have been of use for propulsion when the animal "took off" from the bottom to seize its prey. They may also, and perhaps more importantly, have provided better maneuverability when swimming.

Buffetaut (1982a) has proposed to place the Dyrosauridae in an infraorder of their own within the Mesosuchia, the Tethysuchia. The name is based on the geographical distribution of the group, which is found mainly in shallow marine deposits all along the shores of the Tethys Sea, from the Indian subcontinent through Arabia and northern and western Africa, to the eastern coast of North America, Brazil, and the Andean regions (Buffetaut, 1991) of western South America. This wide distribution certainly suggests a good adaptation to marine life, and may have been favored by marine currents from east to west in the Tethys and across the proto-Atlantic. The absence of dyrosaurids on the northern shore of the Tethys in Europe is difficult to explain in terms of paleogeography or paleoclimatology. It has been suggested (Buffetaut, 1982a) that the extinction of the dyrosaurids during the Eocene may have been the result of the expansion of primitive whales (archaeocetes) and of the resulting competition in the marine realm.

CONCLUSIONS

This introduction has focused on the various groups of mesosuchian crocodilians which, at several periods of the Mesozoic and Cenozoic, became adapted to marine life. The degree of adaptation was different, metriorhynchids being clearly more deeply transformed for life in the sea than the other groups. The modes of adaptation were also different: it is clear that dyrosaurids did not become adapted to life in the sea along the same lines as the earlier teleosaurids and metriorhynchids, but they nevertheless apparently enjoyed a considerable evolutionary success, especially after the disappearance of all other predatory marine reptiles at the Cretaceous-Tertiary boundary. The record of marine mesosuchians thus reflects the emergence of various, not completely convergent, modes of adaptation to marine life at different times in the evolution of the crocodilians. Unfortunately, virtually nothing is known of the nonmarine ancestors of the main groups of marine mesosuchians described above, and their early histories are therefore very obscure.

It should also be mentioned that several groups of eusuchian crocodilians also had marine or partly marine representatives during the latest Cretaceous and Cenozoic, to judge from their occurrence in marine sediments. However, these crocodilians do not differ markedly from their nonmarine counterparts and do not seem to have been much transformed morphologically in relation to their marine habitat. They include the latest Cretaceous and Paleocene longirostrine genus *Thoracosaurus*, the affinities of which are unclear, and various gavialids which have been found in Tertiary marine deposits in Africa and Europe (Buffetaut, 1978b). The occurrence of true gavialids in the Tertiary of South America seems to be explainable only by a crossing of the South Atlantic in the Eocene or Oligocene, at a time when that ocean was already wide enough to form an effective barrier to the dispersal of organisms not adapted to some extent to the marine environment (Buffetaut, 1982b).

Finally, some mention should be made of recent hypotheses about the zoogeography of living crocodiles which involve the crossing of marine barriers. On the basis of Densmore's work (1983) on the molecular phylogeny of crocodilians and the small genetic distance between the different species of the genus *Crocodylus*, Taplin et al. (1985), Taplin (1984), and Densmore (1983) have proposed a hypothesis explaining the zoogeography of the Crocodylidae according to which the species of *Crocodylus* were derived from a marine common ancestor, which was capable of transoceanic crossings during the Plio-Pleistocene. The occurrence of an active salt gland only in the Crocodylidae (among the living crocodilians) is used as evidence in favor of this "Transoceanic Migration Hypothesis." This hypothesis can be discussed on a paleontological basis. First, the "molecular clock" used by its proponents should be regarded with some caution, since it is based on the supposed times of divergence of fossil groups, which are subject to reinterpretation. There is no doubt that the genetic distance between some species of *Crocodylus* is small, as attested by the fact that hybridization between some of them can produce viable offspring (for instance, *C. porosus* x *C. siamensis* or *C. rhombifer* x *C. acutus*). However, the problem of salinity tolerance is more complex than it may seem. As mentioned above, besides the Crocodylidae, some other groups of eusuchians, such as the Gavialidae, have apparently been able to cross oceanic barriers in the course of their histories. A vestigial salt gland is present in gavials and alligators, and its occurrence is therefore not restricted to the Crocodylidae, although it has undergone a regression in other eusuchian families. The main problem with the Transoceanic Migration Hypothesis is that it considers the Crocodylidae as secondarily adapted to fresh water. If one considers marine mesosuchians, however, it becomes apparent that the earliest representatives of the various groups (with the exception of the Metriorhynchidae, the ancestors of which may be found among early teleosaurids) were nonmarine forms. This applies, of course, to the Pholidosauridae, a group represented mainly by freshwater forms, *Teleorhinus* being a marine exception. The early teleosaurid *Peipehsuchus* from the continental Lower Jurassic of China also suggests that the Teleosauridae were originally nonmarine forms. As far as the Dyrosauridae are concerned, the earliest known specimens referable to that family are from nonmarine Cretaceous deposits in Sudan (Buffetaut et al., 1990). It thus seems that the usual trend among crocodilians has been from a freshwater to a marine environment, not the reverse, no matter how the salinity problem was solved in the various groups. In the Eusuchia at least, this problem was apparently solved by the existence of a salt gland which is not restricted to the Crocodylidae, but probably was already present at an early stage of archosaurian evolution and contributed to the dispersal across marine barriers of several taxa whose origin was probably nonmarine.

REFERENCES

Allen, G. R. 1974. The marine crocodile *Crocodylus porosus* from Ponape, Eastern Caroline Islands, with notes on food habits of crocodiles from the Palau archipelago. *Copeia* 2:553.

Andrews, C. W. 1913. *A Descriptive Catalogue of the Marine Reptiles of the Oxford Clay. Part II*. British Museum (Natural History), London, 206 pp.

Benton, M. J. and M. A. Taylor. 1984. Marine reptiles from the Upper Lias (Lower Toarcian, Lower Jurassic) of the Yorkshire coast. *Proceedings of the Yorkshire Geological Society* 44:399-429.

Berckhemer, F. 1929. Beiträge zur Kenntnis der Krokodilier des schwäbischen oberen Lias.

Neues Jahrbuch für Mineralogie und Paläontologie 64, B:1-60.
Buckman, J. 1860. On some fossil reptilian eggs from the Great Oolite of Cirencester. *Quarterly Journal of the Geological Society of London* 16:107-110.
Buffetaut, E. 1977. Sur un Crocodilien marin, *Metriorhynchus superciliosus*, de l'Oxfordien supérieur (Rauracien) de l'Ile de Ré (Charente-Maritime). *Annales de la Société des Sciences Naturelles de la Charente-Maritime* 6, 4:252-256.
Buffetaut, E. 1978a. Crocodilian remains from the Eocene of Pakistan. *Neues Jahrbuch für Geologie und Paläontologie*, Abhandlungen 156, 2:262-283.
Buffetaut, E. 1978b. Sur l'histoire phylogénétique et biogéographique des Gavialidae (Crocodylia, Eusuchia). *Comptes Rendus de l'Académie des Sciences de Paris* 287, D: 911-914.
Buffetaut, E. 1979. A propos d'un crâne de *Metriorhynchus* (Crocodylia, Mesosuchia) de Bavent, Calvados: l'allégement des os crâniens chez les Metriorhynchidae et sa signification. *Bulletin de la Société Géologique de Normandie* 66:77-83.
Buffetaut, E. 1980a. Teleosauridae et Metriorhynchidae: l'évolution de deux familles de Crocodiliens mésosuchiens marins du Mésozoïque. *105e Congrès National des Sociétés Savantes*, pp. 1-12.
Buffetaut, E. 1980b. Position systématique et phylogénétique du genre *Pelagosaurus* Bronn, 1841 (Crocodylia, Mesosuchia) du Toarcien d'Europe. *Geobios* 13, 5:783-786.
Buffetaut, E. 1980c. Les Crocodiliens paléogènes du Tilemsi (Mali): un aperçu systématique. *Palaeovertebrata Mémoire Jubilaire René Lavocat*, pp. 15-35.
Buffetaut, E. 1982a. Radiation évolutive, paléoécologie et biogéographie des crocodiliens mésosuchiens. *Mémoires de la Société Géologique de France* 60, 142:1-88.
Buffetaut, E. 1982b. Systématique, origine et évolution des Gavialidae sud-américains. *Geobios Mémoire Spécial* 6:127-140.
Buffetaut, E. 1987. *A Short History of Vertebrate Palaeontology*. Croom Helm, London, 223 pp.
Buffetaut, E. 1991. Fossil crocodilians from Tiupampa (Santa Lucia Formation, Early Paleocene), Bolivia: a preliminary report. IN R. Suarez-Soruco (Ed.), *Fosiles y Facies de Bolivia*, 1, *Revista Técnica de YPFB*, 12, 3-4:541-544.
Buffetaut, E. and J. Thierry. 1977. Les Crocodiliens fossiles du Jurassique moyen et supérieur de Bourgogne. *Geobios* 10, 2:181-194.
Buffetaut, E. and P. Wellnhofer. 1980. Der Krokodilier *Teleorhinus* Osborn, 1904 (Mesosuchia, Pholidosauridae) im Regensbürger Grünsandstein (Obercenoman). *Mitteilungen der Bayerischen Staatssammlung für Paläontologie und Historische Geologie* 20:83-94.
Buffetaut, E., V. de Buffrénil, A. de Ricqlès, and Z. Spinar. 1982. Remarques anatomiques et paléohistologiques sur *Dyrosaurus phosphaticus*, Crocodilien mésosuchien des phosphates yprésiens de Tunisie. *Annales de Paléontologie* 68, 4:327-341.
Buffetaut, E., R. Bussert, and W. Brinkmann. 1990. A new nonmarine vertebrate fauna in the Upper Cretaceous of northern Sudan. *Berliner Geowissenschaftliche Abhandlungen* 120, 1:183-202.
Bustard, H.R. and B. C. Choudhury. 1980. Long distance movement by a salt-water

crocodile (*Crocodylus porosus*). *British Journal of Herpetology* 6:87.
Cornée, J. J. and E. Buffetaut. 1979. Découverte d'un Téléosauridé (Crocodylia, Mesosuchia) dans le Valanginien supérieur du massif d'Allauch (Sud-Est de la France). *Comptes Rendus de l'Académie des Sciences de Paris* 288, D:1151-1154.
Densmore, L. D. 1983. Biochemical and immunological systematics of the Order Crocodilia. *Evolutionary Biology* 16:397-365.
Dollo, L. 1914. Sur la découverte de Téléosauriens tertiaires au Congo. *Bulletin de la Classe des Sciences de l'Académie Royale de Belgique*, pp. 288-298.
Erickson, B. R. 1969. A new species of crocodile, *Teleorhinus mesabiensis*, from the Iron Range Cretaceous. *Scientific Publications of the Science Museum of Minnesota* 2, 1:1-13.
Frey, E. 1988. Das Tragsystem der Krokodile: eine bio-mechanische und phylogenetische Analyse. *Stuttgarter Beiträge zur Naturkunde* A 426:1-60.
Gasparini, Z. B. 1978. Consideraciones sobre los Metriorhynchidae (Crocodylia, Mesosuchia): su origen, taxonomia y distribucion geografica. *Obra del Centenario del Museo de la Plata* 5:1-9.
Gasparini, Z. B. 1980. South American Mesozoic crocodilians. *Mesozoic Vertebrate Life* 1:66-72.
Gasparini, Z. B. 1981. Los Crocodylia fosiles de la Argentina. *Ameghiniana* 18, 3-4:177-205.
Godefroit, P. 1994. Les reptiles marins du Toarcien (Jurassique inférieur) belgo-luxembourgeois. *Mémoires pour Servir à l'Explication des Cartes Géologiques et Minières de la Belgique* 39:1-98.
Grigg, G. C., L. E. Taplin, P. Harlow, and J. Wright. 1980. Survival and growth of hatchling *Crocodylus porosus* in saltwater without access to fresh drinking water. *Oecologia* 47:264-266.
Hua, S. 1994. Hydrodynamique et modalités d'allégement chez *Metriorhynchus superciliosus* (Crocodylia, Thalattosuchia): implications paléoécologiques. *Neues Jahrbuch für Geologie und Paläontologie,* Abhandlungen 193, 1:1-19.
Hua, S. In press. *Machimosaurus mosae* (Crocodylia, Teleosauridae): ostéologie, paléoécologie et phylogénie. *Palaeontographica*.
Hua, S. and F. Atrops. In press. Un crâne de *Teleidosaurus* cf. *gaudryi* Collot, 1905 (Crocodylia, Metriorhynchidae) dans le Bajocien terminal (Jurassique moyen) des environs de Castellane (Sud-Est de la France). *Bulletin de la Société Géologique de France*.
Hua, S. and V. de Buffrénil. In press. Bone histology as a clue for the interpretation of the functional adaptation in the Thalattosuchia (Reptilia, Crocodylia). *Journal of Vertebrate Paleontology.*
Hutton, J. 1989. Movements, home range, dispersal and separation of size classes in Nile crocodiles. *American Zoologist* 29:1044-1049.
Krebs, B. 1962. Ein *Steneosaurus*-Rest aus dem Oberen Jura von Dielsdorf, Kt. Zürich. *Schweizerische Paläontologische Abhandlungen* 79:1-28.
Krebs, B. 1967. Zwei *Steneosaurus*-Wirbel aus den Birmenstorfer Schichten (Ober-Oxford)

von "Weissen Graben" bei Montal (Kt. Aargau). *Eclogae Geologicae Helvetiae* 60:689-695.

Langston, W. 1973. The crocodilian skull in historical perspective. IN C. Gans and T. S. Parsons (Eds.), *Biology of the Reptilia*, pp. 263-284. Academic Press, London and New York.

Li, J. 1993. A new specimen of *Peipehsuchus teleorhinus* from Ziliujing Formation of Daxian, Sichuan. *Vertebrata Palasiatica* 31, 2:85-94.

Martill, D. M. 1986. The diet of *Metriorhynchus*, a Mesozoic marine reptile. *Neues Jahrbuch für Geologie und Paläontologie*, Monatshefte, pp. 621-625.

Martill, D. M., M. A. Taylor, K. L. Duff, J. B. Riding, and P. R. Bown. 1994. The trophic structure of the biota of the Peterborough Member, Oxford Clay Formation (Jurassic). *Journal of the Geological Society, London* 151:173-194.

Massare, J. A. 1987. Tooth morphology and prey preferences of Mesozoic marine reptiles. *Journal of Vertebrate Paleontology* 7:121-137.

Massare, J. A. 1988. Swimming capabilities of Mesozoic marine reptiles: implications for method of predation. *Paleobiology* 14:187-205.

Mateer, N. J. 1977. Form and function in some Mesozoic reptiles. *Acta Universitatis Upsaliensis* 440:1-39.

Mazzotti, F. J. and W. A. Dunson. 1989. Osmoregulation in crocodilians. *American Zoologist* 29:903-920.

Meyer, C. 1991. Burial experiments with marine turtle carcasses and their paleoecological significance. *Palaios* 6:89-96.

Mook, C. C. 1934. A new species of *Teleorhinus* from the Benton Shales. *American Museum Novitates* 702: 1-11.

Neill, W. T. 1971. *The Last of the Ruling Reptiles. Alligators, Crocodiles and Their Kin.* Columbia University Press, New York and London, 486 pp.

Norell, M. A. and G. W. Storrs. 1989. Catalogue and review of the type fossil crocodilians in the Yale Peabody Museum. *Postilla* 203:1-28.

Osborn, H. F. 1904. *Teleorhinus browni*. A teleosaur in the Fort Benton. *Bulletin of the American Museum of Natural History* 20:239-240.

Rieppel, O. 1981. Fossile Krokodilier aus dem schweizer Jura. *Eclogae Geologicae Helvetiae* 74:735-751.

Storrs, G. W. 1986. A dyrosaurid crocodile (Crocodylia, Mesosuchia) from the Paleocene of Pakistan. *Postilla* 197:1-16.

Swinton, W. E. 1950. On *Congosaurus bequaerti* Dollo. *Annales du Musée du Congo Belge* B 4:1-60.

Taplin, L. E. 1984. Evolution and zoogeography of crocodilians: a new look at an ancient order. IN M. Archer and G. Clayton (Eds.), *Zoogeography and Evolution in Australasia---Animals in Space and Time*, pp. 361-370. Hesperian Press, Perth.

Taplin, L. E. and G. C. Grigg. 1981. Salt glands in the tongue of the estuarine crocodile, *Crocodylus porosus*. *Science* 212:1045-1047.

Taplin, L. E., G. C. Grigg, and L. Beard. 1985. Salt gland function in fresh water crocodiles: evidence for a marine phase in eusuchian evolution? IN G. C. Grigg, R.

Shine, and H. Ehmann (Eds.), *Biology of Australasian Frogs and Reptiles*, pp. 403-410. Surrey Beatly and Sons, Sydney.

Troxell, E. L. 1925. *Hyposaurus*, a marine crocodilian. *American Journal of Science* 9, 54:489-514.

Webb, G. J. W. and H. Messel. 1978. Movement and dispersal patterns of *Crocodylus porosus* in some rivers of Arnhem Land, Northern Australia. *Australian Wildlife Research* 5:263-283.

Webb, P. and J. M. Skadsen. 1979. Reduced skin mass: an adaptation for acceleration in some teleost fishes. *Canadian Journal of Zoology* 57:1570-1575.

Wenz, S. 1968. Contribution à l'étude du genre *Metriorhynchus*. Crâne et moulage endocrânien de *Metriorhynchus superciliosus*. *Annales de Paléontologie, Vertébrés* 54:149-183.

Westphal, F. 1962. Die Krokodilier des deutschen und englischen oberen Lias. *Palaeontographica* A 118:23-118.

Chapter 13

THE MARINE CROCODILIAN *HYPOSAURUS* IN NORTH AMERICA

ROBERT K. DENTON, JR., JAMES L. DOBIE, and DAVID C. PARRIS

INTRODUCTION

The fossil genus *Hyposaurus*, a longirostrine mesosuchian crocodilian, is a characteristic and relatively common fossil from marine sediments of Maastrichtian through Danian ages. Although known from both continents of the western hemisphere as well as Africa, the North American range has been traditionally limited to the Hornerstown Formation of New Jersey.

Owen (1849) described two amphicoelus vertebrae, which he identified as belonging to a fossil crocodilian from the "greensand beds" of New Jersey. He named the species *Hyposaurus rogersii*, in honor of Henry Rogers, who brought the specimen to him for study. The generic name referred to the distinctive hypapophyseal keel that extended along the ventral surface of the centrum. Some years later, Cope (1869-1870) described *Gavialus fraterculus*, often referred to *Hyposaurus*, but which is established herein as belonging to a juvenile thoracosaur. Rogers, Owen, and Cope all believed *Hyposaurus* to be a marine crocodilian related to the teleosaurs, based on similar (i.e., amphicoelus) vertebral morphology and the fact that both came from sediments deposited under marine conditions.

Over the years numerous other specimens referred to *Hyposaurus* were collected from the New Jersey marl pits; however, they generally consisted of isolated vertebrae, and contributed little to a comprehensive knowledge of the anatomy of the species.

The first relatively complete specimen of *Hyposaurus* is that of *H. derbianus* (Cope, 1886) from the Cretaceous of Pernambuco State, Brazil. The presence of an almost complete lower jaw established the longirostrine nature of the species.

Nevertheless, Zittel (1890) placed *Hyposaurus* in the brevirostrine subdivision of the Mesosuchia, a relationship retained by Mook (1925), who believed the genus was synonomous with *Goniopholis*. Ironically, the most complete specimen of *Hyposaurus* yet encountered would be described by Troxell that same year (1925). Because of the elongate snout preserved in that specimen, Troxell questioned Zittel's assignment of *Hyposaurus* to the brevirostrine mesosuchians. Kuhn (1973), following earlier workers, maintained the synonymy of *Hyposaurus* and *Goniopholis*. Huene (1956) first suggested a relationship between *Hyposaurus* and the dyrosaurids, a group of late-surviving, longirostrine, marine mesosuchians. However, it would be left to Buffetaut (1976, 1980) to solidify this relationship, utilizing character analysis, on the basis of both Troxell's specimen and newer (previously undescribed) material in the collections of the New Jersey State Museum. The priority of *H. rogersii* over any other North American species was discussed by Parris (1986). The validity of the type specimen has been questioned (Norell and Storrs, 1989) primarily because the postcranial skeletons of dyrosaurids are undiagnostic. However, there is no evidence that any of the subsequently discovered specimens are unique enough to warrant subdividing the species and erecting new taxa. As *Hyposaurus* is the only valid dyrosaurid genus known from the western hemisphere, even isolated vertebrae bearing the diagnostic hypapophyseal keel, or its remnants, can be compared with the type specimen.

Until recently, the range of the North American species *H. rogersii* has been restricted to the Maastrichtian-Danian sediments of the Middle Atlantic coastal plain (Hornerstown Formation). The biostratigraphic significance of *Hyposaurus* as a marine guide fossil was discussed by Parris (1986).

In 1993, we began our study of a mesosuchian crocodilian from the Clayton Formation of Alabama (early Paleocene, Danian). The remains have since been established as belonging to *Hyposaurus*, which extends the range of the genus to the early Paleocene of the Gulf region. Here we discuss this new specimen and its implications for biostratigraphy. We also review the functional anatomy and paleobiology of *Hyposaurus rogersii*, and discuss several previously unreported specimens in the New Jersey State Museum.

MATERIALS AND METHODS

Most early examples of *Hyposaurus* were collected randomly during the mining of glauconite (greensand) marl. The disarticulated bones were simply picked up and saved by the pit workers while they were engaged in digging. This accounts for the fragmentary, and often damaged, condition of the nineteenth-century specimens.

For several decades, cooperation between the New Jersey State Museum and the

Hungerford family, owners of the Inversand Company Marl Pit (Sewell, New Jersey) has produced a number of relatively complete specimens of *Hyposaurus*. The bones are generally discovered by the marl pit workers; however, upon discovery all mining is suspended until the find can be assessed and removed. The specimens are plaster jacketed in the field, and removed to the museum laboratory for drying, cleaning, and coating with a hydrophobic, polymeric varnish.

The Alabama specimen was collected in bulk samples of a partly indurated Paleocene chalk. The bones were relatively resistant and lent themselves to preparation with dilute acetic acid. Residual insolubles were removed with dental hand tools.

All measurements were taken with an optical micrometer, vernier calipers, or ruler. Illustrations were drawn using scaled photographs and/or direct measurement of specimens.

Samples of gastroliths were identified by x-ray powder diffraction analysis, using a Phillips goniometer and data analyzer. Samples were ground to <80 mesh, back-packed, and run using $CuK\alpha$ radiation.

Institutional Abbreviations

The abbreviations used for the institutions referred to in the text are as follows: ALAM-PI, University of Alabama Museum---Paleoinvertebrate Collection; ALAM-PV, University of Alabama---Paleovertebrate Collection; AMNH, American Museum of Natural History; AUMP, Auburn University Museum, Paleontological Collection; NJSM, New Jersey State Museum; PU, Princeton University Collection, now at Yale Peabody Museum; RMM, Red Mountain Museum, now Discovery 2000; SC, South Carolina State Museum; USNM, United States National Museum; and YPM, Yale Peabody Museum.

SYSTEMATIC PALEONTOLOGY

Order CROCODILIA Gmelin, 1788
Suborder MESOSUCHIA Huxley, 1875
Family DYROSAURIDAE De Stefano, 1903
Genus *HYPOSAURUS* Owen, 1849

Diagnosis: As given by Buffetaut (1980).

HYPOSAURUS ROGERSII Owen, 1849
Figures 1-8

Hyposaurus fraterculus Cope, 1869
Hyposaurus ferox Marsh, 1871
Hyposaurus natator Troxell, 1925

Emended diagnosis: Mandibular symphysis extending to the tenth dentary tooth. Splenial extending to the fifth mandibular tooth. Total dentary tooth count varying from 12 to 15. Teeth laterally compressed, slightly recurved blades, lingually striate, with unserrated carinae. Teeth inflated in largest specimens, with anterior carina displaced lingually. Nasals fused. Limbs well developed. Vertebrae amphicoelus with relatively concave centrum articular surfaces. Enlarged, recurved neural spines present on vertebrae 7-11.

Range: Late Cretaceous through Paleocene, Atlantic and Gulf coasts of North America.

Status of *Hyposaurus fraterculus* Cope: As noted by Parris (1986), no sound basis exists for recognition of more than one species of *Hyposaurus* in North America. Thus, *Hyposaurus rogersii* Owen has priority. Among other proposed and referred species, *Hyposaurus fraterculus* Cope has remained of uncertain status. Originally proposed as *Gavialus fraterculus* (Cope, 1869), it has since been generally referred to *Hyposaurus* without significant reason due to the fragmentary nature of the type specimen, and in spite of teeth distinctly perpendicular to the occlusal plane. Recent collecting has yielded a juvenile specimen of the eusuchian *Thoracosaurus neocesariensis* (NJSM 15437) which includes a dental segment virtually identical to the type specimen of *H. fraterculus* (AMNH 2198), thus demonstrating that *H. fraterculus* is a junior synonym of *Thoracosaurus neocesariensis* (DeKay, 1842).

REFERRED SPECIMENS

NJSM 10861 (Figures 1 and 2)

Locality and horizon: Inversand Company Marl Pit, Sewell, Mantua Township, Gloucester County, New Jersey. Late Cretaceous (Maastrichtian), lower Hornerstown Formation.

Collector and date: D. C. Parris, October 3, 1980.

Material: Nearly complete disarticulated skull including: partial premaxillae, maxillae with teeth, nasals, frontals, lachrymals, prefrontals, postorbitals, squamosals, parietal, supraoccipital, exoccipitals, basioccipital, basicranium, jugals, quadrates, complete mandibles with teeth, miscellaneous skull fragments, 13 vertebrae, rib fragments, and osteoscutes.

Remarks: This specimen comprises the most perfectly preserved skull and jaws of *H. rogersii* yet found. There are 12 preserved alveoli in the left dentary, 11 in the right (Figure 1), but based on other specimens the original tooth count may have been as high as 15 (several of the alveoli in NJSM 10861 are incomplete). A diminutive seventh mandibular tooth is present only in the right dentary; it is completely absent from the left. This reduction of the seventh mandibular tooth is a synapomorphy of the Dyrosauridae. The edge of each dentary is slightly "festooned"; i.e., the labial alveolar borders extend slightly from the lateral surface when viewed from the dorsal (occlusal) plane. The mandibular symphysis varies from quadrangular (anterior) to oval (posterior) in transverse section, and is slightly wider than high. The symphysis extends to the level of the tenth alveolus, and the splenial reaches dorsally to the fifth alveolus.

Both maxillae are preserved (Figure 2), with 11 alveoli preserved in the left and 9 in the right. There were probably 12 teeth in each maxilla. Most of the maxillary teeth are broken, and a number of the tooth bases are present in the alveoli. The few preserved teeth are pointed, laterally compressed, and longitudinally striate on their lingual surfaces. They possess sharp, nonserrated carinae, extending from the tip to the base, along the entire enameled surface of the anterior and posterior edges. The edges of the maxillae are not festooned by the alveolar borders and appear smooth from the dorsal aspect.

No teeth are preserved in the fused premaxillae; however, observations on other skulls suggest that there were a total of 6 in life. There is no evidence of any notch or indentation between the premaxillae and the maxillae, as occurs in the African dyrosaurids *Sokotosuchus* and *Phosphatosaurus*.

The condition of the nasal bones, previously unknown for the genus, is revealed in this specimen. They are completely fused along their entire length, and contact the premaxillae. The premaxillae completely surround the external nares.

NJSM 10861 is the only known specimen of *Hyposaurus* in which the entire dorsal suface of the skull is preserved. Table 1 is a comparison of skull length measurements and their ratios among the better known dyrosaurid genera as well as the generalized mesosuchian genus *Goniopholis* and the eusuchian *Crocodylus*. Although the dyrosaurids are often referred to as "longirostrine," the preorbital vs. overall dorsal length ratios (Table 1, R1) of the various genera do not differ markedly from either *Goniopholis* or the *Crocodylus*, and vary from 0.59 to 0.68. In contrast, the teleosaurids, *Thoracosaurus*, and gavialids all have PreSL/DL ratios of ≥ 0.70. The only dyrosaurids which approach this degree of longirostrine development are the genera *Dyrosaurus*, *Rhabdognathus*, and *Atlantosuchus*.

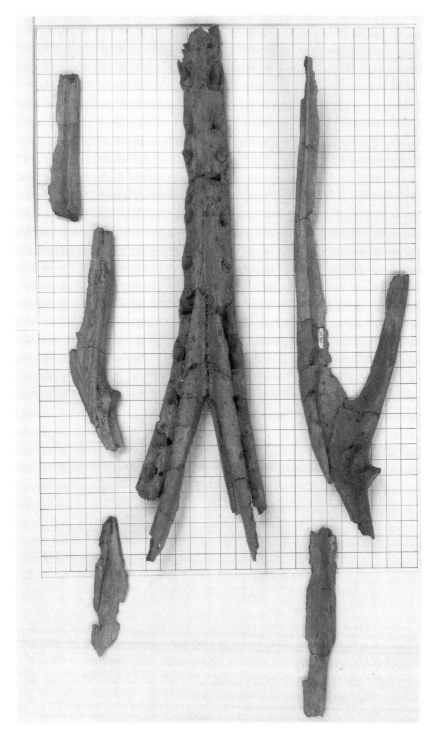

Figure 1. Portions of mandibular elements of NJSM 10861, *Hyposaurus rogersii*.

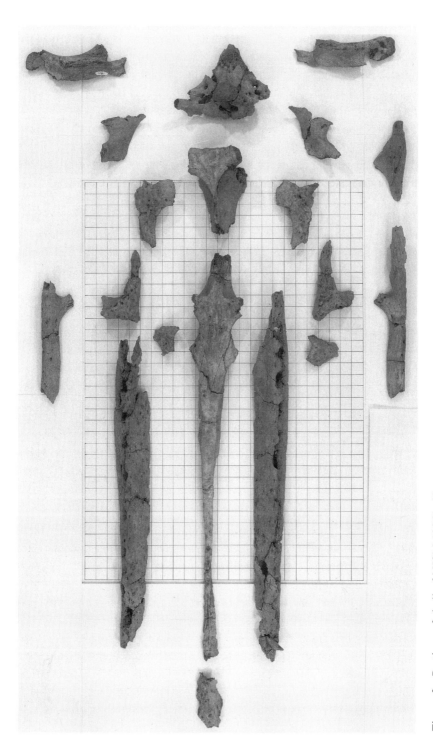

Figure 2. Portions of skull of NJSM 10861, *Hyposaurus rogersii*.

In general, the dyrosaurids are all relatively narrow snouted, and this condition reaches its greatest development in the genera *Dyrosaurus, Hyposaurus* (Table 1, R2), *Rhabdognathus*, and *Atlantosuchus*. This character appears even more exaggerated in articulated skulls, where the widely open upper temporal fenestrae and thin temporal arches give an open, airy look to the rear portion of the skull, in distinct contrast to the narrow, compact maxillofacial region.

This specimen is similar in all other respects to YPM 985, *H. natator* (Troxell, 1925), and conforms to Buffetaut's (1980) diagnosis of the genus, including the presence of well-developed "spurs" on the anterior edge of each postorbital bone, a distinctive character of the family Dyrosauridae.

Table 1. Comparison of dorsal skull length dimensions among selected members of the Dyrosauridae, *Goniopholis*, and *Crocodylus*. Abbreviations: **DL** = total dorsal skull length, **PreSL** = preorbital skull length, **R** = ratio, **SnW** = width at snout mid-length.

Taxon	PreSL (cm)	SnW (cm)	DL (cm)	R1 (PreSL/DL)	R2 (SW/PrSL)
Hyposaurus	28.2	2.7	42.9	0.65	0.09
Dyrosaurus	58.7	6.5	86.5	0.68	0.11
Phosphatosaurus	38.2	7.4	64.2	0.59	0.19
Sokotosuchus	69.5	14.0	108.9	0.63	0.20
Goniopholis	37.0	16.5	60.0	0.61	0.44
Crocodylus	28.0	13.0	44.0	0.63	0.46

After Kuhn (1973) and Buffetaut (1979).

NJSM 12293

Locality and horizon: Inversand Company Marl Pit, Sewell, Mantua Township, Gloucester County, New Jersey. Late Cretaceous (Maastrichtian), lower Hornerstown Formation. Same provenience as NJSM 10861.

Collector and date: D. C. Parris, October 4, 1983.

Material: Twenty-one vertebrae, two femora, two tibiae, osteoscutes, pelvic elements, phalanges, skull fragments, teeth, and gastroliths.

Remarks: This fossil is among the largest known specimens of *Hyposaurus*. The femur is approximately 265 mm in length, slightly exceeding the same bone of Troxell's subspecies *H. natator oweni*, YPM 753 (Troxell, 1925). The centrum of the eighth or ninth vertebra of NJSM 12293 is 65 mm in ventral length, and 42 mm wide (anteriorly), also slightly exceeding the size of Troxell's subspecies. Two well-preserved anterior caudals are 47.5 mm and 47.0 mm in ventral length, and approximately 42 mm high and 34 mm wide (anteriorly). *Hyposaurus n. oweni* was separated by Troxell as a distinct subspecies on the basis of its size, and morphology of the pelvis and femur. The regions of the latter bones which Troxell considered to be anatomically distinct are, however, poorly preserved (Norell and Storrs, 1989), and cannot be considered diagnostic. Furthermore, the recovery of additional specimens of *H. rogersii* since Troxell's time reveals a considerable variation in size within the species. Thus, *H. n. oweni* probably represents nothing more than a fully mature individual of H. *rogersii*.

Aside from its large size, NJSM 12293 possesses two unique features, which are worthy of note. First, a number of well-preserved teeth are associated with the skeleton, and their overall morphology is similar to that of other specimens of *H. rogersii*. They are, for example, lingually ridged and sharp, and possess smooth carinae. Yet, relative to all other specimens of *Hyposaurus*, these teeth appear generally more "robust" and conical, with slightly thickened root walls. Additionally, the anterior carina of each tooth exhibits a slight lingual twist, with the greatest displacement occurring as the carina approaches the root. This "anterior twist" is most exaggerated in the more conical, uncompressed teeth as opposed to the laterally compressed specimens, but it is present to some degree in all of the preserved teeth. The significance of this unique tooth morphology is difficult to interpret, and it may represent a specific ontogenetic stage. Once a certain individual size was reached, dietary requirements may have shifted. A similar tooth morph is seen in the dentition of some mosasaurs, where the conical, ridged teeth also have a twisted and/or displaced anterior carina. Stomach contents analyses suggest that mosasaurs were extremely generalized predators, eating just about anything they could catch, including ammonites (Bukowski and Bond, 1989; Kauffman and Kesling, 1960), birds, fish, and smaller mosasaurs (Martin and Bjork, 1987). Thus, it is possible that as it matured *Hyposaurus* may have shifted from a strictly piscivorous diet to a more generalized one. The robust, conical teeth of NJSM 12293, with their strong lingual ridges and twisted carina, may have added durability and the ability to puncture the integument/shells of larger prey. It is notable that another large *Hyposaurus* (NJSM 11069), from the Paleocene (Danian) upper Hornerstown Formation (Parris, 1986), has teeth which are as large as those of NJSM 12293. Yet all the preserved teeth of the Paleocene specimen are more "gracile" (i.e., relatively longer and thinner), have thinner root walls, and lack any trace of the anterolingually twisted carinae. If the tooth morphology seen in the large Cretaceous specimen of *Hyposaurus* was, in fact, associated with predation on shelled creatures, especially ammonites and large marine reptiles, it would have become "obsolete" with the extinction of the ammonoids and many marine reptiles at the close of the Cretaceous. Crocodilians are highly adaptable, and are known to undergo rapid changes in their dental morphology in response to environmental stress and dietary modification (Langston, 1965). This suggests that the dyrosaurid survivors of the Cretaceous-Tertiary faunal collapse may have reverted to piscivorous habits, even in the largest individuals of *Hyposaurus*, and that this dietary alteration was reflected in their dental morphology.

NJSM 12293 was found with 21 smooth, rounded stones, ranging in mass from 1.0 to 218 g, with only 3 exceeding 30 g and 13 of 6 g or less. They are identical to the gastroliths (stomach stones) often found associated with the skeletons of plesiosaurs. The majority of these supposed gastroliths are composed of a dense, compact quartzite, and resemble river cobbles. They are completely out of place, from a lithological standpoint, in the fine-grained, homogeneous glauconite marl of the Hornerstown Formation. Although the majority are quartzite, one gastrolith is composed of a mixture of quartz and albite (plagioclase feldspar). It is identical in composition to specimens of a finely crystalline, friable feldspar, in the collection of one of the authors (R. Denton). These reference specimens originated from a now inaccessible pegmatite dike located in Upper Darby, Montgomery County, Pennsylvania. We suggest that the source rocks for the white gastrolith may have been the local crystalline piedmont rocks just to the west of the ancient Cretaceous shoreline. The feldspar fragment was probably intermixed with sandstone beach and river cobbles, such as the other gastroliths appear to be. The function, if any, of stomach stones has been controversial. The mass of gastroliths in specimens of *Crocodylus niloticus* has been found to be proportional to the mass of the host (Cott, 1961), and in marine reptiles, the potential of ballast function has been argued (Darby and Ojakangas, 1980). The total mass of the stones found with NJSM 12293 is 680 g, seemingly consistent with ballast function advantageous to a marine reptile.

NJSM 10416 (Figures 3-5)

Locality and horizon: Inversand Company Marl Pit, Sewell, Mantua Township, Gloucester County, New Jersey. Late Cretaceous (Maastrichtian), lower Hornerstown Formation. Same provenience as NJSM 10861.
Collector and date: Collector unknown, 1940s.
Material: Articulated and mounted skull and jaws (maxillae, prefrontals, jugals, postorbitals, nasals, left surangular, left angular, restored).
Remarks: This somewhat enigmatic fossil, identified by E. Buffetaut in 1976, is currently on display in the Natural History Hall of the NJSM, and has been the most frequently illustrated and photographed of any specimen of *Hyposaurus* (Wolfe, 1977; Parris, 1986). Although the locality data for the skull are detailed enough to establish its provenience unequivocally, the exact date it was excavated

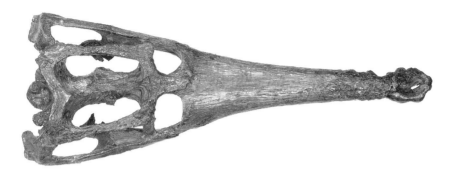

Figure 3. Skull of NJSM 10416, *Hyposaurus rogersii*, dorsal aspect; note restored maxillary region and lack of nasals.

and the identity of the collector remain unknown.

The most noteworthy feature of this specimen is the nearly perfect preservation of the entire posterior face of the skull (Figure 5). This surface shows symmetrically paired and pronounced occipital tuberosities formed by the supraoccipital and exoccipital bones, lying halfway between the foramen magnum and the dorsal surface of the skull. These occipital tuberosities are characteristic of all dyrosaurids, and probably served as attachment points for powerful neck muscles which could swing the skull rapidly in the sagittal plane, as well as from side to side (Buffetaut, 1979). The occipital tuberosities reached their greatest development in the genera *Hyposaurus* and *Dyrosaurus*, among those dyrosaurids in which this region of the skull has been preserved.

SC 83-78-10

Locality and horizon: Spoils piles, Santee rediversion canal, St. Stephen, Berkeley County, South Carolina. Late Paleocene, Williamsburg Formation, Black Mingo Group.
Collector and date: Rudy Mancke, July 1983.
Material: Single midcaudal (or) anterior midcaudal vertebral centrum.
Remarks: The spoils heaps of the Santee rediversion have produced a number of significant specimens, including late Paleocene land mammals, previously unknown from eastern North America (Schoch, 1985). The geology of the Black Mingo Group has been studied by Van Nieuwenhuise and Colquhoun (1982a, b). This specimen is indistinguishable from the anterior midcaudal centra of NJSM

Figure 4. Skull of NJSM 10416, *Hyposaurus rogersii*, right lateral aspect.

Figure 5. Skull of NJSM 10416, *Hyposaurus rogersii*, right oblique view of posteroventral aspect, showing the occipital tuberosities.

12293, and probably represents a slightly larger individual. Table 2 compares the measurements of the equivalent bones of NJSM 12293 and SC 83-78-10.

We tentatively refer this specimen to *Hyposaurus rogersii*, despite its being merely one vertebra, in view of the uniqueness of the species as the only North American dyrosaurid. It is currently indistinguishable from *H. rogersii*, but if the species were reviewed and subdivided, such a specimen might be insufficiently diagnostic.

Table 2. Caudal centrum measurements (in mm) of NJSM 12293 and SC 83-78-10.

	NJSM 12293	SC 83-78-10
Centrum length (ventral)	47.5	52.0
Anterior height	32.0	50.0
Anterior width	34.0	42.0

AUMP 1240 (Figure 6)

Locality and horizon: M. Till farm, east of unmarked road, sec. 17 (T12N, R12E), north of Braggs, Wilcox County, Alabama. Early Paleocene (Danian) Midway Group, Clayton Formation, Pine Barren Sand Member.

Collectors and date: J. Dobie, G. Dobie, J. Owen, and D. Phillips, June 2, 1971.

Material: Nine vertebrae (plus additional fragments), mixed cervicals, thoracics, and caudals. Miscellaneous unidentified bones (unprepared), including possible skull fragments.

Remarks: This specimen was identified in 1993 by the staff of the NJSM. It was until recently the only known crocodilian from the Paleocene of Alabama. Several vertebrae, freed from their chalky matrix, are indistiguishable from those of the New Jersey specimens of *Hyposaurus*.

Among the best preserved is the sixth (or seventh) cervical (Figure 6). It is amphicoelus, with the broken bases of the rib articular processes located laterally. The neural arch is intact, but the dorsal spine is broken off. The hypapophysis is missing, but enough of the peduncle is present to establish its length. The measurements of the centrum are: ventral length = 61.1 mm, anterior height = 41.9 mm, anterior width = 38.9 mm, and length of hypapophyseal peduncle = 18.9 mm.

This fossil extends the known range of *Hyposaurus* to the Paleocene of the Gulf Coast of North America.

Figure 6. Three vetebral centra of AUMP 1240, *Hyposaurus rogersii*, from the Paleocene of Wilcox County, Alabama.

ALAM-PV 990.019

Locality and horizon: M. Till farm, east of unmarked road, sec. 17 (T12N, R12E), north of Braggs, Wilcox County, Alabama. Early Paleocene (Danian) Midway Group, Clayton Formation, Pine Barren Sand Member.

Collectors and date: D. Jones and Oldshue, 1990.

Material: Osteoscutes (fragments), right ischium (proximal fragments), right femur (distal fragment), rib fragments, metapodial fragment (?), bone scraps (unidentifiable).

Remarks: This is the second specimen from the Till farm, collected nearly 18 years after AUMP 1240. It consists primarily of fragmentary bone, including the distal condyles from the (right?) femur, an acetabular fragment of the ischium, a rib fragment, and a broken metapodial (?) shaft. Two broken osteoscutes are thin and have shallow, widely spaced punctations and gently scalloped edges. Their overall morphology is identical to that of *H. rogersii*, and they compare favorably with the osteoscutes of NJSM 12293, NJSM 11069, and YPM 985.

This specimen further demonstrates the presence of *Hyposaurus* in the Paleocene beds of the Gulf Coast of North America.

TENTATIVELY REFERRED SPECIMENS

Materials currently under study include the following specimens which are indistinguishable from the genus *Hyposaurus*, but which are too incomplete for certain reference.

NJSM 13502

Four teeth from the Aquia Formation of Belvidere Beach, King George County, Virginia. These are Thanetian in age, the same as SC 83-78-10.

USNM 23046

Mandibular fragment, from Surfside Beach (Cretaceous age?), South Carolina (E. Buffetaut, personal communication, 1984).

PALEOBIOLOGY OF *HYPOSAURUS*

The adaptation of *Hyposaurus* to the marine environment was established correctly by Troxell (1925), who cited as evidence: eyes located more laterally than dorsally, a characteristic of animals that spend much time below the surface of the water; the long, flattened tail; an elongate, narrow snout with numerous, close-set, homodont teeth; amphicoelus vertebrae; unique pelvic adaptations; and light dorsal armor. All of these observations were based on several Yale specimens, especially YPM 985, the holotype of *Hyposaurus natator*. Since Troxell's description was published a number of specimens of *Hyposaurus* have come to light, which furnish new information about the natural history of the species.

The numerous sizable gastroliths associated with NJSM 12293 suggest that animal may have purposely swallowed cobbles to serve as ballast. This is significant, in that *Hyposaurus* had no special adaptations for diving. The dyrosaurids lacked the reversed heterocercal tail, laterally compressed body, and paddle-like limbs of the Thallatosuchians (e.g., *Geosaurus*). They had unreduced limbs of a unique conformation (Arambourg, 1952; Swinton, 1950). The forelimbs are not significantly smaller than the hindlimbs, and it has been suggested that this may be evidence of the family's descent from an unknown, terrestrial mesosuchian ancestor (Langston, 1995). The overall appendicular development was relatively greater than the extant genus *Alligator* (Troxell, 1925), a character unexpected in a crocodilian adapted for marine life, although the femora are relatively smaller than the corresponding bones in *Crocodylus* (Swinton, 1950). The high vertical angulation of the zygapophyseal facets of the anterior thoracic and posterior cevical vertebrae of *Hyposaurus* suggest its trunk was stiff, almost rod-like, and capable of only limited lateral flexion (Langston, 1995); thus, swimming must have been accomplished by undulations of the tail assisted by some movement of the limbs. They shared with other Mesozoic marine reptiles the habit of swallowing "ballast stones." These stones may have been equally important both as an aid in diving, as well as to prevent rolling in the waves of the open ocean. Although stomach stones are found in many mature crocodiles (Darby and Ojakangas, 1980), they are not so commonly associated with fossil skeletons. We agree with the ballast function interpretation and note that the large size of NJSM 12293 would demand more significant ballast than most crocodilians of lesser dimensions.

Troxell (1925) commented on the large, recurved neural spines of the posterior cervical and anterior thoracic vertebrae of *Hyposaurus* (Figure 7a), while in many other dyrosaurids the neural spines of the entire vertebral column are enlarged. These same vertebrae also exhibit the greatest development of the hypapophyses. Although this character is unique in the dyrosaurids among the Mesosuchia, it is seen with relative frequency in the Eusuchia. It is especially developed among the Alligatorinae (Figure 7b), although never to the same degree as in the dyrosaurids. Troxell (1925) suggested that the enlarged neural spines may have "furnished a place of attachment for the muscles [flexors] which lifted the skull"; however, they are also the origin point for the axial abductor musculature, which turn the head from side to side. Buffetaut (1979) commented on the insertion point of these muscles on the posterior face of the skull, which he called the occipital tuberosities, and discussed their functional significance in aiding the animal to turn or lift its head rapidly. He suggested that the dyrosaurids were pursuit hunters; however, it is just as possible that they were also ambush hunters, who swam slowly through the water and capitalized on their ability to rapidly swing or lift their head to capture large, fast-swimming fish or cephalopods. The stout teeth of large specimens of *Hyposaurus* suggest they hunted strong, active prey. Their teeth are relatively more robust than those of the extant piscivorous gavials of equivalent size, as well as the fossil teleosaurs and metriorhynchids.

The reduced, nonimbricate armor scutes in *Hyposaurus*, many of which are quite frail (Figure 8), are to be viewed as a marine adaptation as well (Troxell, 1925). The scutes do not exhibit the "peg and socket" articulation typical of most mesosuchian crocodilians. We also note that the rarity of scutes among specimens of *Hyposaurus* generally is negative evidence for the presence of relatively light armor. However, in other dyrosaurids, especially *Rhabdognathus* and *Phosphatosaurus*, the armor is partially imbricate and very solidly developed, both

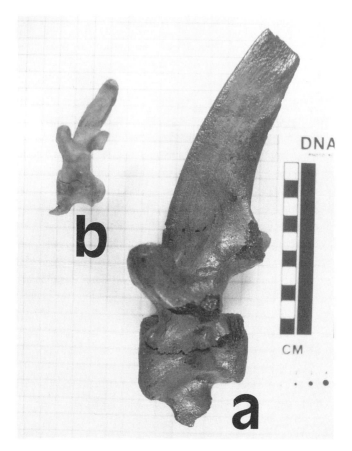

Figure 7. Comparison of dorsal vertebra of **A**) *Hyposaurus rogersii* (YPM 764) with **B**) *Alligator mississipiensis*. Note ventral hypopophyses and relative development of neural spine.

dorsally and ventrally (see Langston, 1995).

Like other marine reptiles, *Hyposaurus* was probably a part-time resident of the strandline environment, at least for purposes of reproduction. Large, well-developed limbs may have enabled *Hyposaurus* to move about on land fairly effectively, but the geological occurrences and details of the anatomy suggest most of the time was spent far from the shore, quite possibly with much submergence. This would also have reduced competition with its contemporary *Thoracosaurus*, a heavily armored, longirostrine eusuchian surface hunter, that could also grow to very large size, but which is known from more nearshore facies. *Hyposaurus* probably fed primarily on fish, although details of the tooth structure of a large Cretaceous specimen suggest they may have preyed on ammonites, or other shelled creatures, as adults.

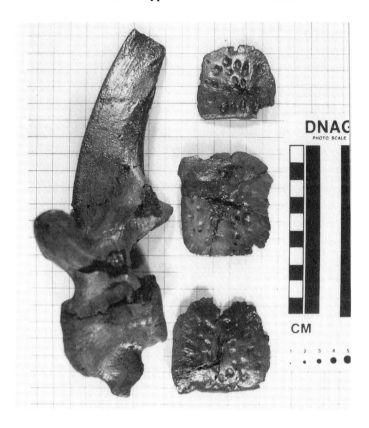

Figure 8. *Hyposaurus rogersii*, YPM 764, dorsal vertebra and three dorsal osteoscutes.

Hyposaurus has no exact ecological equivalent among extant crocodilians, which are all denizens of the lacustrine, fluviatile, estuarine, and nearshore environments. It was probably more thoroughly marine than any other known contemporary crocodilian, except possibly the rare *Bottosaurus*. Some crocodiles are known to venture out into the open sea (e.g., *C. porosus, C. acutus*), possibly to find new mates and territory, but they generally hunt at or near the shoreline (Kuhn, 1973). We speculate that the situation for *Hyposaurus* may have been exactly the opposite: hunting far out to sea, and spending some relatively short amounts of time on or near the land, in a way unable to completely escape the legacy of a currently unknown terrestrial mesosuchian ancestor. This speculation would account for two interesting aspects of the fossil record of the genus. Substantial use of the open marine environment contributed to its wide geographic range (North America, South America, Africa). Occupation of multiple habitats may have enabled *Hyposaurus* (among other crocodilians) to escape the fate that befell many marine reptiles at the end of the Cretaceous; however, their abundance diminishes significantly in the

Paleocene of North America. Yet in other Tethyan regimes the family continued to be numerous and diverse until the end of the Eocene.

BIOSTRATIGRAPHIC CONSIDERATIONS

The correlations of *Hyposaurus* specimens of the New Jersey-Delaware sequences were discussed in detail by Parris (1986) and Buffetaut (1976). Specimens from the Hornerstown Formation have been documented with extensive associated faunas of both Maastrichtian and Danian (Midway) ages. The established stratigraphic range is of limited duration, but spans the Cretaceous-Tertiary boundary. The South Carolina and Alabama specimens referred herein can be correlated with considerable accuracy to the Paleocene span of the range of *Hyposaurus*.

The provenience of AUMP 1240 and ALAM-PV 990.0019 is the Pine Barren Sand Member of the Clayton Formation, which is of unquestioned early Paleocene age (Gibson et al., 1982; Nolf and Dockery, 1993). Numerous whole and fragmentary specimens of *Ostrea pulaskensis* Harris, considered the definitive guide fossil for the early Paleocene of the Gulf region, are present within the matrix of AUMP 1240. As noted by Parris (1986), similar oysters are present in the Upper Fossiliferous Layer at the Inversand Company Marl Pit, site of many New Jersey specimens.

In addition we have recovered one specimen of *Aturia cf. vanuxemi* (ALAM-PI 994.3.1) from the site of AUMP 1240. This distinctive Paleocene nautiloid is well known from New Jersey sites as well, including the upper levels at the Inversand Marl Pit.

A specimen of a chimaeroid taxon new to the Alabama fauna was recovered in the course of field inspection in the present study. We refer this specimen (ALAM-PV 994.3.1) to *Ischyodus williamsae* Case. It is a mandibular element, precisely matching tritor tracts illustrated by Case (1989) for that species and thus further confirming the correlation to the Paleocene. In summary, the correlation of the Alabama specimens to the Danian (Midway) horizons of the Hornerstown Formation is supported by all available faunal evidence.

Although recovered from spoil piles, the faunal associations of the South Carolina specimen indicate Paleocene age, although perhaps later in the epoch than other *Hyposaurus* specimens (Schoch, 1985). As the fauna is under study by others we have not attempted any detailed examination of the correlation, but see no inconsistency with other occurrences of the genus.

SUMMARY

The mesosuchian genus *Hyposaurus* is essentially confined to Maastrichtian and Danian marine sediments. Its geographic range in the United States now extends from New Jersey to Alabama (possibly Mississippi), in glauconite and chalk facies. The type specimen of *Hyposaurus fraterculus* Cope is now known to be a juvenile *Thoracosaurus*, leaving *Hyposaurus rogersii* Owen as the only North American species. It is notable as a marine guide fossil of short stratigraphic range which does, however, span the Cretaceous-Tertiary boundary.

New specimens allow the elongate skull of *Hyposaurus rogersii* to be described and illustrated more completely than before, which is of interest because of published remarks about its degree of aquatic adaptation. Other adaptations to the marine environment have been cited in previous works, notably the light scute armor and enhanced characteristics of propulsion in the tail and pelvic limb. However, one well-preserved specimen has yielded remarkably large gastroliths, prompting comparison to those of elasmosaurs, with potential contributions to stability in marine environments.

ACKNOWLEDGMENTS

We would like to thank James Lamb and W. Brown Hawkins, for their time, patience, and assistance in the field and laboratory studies of the Alabama *Hyposaurus* specimens. We also want to thank the Till family of Braggs, Alabama, on whose property the collection was undertaken, as well as James Knight of the South Carolina State Museum, for his assistance and loan of the specimen under his care.

Additional thanks go to Earl Manning, who originally identified the Alabama fossil as a mesosuchian and suggested sending it to NJSM for study. We also thank Barbara Grandstaff and Bob O'Neill for their assistance throughout the project.

REFERENCES

Arambourg, C. 1952. Les vertébrés fossiles des gisements de phosphates (Maroc-Algérie-Tunisie). *Notes et Memoirs Service Geologique Maroc* 92:295-307.

Buffetaut, E. 1976. Une nouvelle definition de la famile des Dyrosauridae De Stefano, 1903 (Crocodilia, Mesosuchia) et ses consequences: Inclusion des genres *Hyposaurus* et *Sokotosuchus* dans les Dyrosauridae. *Geobios* 9:333-336.

Buffetaut, E. 1979. *Sokotosuchus ianwilsoni* and the evolution of the dyrosaurid Crocodiles. *The Nigerian Field* 1(supplement to Volume 44):31-41.

Buffetaut, E. 1980. Les Crocodiliens Paleogene du Tilemji (Mali), un aperçu systematique. *Palaeovertebrata. Montpellier. Memoire Jubilaire R. Lavocat* pp. 15-35.

Bukowski, F. and P. Bond. 1989. A predator attacks *Sphenodiscus*. *The Mosasaur* 4:69-74.

Case, G. R. 1989. A new species of chimaeroid fish from the Upper Paleocene (Thanetian) of Maryland, U.S.A. *Palaeovertebrata* 21:85-94.

Cope, E. D. 1869-1870. Synopsis of the extinct Batrachia, Reptilia, and Aves of North America. *Transactions of the American Philosophical Society* 14:1-252.

Cope, E. D. 1886. A contribution to the vertebrate paleontology of Brazil. *Proceedings of the American Philosophical Society* 23:15-20.

Cott, H. B. 1961. Scientific results of an inquiry into the ecology and economic status of the Nile crocodile *(Crocodilus niloticus)* in Uganda and northern Rhodesia. *Transactions of the Zoological Society of London* 29:1-211.

Darby, D. G. and R. W. Ojakangas. 1980. Gastroliths from an Upper Cretaceous plesiosaur. *Journal of Paleontology* 54:548-556.

Gibson, T. G., E. A. Mancini, and L. M. Bybell. 1982. Paleocene to middle Eocene stratigraphy of Alabama. *Transactions of the Gulf Coast Association of Geological Societies* 23:449-458.

Huene, F. von. 1956. *Paläontologie und Phylogenie der Niederen Tetrapoden*. Gustav Fischer Verlag, Jena, 716 pp.

Kauffman, E. G. and R. U. Kesling. 1960. An Upper Cretaceous ammonite bitten by a mosasaur. *Contributions From the Museum of Paleontology, University of Michigan* 15(9):193-248.

Kuhn, O. 1973. Crocodylia. Part 16. *Handbuch der Palaoherpetologie*. Gustav Fischer Verlag, Stuttgart, 116 pp.

Langston, W. Jr. 1965. Fossil crocodilians from Columbia and the Cenozoic history of the crocodilia in South America. *University of California Publications in Geological Sciences* 52:1-157.

Langston, W. Jr. 1995. Dyrosaurs (Crocodilia: Mesosuchia) from the Paleocene Umm Himar Formation, Kingdom of Saudi Arabia. *Bulletin of the United States Geological Survey* 2093.

Marsh, O. C. 1871. [A communication on some new reptiles and fishes from the Cretaceous and Tertiary formations.] *Proceedings of the Academy of Natural Sciences, Philadelphia* 1871:103-105.

Martin, J. E. and P. R. Bjork. 1987. Gastric residues associated with a mosasaur from the late Cretaceous (Campanian) Pierre Shale in South Dakota. *Dakoterra* 3:68-72.

Mook, C. C. 1925. A revision of the Mesozoic Crocodilia of North America. *Bulletin of the American Museum of Natural History* 51(9):319-432.

Nolf, D. and D. T. Dockery III. 1993. Fish otoliths from the Matthews Landing Marl Member (Porters Creek Formation), Paleocene of Alabama. *Mississippi Geology* 14:24-39.

Norell, M. A. and G. W. Storrs. 1989. Catalogue and review of the type fossil crocodilians in the Yale Peabody Museum. *Postilla* 203:1-28.

Owen, R. 1849. Notes on the remains of an fossil reptile discovered by Professor Henry

Rogers of Pennsylvania, US, in greensand formations of New Jersey. *Proceedings of the Geological Society of London* 5:380-383.

Parris, D. C. 1986. Biostratigraphy of the fossil crocodile *Hyposaurus* Owen from New Jersey. *Investigations of the New Jersey State Museum* 4:1-16.

Schoch, R. M. 1985. Preliminary description of a new late Paleocene land mammal fauna from South Carolina, U. S. A. *Postilla* 196:1-13.

Swinton, W. E. 1950. On *Congosaurus bequaerti*, Dollo. *Annales du Musée du Congo Belge* 4:1-37.

Troxell, E. L. 1925. *Hyposaurus*, a marine crocodilian. *American Journal of Science* 9:489-514.

Van Nieuwenhuise, D. S. and D. J. Colquhoun. 1982a. Contact relationships of the Black Mingo and Peedee Formations---The Cretaceous-Tertiary boundary in South Carolina, U. S. A. *South Carolina Geology* 26:1-14.

Van Nieuwenhuise, D. S. and D. J. Colquhoun. 1982b. The Paleocene-lower Eocene Black Mingo Group of the east central coastal plain of South Carolina. *South Carolina Geology* 26:47-67.

Wolfe, P. 1977. *Geology and Landscapes of New Jersey*. Rutgers University Press, New Brunswick, New Jersey.

Zittel, R. 1890. *Handbuch der Palaeontologie*. Volume 3. Paleozoologie. Munich and Leipzig, 900 pp.

PART VI
Faunas, Behavior, and Evolution

Part VI: Faunas, Behavior, and Evolution

INTRODUCTION

JUDY A. MASSARE

In today's oceans, reptiles play a very minor role. Sea turtles, a few snakes, the marine iguana, and the saltwater crocodile (*Crocodylus porosus*) represent the diversity of modern marine reptiles. Of these, only the crocodile is a large (>1-m) predator. This was likely the situation for the Cenozoic as well. During most of the Mesozoic, however, the large predators in the oceans were reptiles, although large predaceous fish and sharks became common in the Late Cretaceous. With the exception of sea turtles, all of the Mesozoic marine reptiles were carnivorous. For the most part, they caught active prey from the water column. The predators could choose from such prey as soft cephalopods, belemnoids, thin-shelled ammonoids, a variety of sharks and fish, including the bony-scaled "holostean" fish, as well as other marine reptiles. The main constraints on the kind of prey consumed were what the predator could actually catch (swimming capabilities, size) and what kind of prey the teeth could process (size of the gullet, tooth morphology). The approach taken here focuses on ecological rather than taxonomic types of predators. By looking at marine reptiles from the perspective of ecological roles through time, some interesting patterns emerge which are not as obvious if only one taxonomic group or a single fauna is examined.

SWIMMING BEHAVIOR

Marine reptiles can be characterized by four different body forms, or Baupläne, each with a distinct mode of swimming (Massare, 1994; Figure 1). Bauplan I (post-Triassic ichthyosaurs and some Triassic ichthyosaurs) was a deep, streamlined body, deepest at the pectoral region and tapering posteriorly to the peduncle of a

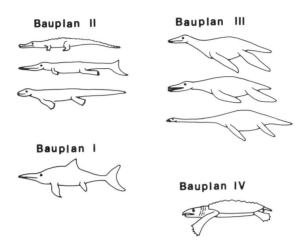

Figure 1. The four body forms, or Baupläne, displayed by Mesozoic marine reptiles.

lunate caudal fin. Propulsion was by oscillation of the tail, with the fin-like limbs used primarily for steering. The general body shape was within the optimum range to minimize total drag for efficient sustained swimming (Massare, 1988). The streamlined shape, fairly narrow caudal peduncle, high dorsal fin, and high aspect-ratio lunate tail are adaptations also found in fast, efficient swimmers in modern oceans. Bauplan I predators were probably the fastest (for their size) sustained swimmers of all of the marine reptiles, and most likely employed a pursuit strategy, actively hunting for prey over large areas (Massare, 1988, 1994).

Bauplan II (mosasaurs, marine crocodiles, thalattosaurs, pachypleurosaurs, some Triassic ichthyosaurs) was a narrow, elongate body with a long, broad tail. Propulsion was by undulation of at least part of the body along with the tail (Massare, 1988; Lingham-Soliar, 1991); thus, the tail could be undulated in a large wave to produce a burst of thrust. This body shape did not minimize total drag, but reduced pressure drag at the expense of friction drag. The added friction of the elongated body provided the resistance for a quick push against the water as the predator lunged for its prey. Bauplan II shows a number of adaptations for rapid acceleration rather than efficient sustained swimming, including an elongated body shape, a broad tail, a broad base to the tail, and an expansion of the distal end of the tail in some species. These adaptations for rapid bursts of speed rather than drag reduction suggest an ambush mode of attack for prey capture (Massare, 1988, 1994).

Bauplan III (plesiosaurs, "nothosaurids"---*sensu* Storrs, 1993) was a stiff, ellipsoidal body with two pairs of elongated limbs. The limbs were wing-shaped in the plesiosaurs. Propulsion was a modified underwater flight using the limbs, somewhat analogous to swimming in extant sea lions (Robinson, 1975; Godfrey, 1984: Halstead, 1989; Storrs, 1993). Subaqueous flying has also been suggested for some ichthyosaurs (Reiss, 1986) and a mosasaur (Lingham-Soliar, 1992), but these hypotheses have been soundly refuted (Klima, 1992; Nicholls and Godfrey, 1994). The plesiosaurs have historically been divided into the plesiosauroids and the pliosauroids (Brown, 1981a), but this distinction may be more of an ecological rather than a taxonomic one (Bakker, 1993; Cruickshank, 1994). The plesiosauroids and nothosaurids had long necks and small heads, whereas the pliosauroids had shorter necks and larger heads, resulting in more compact bodies. This probably resulted in different swimming capabilities. The more streamlined pliosauroids, with relatively larger limbs and larger limb girdles for muscle attachment, had a body shape within the range conducive to reducing total drag, suggesting that they were pursuit predators, although not as fast as the ichthyosaurs of Bauplan I (Massare, 1988, 1994). Because of their long necks, plesiosauroids had a much less compact form, and one which did not minimize the total drag. Furthermore, the long neck would present a problem in that it puts a lot of drag in front of the center of mass. Any movement of the head would cause the animal to veer off course (Alexander, 1989:137). Plesiosauroids probably compensated for this with a slower swimming speed to enable the animal to counteract any inadvertent course changes. The long neck did, however, convey an advantage: the head could encounter a school of small prey before the large body could be detected. Thus, plesiosauroids may have ambushed prey by stealth rather than a sudden lunge. Nothosaurids, with their long necks, were probably more similar to the plesiosauroids than to the pliosauroids in their mode of predation.

Bauplan IV (placodonts, sea turtles) was a dorsoventrally compressed body covered by bony armor. Propulsion was probably achieved by paddling with the limbs, although in some sea turtles, the forelimbs are modified into wings, and they propel themselves by subaqueous flight. The placodonts lacked such adaptations, and they were neither efficient sustained swimmers nor acceleration specialists. Unlike other marine reptiles, they preyed on immobile benthos, so locomotion was not critical in prey capture.

TOOTH FORM AND PREY PREFERENCE

Marine reptilian predators were also constrained by what kinds of prey their teeth could immobilize and process. Except for the placodonts, marine reptiles caught active prey from the water column, seizing it on a long row of usually homodont teeth. Most reptiles probably swallowed their prey whole, as do most marine mammals, but some pliosauroids were capable of dismembering large prey (Taylor, 1992a; Taylor and Cruickshank, 1993; Martill et al., 1994). In spite of their high taxonomic diversity, the common Mesozoic marine predators display a small number of tooth morphologies, many of which are also seen in marine mammals (Massare, 1987). All marine reptiles had fairly simple, conical teeth. The shape of the tooth crown, however, suggests qualitative differences in the diets of marine reptiles. Slender, sharply pointed teeth were used to pierce prey which had a soft exterior and few internal hard parts. Robust, blunt teeth were used to grasp and crush prey with a hard exterior. Robust, pointed teeth with sharp, sometimes serrated, cutting ridges were used for cutting or tearing fleshy prey. These three forms can be viewed as end members of a spectrum of tooth morphologies that reflect the range of morphologies and inferred functions seen in Mesozoic marine reptiles (Massare, 1987). These suggest at least the broad category of prey types which the predator could catch and process.

Six different kinds of predator guilds can be identified on the basis of tooth crown morphology (Table 1). Each guild is occupied by reptiles with a certain range of tooth morphologies and a specific range of preferred prey. The guild name reflects the main function of the teeth. There is a gradation between many of the tooth forms and prey possibilities, so these guilds can be thought of as subdivisions along a continuum of tooth form and prey possibilities. Note that several different morphologies are found within most groups. Ichthyosaurs, for example, had species with morphologies characteristic of five of the six guilds. The geosaurs occupied four of the feeding guilds at various times. Furthermore, most tooth morphologies are found in more than one order of marine reptiles. The Cut and Pierce II guilds, for example, included representatives of four of the marine reptile groups, although they did not necessarily occupy the guild at the same time.

The Smash guild is notable in that only ichthyosaurs display this tooth form, and it is the most widespread morphology among the ichthyosaurs. These teeth are very small with respect to the gullet size, and seem to be adapted for grasping rather than impaling prey (Massare, 1987). Some species with this morphology may have lacked teeth as adults (Huene, 1922, cited by McGowan, 1979:103). This is not unheard of among marine tetrapod predators. The extant beaked whales (family Ziphiidae) have, at most, two pairs of teeth. The best known species prey mainly on squid, although fish, including bottom-dwelling forms, are also consumed (Minasian et al., 1984).

Marine reptiles show a remarkable amount of convergence of tooth form among themselves and also in comparison to modern toothed whales (Massare, 1987). However, as Collin and Janis point out in Chapter 16, there is a notable absence of suspension feeders among the marine reptiles. In over 150 million years of evolution why did none of the reptilian orders evolve a species with a feeding

Faunas, Behavior, and Evolution

strategy similar to that of the very successful baleen whales? The analysis presented by Collin and Janis is an excellent example of the insight that can be achieved when one takes a broader functional approach to the study of marine reptiles.

The convergences observed in tooth form as well as body shape raise further questions. Why do tetrapods seem to be limited in the number of ways of being a large marine predator? Perhaps the tetrapod body plan limits the possible solutions to swimming and feeding in a dense medium. Or perhaps the marine habitat imposes certain physical/ecological constraints on large animals. It is significant, though, that many lineages of reptiles and mammals invaded the marine realm. Because so many unrelated lineages adapted to a marine habitat, the similarity of physical requirements, and thus selection pressures, makes them ideal for studying how evolution operates in transitions between adaptive zones. Carroll explores this in Chapter 17 and discusses what it reveals about processes of large-scale evolution.

FAUNAS THROUGH TIME

Much of our knowledge of marine reptile ecology and evolution comes from a small number of Mesozoic faunas---Lagerstätten that have a very large number of exceptionally well-preserved specimens. These faunas provide us with geologic "snapshots" of the Mesozoic oceans from the mid-Triassic through the Late Cretaceous. Such faunas provide a wealth of information on morphology and mode of life of individual species and on the reptilian community in a particular area, but this is a very limited picture of the Mesozoic seas. Nonetheless, it is the best that we have.

Six Lagerstätten are particularly important in our current understanding of Mesozoic marine reptile communities: (1) the Middle Triassic (Ladinian-Anisian) of the Monte San Giorgio region, Switzerland; (2) the Early Jurassic (Hettangian-Sinemurian) Blue Lias, Lyme Regis, England; (3) the Early Jurassic (Toarcian) Posidonienschiefer, Holzmaden, Germany; (4) the late Middle Jurassic (Callovian) Oxford Clay, Peterborough, England; (5) the Late Cretaceous (Coniacian-Campanian) Smoky Hill Member, Niobrara Chalk of Kansas; and (6) the Late Cretaceous (Campanian-Maastrichtian) Sharon Springs Member, Pierre Shale of the western interior of the United States.

Each of the above faunas represents an averaging in space and time among adjacent or temporally successive environments. The duration of time recorded and the environmental setting varies. What these assemblages have in common is that each is diverse, well preserved, and extensively collected. Although these assemblages are not all from exactly the same kind of environment, the range of predator types reveals interesting patterns that may be independent of local

Table 1. Feeding guilds of the most common Mesozoic marine reptiles.

Guild	Taxa	Tooth Crown Morphology	Prey Attributes
Cut	Ichthyosaurs Pliosauroids Geosaurs Mosasaurs	Pointed, robust cone with two or more sharp longitudinal carinae or ridges, which can be serrated. Tooth breakage is common, and the oblique break is often rounded and polished.	Large, fleshy prey that puts up a fight. Large bones break and polish the teeth.
Pierce II/ General	Pliosauroids Plesiosauroids Teleosaurs Geosaurs	Pointed, somewhat slender, curved tooth of moderate length. Two carinae or fine longitudinal ridges are often present. Worn teeth have a rounded apex or the tip may be broken.	Fleshy prey lacking a hard exterior. Prey is impaled to capture.
Pierce I	Plesiosauroids Ichthyosaurs Teleosaurs	Very slender, sharply pointed, long, delicate tooth which can be smooth or have fine longitudinal ridges. Rarely shows wear.	Very soft prey or very small, bony prey. Prey is impaled to capture, or teeth possibly used as sieve.

Table 1 (continued)

Smash	Ichthyosaurs	Small, straight or slightly curved tooth with a rounded but acute apex. Worn teeth have a blunter apex.	Externally soft prey with some internal hard parts. Prey is grasped rather than impaled.
Crunch	Ichthyosaurs Teleosaurs Geosaurs	Small, robust, straight or slightly curved tooth with a blunt apex. The apex often has a rough surface in new teeth. Older teeth have a smooth, polished apex.	Prey with bony scales or hard, thin exoskeleton Prey is grasped rather than impaled.
Crush	Placodonts Ichthyosaurs Geosaurs Mosasaurs	Very blunt, bulbous or peg-like tooth which often shows abrasion or polishing of the top surface.	Prey with a very hard exterior.

Modified from Massare (1987).

differences in environment.

Although marine reptiles are known from the Early Triassic (Callaway and Massare, 1989; Bardet, 1994), the Middle Triassic fauna of the Monte San Giorgio region, Switzerland, is the earliest well-preserved fauna. The black shales have yielded at least two genera of ichthyosaurs, two genera of placodonts, three genera of thalattosaurs, and five genera of nothosaurs, as well as some other problematic reptiles such as the extremely long-necked *Tanystropheus* (Bürgin et al., 1989; Sander, 1989). The most common ichthyosaur, *Mixosaurus*, was unusual compared to later ichthyosaurs in that it had heterodont dentition. The front teeth were slender, pointed cones, whereas the back teeth were very blunt pegs. The morphologies in a single jaw span several of the guilds previously defined! A large ichthyosaur, *Cymbospondylus*, was most likely in the Smash guild. The thalattosaurs included a fish eater, *Askeptosaurus*, probably in Pierce I or II guild, and *Clarazia* with crushing teeth (Bürgin et al., 1989). The placodonts were in the Crush guild as well. The small pachypleurosaurs were probably in the Pierce I guild, whereas the larger nothosaurs may have occupied either the Pierce I or II guild. It is notable that there was no predator in the Cut guild---no killer whale type of predator (Table 2).

The fauna is dominated by ambush predators. *Cymbospondylus* was a very elongated, narrow-bodied ichthyosaur whose gross body shape is within Bauplan II (Massare, 1988; Massare and Callaway, 1990). The thalattosaurs, with a similar body form, and the long-necked nothosaurs were also ambush predators. The placodonts were slow swimmers that caught benthic, immobile prey. The only pursuit predator was the small (1.5-m) ichthyosaur, *Mixosaurus*. Although *Mixosaurus* was somewhat elongated for an ichthyosaur (length/depth ratio of 6.6), its body shape was within the range for efficient sustained swimming (Massare, 1988). *Mixosaurus*, however, lacked the lunate tail typical of the Jurassic ichthyosaurs. Elongated neural spines toward the end of the tail suggest that it had an asymmetric caudal fin. Thus, it may not have been as fast for its size or as efficient a pursuit predator as many of the later ichthyosaurs. Its heterodont dentition may have enabled it to be more of a generalist in its feeding than were many of the later ichthyosaurs.

The marine reptilian predators underwent changes in the Late Triassic, but a clear picture of this time has yet to be developed. The largest ichthyosaurs known lived in the Late Triassic (Camp, 1980; Kosch, 1990), as did some of the smallest species (*Merriamia, Toretocnemus*). The long-bodied ichthyosaur *Cymbospondylus* persisted through the Carnian. We lack good material, however, on the smaller species, so it is not possible to determine their body forms. The placodonts, thalattosaurs, and nothosaurs were much less diverse after the Ladinian, and only one family of thalattosaurs and one family of placodonts survived past the Carnian

Table 2. Monte San Giorgio fauna, Middle Triassic. *The ichthyosaur *Mixosaurus* had heterodont teeth which span the tooth morphologies of several guilds.

Feeding Guild	Pursuit Predators	Ambush Predators
Cut	---	---
Pierce II	---	Nothosaur
		?Thalattosaur
Pierce I	*	?Nothosaur
		?Thalattosaur
		Pachypleurosaur
Smash	*	Ichthyosaur
Crunch	*	---
Crush	*	Thalattosaur
	?Placodont	?Placodont

(Bardet, 1994). The Late Triassic appears to have been a time of high diversity for ichthyosaurs (Callaway and Massare, 1989), although ambush predator diversity decreased through the Late Triassic.

The Early Jurassic fauna of Lyme Regis, England, has yielded about a half-dozen ichthyosaur species, ranging in size from the 1.5-m-long *Ichthyosaurus breviceps* to the 9-m-long *Temnodontosaurus platyodon* (McGowan, 1974a, b, 1989a, 1994a). Two other species, *Excalibosaurus costini* and *Leptopterygius solei* are also known from the lower Liassic of the region (McGowan, 1986, 1989b, 1993). The ichthyosaurs occupied four of the six feeding guilds (Massare, 1987). There was some variation in tooth form within a genus. The small *Ichthyosaurus breviceps* and *I. conybeari* occupied the Smash guild, whereas the larger *I. communis* occupied the Crunch guild. Similarly, *Temnodontosaurus platyodon* occupied the Cut guild, whereas *T. longirostris* probably occupied the Pierce I guild. Plesiosaurs were represented by two long-necked plesiosauroids and three pliosauroids (Massare, 1987). The plesiosaurs were much less diverse than the ichthyosaurs, occupying only two feeding guilds: *Plesiosaurus dolichodeirus* was in the Pierce I guild, and the others, probably including the plesiosauroid *Attenborosaurus* (Bakker, 1993), occupied the Pierce II/General guild (Massare, 1987; Cruickshank, 1994). Interestingly, there was not much difference in tooth morphology between plesiosauroids and pliosauroids at this time. The teeth of pliosauroids were more robust and had more prominent and more numerous longitudinal ridges, but these differences were minor compared to differences seen in later Jurassic forms. In fact, the plesiosauroid-pliosauroid distinction is breaking down as more Early Jurassic forms become known (Cruickshank, 1994). The Early Jurassic may have been a time of ecological divergence between the two plesiosaur lineages.

As a whole, the Lyme Regis reptilian fauna was fairly diverse, with five of six feeding guilds occupied (Table 3). Ichthyosaurs were at the top of the trophic pyramid, with a killer whale type of predator, *Temnodontosaurus platyodon*, in the Cut guild. The fauna, however, was almost entirely pursuit predators, mainly ichthyosaurs. The only ambush predators were the long-necked plesiosauroids, which could take advantage of their long necks to sneak up on prey. There were no true ambush predators (Bauplan II) that relied on rapid acceleration and a quick lunge to capture prey.

Table 3. Lyme Regis fauna, Early Jurassic.

Feeding Guild	Pursuit Predators	Ambush Predators
Cut	Ichthyosaur	---
Pierce II	Pliosauroid	?Plesiosauroid
Pierce I	Ichthyosaur	Plesiosauroid
Smash	Ichthyosaur	---
Crunch	Ichthyosaur	---
Crush	---	---

Data from Massare (1987, 1988).

The next glimpse of Mesozoic oceans are from slate quarries of southern Germany, although many specimens of comparable age are known from England (Benton and Taylor, 1984; Taylor, 1992a, b; McGowan, 1989a, 1994b). For more than 100 years, workers have unearthed a spectacular assemblage of marine reptiles from the Lias Epsilon (Toarcian) Posidonia Shales of Holzmaden (Fraas, 1902, 1910; Huene, 1922; Hauff, 1953). Nearly a dozen ichthyosaur species have been identified (Huene, 1922; McGowan, 1979, 1994b), and their taxonomic diversity is reflected in their feeding diversity. Ichthyosaurs again occupied four of the six guilds and this excludes three species (*Stenopterygius cuneiceps*, *S. macrophasma*, and *S. longipes*; McGowan, 1979) for which I lack data needed in this analysis. *Eurhinosaurus* occupied the Pierce I guild, *Leptopterygius disinteger* occupied the Crunch guild, and *L. burgundiae* (*L. acutirostris* in older literature; see McGowan, 1979) occupied the Cut guild. The remaining species, all of the genus *Stenopterygius*, occupied the Smash guild, although small *S. quadriscissus* may have occupied the Pierce I guild (Massare, 1987). Plesiosaurs were represented by one plesiosauroid, *Plesiosaurus brachypterygius* and two pliosauroids, *Rhomaleosaurus* (*Thaumatosaurus* in older literature; Tarlo, 1960) and *Macroplata* (specimen on display at Museum Hauff, Holzmaden, identified as *T. victor*). The teeth of the pleisosauroid were very long and slender with pointed, unworn tips, typical of later

Jurassic pleisosauroids and distinct from the more robust teeth of the Holzmaden pliosauroids. The plesiosaurs as a whole occupied only two feeding guilds, but a divergence between the plesiosauroids and pliosauroids had already occurred: they occupied different guilds for the remainder of the Jurassic. The biggest difference between the Holzmaden and Lyme Regis faunas was the appearance of three species of teleosaurs, marine mesosuchian crocodiles: two large species (skull length approximately 1 m) and one small species (total length 1.5 m). They occupied the Pierce II and Pierce I feeding guilds, respectively.

In the Holzmaden fauna, we again see a high diversity of pursuit predators, dominated by ichthyosaurs (Table 4). Once again, ichthyosaurs were at the top of the trophic pyramid. In general, the structure of the marine predator community was similar for both Lyme Regis and Holzmaden times, although the latter showed an expansion in diversity by the addition of ambush predators in two guilds. Guilds appear to have been stable for 10-15 million years, although there was a species level replacement within guilds during this time. It is interesting that true ambush predators did not appear for more than 10 million years, and when they finally appeared, they did not immediately radiate to fill all of the feeding guilds.

Table 4. Holzmaden fauna, Early Jurassic.

Feeding Guild	Pursuit Predators	Ambush Predators
Cut	Ichthyosaur	---
Pierce II	Pliosauroid	Teleosaur
Pierce I	Ichthyosaur	Plesiosauroid
		Teleosaur
Smash	Ichthyosaur	---
Crunch	Ichthyosaur	---
Crush	---	---

Data from Massare (1987, 1988).

In the Middle Jurassic, a major reorganization of reptilian predator guilds occurred. Reptiles within each guild changed at the ordinal level. Unfortunately, the Middle Jurassic marine fossil record is poor, but when a diverse reptile assemblage appears again, in the Middle Callovian, the ichthyosaur-dominated fauna had been replaced by a pliosauroid/crocodile-dominated one.

The Peterborough Member of the Oxford Clay Formation has produced a large assemblage of marine reptiles over the last 100 years or more of collecting (Martill et al., 1994). The fauna includes a single species of ichthyosaur, ten plesiosaurs, and at least four crocodiles (Andrews, 1910-1913; Tarlo, 1960; Adams-Tresman, 1987a, b; Brown, 1981, 1993; Martill et al., 1994). The lone ichthyosaur,

Ophthalmosaurus, was a large species, with a skull length of just over 1 m (Andrews, 1910-1913). Its small teeth suggest a Smash guild predator (Massare, 1987), although teeth are rarely preserved, leading to the suggestion that *Ophthalmosaurus* was edentulous as an adult (Andrews, 1910-1913; McGowan, 1979). Its long, narrow snout precludes the possibility that it occupied the Crunch or Crush guild. The plesiosauroids include *Muraenosaurus*, *Cryptoclidus*, and *Tricleidus* (Brown, 1981a). Although they differed in the relative proportion of the neck, all occupied the Pierce I guild. *Cryptoclidus*, however, may have used its teeth to sieve small organisms from the water (Brown, 1981b; Brown and Cruickshank, 1994; Martill et al., 1994). The four pliosauroids were larger than their Early Jurassic precursors. For the most part, their necks were shorter and their heads were larger, resulting in more compact and powerful bodies. The larger forms, *Liopleurodon* and *Pliosaurus*, occupied the Cut guild, whereas the smaller forms, *Peloneustes* and *Simolestes*, occupied the Pierce II/General guild (Massare, 1987). The marine crocodiles include members of two families, the teleosauridae and the metriorhynchidae (geosaurs). Both the geosaurs and the teleosaurs occupied the Pierce II and Crunch guilds (Massare, 1987).

The Oxford Clay fauna shows a reduction in ichthyosaur diversity compared to the Early Jurassic. Pliosaurs had become more diverse, and occupied two feeding guilds, most notably having replaced the ichthyosaurs as top predators in the Cut guild. The plesiosauroids, although represented by more species, had not changed in feeding diversity, still occupying the Pierce I guild. The major change since the Early Jurassic was an expansion in the taxonomic diversity of the ambush predators, now represented by two families of crocodiles. Thus, the ichthyosaur-dominated pursuit predator fauna of the Early Jurassic was replaced by a more diverse pliosaur/crocodile-dominated fauna with a mix of pursuit and ambush predators (Table 5).

Table 5. Oxford Clay fauna, Middle Jurassic.

Feeding Guild	Pursuit Predators	Ambush Predators
Cut	Pliosauroid	---
Pierce II	Pliosauroid	Teleosaur
		Geosaur
Pierce I	Plesiosauroid	Plesiosauroid
Smash	Ichthyosaur	---
Crunch	---	Teleosaur
		Geosaur
Crush	---	---

Data from Massare (1987, 1988).

This pattern probably continued through the Late Jurassic, although there are no Lagerstätten of this age that provide a good assemblage for comparison. There are, however, specimens that provide some clues. A large ichthyosaur, *Grendelius* (McGowan, 1976), has teeth suggestive of the Crunch guild. Another ichthyosaur, *Nannopterygius*, resembles *Ophthalmosaurus* in gross skull morphology and so may have similarly occupied the Smash guild. Plesiosauroids include *Kimmerosaurus*, with even more slender, delicate teeth than the earlier plesiosauroids. It has been suggested that *Kimmerosaurus* used its long, recurved teeth to sieve small prey from the water (Brown, 1981b). Another large plesiosauroid, *Colymbosaurus*, is known only from postcranial material, and may be synonymous with *Kimmerosaurus* (Brown et al., 1986; Brown, 1993). In either case, plesiosauroids seem to be quite conservative in their feeding morphology and gross body shape, so it is likely that these plesiosauroids occupied the same guild as the Oxford Clay genera. Kimmeridgian pliosauroids were huge, even in comparison to their Oxford Clay precursors (Tarlo, 1959, 1960; Taylor and Cruickshank, 1993; Taylor et al., 1995). All occupied the Cut guild, and probably had streamlined, compact bodies similar to those of the larger Oxford Clay species. Crocodiles continued to increase in diversity (Fraas, 1902). The large, Late Jurassic geosaur, *Dakosaurus*, with its massive skull and robust, serrated teeth was clearly capable of devouring large prey, and therefore occupied the Cut guild as one of the highest-order predators.

Thus, the reorganization of guilds that occurred by the Callovian seems to have remained stable through the Late Jurassic. Pliosauroids continued as the highest-order predators. Most of the feeding guilds were occupied by both ambush and pursuit predators; thus, faunas were diverse not only in terms of feeding types but also in terms of prey capture strategies. The expansion in diversity of ambush predators, coupled with the continued high diversity of the pursuit predators, must have caused an overall increase in predation pressure by reptiles on the various prey species. Different attack methods, and their implied differences in hunting territories, permitted animals with similar prey preferences to coexist.

The Early Cretaceous record of marine reptiles is even spottier than the Late Jurassic record. During this time there must have been a major reorganization of the large marine predator guilds. When we next see a well-preserved, diverse assemblage of marine reptiles in the Late Cretaceous, the number of guilds occupied by reptiles was greatly reduced, and reptilian pursuit predators had nearly disappeared. Exactly when or how this occurred is not clear. Ichthyosaurs are even less diverse in the Cretaceous than in the Late Jurassic, with only one known genus (McGowan, 1972). The last reliable record of ichthyosaurs is from the Cenomanian (Baird, 1984; Bardet, 1992). Large pliosauroids are known from the Early Cretaceous: *Kronosaurus* from Australia (White, 1940) and *Brachauchenius* (Williston, 1907) from the United States, but the latter at least has teeth which lack the sharp, serrated cutting ridges seen on teeth of the Late Jurassic pliosauroids. Although large pliosaurs persisted into the Cretaceous, they may not have continued in the Cut guild. Cretaceous long-necked plesiosaurs are represented by the elasmosaurs, whose relationship to the long-necked Jurassic forms is problematic (Bakker, 1993; Brown, 1993; Brown and Cruickshank, 1994; Carpenter, Chapter 7). Many elasmosaurs had more robust teeth than the Late Jurassic plesiosauroids, suggesting a shift from the Pierce I guild into the Pierce II guild. The geosaurs and teleosaurs became extinct early in the Cretaceous (Neocomian), but their diversity is greatly reduced after the Jurassic (Bardet, 1994).

The lack of an extensively collected Lagerstätten of the appropriate age makes it difficult to document when and how a reorganization of predator guilds occurred in the Cretaceous. Bardet (1994) and Bakker (1993), however, recognize two extinctions that affected marine reptiles during this interval. The latest Jurassic saw familial-level extinctions of ichthyosaurs and plesiosauroids, and a reduction in diversity of the marine crocodiles (Bardet, 1994). Ecologically, it affected a wide range of predator types. The Cenomanian-Turonian boundary records a second extinction of marine reptiles (Bakker, 1993; Bardet, 1994). Ichthyosaurs and a marine crocodile family (Pholidosauridae) became extinct, and large pliosauroids became rare (Bakker, 1993; Bardet, 1994). The mid-Cretaceous extinction, however, did not affect as many guilds as did the Late Jurassic extinction. Neither the ichthyosaurs, pliosauroids, nor crocodiles were as ecologically or taxonomically diverse in the Early Cretaceous as they had been in the Late Jurassic.

The reptilian fauna of the Smoky Hill Member of the Niobrara Chalk (Coniacian-Campanian) documents a major guild reorganization (Table 6). Mosasaurs dominate the fauna, the most common species being from the genera *Platecarpus*, *Tylosaurus*, and *Clidastes* (Russell, 1967; Carpenter, 1990). Most mosasaurs occupied the Cut guild, although the rare *Ectenosaurus* occupied the Pierce II guild (Massare, 1987). As ambush predators, mosasaurs may have been opportunistic feeders, generalists, as well as being at the top of the trophic pyramid. Mosasaurs shared the seas with the elasmosaurs *Styxosaurus*, *Elasmosaurus*, and possibly another genus (Welles, 1952; Carpenter, 1990 and personal communication, 1995). Their teeth were much more robust than those of the Oxford Clay plesiosauroids, suggesting that they occupied the Pierce II guild, but the lack of information on some species does not permit ruling out the Pierce I guild as a possibility. A small pliosauroid, *Dolichorhynchops*, occupied the Pierce II guild. A larger pliosauroid, *Polycotylus*, is not well known, but its teeth suggest that it shared the Pierce II guild with the smaller pliosauroid (Massare, 1987).

The Sharon Springs Member of the Pierre Formation (Campanian-Maastrichtian) is similar to the Niobrara Chalk in the kinds of reptilian predators (Table 7). Five mosasaur species have been found (Carpenter, 1990 and

Table 6. Niobrara Chalk fauna, Late Cretaceous.

Feeding Guild	Pursuit Predators	Ambush Predators
Cut	---	Mosasaur
Pierce II	Pliosauroid	Mosasaur
		Elasmosaur
Pierce I	---	?Elasmosaur
Smash	---	---
Crunch	---	---
Crush	---	---

Data from Massare (1987, 1988).

Table 7. Pierre Shale fauna, Late Cretaceous.

Feeding Guild	Pursuit Predators	Ambush Predators
Cut	---	Mosasaur
Pierce II	Pliosauroid	?Mosasaur
		Elasmosaur
Pierce I	---	?Elasmosaur
Smash	---	---
Crunch	---	---
Crush	---	Mosasaur

Modified from Massare (1983).

personal communication, 1995), including *Globidens*, which marks the first appearance of reptilian predators in the Crush guild since the Triassic. There were two elasmosaurs, *Elasmosaurus* and *Styxosaurus,* and one small pliosauroid, *Dolichorhynchops* (Carpenter, 1990 and personal communication, 1995). *Elasmosaurus* was so microcephalic that it must have occupied the Pierce I guild. *Styxosaurus* and *Dolichorhynchops* occupied the Pierce II guild. As in the Niobrara fauna, the pliosauroid was the only reptilian pursuit predator. Note that the Late Cretaceous plesiosaurs did not show a marked difference in feeding guilds as they did in the Middle and Late Jurassic.

The Late Cretaceous faunas were dominated by ambush predators. The top reptilian carnivores were ambush predators, namely, mosasaurs. Reptilian pursuit predators occupied only one guild, and they were much less common than the ambush predators. In both the Niobrara and Pierre formations, however, large (>1-m) fish and large sharks were much more abundant and clearly a more important component of the large predator community than previously in the Mesozoic. It may well be that the fish and sharks replaced the reptiles in many of the pursuit predator guilds. The terminal Cretaceous extinction eliminated only a few ecological types of reptilian predators, but it ended the reign of reptiles in the oceans.

An interesting pattern seems to be emerging. During certain times in the Mesozoic, ambush predators dominate the large predator communities (Middle Triassic, Late Cretaceous). At other times (Early Jurassic), pursuit predators dominate as they do in today's oceans. At still other times (late Middle-Late Jurassic), there is an almost even mixture of ambush and pursuit predators, filling nearly all of the feeding guilds (Table 8). Does this reflect differing conditions in the world's oceans? Whether it does or not, the differences point out the biostratigraphic potential of marine reptiles, at least on a very gross scale. As Lucas argues in Chapter 14, marine reptiles may be of biostratigraphic value on an even finer scale that has not yet been developed to its potential.

Table 8. Change in types of reptilian predators throughout the Mesozoic. "Top Predator" refers to the highest-order predator in the Cut guild.

Fauna	Age	Pursuit Guilds	Top Pursuit Predator	Ambush Guilds	Top Ambush Predator
Monte San Giorgio	Middle Triassic	?1	none	4	none
Lyme Regis	Early Jurassic	5	Ichthyosaur	1-2	none
Holzmaden	Early Jurassic	5	Ichthyosaur	2	none
Oxford Clay	Middle Jurassic	4	Pliosauroid	3	none
Kimmeridgian of Europe	Late Jurassic	5	Pliosauroid	4	Geosaur
Niobrara Chalk	Late Cretaceous	1	(Shark)	2-3	Mosasaur
Pierre Shale	Late Cretaceous	1	(Shark)	3-4	Mosasaur

How real is the observed variation in diversity? The observed pattern of predator types might be an artifact of sampling biases. Although the analysis focuses on the range of predator types rather than on the proportion of each type, biases can arise from the duration of time represented, the geographic area sampled, and the total number of specimens collected. As it happens, however, the least diverse faunas from the Pierre Shale and Niobrara Chalk represent the longest duration of time and a very wide sampling area (at least on the order of tens of square miles). One of the most diverse faunas, the Oxford Clay fauna, is the one of the most geographically and temporally restricted. So perhaps these kinds of sampling biases are less critical than it might initially appear.

One bias cannot be ignored: all of these faunas are from the northern hemisphere. Specimens and faunas have been described from the southern continents but we lack well-known, extensively collected faunas comparable to those discussed here. What was happening in the southern oceans? Studies such as the one in Chapter 15 by Gasparini and Fernandez are very important to our understanding of marine reptile diversity, biogeography, and evolution. Data from the southern oceans will be critical in evaluating how well our long-known Lagerstätten from Europe and North America actually represent the diversity of predator types in the oceans.

It is evident that in spite of numerous specimens, some spectacular preservation, and dedicated collecting there are tremendous geographic and stratigraphic gaps in our knowledge of marine reptiles. The Early Triassic, early Middle Jurassic, and the Early Cretaceous are especially problematic (Bardet, 1994). It is particularly important to realize that most of our knowledge and most of our specimens come from a very limited number of locations and horizons.

Given these limitations, what can be concluded about the distribution and composition of guilds? Our present knowledge indicates that morphological types are not "evenly spaced" within the range of possibilities. At times there appear to be many species in some guilds and none in others. This suggests that competition for food is not a critical factor, at least at this large scale. Furthermore, frequently there is room ecologically for more kinds of predators than are actually present. Guild composition may be largely a matter of chance---which taxa happened to evolve what adaptations at a particular time, influenced, of course, by what survived the last extinction. On the other hand, occurrences may be tempered by environmental or ecological conditions favoring certain kinds of predators at different times. Clearly, more work is needed.

ACKNOWLEDGMENTS

I thank Drs. M. A. Taylor, D. S. Brown, and E. L. Nicholls for their help in locating references and K. Carpenter for information on the Pierre and Niobrara plesiosaurs. An earlier version of this research was part of a Ph. D. dissertation submitted to the Johns Hopkins University, and I again thank the many individuals and museums that helped make that work possible. Thanks also go to Dr. M. A. Taylor and Dr. J. M. Callaway for their very helpful reviews of this introduction.

REFERENCES

Adams-Tresman, S. M. 1987a. The Callovian (Middle Jurassic) marine crocodile *Metriorhynchus* from central England. *Palaeontology* 30:179-194.

Adams-Tresman, S. M. 1987b. The Callovian (Middle Jurassic) teleosaurid marine crocodiles from central England. *Palaeontology* 30:195-206.

Alexander, R. M. 1989. *Dynamics of Dinosaurs and Other Extinct Giants*. Columbia University Press, New York.

Andrews, C. W. 1910-1913. *A Descriptive Catalogue of the Marine Reptiles of the Oxford Clay*, 2 vols. British Museum, London, 411 pp.

Baird, D. 1984. No ichthyosaurs in the Upper Cretaceous of New Jersey... or Saskatchewan. *The Mosasaur* 2:129-133.

Bakker, R. T. 1993. Plesiosaur extinction cycles---events that mark the beginning, middle, and end of the Cretaceous. IN W. G. E. Caldwell and E. G. Kauffman (Eds.), *Evolution of the Western Interior Basin*, pp. 641- 664. Geological Association of Canada Special Paper 39.

Bardet, N. 1992. Stratigraphic evidence for the extinction of the ichthyosaurs. *Terra Nova* 4:649-656.

Bardet, N. 1994. Extinction events among Mesozoic marine reptiles. *Historical Biology* 7:313-324.

Benton, M. J. and M. A. Taylor. 1984. Marine reptiles from the Upper Lias (Lower Toarcian, Lower Jurassic) of the Yorkshire coast. *Proceedings of the Yorkshire Geological Society* 44:399-429.

Brown, D. S. 1981a. The English Upper Jurassic Plesiosauria (Reptilia) and a review of the phylogeny and classification of the Plesiosauria. *Bulletin of the British Museum (Natural History)*, Geology Series 35:253-347.

Brown, D. S. 1981b. Dental morphology and function of plesiosaurs [Abstract 146]. *Journal of Dental Research* 60:1114.

Brown, D. S. 1993. A taxonomic reappraisal of the families Elasmosauridae and Cryptoclididae (Reptilia: Plesiosauria). *Revue de Paléobiologie, Vol. Spéciale* 7:9-16.

Brown, D. S. and A. R. I. Cruickshank. 1994. The skull of the Callovian plesiosaur *Cryptoclidus eurymerus* and the sauropterygian cheek. *Palaeontology* 37:941-953.

Brown, D. S., A. C. Milner, and M. A. Taylor. 1986. New material of the plesiosaur *Kimmerosaurus langhami* Brown from the Kimmeridge Clay of Dorset. *Bulletin of the British Museum (Natural History)*, Geology Series 40:225-234.

Bürgin, T., O. Rieppel, P. M. Sander, and K. Tschanz. 1989. The fossils of Monte San Giorgio. *Scientific American* 260 (6):64-81.

Callaway, J. M. and J. A. Massare. 1989. Geographic and stratigraphic distribution of the Triassic Ichthyosauria (Reptilia, Diapsida). *Neues Jahrbuch für Geologie und Paläontologie*, Abhandlungen 178:37-58.

Camp, C. 1980. Large ichthyosaurs from the Upper Triassic of Nevada. *Palaeontographica* A 170:139-200.

Carpenter, K. 1990. Upward continuity of the Niobrara fauna with the Pierre Shale fauna.

IN S. C. Bennett (Ed.), *Niobrara Chalk Excursion Guidebook*, pp. 73-81. Society of Vertebrate Paleontology.

Cruickshank, A. R. I. 1994. A juvenile plesiosaur (Plesiosauria: Reptilia) from the Lower Lias (Hettangian: Lower Jurassic) of Lyme Regis, England: a pliosauroid-plesiosauroid intermediate? *Zoological Journal of the Linnean Society* 112:151-178.

Fraas, E. 1902. Die Meer-Krokodilier (Thalattosuchia) des oberen Jura unter specieller Berucksichtigung von *Dacosaurus* und *Geosaurus*. *Palaeontographica* A 49, pt.1:1-72.

Fraas, E. 1910. Plesiosaurier aus dem oberen Lias von Holzmaden. *Palaeontographica* A 57:105-140.

Godfrey, S. J. 1984. Plesiosaur subaqueous locomotion: a reappraisal. *Neues Jahrbuch für Geologie und Paläontologie*, Monatshefte 11:661-672.

Halstead, L. B. 1989. Plesiosaur locomotion. *Journal of the Geological Society of London* 146:37-40.

Hauff, B. 1953. *Das Holzmadenbuch*. Verlag der Hohenlohe'schen Buchhandlung, F. Rau Öhringen, 45 pp.

Huene, F. von. 1922. Die Ichthyosaurier des Lias und ihre Zusammenhange. *Monographien zur Geologie und Paläontologie*, Serie 1, Heft 1, 114 pp.

Klima, M. 1992. Schwimmbewegungen und Auftauchmodus bei Whalen und bei Ichthyosaurien. I. Anatomische Grundlagen der Schwimmbewegungen. *Natur und Museum, Frankfurt* 122:1-17.

Kosch, B. F. 1990. A revision of the skeletal reconstruction of *Shonisaurus popularis* (Reptilia; Ichthyosauria). *Journal of Vertebrate Paleontology* 10:512-514.

Lingham-Soliar, T. 1991. Locomotion in mosasaurs. *Modern Geology* 16:229-248.

Lingham-Soliar, T. 1992. A new mode of locomotion in mosasaurs: subaqueous flying in *Plioplatecarpus marshi*. *Journal of Vertebrate Paleontology* 12:405-421.

Martill, D. M., M. A. Taylor, and K. L. Duff with contributions by J. B. Reading and P. R. Brown. 1994. The trophic structure of the biota of the Peterborough Member, Oxford Clay Formation (Jurassic), UK. *Journal of the Geological Society, London* 151:173-194.

Massare, J. A. 1983. *Ecology and Evolution of Mesozoic Marine Reptiles*. Unpublished Ph. D. dissertation, The Johns Hopkins University, Baltimore, Maryland.

Massare, J. A. 1987. Tooth morphology and prey preference of Mesozoic marine reptiles. *Journal of Vertebrate Paleontology* 7:121-137.

Massare, J. A. 1988. Swimming capabilities of Mesozoic marine reptiles: implications for method of predation. *Paleobiology* 14:187-205.

Massare, J. A. 1994. Swimming capabilities of Mesozoic marine reptiles: a review. IN L. Maddock, Q. Bone, and J. M. V. Rayner (Eds.), *Mechanics and Physiology of Animal Swimming*, pp. 133-149. Cambridge University Press, Cambridge.

Massare, J. A. and J. M. Callaway. 1990. The affinities and ecology of Triassic ichthyosaurs. *Bulletin, Geological Society of America* 102:409-416.

McGowan, C. 1972. The systematics of Cretaceous ichthyosaurs with particular reference to the material from North America. *Contributions to Geology, University of Wyoming* 11:19-29.

McGowan, C. 1974a. A revision of the longipinnate ichthyosaurs of the Lower Jurassic of England, with descriptions of two new species (Reptilia: Ichthyosauria). *Life Sciences Contributions, Royal Ontario Museum* 97, 37 pp.

McGowan, C. 1974b. A revision of the latipinnate ichthyosaurs of the Lower Jurassic of England. *Life Sciences Contributions, Royal Ontario Museum* 100, 30 pp.

McGowan, C. 1976. The description and phenetic relationships of a new ichthyosaur genus from the Upper Jurassic of England. *Canadian Journal of Earth Sciences* 13:668-683.

McGowan, C. 1979. A revision of the Lower Jurassic ichthyosaurs of Germany, with descriptions of two new species. *Paleontographica* A 166:93-135.

McGowan, C. 1986. A putative ancestor for the swordfish-like ichthyosaur *Eurhinosaurus*. *Nature* 322:454-456.

McGowan, C. 1989a. Computed tomography reveals further details of *Excalibosaurus*, a putative ancestor for the swordfish-like ichthyosaur *Eurhinosaurus*. *Journal of Vertebrate Paleontology* 9:269-281.

McGowan, C. 1989b. *Leptopterygius tenuirostris* and other long-snouted ichthyosaurs from the English Lower Lias. *Palaeontology* 32:409-427.

McGowan, C. 1993. A new species of large, long-snouted ichthyosaur from the English lower Lias. *Canadian Journal of Earth Sciences* 30:1197-1204.

McGowan, C. 1994a. *Temnodontosaurus risor* is a juvenile of *T. platyodon* (Reptilia: Ichthyosauria). *Journal of Vertebrate Paleontology* 14:472-479.

McGowan, C. 1994b. The taxonomic status of the Upper Liassic ichthyosaur *Eurhinosaurus longirostris*. *Palaeontology* 37:747-753.

Minasian, S. M., K. C. Balcomb III, and L. Foster. 1984. *The World's Whales*. Smithsonian Books, Washington, D. C., 223 pp.

Nicholls, E. L. and S. J. Godfrey. 1994. Subaqueous flight in mosasaurs---a discussion. *Journal of Vertebrate Paleontology* 14:450-452.

Reiss, J. 1986. Fortbewegungsweise, Schwimmbiophysik, und Phylogenie der Ichthyosaurier. *Palaeontographica* A 192:93-155.

Robinson, J. A. 1975. The locomotion of plesiosaurs. *Neues Jahrbuch für Geologie und Paläontologie*, Abhandlungen 153:86-128.

Russell, D. 1967. Systematics and morphology of North American mosasaurs. *Peabody Museum of Natural History, Bulletin* 23:240 pp.

Sander, P. M. 1989. The pachypleurosaurids (Reptilia: Nothosauria) from the Middle Triassic of Monte San Giorgio (Switzerland) with the description of a new species. *Philosophical Transactions of the Royal Society of London*, Series B 325:561-666.

Storrs, G. 1993. Function and phylogeny in sauropterygian (Diapsida) evolution. *American Journal of Science* 293A:63-90.

Tarlo, L. B. 1959. *Stretosaurus* gen. nov., a giant pliosaur from the Kimmeridge Clay. *Palaeontology* 2:39-55.

Tarlo, L. B. 1960. A review of the Upper Jurassic pliosaurs. *Bulletin of the British Museum (Natural History), Geology Series* 4:145-189.

Taylor, M. A. 1992a. Functional anatomy of the head of the large aquatic predator *Rhomaleosaurus zetlandicus* (Plesiosauria, Reptilia) from the Toarcian (Lower Jurassic) of Yorkshire, England. *Philosophical Transactions of the Royal Society of London* B 335:247-280.

Taylor, M. A. 1992b. Taxonomy and taphonomy of *Rhomaleosaurus zetlandicus* (Plesiosauria, Reptilia) from the Toarcian (Lower Jurassic) of the Yorkshire coast. *Proceedings of the Yorkshire Geological Society* 49:49-55.

Taylor, M. A. and A. R. I. Cruickshank. 1993. Cranial anatomy and functional morphology of *Pliosaurus brachyspondylus* (Reptilia: Plesiosauria) from the Upper Jurassic of Westbury, Wiltshire. *Philosophical Transactions of the Royal Society of London* B 341:399-418.

Taylor, M. A., D. B. Hill, and A. R. I. Cruickshank. 1995. The first Westbury pliosaur, *Pliosaurus brachyspondylus*, from the Kimmeridge Clay of Westbury, Wiltshire. *Wiltshire Archaeological and Natural History Magazine* 88:141-146.

Welles, S. P. 1952. A review of North American Cretaceous elasmosaurs. *University of California Publications in Geological Sciences* 29(3):47-144.

White, T. 1940. On the skull of *Kronosaurus queenslandicus* Longman. *Boston Society for Natural History, Occasional Papers* 8:219-228.

Williston, S. W. 1907. The skull of *Brachauchenius*. *United States National Museum Proceedings* 32:477-489.

Chapter 14

MARINE REPTILES AND MESOZOIC BIOCHRONOLOGY

SPENCER G. LUCAS

INTRODUCTION

The standard global chronostratigraphic scale (SGCS) for the Mesozoic is strongly rooted in a robust marine biochronology developed principally for ammonoids, bivalves, conodonts (in the Triassic), foraminiferans, dinoflagellates, and calcareous nannoplankton (e.g., Harland et al., 1990). The fossil record of marine reptiles spans nearly the entire Mesozoic (Figure 1), but has played virtually no role in correlation and biochronology. This omission is surprising when one considers the broad geographic distribution, short stratigraphic ranges, and relative ease of identification of some Mesozoic marine reptile taxa. These characteristics should make them ideal index taxa, but little effort has been made to utilize them biochronologically. Indeed, in Mesozoic biochronology marine reptiles do not play second fiddle to the marine invertebrates; they are not even in the orchestra!

This chapter explores the problems of the Mesozoic marine reptile record that have limited its biochronological utility. These are problems of taxonomy, distribution, and neglect. Despite them, marine reptiles can be used in correlation and thereby contribute to the Mesozoic SGCS, as the two examples that conclude this chapter demonstrate.

MESOZOIC MARINE REPTILES AND BIOCHRONOLOGY

Marine reptiles of the Mesozoic were the Placodontia, Nothosauria, Mosasauria,

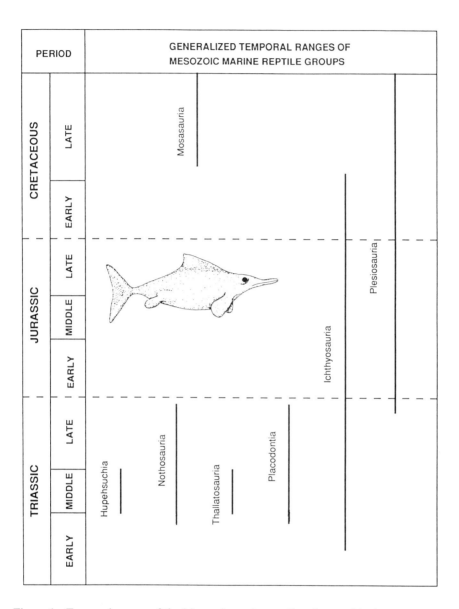

Figure 1. Temporal ranges of the Mesozoic marine reptiles discussed in the text.

Thallatosauria, Ichthyosauria, Plesiosauria, and Hupehsuchia (Figure 1). Some Mesozoic turtles and crocodilians also had a marine habitus, but they are not considered here.

Hupehsuchia consists of two genera from a single Middle Triassic horizon in Hubei Province, People's Republic of China (Carroll and Dong, 1991). Until hupehsuchians are discovered elsewhere, they are of no biochronological significance.

Nothosaurs are best known from the Middle Triassic of Europe and China, although their stratigraphic range extends into the Upper Triassic, and two genera *are* known from the late Early Triassic (Storrs, 1991). Nothosaur genera are endemic to one continent, but the stratigraphic ranges of European nothosaurs are well established and of some regional biochronological value (Figure 2).

Thallatosaurs are known only from the Middle Triassic of Switzerland and western North America (e.g., Kuhn-Schnyder, 1974). Genera are endemic to both regions, so the biochronological significance of thallatosaurs is limited.

Placodonts are known only from the uppermost Lower, Middle, and Upper Triassic of Europe, North Africa, and the Middle East (Pinna and Mazin, 1993); their distribution was restricted to the western periphery of Triassic Tethys.

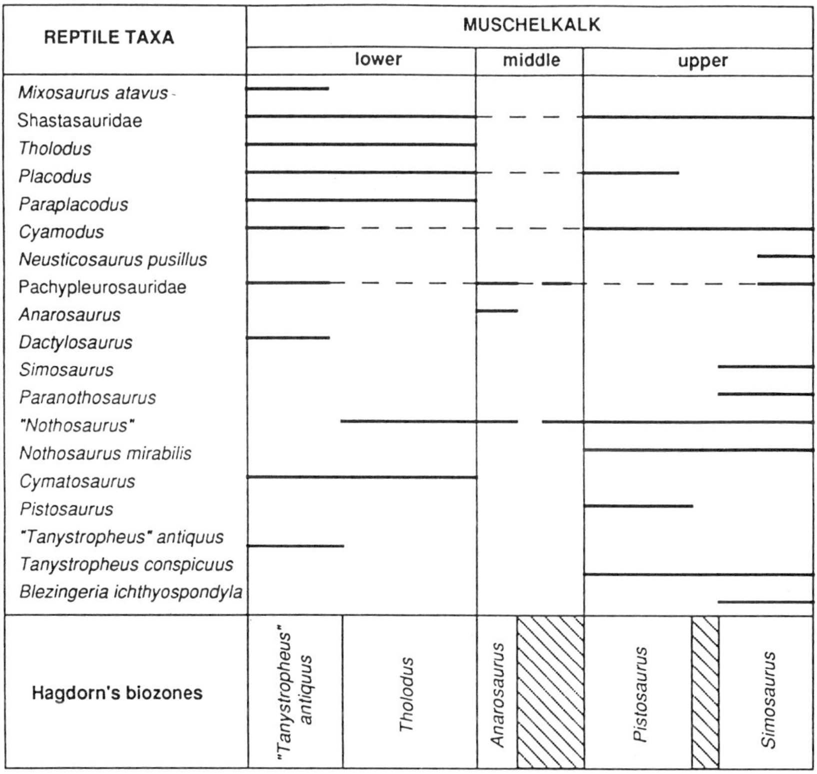

Figure 2. Stratigraphic ranges and reptile biozones from the German Middle Triassic Muschelkalk (after Hagdorn, 1993).

Virtually all the genera are from Europe, where they have established stratigraphic ranges of regional significance (Figure 2). However, the geographic restriction of placodonts makes them insignificant to global biochronology.

Ichthyosaurs appeared during the Early Triassic and persisted until the beginning of the Late Cretaceous (late Cenomanian) (Bardet et al., 1994). These most specialized and fishlike of the marine reptiles achieved a global distribution by the Middle Triassic (see below). Many ichthyosaur genera are well known from widespread localities, and taxonomy relies not just on cranial features but on the highly distinctive and durable vertebral centra. Ichthyosaurs thus have strong potential for Mesozoic biochronology, the highest of all the Mesozoic marine reptile groups.

Plesiosaurs first appeared during the Late Triassic (Rhaetian) and persisted until the end of the Cretaceous (Figure 1). During the Jurassic and Cretaceous, plesiosaurs had an essentially global distribution, but most genera as now delineated were endemic to one continent (e.g., Welles, 1962; Brown, 1981). Plesiosaur taxonomy has long relied on features of the durable elements of the limb girdles, in addition to cranial and vertebral characters. This means isolated elements can often be identified precisely, and the potential for a plesiosaur-based biochronology is high, though limited to regional correlation because of endemicity.

Mosasaurs were marine lizards of the Late Cretaceous that first appeared during the Cenomanian (Russell, 1967, 1993). Like plesiosaurs, most mosasaur taxa as now delineated were endemic to one region. Yet, in the North American Western Interior, some mosasaur taxa have temporal ranges shorter than a stage/age (Bell, 1985 and personal communication, 1994). This demonstrates that mosasaurs have strong regional potential in biochronology.

To summarize, the following ranking (highest to lowest) of the potential of the different Mesozoic marine reptiles for global and regional biochronology can be made: ichthyosaurs, plesiosaurs, mosasaurs, placodonts, nothosaurs, thallatosaurs, and hupehsuchians. Ichthyosaurs have the highest global and regional potential. Plesiosaurs have strong regional potential, as do mosasaurs. Placodonts have strong potential in only one region. Nothosaurs have some global and regional potential, but thallatosaurs and hupehsuchians are of no biochronological utility.

TAXONOMIC PROBLEMS

A good index fossil must be easy to identify. Therefore, it must not only be morphologically distinctive, but there must be a sound taxonomy with which to identify it. This taxonomy must distinguish taxa strictly by unique morphological and metric features. Stratigraphic position cannot be used to differentiate taxa.

Marine reptile taxa are typically most readily identified from whole skeletons or complete skulls. Yet, most fossils of marine reptiles are fragments of skulls or isolated postcranial bones. For some groups (e.g., ichthyosaurs, plesiosaurs), genus-level identifications can be made from certain isolated postcrania, but for the most part precise identifications require more complete material, especially skulls.

This problem afflicts much of fossil vertebrate taxonomy, except for mammals, which can usually be identified precisely from their isolated cheek teeth. Yet, it has not hindered the development of detailed terrestrial reptile biochronologies for parts of the Mesozoic and so should not hinder a biochronology based on marine reptiles.

A maximum effort is needed by paleontologists revising the taxonomy of Mesozoic marine reptiles to extract taxonomically useful features from postcrania. Head-hunting taxonomy---a strict reliance on cranial features---must be superseded. Also needed are broad-based taxonomic revisions of marine reptile groups. For example, a mosasaur taxonomy that is based on a study of all the mosasaurs, not just those from the North American Western Interior or from Belgium, is necessary to establish endemism and cosmopolitanism. It is mandatory that stratigraphic position does not figure---explicitly or implicitly---into taxonomic assignments.

DISTRIBUTIONAL PROBLEMS

Good index fossils have two key distributional attributes---they are widespread (geography) and they represent short-lived taxa (stratigraphy).

Geography

Restricted geographic distribution is the single greatest problem facing those who will use the marine reptile record for Mesozoic biochronology. Most marine reptiles lived in both epeiric seas and the oceanic basins, or at least some must have crossed those basins. Yet, most of the marine rock record exposed on the continents is that of relatively shallow epeiric seas, not of abyssal oceanic sediments. This means a relatively high percentage of marine reptile fossils will never be found. For the Mesozoic prior to the Middle Jurassic, subduction has eliminated virtually all the oceanic crust, and during this interval epeiric seas were at a minimum during the Pangaean lowstand. Therefore, the Triassic and Early Jurassic record of marine reptiles must be less complete than that of the later Mesozoic.

Nothing can be done to alleviate this problem. Real endemism and provinciality do exist for some Mesozoic marine reptiles and will, as in other fossil groups, limit their use in global biochronology. However, paleontologists can avoid false endemism and provinciality based on what Cooper (1982) aptly termed "provincial taxonomy." The geographic location of a Mesozoic marine reptile fossil should not affect taxonomic assignment. Taxonomic revisions of marine reptiles need to be based broadly on a consideration of their complete geographic record. Such revisions are fraught with practical problems (they require much funding, time, and travel), but only they can produce a taxonomy with which to gauge accurately endemism and provinciality.

Stratigraphy

A good index fossil has a short stratigraphic range so that its stratigraphic distribution defines a correspondingly short interval of geologic time. Some Mesozoic marine reptile taxa have short stratigraphic ranges, some do not, and we do not know the precise stratigraphic ranges of many taxa. Careful stratigraphic organization of much of the fossil record of Mesozoic marine reptiles needs to take place in order to establish stratigraphic ranges.

NEGLECT

Mesozoic marine reptiles are among the most unusual and fascinating animals to have ever lived. They have an extensive fossil record. This record includes outstanding Lagerstätten, such as the Jurassic Holzmaden ichthyosaurs of Germany, that present the paleontologist with complete and exquisitely preserved fossils. For these reasons, much attention has been lavished on the study of Mesozoic marine reptiles, particularly on their taxonomy, phylogeny, and paleobiology. However, the biochronological value of Mesozoic marine reptiles has been much less studied. Such neglect, of course, is easily remedied by careful taxonomic revisions and stratigraphic organization aimed at the biochronological exploitation of the record of Mesozoic marine reptiles.

EXAMPLES

Despite the problems just discussed, Mesozoic marine reptiles have been used in biochronology by some workers, and that use is increasing. For example, Russell (1993) recently defined four marine vertebrate ages (Trinitian, Woodbinian, Niobraran, and Navesinkian), noting biostratigraphically significant horizons based on marine reptile occurrences that help to define these biochronological units. Thus,

Trinitian is in part defined by the last North American ichthyosaurs; Woodbinian, by the absence of marine reptiles; Niobraran, by the first mosasaurs; and Navesinkian, by the index fossils *Mosasaurus maximus* and *Plioplatecarpus*, its beginning by the first appearance of *M. conodon* and *M. missouriensis*, and its end by the last mosasaur.

I applaud this work, and that of Bell (1985), among others, who are showing the way to use Mesozoic marine reptiles in biochronology. The following two examples demonstrate the biochronological utility of Mesozoic marine reptiles at the regional and the global scale.

Muschelkalk Marine Reptiles

Hagdorn (1993) recently published the stratigraphic ranges of all reptiles known from the marine Middle Triassic Muschelkalk of Germany (Figure 2). These reptiles have been collected and studied for well over a century and can be directly correlated to conodont and ammonite (ceratite) zonations. Hagdorn's range chart thus represents a robust regional biostratigraphic record.

Hagdorn (1993) defined five "mögliche biozonen mariner Reptilien" from these data, but his *Tholodus* zone does not correspond to the stratigraphic range of *Tholodus* and can be abandoned. Nevertheless, four obvious range zones, a variety of interval zones, and other biostratigraphically useful horizons can be identified from Hagdorn's compilation. The temporal resolution of the Muschelkalk marine reptiles does not rival that of conodonts, which define 10 biozones in the Muschelkalk (Kozur, 1974), but it does produce temporal resolution below the level of stage/age for regional correlations.

Triassic Ichthyosaur Biochronology

Using data compiled by Callaway and Massare (1989), with recent additions from Mazin et al. (1991), Brinkman et al. (1992), Nicholls and Brinkman (1993), Mazin and Sander (1993), Sander and Mazin (1993), Massare and Callaway (1994), and Lucas and González-Léon (1995), the temporal ranges of Triassic ichthyosaur genera can be plotted (Figure 3). This biochronological organization identifies three intervals of Triassic time (labeled A, B, and C), as was long the case with Triassic terrestrial tetrapods (Lucas, 1990).

The oldest known ichthyosaurs are generically indeterminate fossils from early Olenekian (Smithian) strata in British Columbia and Svalbard (Cox and Smith, 1973; Callaway and Brinkman, 1989). The oldest generically identifiable material

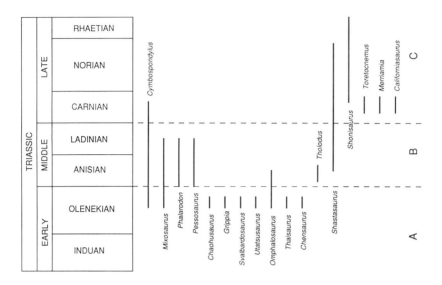

Figure 3. Temporal ranges of Triassic ichthyosaur genera (from data in Callaway and Massare, 1989; Mazin et al., 1991; Nicholls and Brinkman, 1993; Mazin and Sander, 1993; Sander and Mazin, 1993; Massare and Callaway, 1994; Lucas and González-León, 1995). Note that the Chinese Late Triassic giant shastasaurid genera *Tibetosaurus* and *Himalayasaurus* are not included because both are nomina dubia. Also note that contrary to Brinkman et al. (1992), *Phalarodon* is a valid taxon (J. M. Callaway, personal communication, 1994).

is of late Olenekian (Spathian) age from widespread localities---Anhui, China (*Chaohusaurus*); Svalbard (*Grippia, Svalbardosaurus, Omphalosaurus*); Honshu, Japan (*Utatsusaurus*); Idaho and Nevada (*Omphalosaurus, Cymbospondylus*); British Columbia (*Grippia, Utatsusaurus*); and Thailand (*Thaisaurus*). The oldest *Mixosaurus* is from the late Olenekian of British Columbia (Callaway and Brinkman, 1989). The lack of cosmopolitanism of these primitive ichthyosaur genera---most are known from only a single location---renders them of little biochronological utility, although the Early Triassic can be identified as a time of primitive ichthyosaurs. However, if Motani's (1994) claim (in an abstract) that *Grippia* is a senior subjective synonym of *Utatsusaurus* and *Chaohusaurus* is correct, then a *Grippia* biochron of Olenekian age can be widely recognized.

This changes dramatically during the Middle Triassic (interval B) when the cosmopolitan genus *Mixosaurus* is known from Alaska, the Northwest Territories, British Columbia, Nevada, Timor, Turkey, China, Italy, Switzerland, Germany, France, Poland, and Svalbard (Callaway and Massare, 1989). Like *Mixosaurus*, *Cymbospondylus* has its earliest (single-locality) occurrence during the late

Olenekian (Massare and Callaway, 1994), but has a cosmopolitan distribution during the Middle Triassic. This parallels the greater cosmopolitanism of ammonoids during the Middle Triassic (Dagys, 1988). Most of the other Middle Triassic ichthyosaurs are restricted to the Germanic basin Muschelkalk and/or Svalbard. The Middle Triassic thus corresponds to a *Mixosaurus-Cymbospondylus* abundance biochron; species-level revisions of these genera might produce even more refined correlations within this interval (Brinkman et al., 1992).

The Late Triassic shastasaurids were also a cosmopolitan group with principal occurrences in Oregon, Nevada, California, Sonora, New Zealand, New Caledonia, Russia, Tibet, Switzerland, and Austria (Callaway and Massare, 1989). Most of these are occurrences of *Shastasaurus* or of generically indeterminate large shastasaurid fossils. A shastasaurid biochron---interval C of Figure 3---encompasses the Late Triassic, one which can hopefully be refined and subdivided with further collecting and more precise taxonomy.

In conclusion, Triassic ichthyosaurs discriminate three time intervals: (A) Early Triassic---primitive ichthyosaurs;(B) Middle Triassic---*Mixosaurus-Cymbospondylus* abundance biochron; and (C) Late Triassic---shastasaurid biochron. Ichthyosaurs resolving three intervals of Triassic time equivalent to the three Triassic Series may not seem like impressive resolution, but paleontologists should well remember that as recently as 1975 terrestrial reptile biochronology only resolved the same three intervals of Triassic time globally (Lucas, 1990). Twenty years of patient work by terrestrial biostratigraphers has improved this resolution, and the Triassic ichthyosaur record is ripe for such improvement.

SUMMARY

The marine Mesozoic biochronology of the standard global chronostratigraphic scale is based primarily on the stratigraphic succession of ammonoid cephalopods and several microfossil groups, especially conodonts (in the Triassic), calcareous nannoplankton, dinoflagellates, and planktic foraminiferans. Marine reptiles have played essentially no role in Mesozoic biochronology. Three Mesozoic marine reptile groups---Olenekian-Cenomanian Ichthyosauria, Rhaetian-Maastrichtian Plesiosauria, and Cenomanian-Maastrichtian Mosasauria---have the greatest biochronological potential by virtue of their temporal and geographic distribution, relative abundance, and intensity of study. Nevertheless, the use of ichthyosaurs, plesiosaurs, and mosasaurs in Mesozoic biochronology is greatly diminished by the general incompleteness of their fossils in a taxonomic context that relies on a whole skull or a large part of the postcrania for precise identification. Endemism of Mesozoic marine reptiles also hinders their biochronological utility. Furthermore, paleontologists have generally neglected to use Mesozoic marine reptiles in biochronology. Yet, despite these problems, Mesozoic marine reptiles do provide some robust regional and low-resolution global biochronology. For example, Triassic marine reptiles from the European Muschelkalk provide some correlations below the stage/age level, and ichthyosaurs resolve three intervals of Triassic time. Further collecting, more whole-animal taxonomy, and more precise stratigraphic data should increase the biochronological utility of Mesozoic marine reptiles.

ACKNOWLEDGMENTS

I thank G. Bell, J. Callaway, and S. Welles for freely sharing information with me. C. González-Léon, B. Kues, and P. Reser worked with me in the field collecting and stratigraphically organizing some marine reptile fossils. Jack Callaway, Jeff Eaton, and Bill Orr provided helpful reviews of the manuscript.

REFERENCES

Bardet, N., P. Wellnhofer, and D. Herm. 1994. Discovery of ichthyosaur remains (Reptilia) in the upper Cenomanian of Bavaria. *Mitteilungen der Bayerische Staatssammlung für Paläontologie und Historische Geologie* 34:213-220.

Bell, G. L. 1985. Vertebrate faunal zones in the Upper Cretaceous of west-central Alabama. *Geological Society of America, Abstracts with Programs* 17(2):80.

Brinkman, D. B., E. L. Nicholls, and J. M. Callaway. 1992. New material of the ichthyosaur *Mixosaurus nordenskioeldii* from the Triassic of British Columbia, and the interspecific relationships of *Mixosaurus*. *The Paleontological Society Special Publication* (Fifth North American Paleontological Convention, Abstracts with Programs) 6:37.

Brown, D. S. 1981. The English Upper Jurassic Plesiosauroidea (Reptilia), and a review of the phylogeny and classification of the Plesiosauria. *Bulletin of the British Museum (Natural History), Geology* 35: 253-347.

Callaway, J. M. and D. B. Brinkman. 1989. Ichthyosaurs (Reptilia, Ichthyosauria) from the Lower and Middle Triassic Sulphur Mountain Formation, Wapiti Lake area, British Columbia, Canada. *Canadian Journal of Earth Sciences* 26:1491-1500.

Callaway, J. M. and J. A. Massare. 1989. Geographic and stratigraphic distribution of the Triassic Ichthyosauria (Reptilia, Diapsida). *Neues Jahrbuch für Geologie und Paläontologie*, Abhandlungen 178:37-58.

Carroll, R. L. and Z. Dong. 1991. *Hupehsuchus*, an enigmatic aquatic reptile from the Triassic of China, and the problem of establishing relationships. *Philosophical Transactions of the Royal Society of London* B 331:131-153.

Cooper, M. R. 1982. A mid-Permian to earliest Jurassic tetrapod biostratigraphy and its

significance. *Arnoldia Zimbabwe* 9:77-101.
Cox, C. B. and D. G. Smith. 1973. A review of the Triassic vertebrate faunas of Svalbard. *Geological Magazine* 110:405-418.
Dagys, A. 1988. Major features of the geographic differentiation of Triassic ammonoids. IN J. Wiedmann and J. Kullman (Eds.), *Cephalopods---Present and Past*, pp. 341-349. Schweizerbart'sche Verlagsbuchhandlung, Stuttgart.
Hagdorn, H. 1993. Reptilien-Biostratigraphie des Muschelkalks. IN H. Hagdorn and A. Seilacher (Eds.), *Muschelkalk, Schontaler Symposium 1991*, p. 186. Korb (Goldschenk), Stuttgart.
Harland, W. B., R. L. Amstrong, A. V. Cox, L. E. Craig, A. G. Smith, and D. G. Smith. 1990. *A Geologic Time Scale 1989*. Cambridge University Press, Cambridge, 263 pp.
Kozur, H. 1974. Biostratigraphie der germanischen Mitteltrias. *Frieberger Forschungen C* 280(1):1-56, 280(2):1-71, 280(3):9 plates.
Kuhn-Schnyder, E. 1974. Die Triasfauna der Tessiner Kalkalpen. *Neujahrsbatt Naturforschung Gessellschaft Zürich* 176:1-119.
Lucas, S. G. 1990. Toward a vertebrate biochronology of the Triassic. *Albertiana* 8:36-41.
Lucas, S. G. and C. González-Léon. 1995. Ichthyosaurs from the Upper Triassic of Sonora and the biochronology of Triassic ichthyosaurs. IN C. Jacques-Ayala, C. M. González-Léon, and J. Roldán-Quintana (Eds.), Studies on the Mesozoic of Sonora and Adjacent Areas, pp. 17-20. *Geological Society of America Special Paper* 301.
Massare, J. A. and J. M. Callaway. 1994. *Cymbospondylus* (Ichthyosauria, Shastasauridae) from the Early Triassic Thaynes Formation of southeastern Idaho. *Journal of Vertebrate Paleontology* 14:139-141.
Mazin, J.-M. and P. M. Sander. 1993. Paleobiogeography of the Early and Late Triassic Ichthyopterygia. *Paleontologia Lombarda Nuova Serie* 2:93-107.
Mazin, J.-M., Y. Suteethorn, E. Buffetaut, J.-J. Jäger, and H.-I. Rucha. 1991. Preliminary description of *Thaisaurus chonglakmanni* n. g., n. sp., a new ichthyosaurian (Reptilia) from the Early Triassic of Thailand. *Compte Rendus Académie Science Paris II* 313:1207-1212.
Motani, R. 1994. Computer aided comparisons among Early Triassic ichthyosaurs reveal smaller taxonomic diversity than was believed. *Journal of Vertebrate Paleontology* 14, supplement:39A.
Nicholls, E. L. and D. B. Brinkman. 1993. A new specimen of *Utatsusaurus* (Reptilia: Ichthyosauria) from the Lower Triassic Sulphur Mountain Formation of British Columbia. *Canadian Journal of Earth Sciences* 30: 486-490.
Pinna, G. and J.-M. Mazin. 1993. Stratigraphy and paleobiogeography of the Placodontia. *Paleontologia Lombarda Nuova Serie* 2:125-130.
Russell, D. A. 1967. Systematics and morphology of American mosasaurs. *Peabody Museum of Natural History, Yale University Bulletin* 23:1-241.
Russell, D. A. 1993. Vertebrates in the Cretaceous Western Interior sea. *Geological Association of Canada Special Paper* 39:665-680.
Sander, P. M. and J.-M. Mazin. 1993. The paleobiogeography of Middle Triassic ichthyosaurs: the five major faunas. *Paleontologia Lombarda Nuova Serie* 2:145-152.

Storrs, G. W. 1991. Anatomy and relationships of *Corosaurus alcovensis* (Diapsida: Sauropterygia) and the Triassic Alcova Limestone of Wyoming. *Peabody Museum of Natural History, Yale University Bulletin* 44:1-151.

Welles, S. P. 1962. A new species of elasmosaur from the Aptian of Columbia and a review of the Cretaceous plesiosaurs. *University of California Publications in Geological Sciences* 44:1-89.

Chapter 15

TITHONIAN MARINE REPTILES OF THE EASTERN PACIFIC

ZULMA GASPARINI and MARTA FERNANDEZ

INTRODUCTION

South American Jurassic marine reptiles are found only in the Pacific margin, mainly in Argentina and Chile. An important collection of marine reptiles has been gathered during the last 20 years. The search was carried out mainly in the Tithonian of the Vaca Muerta Formation, which is exposed in several localities of the Neuquén Basin, west-central Argentina. The specimens discovered represent a rich Tithonian fauna of ichthyosaurs, crocodiles, turtles, and plesiosaurs. Until now, Liassic records have been very scarce; the fragmentary materials are determinable only at higher taxonomic levels (Gasparini, 1985). The Middle Jurassic is richer than the Lias. Recent fieldwork in the Neuquén Basin has discovered Bajocian pliosaurids, crocodiles, and ichthyosaurs (Fernández, 1994; Spalletti et al., 1994) and Callovian pliosaurids, elasmosaurids, and cryptoclidids (Gasparini and Spalletti, 1993). To this list must be added Callovian metriorhynchids from northern Chile (Gasparini and Chong Diaz, 1977; Gasparini, 1980). Nevertheless, the most important marine herpetofauna from the eastern Pacific, both in abundance as well as in diversity, was found in the Tithonian of the Neuquén Basin.

The aims of this chapter are to present an up-to-date analysis of the South American Tithonian reptile fauna in a biogeographical context and interpret the pattern of the marine reptile distribution during the Late Jurassic.

GEOGRAPHIC AND GEOLOGIC OCCURRENCE

Except for one thalattosuchian crocodile vertebra from the Lo Valdés Formation (late Tithonian) of central Chile (Gasparini, 1985), all the South American Tithonian reptiles were discovered in the Neuquén Basin (Figure 1). This basin is located in west-central Argentina (32°-41° S, 68°-72° W) and includes an almost complete Upper Triassic-Lower Tertiary sedimentary succession. The Neuquén Basin is a back-arc basin with the axis roughly coincident with that of the Andes, and with a southeastward expansion known as the Neuquén embayment (Riccardi et al., 1992). This basin is limited by two cratonic areas (Sierra Pintada Block to the northeast, and Northpatagonian Massif to the southeast) and by the volcanic arc of the Cordillera Principal to the west. The Jurassic sedimentary succession has been included by Riccardi and Gulisano (1992) in the Araucanian Synthem (Rhaetian-upper Oxfordian) and in the lower part of the Andean Cycle (Kimmeridgian-lower Valanginian). The Jurassic-Early Cretaceous evolution of the Neuquén Basin has been analyzed by several recent workers (e.g., Digregorio and Uliana, 1980; Riccardi, 1983; Gulisano, 1984; Gulisano et al., 1984; Mitchum and Uliana, 1985; Legarreta and Gulisano, 1989; Legarreta and Uliana, 1991; Riccardi and Gulisano, 1992; Riccardi et al., 1992).

All the Late Jurassic marine reptiles of the Neuquén Basin are from the Vaca Muerta Formation, which is Tithonian-Valanginian (Leanza, 1980, 1981). The record covers the early-late Tithonian lapse, but not all of them have biostratigraphic control based on ammonites. In this chapter the only localities mentioned are those where the reptiles are associated with datable invertebrates.

During the middle early Tithonian, a relative sea-level rise caused the widespread deposition of black shales, marls, and bituminous limestones of the lower Vaca Muerta Formation (Riccardi et al., 1992). Consequently, the early and middle Tithonian reptiles are found in two areas (Cerro Lotena and Los Catutos) in the east-central part of the basin (Figure 1). During the late Tithonian, a strong progradation of clastic carbonates from the south and east (Neuquén, Río Negro, and La Pampa provinces) caused the area of anoxic sedimentation (Vaca Muerta Formation) to be restricted (Riccardi et al., 1992). The reptile-bearing localities, (Chacay Melehue, Cajón de Almanza, Bardas Blancas, Poti Malal) consequently appear aligned along the western sector of the basin (Figure 1).

The temporal sequence of the Tithonian marine reptiles record begins in the early Tithonian levels of Cerro Lotena (*Virgatosphinctes mendozanus* zone: Leanza, 1980, 1981). In this locality, the middle section of the Vaca Muerta Formation is composed of light green calcareous shales interspersed with light gray concretional limestones up to 1 m thick. The vertebrates, quite well preserved, were found in the lower part of this section (Gasparini and Dellapé, 1976). The herpetofauna

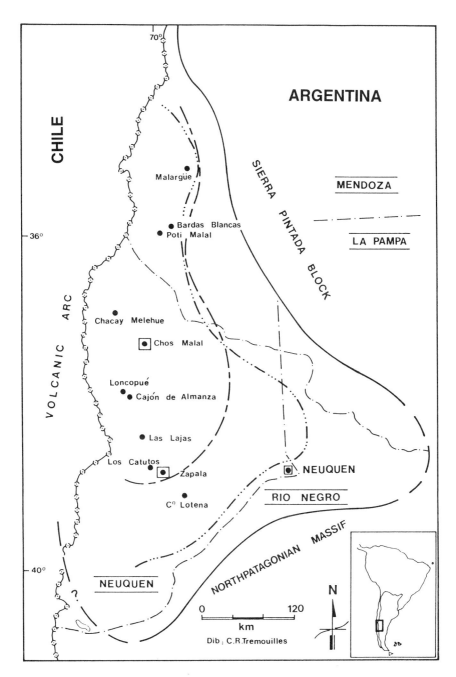

Figure 1. Map of the Neuquén Basin showing principal localities where marine reptiles were found. Simplified after Riccardi et al. (1992). Selected symbols: **basin margin** = ——— (dashed where inferred), **position of the coast line during the early Tithonian** = — ··· —, **position of the coast line during the late Tithonian** = — — —.

(Figure 2) consists of crocodiles (*Geosaurus araucanensis* Gasparini and Dellapé, 1976; *Metriorhynchus* sp.), ichthyosaurs (*Ophthalmosaurus* sp.), pliosaurids (*Liopleurodon* cf. *L. macromerus* Tarlo, 1959), and turtles (*Notoemys laticentralis* Cattoi and Freiberg and *Neusticemys neuquina* Fernández and Fuente).

The middle Tithonian reptiles were found in the lithographic limestone of the Los Catutos area, which crops out close to Zapala, Neuquén Province, in the south-central region of the Neuquén Basin (Figure 1). This lithographic limestone belongs to the Los Catutos Member of the Vaca Muerta Formation, *Windhauseniceras interspinosum* zone, of the uppermost middle Tithonian (Leanza and Zeiss, 1990). The herpetofauna consists of ichthyosaurs (*Ophthalmosaurus monocharactus, Ophthalmosaurus* sp.), crocodiles (*Geosaurus* sp.), turtles (*Notoemys laticentralis* and *Neusticemys neuquina*), an indeterminate plesiosaur, and a Pterodactyloidea indet. (Cione et al., 1987; Gasparini et al., 1987; Leanza and Zeiss, 1990).

The greatest diversity of late Tithonian reptiles is found in the Cajón de Almanza area, near Loncopué (Neuquén Province) in the west-central region of the Neuquén Basin (Figure 1). Here, the Vaca Muerta Formation is composed of a thick succession of black shales and marls with light authigenic carbonate intercalations. The fossil-bearing levels, at the base of *Substeueroceras koeneni* zone, belongs to the uppermost Tithonian (H. Leanza, personal communication, 1994). Most of the material was recently collected, and is still in preparation and/or study. However, preliminary studies support the presence of ichthyosaurs, *Pliosaurus* sp., *Geosaurus* sp., and the turtle *Neusticemys neuquina*. Late Tithonian

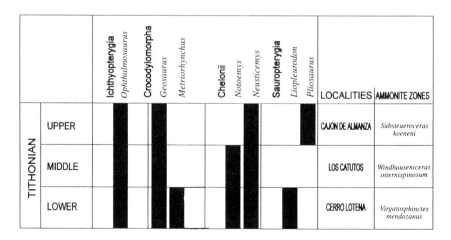

Figure 2. Geographic and temporal distribution of marine reptiles in the Tithonian of the Neuquén Basin.

reptiles were found in several localities of the northwestern Neuquén Basin in Mendoza Province (Bardas Blancas, Poti Malal, Figure 1). Most of them were studied by Rusconi (1947, 1948a, b), and are now referred to *Ophthalmosaurus* sp., *Metriorhynchus potens,* and *Dakosaurus* sp.

SYSTEMATICS

Crocodylomorpha

All the Tithonian thalattosuchians from the eastern Pacific are Metriorhynchidae. Several skulls of *Geosaurus araucanensis* Gasparini and Dellapé, 1976 (Figure 3A), are known from the early Tithonian of Cerro Lotena, along with one skull, currently under study, of *Metriorhynchus*. Recently, an almost complete specimen of *Geosaurus* (probably *G. araucanensis*) was found in the middle Tithonian of Los Catutos, and, in 1994, two new specimens were exhumed in the late Tithonian of Cajón de Almanza. One of these, referable to *Geosaurus* sp. *Metriorhynchus potens* (Rusconi, 1948a), was found in the south-central part of Mendoza Province, in the Malargüe area, in levels assigned to the late Tithonian (*Substeueroceras koeneni* zone). It was originally identified as a plesiosaur, assigned to ?*Metriorhynchus* by Kuhn (1968), to ?*Purranisaurus potens* by Gasparini (1973, 1981), and finally to *M. potens* by Gasparini (1985, 1992). Taxonomic confusion surrounding this metriorhynchid with a low skull and narrow rostrum is due to preservation and the complex generic taxonomy (Wenz, 1968; Adams-Tresman, 1987; Mazin et al., in press). Another metriorhynchid from the late Tithonian (Figure 3B and C), in Poti Malal, Mendoza, was referred by Goñi (1987) to *Metriorhynchus* aff. *M. durobrivensis* (Andrews). However, its short and very high rostrum are consistent with *Dakosaurus* Quenstedt, a Kimmeridgian-Tithonian genus from western Europe.

Sauropterygia

Tithonian sauropterygians have been recorded in Cerro Lotena, Los Catutos, and Cajón de Almanza, but they are proportionally scarcer than other Tithonian reptiles from the Neuquén Basin. A mandible and part of the palate of *Liopleurodon* cf. *L. macromerus* (Tarlo, 1960) were found in the early Tithonian of Cerro Lotena. This specimen (Figure 3D) was originally assigned to *Stretosaurus* (= *Liopleurodon*; Halstead, 1989) by Gasparini et al. (1982), who pointed out its close affinities with *S. macromerus*. A single, strongly striated tooth was found in the lithographic limestone of Los Catutos (middle Tithonian) and was referred to an indeterminate plesiosaur (Leanza and Zeiss, 1990). Remains belonging to three specimens have been found in the late Tithonian of the Cajón de Almanza area. One of them may be referred to *Pliosaurus* sp.; the other two are indeterminate Pliosauridae.

Ichthyosauria

In spite of their relative abundance, the diversity of ichthyosaurs in the Tithonian of South America is low. All the specimens found in the Neuquén Basin are referable to *Ophthalmosaurus* as defined by Appleby (1956) and McGowan (1976) (Figure 3E). One of the specimens found in the lithographic limestone of Los Catutos, almost complete and excellently preserved, was identified by Gasparini (1988) as *Ophthalmosaurus monocharactus* Appleby (= *O. icenicus* Seeley).

Only two specimens discovered in the early Tithonian of Cerro Lotena are not referable to *Ophthalmosaurus*. One rostrum, characterized by its gracility, the lack of teeth, the maxillary taking part in the external naris, and the relatively small orbit, was identified as ?*Platypterygius* Huene, 1922 by Gasparini and Goñi (1990). Recently, Dr. José Bonaparte discovered, also in the early Tithonian of Cerro Lotena, a large, well-preserved specimen. Both specimens are under study and share a gracile and edentulous rostrum, and a rather small orbit. This combination, together with other characters, distinguishes them from other taxa described from the Late Jurassic and suggests that Tithonian diversity was higher than previously supposed.

Chelonia

The Jurassic chelonian record in South America is restricted to the Late Jurassic of the Neuquén Basin, mainly Cerro Lotena and Los Catutos. This record ranges in age from the early Tithonian of Cerro Lotena to the late Tithonian of Cajón de Almanza.

Figure 3. **A)** *Geosaurus araucanensis*, MLP 72-IV-7-1, dorsal view. **B)** *Dakosaurus*, MSR 344, ventral view. **C)** *Dakosaurus*, MSR 344, lateral view. **D)** *Liopleurodon* cf. *L. macromerus*, MLP 80-V-29-1, ventral view. **E)** *Ophthalmosaurus*, MLP 85-I-15-1, lateral view. **F)** *Neusticemys neuquina*, MOZ S/N, ventral view. **G)** *Notoemys laticentralis*, MOZ 2487, dorsal view. Abbreviations: **MLP** = Museo de La Plata, Argentina; **MOZ** = Museo Prof. Olsacher, Zapala, Argentina; **MSR** = Museo de San Rafael, Mendoza, Argentina. Scale bars = 5 cm.

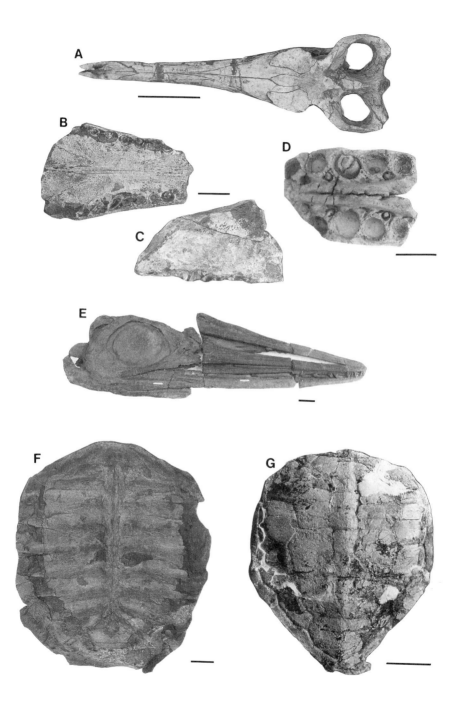

Cryptodira and Pleurodira, the two groups of extant turtles, are represented in the Tithonian of the Neuquén Basin by *Neusticemys neuquina* (Fernández and Fuente, 1988) (Figure 3F) and *Notoemys laticentralis* Cattoi and Freiberg, 1961 (Figure 3G), respectively. All the remains found in this basin can be referred to the these taxa.

Neusticemys neuquina was tentatively included in the Plesiochelyidae Baur, 1888 (Fernández and Fuente, 1993). The phylogenetic analysis of this new genus has been hampered by the lack of a systematic review of the Plesiochelyidae-Thalassemydidae. *Neusticemys* shows the same transformation of the forelimb into a paddle as the Cheloniidae. However, the hypothesis that *Neusticemys* is an early representative of this taxon has been rejected. Fernández and Fuente (1993) interpreted this transformation of the forelimb as a homoplasy resulting from pelagic marine life.

Notoemys was described as a Plesiochelyidae by Wood and Freiberg (1977). Years later, specimens more complete than the holotype were discovered in the Tithonian of Cerro Lotena and Los Catutos. Based on the new specimens *Notoemys* was referred to the infraorder Pleurodira (Fuente and Fernández, 1989). Fernández and Fuente (1994) analyzed the phylogenetic relationships of this taxon within the infraorder and proposed the inclusion of *Notoemys* in a separate family. According to their hypothesis, *Platychelys* Wagner, 1853, from the Kimmeridgian of Solothurn, is the sister-group of *Notoemys*, Cheliidae, and pelomedusoids.

TITHONIAN MARINE REPTILE ASSEMBLAGE IN THE NEUQUEN BASIN

The Tithonian marine reptiles of the Neuquén Basin are distributed in sedimentary facies assigned to basinal environments (Gulisano et al., 1984). The marine reptile fauna suggests an offshore assemblage throughout the Tithonian (Figure 4).

Ichthyosaurs are the most common reptiles, comprising 50% of the fauna. The most frequently found taxon, *Ophthalmosaurus*, not only is abundant, but also has a worldwide distribution (McGowan, 1978).

Crocodiles and turtles make up 40% of the fauna. Although the association of these two groups has been interpreted as one characteristic of nearshore and lagoonal environments in the Late Jurassic of the West Tethys (Mazin et al., in press), this is not the case for the Neuquén Basin. The only Tithonian crocodiles identified in this basin are the pelagic metriorhynchids---*Geosaurus*, *Metriorhynchus*, and *Dakosaurus*---characterized by a naked body, transformed limbs, and dorsal and caudal fins. Gasparini (1981) suggested that South American metriorhynchids were

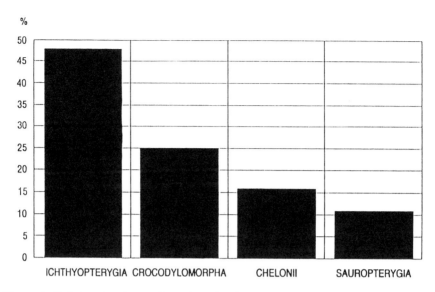

Figure 4. Relative abundance of remains of each taxon expressed as a percentage of total fossils.

not only highly adapted to marine life, as has been previously proposed (Buffetaut and Thierry, 1977), but that they had also probably developed independence from land by acquiring ovoviviparity.

The most abundant turtle in the Vaca Muerta Formation is *Neusticemys*. Its habits were probably different from those inferred for the other plesiochelyids of Europe, China, and the former Soviet Union (Bräm, 1965; Nessov, 1988; Mazin et al., in press). Tethyan turtles were supposed to be estuarine or lagoonal. Some features of the appendicular skeleton and carapace of *Neusticemys*, however, suggest that this turtle inhabited open marine environments. Its forelimbs are transformed into paddles. The proximal elements of the carpus are round and metacarpals and phalanges are conspicuously elongated. This paddle-like forelimb is unique among Jurassic turtles.

The other turtle known from the Neuquén Basin is the pleurodire *Notoemys*. This taxon has retained the appendicular skeleton of a continental form similar to the extant Pleurodira. Despite these features, it is possible that *Notoemys* was not an estuarine or continental form, as suggested by Wood and Freiberg (1977). All specimens were found in sediments deposited in basinal and offshore environments at three different localities: Cerro Lotena, Las Lajas, and Los Catutos. The remains are well preserved and there is no evidence that they are reworked from other continental units, or that they have been transported postmortem from littoral environments (Fernández and Fuente, 1993).

Although sauropterygians are poorly represented within this fauna, it is noteworthy that the remains found can all be referred to the large Pliosauridae.

PALEOBIOGEOGRAPHIC DISTRIBUTION

The Tithonian marine reptile fauna from the Neuquén Basin resembles those from the Late Jurassic of the West Tethys. Except for the turtles, all the Tithonian marine reptiles of South America correspond to already known genera of the European Middle-Late Jurassic: *Ophthalmosaurus, Metriorhynchus, Geosaurus, Dakosaurus, Liopleurodon,* and *Pliosaurus* (Bardet, 1992; Gasparini, 1992; Mazin et al., in press). When faunal comparisons are made with the Late Jurassic, taxonomic affinities indicate close paleobiogeographic relationships (Gasparini, 1992). However, when the comparison is restricted to the Tithonian, many differences are noted. These differences are reflected not only in faunal composition but also in the relative abundance of each taxon within the assemblage.

In European Tithonian basins, turtles, teleosaurids, and metriorhynchid crocodiles are characteristic of nearshore and lagoonal environments (Mazin et al., in press). More pelagic groups such as ichthyosaurs and plesiosaurs are scarce and low in diversity. Ichthyosaurs represent only 17% of all species recorded in the Britain-Normandy Basin. Sauropterygians, represented only by the elasmosaurid *Colymbosaurus trochanterius* (Owen), also make up 17% of the fauna (Bardet, 1992; Mazin et al., in press). In contrast, in the Neuquén Basin Tithonian, despite comparatively little collecting, the general pattern suggests an offshore assemblage with a high abundance of ichthyosaur remains. These assemblage differences could be due to different environmental conditions established during the Tithonian. In European basins, the Tithonian regression probably produced adverse conditions for pelagic marine reptiles (Mazin et al., in press). In the Neuquén Basin, however, the transgressive Tithonian sea was probably an adequate environment for these reptiles.

Some well-represented taxa in the West Tethys Callovian-Kimmeridgian basins (*Ophthalmosaurus, Liopleurodon, Pliosaurus,* and *Metriorhynchus*) disappeared or were poorly represented during the Tithonian. In the Eastern Pacific, however, these genera probably persisted throughout the Tithonian. Some other genera shared by both faunas, such as *Geosaurus*, showed an endemism at the specific level (*G. suevicus* in Europe and *G. araucanensis* in the Neuquén Basin). These taxonomic differences suggest that exchange events occurred between both bioprovinces before the7XTithonian.

The swimming capacity of pelagic reptiles (Massare, 1988; Halstead, 1989; McGowan, 1991), the opportunistic prey capture, and the ovoviviparity, demonstrated in ichthyosaurs and supposedly in metriorhynchids and plesiosaurs, must have allowed their Jurassic dispersion through the warm currents which circumnavigated the globe (Parrish, 1992). Taking into account the similarities between West Tethys and Eastern Pacific faunas during the Late Jurassic, and the position of the macrocontinents (Scotese, 1991), the most probable seaway must have been the Hispanic Corridor and the passage between the southeast of Africa and the southwest of South America (Figure 5).

The hypothesis of the Hispanic Corridor acting as the principal seaway for the marine reptile dispersal in Europe and the East Pacific, at least since the Middle Jurassic, has already been proposed (Gasparini, 1978, 1985, 1992; Gasparini and Spalletti, 1993). *Metriorhynchus* in the Chilean Callovian, cf. *Muraenosaurus* and cf. *Cryptoclidus* in the early Callovian of the Neuquén Basin, and a thalattosuchian in the early Callovian of Cualac, western Mexico (Gasparini, 1992), presuppose that the Hispanic Corridor was used during the Middle Jurassic. Other evidence, mainly the affinities of the ammonoid and brachiopod faunas from west-central South America and the West Tethys, demonstrate that such a seaway existed, perhaps in an intermittent manner during global highstands of sea level since the Middle Jurassic, or even earlier (Riccardi, 1991; Manceñido and Dagys, 1992; Hillebrandt et al., 1992; Parrish, 1992). Oxfordian plesiosaurs and crocodiles of Cuba (Torres and Rojas, 1949; Iturralde-Vinent and Norell, 1992) support the hypothesis of the Hispanic Corridor acting as a seaway in the Late Jurassic.

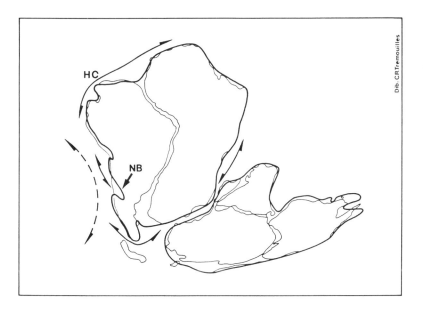

Figure 5. Possible dispersal routes of Late Jurassic marine reptiles. Abbreviations: **HC** = Hispanic Corridor, **NB** = Neuquen Basin.

The breakup of Gondwanaland and marine transgressions in the Late Jurassic resulted in new dispersal routes for pelagic reptiles. A marine transgression during the latest Tithonian-Berriasian produced a shallow intermittent epicontinental seaway between east Africa and southern Patagonia (Riccardi, 1991). The "ichthyosaurid" (sic) found in late Tithonian or early Berriasian (Whitham and Doyle, 1989) in the northeast of the Antarctic Peninsula supports this hypothesis.

Turtles may have had a different history. During the Late Jurassic cryptodires of the Neuquén Basin were endemic at the generic level, while pleurodires were endemic at the familial level. Although *Notoemys* inhabits relatively open marine environments, its limbs were not paddle-like, suggesting that *Notoemys* was not pelagic, and probably did not use the seaways proposed for the other marine reptiles. A vicariance event probably took place before the Late Jurassic and exchanges between the European Tethys and the Eastern Pacific no longer occurred during the Late Jurassic, resulting in endemism.

Many unknowns remain to be elucidated, including which taxa were present in the Eastern Pacific during the Oxfordian and Kimmeridgian because some isolated and indeterminate remains were found in Chile (Gasparini, 1985). However, this requires further collecting, in the same way as the search in the Occidental Pacific Jurassic basins revealed paleobiogeographic affinities, such as those observed in nekto-pelagic invertebrates (Riccardi, 1991).

SUMMARY

Jurassic marine reptiles are found in South America only along the Pacific margin. The largest collection of these reptiles comes from the Tithonian of the Neuquén Basin (Argentina). Analysis of this fauna indicates some biogeographic trends: (1) turtles (*Notoemys, Neusticemys*) are endemic at familial and generic levels; (2) crocodiles (*Geosaurus, Metriorhynchus, Dakosaurus*), plesiosaurs (*Liopleurodon, Pliosaurus*), and ichthyosaurs (*Ophthalmosaurus*) are the same genera recorded in the West Tethys. We propose that during the Late Jurassic, and probably before, during the Middle Jurassic, exchange events between reptiles from the Andean region and the West Tethys were favored by the opening of the Hispanic Corridor. Callovian crocodiles in Chile and southwestern Mexico, Callovian plesiosaurs (cf. *Muraenosaurus*; cf. *Cryptoclidus*) in the Neuquén Basin, and Oxfordian plesiosaurs and crocodiles in Cuba support this hypothesis. The invertebrate fauna also supports this idea. The links to the Tethyan fauna could also be favored during the Late Jurassic by the opening of a new seaway between South America---Africa and Antarctica. The discovery of ichthyosaur remains and fishes from the Upper Jurassic outcrops in the northeast of the Antarctic Peninsula is the first evidence of marine vertebrates in this seaway.

ACKNOWLEDGMENTS

We thank Dr. L. Spalletti (Centro de Investigaciones Geológicas, La Plata) for constructive comments on the original manuscript. The junior author thanks Dr. J. Bonaparte (Museo B. Rivadavia, Argentina) for permission to examine a large Tithonian ichthyosaur. This research was supported by National Geographic Society Grant 5178 (to ZG), and is a contribution to Jurassic Events in South America (IGCP 322).

REFERENCES

Adams-Tresman, S. 1987. The Callovian (Middle Jurassic) marine crocodile *Metriorhynchus* from central England. *Palaeontology* 30(10):179-194.

Appleby, R. 1956. The osteology and taxonomy of the fossil reptile *Ophthalmosaurus*. *Proceedings of the Zoological Society of London* 126(3):403-447.

Bardet, N. 1992. *Évolution et Extinction des Reptiles Marins au cours du Mésozoïque*. Thèse de Doctorat, Université Paris 6 (9230), 212 pp. (unpublished).

Bräm, H. 1965. Die Schildkröten aus dem oderem Jura (Malm) der Gegend von Solothurn. *Schweizerische Paläontologische Abhandlungen* 83:1-190.

Buffetaut, E. and J. Thierry. 1977. Les Crocodiliens fossiles du Jurassique moyen et supérieur de Bourgogne. *Géobios* 10(2):151-194.

Cione, A., Z. Gasparini, H. Leanza, and A. Zeiss. 1987. Marine oberjurassiche Plattenkalke in Argentina (ein erster Forschungsbericht). *Archaeopteryx* 5:13-22.

Digregorio, J. and M. Uliana. 1980. Cuenca Neuquina. IN *Segundo Simposio Geología Regional Argentina*, pp. 985-1032. Academia Nacional de Ciencias de Córdoba 2.

Fernández, M. 1994. A new long-snouted ichthyosaur from the Early Bajocian of the Neuquén Basin (Argentina). *Ameghiniana* 31(2):291-297.

Fernández, M. and M. de la Fuente. 1988. Nueva tortuga (Cryptodira: Thalassemydidae) de la Formación Vaca Muerta (Jurásico, Tithoniano) de la provincia del Neuquén, Argentina. *Ameghiniana* 25(2):129-138.

Fernández, M. and M. de la Fuente. 1993. Las tortugas Casiquelidias de las calizas litográficas del área Los Catutos, Neuquén, Argentina. *Ameghiniana* 30(3):283-295.

Fernández, M. and M. de la Fuente. 1994. Redescription and phylogenetic position of *Notoemys*: the oldest gondwanian pleurodira turtle. *Neues Jahrbuch für Geologie und Paläontologie* 193(1):81-105.

Fuente, M. de la and M. Fernández. 1989. *Notoemys laticentralis* Cattoi and Freiberg, 1961, from the Upper Jurassic of Argentina: a member of the Infraorder Pleurodira (Cope, 1868). *Studia Geológica Salamanticensia. Studia Palaeocheloniologica* 3(2):25-32.

Gasparini, Z. 1973. Revisión de ?*Purranisaurus potens* Rusconi, 1948 (Crocodilia, Thalattosuchia). Los Thalattosuchia como un nuevo infraorden de los Crocodilia. *Actas V Congreso Geológico Argentino* (Villa Carlos Paz, 1972) 3:423-431.

Gasparini, Z. 1978. Consideraciones sobre los Metriorhynchidae (Crocodylia, Mesosuchia): su origen, taxonomía y distribución geográfica. Obras del Centenario del museo de La Plata. *Paleontología* 5:1-9.

Gasparini, Z. 1980. Un nuevo cocodrilo marino (Crocodylia, Metriorhynchidae) del Caloviano del norte de Chile. *Ameghiniana* 17(2):97-103.

Gasparini, Z. 1981. Los Crocodylia fósiles de la Argentina. *Ameghiniana* 18(3-4):177-205.

Gasparini, Z. 1985. Los reptiles marinos jurásicos de América del Sur. *Ameghiniana* 22(1-2):23-34.

Gasparini, Z. 1988. *Ophthalmosaurus monocharactus* Appleby (Reptilia, Ichthyopterygia), en las calizas litográficas titonianas del area Los Catutos, Neuquén, Argentina. *Ameghiniana* 25(1):3-16.

Gasparini, Z. 1992. Marine reptiles of the Circum-Pacific region. IN G. E. G. Westermann (Ed.), *The Jurassic of the Circum-Pacific, World and Regional Geology* 3, pp. 361-364. Cambridge University Press, London.

Gasparini, Z. and G. Chong Diaz. 1977. *Metriorhynchus casamiquelai* n. sp. (Crocodylia, Thalattosuchia), a marine crocodile from the Jurassic (Callovian) of Chile, South America. *Neues Jahrbuch für Geologie und Paläontologie* 153(3):341-360.

Gasparini, Z. and D. Dellapé. 1976. Un nuevo cocodrilo marino (Thalattosuchia, Metriorhynchidae) de la Formación Vaca Muerta (Jurásico, Tithoniano) de la provincia de Neuquén. *Actas I Congreso Geológico Chileno*, Santiago 1:C1-C21.

Gasparini, Z. and R. Goñi. 1990. Los ictiosaurios jurásico-cretácicos de la Argentina. IN W. Volkheimer (Ed.), *Bioestratigrafía de los Sistemas Regionales del Jurásico y Cretácico de América del Sur* 2, pp. 299-311. Comité Sudamericano del Jurásico y Cretácico IGCP 242, Mendoza.

Gasparini, Z. and L. Spalletti. 1993. First Callovian plesiosaurs from the Neuquén Basin, Argentina. *Ameghiniana* 30(3):245-254.

Gasparini, Z., R. Goñi, and O. Molina. 1982. Un plesiosaurio (Reptilia) tithoniano de Cerro Lotena, Neuquén, Argentina. *V Congreso Latinoamericano de Geología, Argentina* 5:33-47.

Gasparini, Z., H. Leanza, and J. Garate Zubillaga. 1987. Un pterosaurio de las calizas litográficas titonianas del área de Los Catutos, Neuquén, Argentina. *Ameghiniana* 24(1-2):141-143.

Goñi, R. 1987. *Metriorhynchus* aff. *M. durobrivensis* (Crocodylia, Thalattosuchia), un cocodrilo marino de la Formación Vaca Muerta, Mendoza, Argentina. Revista del Museo Argentino de Ciencias Naturales "Bernardino Rivadavia." *Paleontología* 4(1):1-8.

Gulisano, C. 1984. Esquema estratigráfico de la secuencia jurásica del oeste de la provincia del Neuquén. *Actas IX Congreso Geológico Argentino* 1:236-259.

Gulisano, C., A. Gutierrez Pleimling, and R. Digregorio. 1984. Análisis estratigráfico del intervalo Tithoniano-Valanginiano (Formaciones Vaca Muerta, Quintuco y Mulichinco) en el suroeste de la provincia del Neuquén. *Actas IX Congreso Geológico Argentino*

1:221-235.

Halstead, B. 1989. Plesiosaur locomotion. *Journal of the Geological Society* 146:37-40.

Hillebrandt, A. von, G. Westermann, J. Callomon, and R. Detterman. 1992. Ammonites of the circum-Pacific region. IN E. G. Westermann (Ed.), *The Jurassic of the Circum-Pacific, World and Regional Geology* 3, pp. 342-359. Cambridge University Press, London.

Iturralde-Vinent, N. and M. Norell. 1992. Los saurios del Jurásico (Oxfordiano) de Cuba Occidental. *IX Convención Geológica Nacional*, p. 96. Veracruz, México.

Kuhn, O. 1968. *Die vorzeilichen Krokodile.* Verlag Oeben München, 124 pp.

Leanza, H. 1980. The Lower and Middle Tithonian ammonite fauna from Cerro Lotena, province of Neuquén, Argentina. *Zitteliana* 5:1-49.

Leanza, H. 1981. Faunas de Ammonites del Jurásico superior y del Cretácico inferior de América del Sur, con especial consideración de la Argentina. Cuencas Sedimentarias. *Jurásico Cretácico de América del Sur* 2:559-97.

Leanza, H. and A. Zeiss. 1990. Upper Jurassic lithographic limestones from Argentina (Neuquén Basin): stratigraphy and fossils. *Facies* 22(4):169-186.

Legarreta, L. and C. Gulisano. 1989. Análisis estratigráfico secuencial de la Cuenca Neuquina (Triásico superior-Terciario inferior), Argentina. IN G. Chebli and L. Spalletti (Eds.), Cuencas Sedimentarias Argentinas, pp. 221-243. *Serie Correlaciones Geológicas* 6. Universidad Nacional de Tucumán, Tucumán.

Legarreta, L. and M. Uliana. 1991. Jurassic-Cretaceous marine oscillations and geometry of Backarc Basin Fill, Central Argentine Andes. *ASTRA C.A.P.S.A. Contracts and New Ventures*. Buenos Aires, 50 pp.

Manceñido, M. and A. Dagys. 1992. Brachiopods of the Circum-Pacific region. IN E. G. Westermann (Ed.), *The Jurassic of the Circum-Pacific, World and Regional Geology* 3, pp. 328-333. Cambridge University Press, London.

Massare, J. 1988. Swimming capabilities of Mesozoic marine reptiles: implications for method of predation. *Paleobiology* 14(2):187-205.

Mazin, J.-M., N. Bardet, P. Vignaud, and S. Hua. In press. Marine reptiles from the Middle and Upper Jurassic of western Europe. *Palaeovertebrata*.

McGowan, C. 1976. The description and phenetic relationships of a new ichthyosaur genus from the Upper Jurassic of England. *Canadian Journal of Earth Sciences* 13(5):668-683.

McGowan, C. 1978. Further evidence for the wide geographical distribution of ichthyosaur taxa (Reptilia: Ichthyosauria). *Journal of Paleontology* 52:1155-1162.

McGowan, C. 1991. *Dinosaurs, Spitfires and Sea-Dragons.* Harvard University Press, Cambridge, Massachusetts, 365 pp.

Mitchum, R. and M. Uliana. 1985. Seismic stratigraphy of carbonate depositional sequences, Upper Jurassic-Lower Cretaceous, Neuquén Basin, Argentina. IN O. Berg and D. Woolverton (Eds.), Seismic Stratigraphy II, an Integrated Approach. *American Association of Petroleum Geologists, Memoir* 39:255-274.

Nessov, L. 1988. Some Late Mesozoic and Paleocene turtles of Soviet Middle Asia. *Studia Geológica Salamanticensia. Studia Palaeocheloniologica* 2(2):7-22.

Parrish, J. 1992. Jurassic climate and oceanography of the Pacific region. IN E. G. Westermann (Ed.), *The Jurassic of the Circum-Pacific, World and Regional Geology* 3, pp. 365-379. Cambridge University Press, London.

Riccardi, A. 1983. The Jurassic of Argentina and Chile. IN M. Moullade and A. Narin (Eds.), *The Phanerozoic Geology of the World. II: The Mesozoic* B, pp. 201-263. Elsevier Press, Amsterdam.

Riccardi, A. 1991. Jurassic and Cretaceous marine connections between the Southeast Pacific and Tethys. *Palaeogeography, Palaeoclimatology, Palaeoecology* 87:155-189.

Riccardi, C. and C. Gulisano. 1992. Unidades limitadas por discontinuidades. Su aplicación al Jurásico andino. *Asociación Argentina de Geología Revista* 45:346-364.

Riccardi, A., C. Gulisano, J. Mojica, O. Palacios, C. Schubert, and M. Thomson. 1992. Western South America and Antarctica. IN E. G. Westermann (Ed.), *The Jurassic of the Circum-Pacific, World and Regional Geology* 3, pp. 122-155. Cambridge University Press, London.

Rusconi, C. 1947. Plesiosaurios del Jurásico de Mendoza. *Anales de la Sociedad Científica* 146(5):327-351.

Rusconi, C. 1948a. Nuevo plesiosaurio, pez, langosta del mar jurásico de Mendoza. *Revista del Museo de Historia Natural de Mendoza* 2(1-2):3-12.

Rusconi, C. 1948b. Ictiosaurios del Jurásico de Mendoza. *Revista del Museo de Historia Natural de Mendoza* 2(1-2):17-160.

Scotese, C. 1991. Jurassic and Cretaceous plate tectonic reconstructions. *Palaeogeography, Palaeoclimatology, Palaeoecology* 87:493-501.

Spalletti, L., Z. Gasparini, and M. Fernández. 1994. Facies, ambientes y reptiles marinos de la transición entre las Formaciones Los Molles y Lajas (Jurásico medio), Cuenca Neuquina, Argentina. *Acta Geológica Leopoldesia* 39 (1):329-344.

Tarlo, L. 1960. A review of the Upper Jurassic pliosaurs. *Bulletin of the British Museum (Natural History), Geology* 4(5):147-189.

Torres, R. and L. Rojas. 1949. Una nueva especie y dos subespecies de Ichtyosauria del Jurásico de Viñales, Cuba. *Memorias de la Sociedad Cubana de Historia Natural* 19(2):197-200.

Wenz, S. 1968. Contribution à l'étude du genre *Metriorhynchus*. Crâne et moulage endocrânien de *Metriorhynchus superciliosus*. *Annales Paléontologie* (Vertébrés) 54(2):149-183.

Whitham, A. and P. Doyle. 1989. Stratigraphy of the Upper Jurassic-Lower Cretaceous Nordenskjöld Formation of eastern Graham Land, Antarctica. *Journal of South American Earth Sciences* 2(4):371-384.

Wood, R. and M. Freiberg. 1977. Redescription of *Notoemys laticentralis*, the oldest fossil turtle from South America. *Acta Geológica Lilloana* 13(6):187-204.

Chapter 16

MORPHOLOGICAL CONSTRAINTS ON TETRAPOD FEEDING MECHANISMS: WHY WERE THERE NO SUSPENSION-FEEDING MARINE REPTILES?

RACHEL COLLIN and CHRISTINE M. JANIS

INTRODUCTION

Although tetrapods are primarily designed for a terrestrial existence, tetrapod history is rife with secondarily aquatic forms. Semiaquatic taxa are usually freshwater, and more fully aquatic taxa are usually marine. With the reinvasion of the aquatic environment, modification of the tetrapod body plan originally designed for feeding and locomotion on land would be necessary to accommodate these functions in the very different medium of water. Both the diversity of living marine tetrapods and the additional diversity afforded by the fossil record demonstrate numerous examples of parallelisms and convergences in the various lineages (Massare, 1987, 1988; Carroll, 1985). Yet aquatic suspension feeding, common among fish, is conspicuously absent in marine nonmammalian tetrapods: baleen whales appear to be the only group that has adopted this form of feeding in past or present marine environments. The notion of morphological or phylogenetic constraint is often invoked when evolutionary patterns differ between groups of organisms (McKitrick, 1993; Carroll, 1985). In this chapter we review the morphological requirements for suspension feeding, and address the issue of whether differences in design between mammals and other tetrapods have resulted in morphological constraints that can explain the apparent absence of suspension-feeding marine reptiles.

During the Paleozoic, when oceanic productivity is thought to have been comparatively low (Bambach, 1993 and personal communication), marine tetrapods were rare. Those tetrapods that did venture into the water, such as the Permian mesosaurs, were small (1 m or so in length) and limited to inland saltwater basins (Carroll, 1988). As productivity increased, more groups of tetrapods diversified in the oceans. During the Mesozoic there were invasions and significant marine radiations of at least 16 diapsid lines (Carroll, 1985), as well as radiations of several groups of turtles, birds, and amphibians. Benton (in Cruickshank, 1993) estimates that as many as 31 groups of tetrapods have undergone marine radiations during the Mesozoic and Cenozoic combined (including three Cenozoic episodes of mammalian invasion: whales, sea cows, and pinnipeds). Radiations within these groups produced an array of high-level marine carnivores preying on a variety of other marine reptiles, fish, and invertebrates.

All Mesozoic marine tetrapods, except turtles, became extinct at or before the end of the Cretaceous. Turtles, sea snakes, and marine iguanas are reptiles found in today's oceans, but birds and mammals have been the dominant marine tetrapods throughout the Cenozoic. Whales represent the only radiation of marine tetrapods that includes suspension feeders. The crabeater seal, *Lobodon carcinophagus*, is sometimes considered to be a suspension feeder: however, it probably captures macroplanktonic food items individually (Sanderson and Wassersug, 1993).

MARINE TETRAPOD PREDATORY GUILDS

With the exception of sea cows, marine iguanas, and a few species of turtles, all of which have limited coastal distributions, marine tetrapods are, or were, carnivorous. The prey preference and predatory mode of extinct marine tetrapods can be inferred from tooth morphology and swimming capabilities. Most fossil and Recent marine tetrapods can be assigned to one of the predatory guilds Massare (1987) described for Mesozoic marine reptiles. All of these guilds are generally applicable to, and include, members from several radiations of marine reptiles and mammals, but do not include a suspension-feeding mode of life.

Suspension feeding, defined as the "capture (of) planktonic prey as water flows past the feeding apparatus" (Sanderson and Wassersug, 1993), implies microphagy and nonselectivity. Among tetrapods in today's oceans this feeding method is used only by baleen whales. However, suspension feeding is not particularly unusual in fish. It has evolved multiple times in actinopterygians (e.g., in the paddlefish, family Polydontidae, and within many teleost families, especially in the Clupeidae, Cyprinidae, and Cichlidae) and elasmobranchs (in the manta ray, the whale shark, the basking shark, and the megamouth shark) (Sanderson and Wassersug, 1993). Suspension feeding is also practiced by amphibian tadpoles, which possess gills and are thus not subject to typical tetrapod constraints on aquatic feeding (Sanderson and Wassersug, 1993).

Given the diversity of predatory guilds represented by the Mesozoic marine reptiles, and their prominence in the oceans for at least 150 million years, it is surprising that there is no evidence of suspension-feeding marine reptiles. A few fossil reptiles have been claimed to be suspension feeders, but on closer examination these interpretations are dubious at best. The Permian mesosaurs in the genus *Mesosaurus* are sometimes described as suspension feeders (Carroll, 1988). However, this reconstruction depends on an erroneous interpretation of the mandibular teeth as small marginal upper teeth, and a more detailed examination of mesosaur functional morphology suggests that they probably captured individual prey selectively rather than processing large volumes of water nonselectively as do true suspension feeders (S. Modesto, personal communication). The plesiosaur *Kimmerosaurus* has also been proposed to be a suspension feeder (Brown, 1981), but Sanderson and Wassersug (1993) consider its teeth to be too short and too widely spaced for this interpretation to be plausible. Based on its lack of teeth, Carroll and Zhi-Ming (1991) suggest the Triassic reptile *Hupehsuchus* may have been a suspension feeder. Not only is there no positive evidence to support this interpretation of its feeding biology, but the authors themselves point out that the small narrow skull and relatively long neck make suspension feeding unlikely (Carroll and Zhi-Ming, 1991).

Why is there no evidence of suspension-feeding marine reptiles in the Mesozoic? The fossil record of Mesozoic marine reptiles and large Cenozoic marine vertebrates, such as whales and seals, is by no means complete. However, we think it highly unlikely that preservational bias could account for the absence of suspension-feeding marine reptiles from the record. Fossils of other marine tetrapods show considerable diversity in feeding mechanisms (Massare, 1987; Carroll, 1988), and baleen whales and suspension-feeding sharks and rays from the Cenozoic have been preserved (Carroll, 1988).

The size attained by some baleen whales might suggest that suspension feeding is restricted to very large animals. However, suspension-feeding fish range from a few centimeters to several meters in length. Many early baleen whales were only 3 or 4 m long (Pivorunas, 1979), and modern baleen whales range in length from 6 m (the pygmy right whale, *Caperea marginata*) to 30 m (the blue whale, *Balaenoptera musculus*) (Leatherwood et al., 1983). This size range easily includes many marine reptiles, although at small body sizes suspension feeding is apparently limited to animals that possess gills (Sanderson and Wassersug, 1993).

We propose instead that neomorphies of the mammalian mouth and pharynx, especially those that allowed for a seal to be formed between the nasopharynx and the buccal cavity, were the key adaptive features in the origination and diversification of suspension-feeding mammals. All other fully marine tetrapods were constrained to feed on larger prey items. We first review the feeding methods

PLANKTON AVAILABILITY AND UPWELLING ZONES

employed by modern suspension-feeding vertebrates, and then discuss the morphological constraints that prevent nonmammalian tetrapods from feeding in these ways.

High productivity of nearshore waters during times of intense upwelling are assumed to have played an important role in the multiple origins and diversifications of marine tetrapods since the Permian (Lipps and Mitchell, 1976). Upwelling supplies surface waters with enough nutrients to support high densities of phytoplankton that, in turn, support high standing crops of zooplankton. Aquatic planktivores depend on the dense swarms of zooplankton that commonly occur near areas of high primary productivity, as capture rates of suspension-feeding organisms are closely related to prey density. Huge numbers of small suspension-feeding fish, or smaller numbers of large suspension feeders like whales and whale sharks, feed on these zooplankton. Although Mesozoic plankton communities were undoubtedly taxonomically different from those in Recent oceans (Tappan, 1969), there is no evidence that their productivity would not result in enough zooplankton to support large suspension feeders. In fact, areas of high productivity produced by upwelling and current divergences seem to have been present throughout the Mesozoic (Lipps and Mitchell, 1976). Parrish and Curtis (1982) used a paleoclimatological model to retrodict areas of upwelling from the deposition of organic-rich sediments, and showed that upwelling was especially prominent during the Triassic and Cretaceous, coinciding with the periods of greatest diversifications of marine reptiles (Bardet, 1994). Similarly, high levels of Cenozoic upwelling coincide with radiations of marine mammals (Lipps and Mitchell, 1976).

Although the physical conditions conducive to high primary productivity and large standing crops of zooplankton were probably common at some time during the Mesozoic, the absolute intensity and productivity of these areas cannot be measured. However, despite a sparse fossil record, the diversity and abundance of zooplankton can be inferred from fossilized phytoplankton and the maturity of the ecosystem. Tappan (1969) traced the stages of oceanic ecosystems through the Mesozoic and found that high phytoplankton diversity and productivity began during the Early Jurassic. The abundance of fossil coccolithophorids and dinoflagellates from this time, as well as radiations of herbivorous pelagic animals that rely on high levels of phytoplankton, implies high oceanic productivity. The fossil evidence indicates that Mesozoic oceanic productivity was able to support a diversity of trophic groups similar to that found during the Cenozoic. Therefore, limited ecological opportunity cannot account for the absence of suspension-feeding marine reptiles, especially as suspension-feeding actinopterygians were common during the Mesozoic (B. Chernoff, personal communication).

PREY CAPTURE IN SUSPENSION FEEDERS

Aquatic suspension feeders can collect zooplankton either by creating suction to draw prey into their mouths (suction feeding) or by relying on their momentum to force water into their mouths (ram feeding) (Sanderson and Wassersug, 1993). Suction feeding is used primarily by some suspension-feeding fish, which take advantage of opercular movements to draw water into the mouth and through the gill slits. Tadpoles and some birds may use other buccal structures to create suction. Although suspension-feeding birds may seem relevant to this discussion, only a bird that fed while fully submerged would be confronted with the same functional problems as other marine tetrapods, and none are known to have existed (Feduccia, 1980). Ducks and flamingos, which feed with their heads out of the water or only partially submerged, use their tongue in a suction-feeding fashion to pump small amounts of water through their mouth, where they collect algal cells on the lamellae of their partially submerged beak (Sanderson and Wassersug, 1993). The fringe-billed pterosaur *Pterodaustro* may have fed in an analogous fashion (Bakker, 1986). This use of the tongue to strain food may only be possible if the action is essentially performed out of the water, where large amounts of water will not be forced into the mouth and where presumably the action of gravity can also aid in expelling the water from the mouth. The bird mode of suspension feeding is a very different functional issue from engulfing a mouthful of water and expelling it in a completely submerged situation.

Suction is not usually employed by large suspension-feeding vertebrates: most whales and large chondrichthyans obtain food by ram feeding (Sanderson and Wassersug, 1993). However, suction feeding involving the capture of large prey items has been reported for a number of aquatic tetrapods, including salamanders, turtles, walruses, seals, and toothed (odontocete) whales (Werth, 1992), and intermittent suction feeding may be the mode of suspension feeding in the megamouth shark (Sanderson and Wassersug, 1993). The only baleen (mysticete) whales that use suction to feed are the gray whales (family Eschrichtidae), which are not planktonic suspension feeders, but instead suck up mud and water from the sea floor, and then strain it through the baleen, entrapping small food items (Nowak, 1991).

The majority of baleen whales are suspension feeders that feed by continuous or intermittent ram feeding. Continuous ram feeders, also known as skimmers or tow-net feeders (including large animals such as whale sharks and right whales as well as some small fish such as menhaden and anchovies), take advantage of the unidirectional flow produced by their movement through the water. Swimming with the mouth open, they continually force water into the mouth anteriorly; the water then flows out of the buccal cavity laterally through the baleen (in the case of the whales), or exits through the gill slits in fish (where the particles are either trapped on gill rakers or entrained in mucus on the roof of the mouth) (Sanderson et al., 1991). Several species of whales in the genus *Balaena* (right or bowhead whales) feed continuously by trapping particles as water exits through baleen filters along the side of the mouth.

Rorqual whales (Balaenopteridae), the only known intermittent ram feeders, collect food by engulfing plankton-laden water. To begin the feeding process, the animal swims forward with its mouth open, engulfing a tremendous amount of water. The mouth is subsequently closed until the baleen covers the gape between the upper and lower jaws; water is then forcefully expelled, trapping plankton on the baleen. The expulsion of water is possibly by the compression of the buccal cavity and the protrusion of the tongue, although the precise mechanism is unknown (Sanderson and Wassersug, 1993). In order to generate enough momentum to feed in this way, intermittent ram feeders must be of large body size (Sanderson and Wassersug, 1993); the smallest rorqual is the Minke whale (*Balaenoptera acutororostrata*) of around 10,000 kg, and the other four species are considerably larger than this, with the blue whale (*Balaenoptera musculus*) attaining body masses of up to 190,000 kg (Nowak, 1991). Note that there is no biomechanical reason for a similar size constraint on continuous ram feeders or on suction feeders.

RECOGNIZING A FOSSIL SUSPENSION-FEEDING TETRAPOD

How might one identify a suspension-feeding marine reptile from its fossil remains? Sanderson and Wassersug (1993) have identified a number of morphological features, many of which could be preserved in fossil specimens, that are functionally correlated with various modes of suspension feeding, and the discussion below is modified from their work.

First, the head is large, from a quarter to a third of the total length, with the elongation of the rostrum to allow for a large buccal volume. The dentition is reduced: chondrichthyans have many rows of small, reduced teeth, and baleen whales have lost the teeth entirely and replaced them with baleen. The eyes are small, and the orbits are reduced in all large living suspension feeders, presumably because vision is not an important sense for this mode of feeding.

Baleen whales have a large, anteriorly sloping, occipital area for the expansion of the area of attachment of the epaxial muscles. These muscles probably act to resist the downward torque on the head produced when the mouth is opened and to resist sideways movement of the head. They also have a very short cervical region, to stabilize the head by limiting the movement of the head on the body. The jaw bones are elongated and broadened, but are not heavily ossified, as they function as regulators of water flow rather than organs of prey capture. In the intermittent ram-feeding rorquals, there are additional morphological features that allow the jaws to be rapidly closed around a mouthful of water, such as a small coronoid process reflecting the presence of somewhat more powerful jaw adductor muscles than in right whales.

How might these criteria be applied to a marine tetrapod of unknown feeding mode? A possible candidate is *Shonisaurus* from the Late Triassic of Nevada (Camp, 1980), an animal not discussed by Massare (1987). This is the largest known ichthyosaur, with a total length of up to 15 m (Camp, 1980; Kosch, 1990), and is thus of comparable size to many baleen whales. It has a relatively large head, long slender jaws with a reduced dentition, and a peculiarly deep body. Its limbs are unusual in being greatly elongated through the manus and the pes, with the hindlimb of equal size to the forelimb. Carroll (1988) points out that the limbs are unusual in only having three rows of phalanges, where hyperphalangy is the norm for ichthyosaurs, and suggests that the limbs functioned more like paddles than like fins. *Shonisaurus* certainly presents the appearance of an animal that gently skulled through the water, rather than one that actively pursued its prey. But might this animal have been a suspension feeder?

Although the head of *Shonisaurus* is relatively large and slender-jawed, its head is only about 20% of its total length (Kosch, 1990), which is somewhat smaller than the proportions of living large suspension feeders. The orbit is large, suggesting a large eye which is not typical of present-day suspension feeders. However, most importantly, *Shonisaurus* does not possess any morphological features that could be interpreted as a straining device, or as supporting a soft-tissue straining device. Although the teeth are small, they are fairly widely spaced, and are certainly not closely packed in any fashion that might act as a sieve to strain plankton. The upper and lower jaws are appressed against each other: there is no arching of the upper jaw to accommodate a soft-tissue filtering device, as in continuous-feeding right whales. An intermediate-feeding mode (as in rorquals), where the filtering tissue might be less extensive and not evidenced by a profound arching of the jaws, is precluded by the small coronoid bone, suggesting the lack of sufficient jaw adductor musculature to rapidly close the jaws around a mouthful of water.

The form of the teeth best seem to approximate to the type that Massare (1987) interpreted as generalized grasping organs, possibly indicative of a diet of soft-bodied prey items such as cephalopods. The reduced dentition is reminiscent of that of present-day beaked whales (Ziphiidae) which employ suction to ingest squid and small fish (Nowak, 1991; Werth, 1992): perhaps *Shonisaurus* had developed some analogous mode of feeding, a dietary interpretation that would be consistent with the skeleton which suggests slow swimming rather than active pursuit. However, the available evidence does not indicate that *Shonisaurus* was a suspension feeder.

NEOMORPHIES OF THE MAMMALIAN PHARYNX

We suggest that the loss of gills in adult tetrapods and the design of the tetrapod oropharyngeal region severely constrains possible aquatic feeding mechanisms. However, neomorphies associated with the evolution of mastication and suckling may have allowed mammals to develop suspension feeding. These features include the ability to separate different portions of the oropharyngeal cavity from each other by the means of muscular seals acting against the posterior soft palate (Smith, 1992; Crompton, 1995) (see Figure 1B). These seals can act to contain food being masticated in the oral cavity from the pharyngeal region until swallowing occurs, and can also allow liquids to be swallowed while maintaining a patent airway.

These seals are formed in the following fashion (Crompton, 1995). The anterior seal (seal #1) is formed between the oral cavity and the oropharynx by the tensing of the tensor veli palatini muscle, and the drawing of the tongue dorsally by the palatoglossus muscle. The intrinsic tongue musculature can then shape the tongue so that its dorsal surface is forced up against the anterior portion of the soft palate. The posterior seal (seal #2), separating the oropharynx from the nasopharynx, results from the contraction of the palatopharyngeal muscle within the posterior portion of the soft palate, which grips the epiglottis and the larynx so that they form a seal against the back of the soft palate. The typical mammalian pharyngeal elevator and constrictor muscles, which include the palatoglossus and the palatopharyngeus, represent a differentiation of the branchiomeric musculature unique to mammals, especially in the fact that they are largely innervated by cranial nerve X (as opposed to the muscles used in swallowing in nonmammalian amniotes, which are primarily innervated by nerve VII) (Smith, 1992).

The action of these two seals can best be seen in suckling: first, with seal #1 applied, depression of the tongue behind the nipple allows milk to accumulate in the oral cavity; next, seal #1 is broken by lowering the tongue and releasing the tension on the anterior soft palate, and milk accumulates in the oropharynx, where seal #2 prevents it from entering the nasopharynx; finally, with the reapplication of seal #1 and the drawing backward of the tongue, milk is forced backward around the epiglottis into the esophagus. The continued application of seal #2 as the milk is

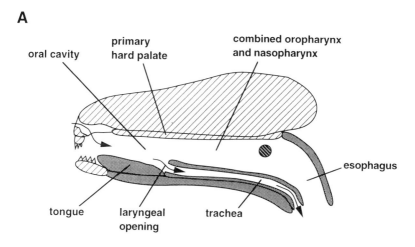

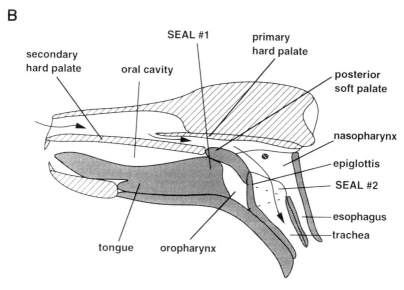

Figure 1. Schematic cutaway views of tetrapod oropharyngeal cavities: **A)** Condition in a nonmammalian amniote (modified from Smith, 1992, based on *Ctenosaura similis*, and from Kardong, 1995). **B)** Condition in mammal (modified from Crompton, 1995, based on *Didelphis virginianus*). Explanation: **arrows** = passage of air, **lightly hatched areas** = bone, **darkly hatched areas** = middle ear opening, **shaded areas** = soft tissue.

swallowed allows for a patent airway to be maintained during swallowing, so that suckling and breathing can occur at the same time. However, when mammals swallow solid food, seal #2 is invariably broken to allow its passage into the esophagus.

In adult mammals seal #1 acts to retain the food in the oral cavity until it has been masticated to a suitable consistency. The food then moves past seal #1 into the oropharynx, where a bolus is formed, and the bolus subsequently moves over the epiglottis, across the lower part of the nasopharynx, into the esophagus. During the swallowing of the food bolus seal #2 prevents the passage of food into the trachea. Note that in adult humans the larynx is moved posteriorly; the oropharynx and nasopharynx are confluent, and seal #2 is absent. Humans are the only mammals that are at risk of choking by accidentally inhaling food into the trachea, and are also the only ones that can voluntarily breathe through the mouth (see Lieberman, 1984).

Baleen whales invert the tongue to form a large pocket during feeding, presumably allowing the confluence of the oral cavity and oropharynx for the collection of plankton-laden water with the loss of seal #1 (F. W. Crompton, personal communication, following illustrations in Lambertsen, 1983). In this case the tight application of seal #2, preventing the mass of water from entering the trachea or esophagus, is clearly essential.

Despite the obvious utility of these seals in suckling, they probably originally evolved as a component of the mammalian swallowing reflex (Smith, 1992; F. W. Crompton, personal communication), where mastication reduces food to a bolus of small particles that is then swallowed in a single action down a relatively narrow food passageway. The morphology of the hard palate of derived cynodonts such as *Pachygenelus* suggests that it had a tensor veli palatini muscle (essential for forming seal #1) (Crompton, 1995). Presumably seal #2 was also present; the presence of seal #1 suggests that suckling may have occurred (Crompton, 1995), and the mammalian suckling reflex requires the interaction of the two seals. The inferred presence of these pharyngeal seals in derived cynodonts implies that they are mammalian symplesiomorphies.

SUSPENSION FEEDING IN MARINE TETRAPODS

Although the detailed mechanisms of prey capture in modern suspension feeders are poorly understood (Sanderson and Wassersug, 1993), a few general constraints may be inferred from the process. To be a successful suspension feeder without gill slits, an animal must be able to seal the back of the oral cavity to prevent engulfed water from entering the pharynx. It must also possess the appropriate facial musculature (lips and cheeks in at least the primitive mammal condition) that can resist the intraoral water pressure produced during feeding, and a way to collect and swallow small particulate food items. Unfortunately, many of the structures that confer these abilities on modern suspension feeders are composed of soft tissues that are not easily characterized in the fossil record. In the absence of contradictory evidence, we consider the characteristics of modern nonavian diapsids representative of diapsid marine reptiles. All modern nonmammalian amniotes and adult terrestrial amphibians lack the features that would facilitate the transition from a carnivorous to a nonselective microphagous way of life.

When water is forced into the oral cavity during any type of suspension feeding, food particles must be retained and directed toward the pharynx while water is expelled from the oral cavity. Food may be caught in mucus or trapped in strainers that cover the area through which water exits the mouth (gills in fish and the sides and/or front of the mouth in whales). In animals without gills, a primary alternate route of escape for water is through the rear of the oral cavity. Separation of the oral and nasal cavities by hard and soft palates and the associated pharyngeal musculature are mammalian neomorphies that allow the formation of seals at the back of the mouth. The critical feature for suspension-feeding mammals is seal #2, which grips the epiglottis and larynx against the posterior pharynx, preventing the animal from inhaling or ingesting large volumes of water, and allowing it to collect and swallow food without disrupting the airway.

Among modern aquatic reptiles, turtles have been reported to suction feed, although they cannot seal the back of their oral cavity. Turtles apparently swallow water together with the food (D. Bramble, personal communication), and it seems unlikely that they could evolve suspension feeding from this mode of food intake. Thus, it is clear that nonmammalian amniotes are not incapable of suction feeding of some sort, although the separation of the oropharynx from the trachea by seal #2 appears to be a critical feature in suction feeding toothed whales (Werth, 1992). As ectotherms only require about one tenth the amount of food per day as endotherms, the actual amounts of water that might be accidentally ingested or inspired along with engulfed food would similarly be less.

Crocodiles might be potential candidates for developing a form of suspension feeding analogous to that seen in mammals. Crocodiles alone among modern reptiles can elevate the base of the tongue to form a seal between a hard palate and the tongue, and have a separation of the oral, nasal, and tympanic cavities, paralleling the mammalian condition (Smith, 1992). Although an anterior hard palate consisting of the maxilla and vomer is primitive for crocodiles, a posterior hard palate, including extensions of the pterygoids, is an apomorphy of the Eusuchia, the Cenozoic clade of primarily freshwater crocodiles (Benton and Clark, 1988). The oral, nasal, and pharyngeal cavities are continuous with each other in all other modern tetrapods, except some lacertilians (Smith, 1992) (see Figure 1A). In these groups, water pushed toward the back of the mouth could easily be forced through the pharynx and into the esophagus. Large volumes of accidentally ingested seawater could potentially cause serious osmotic stress to an animal, although living marine diapsids do possess extrarenal salt glands which can excrete excess salt. Ingested salt might especially be a problem for suspension feeders, as they must process a greater number of mouthfuls containing a greater volume of water than a macrophagous animal.

Water may also be forced from the oropharyngeal cavity into the lungs. In modern baleen whales the epiglottis is permanently positioned to separate the entrance to the trachea completely from the oropharynx (Werth, 1992), but in the typical reptilian condition the larynx opens directly into the oral cavity at the base of the tongue. The laryngeal constrictor muscles may help guard against accidental inhalations; however, they are probably not as effective as complete separation of the two passages. The oropharyngeal cavity is also continuous with the choana and the tympanic area (Smith, 1992), both of which could be stressed by high internal water pressure associated with suspension feeding. Thus, the divisions of the mammalian oropharyngeal region may be vital to successful suspension feeding.

Facial muscles may also be needed for efficient plankton capture in the absence of gills. During continuous ram feeding, the sides of the face must resist the lateral component of the water pressure to prevent disruption of the flow past the straining devices. During intermittent ram feeding, contraction of facial muscles may be necessary to complete expulsion of water from the mouth. In whales, expulsion may be brought about by contraction of a sheet of muscles surrounding the buccal area and face (Lambertsen, 1983). Since facial muscles are not present in nonmammalian tetrapods, they lack convenient precursors to be modified for this function. Lips, which form a seal at the front of the mouth, allowing mammals to exclude water from the oral cavity, are also a uniquely mammalian character. Although whales do not have typically mammalian "lips" or "cheeks," their sheet of facial musculature is presumably homologous with the more generalized mammalian condition of facial muscles. Again, both cheeks and lips probably evolved in association with mastication and suckling.

After water is expelled from the mouth, the food must be collected and swallowed. In contrast with the mammalian condition, the primitive tetrapod buccal apparatus is not well suited for ingesting small particles. The unmodified tetrapod tongue does not have a mobile fleshy tip and can produce less movement independent of the hyoid than can a mammalian tongue (Smith, 1992). This may severely hinder an animal's ability to collect many small, widely distributed particles, and to form a bolus from them. Furthermore, nonmammalian tetrapods are poorly equipped to swallow small particulate food items; they generally ingest large lumps of food that they slowly force from the buccal cavity to the esophagus. Rather than swallowing their food with a quick swallowing reflex, as do mammals, nonmammalian tetrapods rely on the slow squeezing of the constrictor coli muscle, and often on head movements and gravity, to force food from the back of the mouth into the esophagus, and swallowing is a more gradual process (Smith, 1992). This method of swallowing would not work well for collections of small particles: the action of the constrictor coli would simply squash such a food bolus, effectively redistributing the food throughout the mouth. Since the nonmammalian nasal and auditory cavities are continuous with the oropharyngeal cavity, they could also potentially become coated with food particles (Smith, 1992). Mammals do not encounter any of these problems when processing small particulate food since the oropharyngeal cavity is well divided and they have a sophisticated swallowing reflex ideally suited to the ingestion of small particles. The mammalian swallowing reflex probably arose in conjunction with teeth suitable for chewing (Smith, 1992).

CONCLUSION

In the absence of gill slits, the primitive tetrapod characters of an undivided oropharyngeal cavity, the absence of muscular lips and cheeks, a relatively immobile tongue, and lack of a coordinated swallowing reflex were not conducive to the evolution of suspension feeding. However, it is not inconceivable that a nonmammalian tetrapod could have evolved mammal-like features of the orobuccal region in a convergent fashion. For example, a posterior secondary palate could have developed earlier within the crocodilian line, in which case it might have been a feature of the Mesozoic marine crocodiles. Additionally, large, scaly plates could have acted in an analogous fashion to facial muscles in adding the necessary resistance to the cheeks of a potential continuous ram feeder. Certain birds and amphibians have also evolved a more mobile tongue (Smith, 1992; Bramble and Wake, 1985), and ornithischian dinosaurs probably had muscular cheeks (Norman and Weishampel, 1985). However, it would require the modification of a whole complex of primitive tetrapod characters, rather than any one key morphological feature, to permit the capture and ingestion of small particulate food items necessary for successful suspension feeding in a nonmammalian tetrapod.

Mammals, on the other hand, are already equipped with a suite of characters that allow them to perform many of the steps associated with successful suspension feeding. The ability to form a seal between the nasopharynx and the buccal cavity, and the possession of a fleshy mobile tongue and muscular lips and cheeks, are all part of the complex morphological changes associated with the evolution of mammalian mastication, swallowing, and suckling (Smith, 1992; Crompton, 1995). These features are ideally suited to the demands of suspension feeding, and they probably needed little modification to function effectively in this new context.

Marine reptiles also show a diversity of body sizes that overlap with modern vertebrate suspension feeders, so it is unlikely that their diet was constrained by size. However, it is also true that there are no known aquatic reptiles that attained the very large body sizes seen in some present-day baleen whales, especially of those that feed by intermittent ram feeding, for which large size is a requirement (Sanderson and Wassersug, 1993). The largest Mesozoic marine reptiles include the pliosaur (short-necked plesiosaur) *Kronosaurus* and the ichthyosaur *Shonisaurus*, which both reached lengths of around 15 m (Carroll, 1988). Coincidentally, this is a similar size to that of the largest living toothed whale, the sperm whale (Nowak, 1991), so perhaps the lack of a suspension-feeding mode among Mesozoic marine reptiles may also explain why they did not reach sizes attained by the largest of the present-day baleen whales.

We believe that ecological constraints do not adequately account for the absence of suspension-feeding marine reptiles. Although the planktonic communities were different from those in Recent oceans, it is unlikely that there was too little productivity to support large-bodied suspension feeders at any time during the Mesozoic. In conclusion, we believe that morphological phylogenetic constraints and not ecological conditions may have been insurmountable obstacles in the evolution of nonmammalian tetrapod suspension feeders.

SUMMARY

Diapsid reptiles were common free-swimming marine predators during the Mesozoic. Although there was considerable diversity in their modes of predation, there were no suspension-feeding species. We suggest a morphological constraint prevented the evolution of this feeding type among Mesozoic marine reptiles, rather than scarcity of suitable planktonic prey. Aquatic tetrapod suspension feeders cannot use gill slits to create a unidirectional current through the buccal cavity, as do suspension-feeding fish. Instead they must feed by engulfing water containing many small planktonic animals and expelling water from the oral cavity via the mouth. The characters that enable baleen whales to feed in this way are all mammal neomorphies. Without the mammalian ability to form a tight seal at the back of the mouth, and to collect and swallow small particulate food, nonmammalian tetrapods, including Mesozoic marine reptiles, may have been limited to capturing larger individual prey items.

ACKNOWLEDGMENTS

We would like to thank the other members of GRIPS (S. D'Hont, D. Fastovsky, J. Sepkoski, and T. Webb), and L. Adler, A. W. Crompton, B. Crowley, M.

Fortelius, J. Hopson, J. Massare, C. McGowan, S. Modesto, D. Morse, and R. Strathmann for their thoughtful suggestions on earlier drafts of the manuscript. We also thank R. Bambach, D. Bramble, B. Chernoff, and S. Modesto for personal communications, and A. W. Crompton both for extensive personal communications and for access to a manuscript in press. Special thanks to C. McGowan for setting us the delightful challenge to demonstrate that *Shonisaurus* was not a suspension feeder. This is a GRIPS (Greater Rhode Island Paleontological Society) contribution.

REFERENCES

Bakker, R. T. 1986. *The Dinosaur Heresies: New Theories Unlocking the Mystery of the Dinosaurs and Their Extinctions.* William Morrow and Company, New York, 481 pp.

Bambach, R. K. 1993. Seafood through time: changes in biomass, energetics, and productivity in the marine ecosystem. *Paleobiology* 19:372-397.

Bardet, N. 1994. Extinction events among Mesozoic marine reptiles. *Historical Biology* 7:313-324.

Benton, M. J. and J. M. Clark. 1988. Archosaur phylogeny and the relationships of the Crocodilia. IN M. J. Benton (Ed.), *Phylogeny and Classification of Tetrapods, Volume 1: Amphibians, Reptiles, Birds,* pp. 295-338. Systematics Association Special Publication 35A. Clarendon Press, Oxford.

Bramble, D. M. and D. B. Wake. 1985. Feeding mechanisms of lower tetrapods. IN M. Hildebrand, D. M. Bramble, K. F. Liem, and D. B. Wake (Eds.), *Functional Vertebrate Morphology,* pp. 230-262. Belknap Press, Cambridge.

Brown, D. S. 1981. The English Upper Jurassic Plesiosauroidea (Reptilia) and a review of the phylogeny and classification of the Plesiosauria. *Bulletin of the British Museum of Natural History, Geology* 35:253-347.

Camp, C. L. 1980. Large ichthyosaurs from the Upper Triassic of Nevada. *Palaeontographica,* Abteilung A 170:139-200.

Carroll, R. L. 1985. Evolutionary constraints in aquatic diapsid reptiles. *Special Papers in Palaeontology* 44:145-155.

Carroll, R. L. 1988. *Vertebrate Paleontology and Evolution.* W. H. Freeman and Co., New York, 698 pp.

Carroll, R. L. and D. Zhi-Ming. 1991. *Hupehsuchus,* an enigmatic aquatic reptile from the Triassic of China, and the problem of establishing relationships. *Philosophical Transactions of the Royal Society of London* B 331:131-153.

Crompton, A. W. 1995. Masticatory function in nonmammalian cynodonts and early mammals. IN J. J. Thomason (Ed.), *Functional Morphology in Vertebrate Paleontology,* pp. 55-75. Cambridge University Press, Cambridge.

Cruickshank, A. R. I. 1993. Meeting Reports. Review seminar: marine tetrapods (University of Leicester, 24th February 1993). *Palaeontology Newsletter,* The

Palaeontological Association 18 (Spring 1993):6.

Feduccia, A. 1980. *The Age of Birds*. Harvard University Press, Cambridge, Massachusetts, 196 pp.

Kardong, K. V. 1995. *Vertebrates: Comparative Anatomy, Function, Evolution*. Wm. C. Brown Publishers, Dubuque, Iowa, 777 pp.

Kosch, B. F. 1990. A revision of the skeletal reconstruction of *Shonisaurus popularis* (Reptilia: Ichthyosauria). *Journal of Vertebrate Paleontology* 10:512-514.

Lambertsen, R. H. 1983. Internal mechanism of rorqual feeding. *Journal of Mammalogy* 64:76-88.

Leatherwood, S., R. R. Reeves, and L. Foster. 1983. *The Sierra Club Handbook of Whales and Dolphins*. Sierra Club Books, San Francisco, 302 pp.

Lieberman, P. 1984. *The Biology and Evolution of Language*. Harvard University Press, Cambridge, Massachusetts, 399 pp.

Lipps, J. and E. Mitchell. 1976. Trophic model for the adaptive radiations and extinctions of pelagic marine mammals. *Paleobiology* 2:147-155.

Massare, J. 1987. Tooth morphology and prey preference of Mesozoic marine reptiles. *Journal of Vertebrate Paleontology* 7:121-137.

Massare, J. 1988. Swimming capabilities of Mesozoic marine reptiles: implications for method of predation. *Paleobiology* 14:187-205.

McKitrick, M. C. 1993. Phylogenetic constraint in evolutionary theory: has it any explanatory power? *Annual Review of Ecology and Systematics* 24:307-330.

Norman, D. B. and D. B. Weishampel. 1985. Ornithopod feeding mechanisms: their bearing on the evolution of herbivory. *American Naturalist* 126:151-164.

Nowak, R. M. 1991. *Walker's Mammals of the World*, 5th edition. The Johns Hopkins University Press, Baltimore, 2 vols., 1629 pp.

Parrish, J. T. and R. L. Curtis. 1982. Atmospheric circulation, upwelling and organic-rich rocks in the Mesozoic and Cenozoic eras. *Palaeogeography, Palaeoclimatology, Palaeoecology* 40:31-66.

Pivorunas, A. 1979. The feeding mechanisms of baleen whales. *American Scientist* 67:432-440.

Sanderson, S. L. and R. Wassersug. 1993. Convergent and alternative designs for vertebrate suspension feeding. IN J. Hanken and B. K. Hall (Eds.), *The Skull, Vol. 3: Functional and Evolutionary Mechanisms*, pp. 37-112. Chicago University Press, Chicago.

Sanderson, S. L., J. J. Cech, and M. R. Patterson. 1991. Fluid dynamics in suspension-feeding blackfish. *Science* 251:1346-1348.

Smith, K. K. 1992. The evolution of the mammalian pharynx. *Zoological Journal of the Linnean Society* 104:313-349.

Tappan, H. 1969. Microplankton, ecological succession and evolution. *North American Paleontological Convention Proceedings*, Part H 1058-1103.

Werth, A. J. 1992. *Anatomy and Evolution of Odontocete Suction Feeding*. Unpublished Ph. D. Thesis, Harvard University, Cambridge, Massachusetts, 312 pp.

Chapter 17

MESOZOIC MARINE REPTILES AS MODELS OF LONG-TERM, LARGE-SCALE EVOLUTIONARY PHENOMENA

ROBERT L. CARROLL

INTRODUCTION

There is less consensus among biologists today than at any time since the early years of the 20th century as to the patterns and processes of evolution. It is clear that the pattern of large-scale, long-term evolution hypothesized by Darwin (1859), on the basis of direct extrapolation from patterns seen in modern populations, was wrong in suggesting a gradual and progressive unfolding over hundreds of millions of years.

The one illustration in the first edition of *On the Origin of Species* diagrams a continuous and essentially isotropic distribution of radiating lineages and a similar range of evolutionary rates over all time scales. This contrasts strongly with the disjunct distribution of morphological patterns that is shown by the phylogenies of all major taxonomic groups known from the fossil record. Vascular plants, nonvertebrate metazoans, and vertebrates all show a pattern in which there are a relatively few major taxonomic groups, all of which are clearly distinct from one another, both morphologically and adaptively. Most appear suddenly, already highly derived relative to any possible predecessors, and retain a basically similar anatomical pattern for tens or hundreds of millions of years. Very few intermediate forms, linking the major groups, are known.

Gould and Eldredge (1977, 1993), Stanley (1975, 1979), and Eldredge and Gould (1972) specifically emphasized the importance of stasis within species and the significance of intrinsic constraints at all taxonomic levels in accounting for the limited capacity for change within groups. They argued that major changes between groups are the result of factors that were not discussed by Darwin, nor adequately considered within the evolutionary synthesis. The importance of stasis was illustrated by Eldredge and Stanley's (1984) book, *Living Fossils*, in which were discussed the case histories of many taxa that have undergone very little evolutionary change over millions of years. In view of the impression given by the writings of Darwin and many subsequent textbooks, that all species are continually undergoing progressive evolutionary change, it is necessary to emphasize that both species and larger taxonomic groups retain basically similar anatomical, physiological, and behavioral characteristics for very long periods of time.

On the other hand, concentration on stasis provides only a very limited view of a particular category of evolutionary rates. If one is concerned with the patterns and mechanisms of evolutionary change, it is necessary to study groups within which major anatomical and adaptive changes have occurred.

An excellent example is provided by the aquatic adaptation of many groups of reptiles during the Mesozoic (Carroll, 1987). This assemblage provides a very informative model for the study of anatomical changes in a multitude of lineages that shifted from one major adaptive zone to another. Leaving aside marine turtles, most aquatic reptiles are derivatives of the vast radiation of diapsid reptiles that began in the Late Carboniferous. We have detailed knowledge of the anatomy of early diapsids (Reisz, 1981, 1995; Laurin, 1991) and of the primitive archosauromorphs and lepidosauromorphs that make up the major terrestrial groups (Benton, 1985; Carroll and Currie, 1991). Aquatic diapsids do not constitute a monophyletic assemblage, but are represented by a number of distinct lineages, each of which evolved independently from different subgroups of terrestrial diapsids (Figure 1). Nevertheless, all early diapsid lineages share a generally similar anatomical pattern and presumably similar developmental processes that provide a common basis for comparison with the many different ways in which derived groups have adapted to life in the water.

Approximately 20 lineages have independently adapted to an aquatic way of life. The major groups are represented by many superbly preserved fossils, frequently showing every bone in the body in nearly perfect articulation. Even where specific ancestry cannot be established, the polarity of character changes and their functional association with the physical properties of water can be readily determined. All groups show clearly recognizable anatomical changes associated with locomotion, buoyancy control, and feeding in the water. Several exhibit modifications associated with hearing and respiration. The ichthyosaurs provide the most extreme example of aquatic adaptation, with a body form comparable to that of the fastest swimming modern fish, and were apparently unique in giving birth to live young in the water.

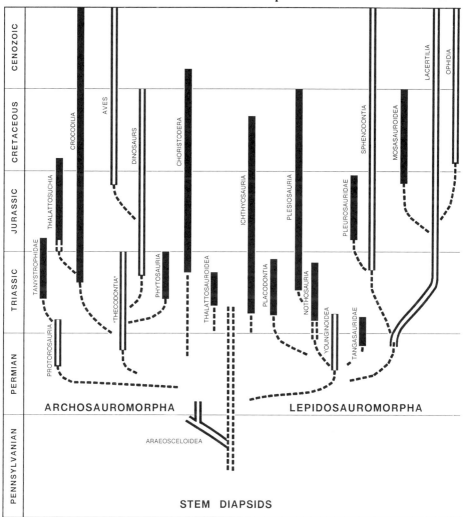

Figure 1. Phylogeny of diapsid reptiles, emphasizing lineages that have secondarily adapted to life in the water. Explanation: **open lines** = terrestrial lineages, **solid lines** = aquatic lineages, **dashed lines** = assumed relationship or range extensions.

Mesozoic marine reptiles provide a model for evaluating many aspects of macroevolutionary change (Carroll, 1985b; deBraga and Carroll, 1993; Bardet, 1994). This chapter deals with the degree to which it is necessary to hypothesize special processes, not considered by Darwin or discussed in the evolutionary synthesis, to explain the origin of major groups, novel structures, and major habitat changes.

Evolutionary biologists have long argued that special factors must be involved in such significant changes. Simpson (1944, 1953) coined the term *quantum evolution* to refer to the special circumstances that might explain these phenomena.

Thomson (1988) and Hall (1992) recently argued that such transitions may be associated with major reorganization of developmental processes. Each of the groups of secondarily aquatic reptiles provides a separate example of major changes in structure and way of life from their primitive, terrestrial ancestors. This assemblage may thus provide considerable data as to the phenomena involved.

COMPLETENESS OF THE FOSSIL RECORD

Before discussing the individual cases, it is necessary to review the nature of the fossil record in order to determine the level of resolution that is possible. Simpson (1944, 1953) stated that there were few, if any, examples in which transitions between major groups were adequately documented in the fossil record. Unfortunately, this is still largely true. All of the postulated transitions between primitive terrestrial diapsids and the many aquatic lineages lack significant intervals. Most of the descendant groups are markedly different from their putative ancestors when they first appear in the fossil record. Intermediates are known, but in no lineage is a step-by-step transition recorded.

A question that is seldom asked is whether the nearly universal rarity of intermediate forms results from biological aspects of major transitions, or is simply a result of the general incompleteness of the fossil record. Transitions between major groups are nearly always represented in pictorial phylogenies as gaps or dotted lines, while the duration of the resulting lineage is presented as a solid line or an outline implying a continuous fossil record. This can be quit misleading. McGowan (1983) emphasized that Jurassic ichthyosaurs are known from only a small number of horizons of limited temporal and geographical extent. He estimated that more than 600 specimens have been recovered from Holzmaden, but there are only a handful of productive horizons throughout the remainder of the Jurassic. His global description of our knowledge of this group is expressed thus: "We therefore catch only glimpses of their evolutionary history, as if viewed through a series of rents in the curtain." These gaps in the fossil record have nothing to do with biological processes, but are simply the result of the absence of strata representing the environment in which these animals lived, as a result of nondeposition or subsequent erosion.

The nothosaurs (Figure 2) also show a very patchy temporal distribution, with nearly all described species and most genera known from only a single horizon. Significant gaps not only mark their origin, but also occur during the time of their early radiation and within all individual lineages. The current knowledge of the Choristodera provides an even more dramatic picture (Figure 3) in which the fossil record is so incomplete that no biological significance can be attached to gaps at any

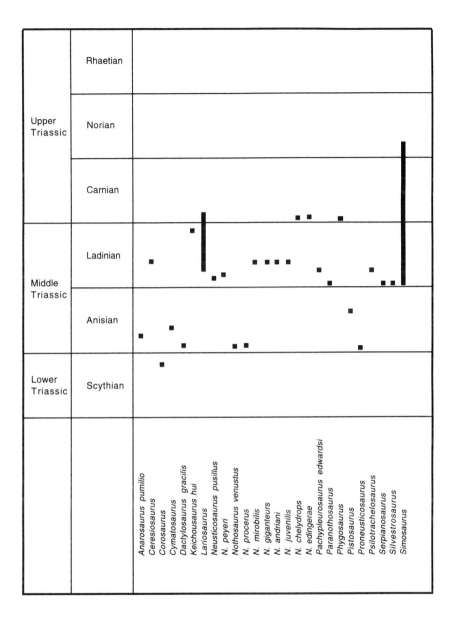

Figure 2. Stratigraphic record of nothosaurs showing gaps in the fossil record during the period of their origin from primitive terrestrial diapsids, during the early radiation within the group, and within individual lineages. Systematics of *Lariosaurus* and *Simosaurus* are not sufficiently well known to establish ranges of included species.

stage of their evolution (Evans and Hecht, 1993).

Despite the very incomplete fossil record, one can say that the period of

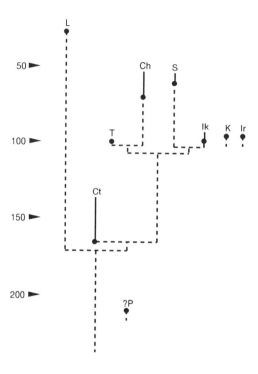

Figure 3. Fossil record of the Choristodira, from Evans and Hecht (1993). Abbreviations: **Ch** = Champsosaurus, **Ct** = Cteniogenys, **Ik** = Ikechosaurus, **Ir** = Irenosaurus; **K** = Khurendukhosaurus, **L** = Lazarossuchus, **P** = Pachystropheus, **S** = Simoedosaurus, **T** = Tchoiria, **P** = Pachystropheus.

transition between the putative terrestrial ancestor and the emergence of each aquatic lineage is substantially shorter than the duration of the descendant group, but this can rarely be quantified. Nevertheless, the gap is sufficiently long that almost any pattern of transformation might be hypothesized. Evolution of some characters might be more rapid than at other stages, but even this is subject to debate (see Gingerich, 1983, 1993, on difficulties of comparing evolutionary rates over different time scales).

The most serious gap in the fossil record of Mesozoic aquatic reptiles is between the Late Permian and Early Triassic. This is the period of origin of two major groups, the ichthyosaurs and the sauropterygians, as well as the assemblage including the Claraziidae, Thalattosauridae, and Askeptosauridae (Rieppel, 1987). All of these groups are very distinct from any putative ancestor when they first appear in the fossil record, suggesting a significant period of prior evolution. Such gaps preclude not only analysis of evolutionary processes associated with their origin, but even determination of their phylogenetic relationships within the Diapsida. The large gap at the base of these lineages might in some way be related to the major extinction at the end of the Permian and the resulting reorganization of the marine biota (Erwin, 1993), but two lineages are known during this time from East Africa and Madagascar (Carroll, 1981; Currie, 1982). The tangasaurids were younginoid derivatives that are not known beyond the Early Triassic, but *Claudiosaurus* has been proposed as a sister taxon to the sauropterygians (Storrs, 1993). In the absence of any information on groups specifically ancestral to the ichthyosaurs, sauropterygians, and thalattosaurs during this period, nothing can be said of the evolutionary processes involved in their origin.

In contrast with the arguments of Eldredge and Gould (1972) that the fossil record is sufficiently well known to establish patterns of evolution at the species level throughout the Phanerozoic, enormous gaps in time separate productive horizons of most vertebrate groups during the Paleozoic and Mesozoic. Even within the nearshore, aquatic environment where the antecedents of later marine reptiles may have lived, and for which a higher than average chance of fossilization would be expected, our knowledge of the fossil record is still too incomplete to document the origin of many of the major lineages. On the other hand, evidence of several transitions is informative.

BETTER KNOWN TRANSITIONS

One of the simplest of the transitions is between primitive terrestrial sphenodontids (resembling the living genus *Sphenodon*) from the Late Triassic and the Late Jurassic pleurosaurs (Figure 4). The pleurosaurs had greatly elongate trunks, with up to 60 presacral vertebrae and reduced limbs. The skulls were also very elongate. The size of the limbs was greatly reduced, as was the degree of ossification and the number of carpals and tarsals. The fossil record of pleurosaurs is limited to one locality in the Upper Jurassic of France, two horizons in the Upper Jurassic of the Solenhofen area, and Holzmaden, in the Lower Jurassic (Carroll, 1985a; Carroll and Wild, 1994). The Holzmaden genus, *Palaeopleurosaurus*, is intermediate in form between Late Triassic terrestrial sphenodontids and the very distinctive Late Jurassic genus *Pleurosaurus*. The differences in skeletal anatomy can be associated almost entirely with three processes: (1) progressive reduction in ossification, (2) changes in proportions, and (3) increase in vertebral number, all of which may be attributed to long-term phyletic evolution. The changes are closely comparable to those that must have occurred in the many lineages of quadrupedal lizards that have greatly increased the length of the trunk and much reduced or completely lost their limbs (Greer, 1991). The really striking feature of pleurosaur

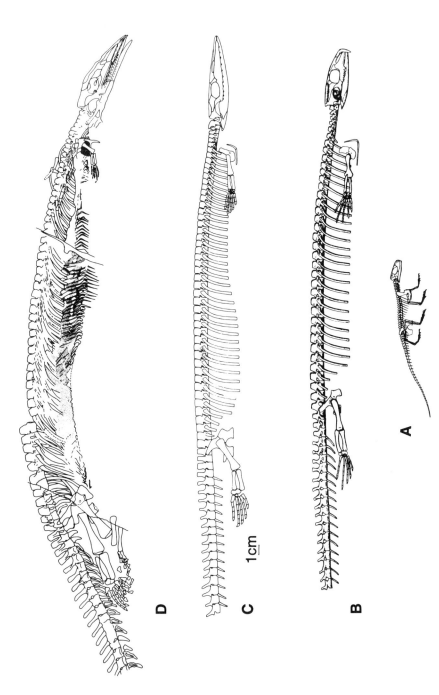

Figure 4. Skeletons of an early sphenodontid and pleurosaurs. **A)** The Upper Triassic sphenodontid *Planocephalosaurus robinsonae* (redrawn from Fraser and Walkden, 1984). **B)** *Palaeopleurosaurus*, Lower Jurassic of Holzmaden. **C)** *Pleurosaurus goldfussi*, Upper Jurassic of Solenhofen. **D)** *Pleurosaurus ginburgi*, Upper Jurassic of France. B and D from Carroll (1985a). All drawn to same scale.

evolution was how long it went on, at what appears to have been a very slow place. It must be emphasized, however, that no fossils are known between the Early and Late Jurassic, a period of approximately 30 million years. Unfortunately, it is difficult to quantify the rate of pleurosaur evolution. The most significant differences are in body proportions, which cannot be directly expressed in darwins. Size increase is significant, but we don't know the size of the particular early sphenodontid lineage from which the pleurosaurs evolved. The number of presacral vertebrae increased from approximately 23 in Late Triassic terrestrial sphenodontids to 60 in Late Jurassic pleurosaurs, which amounts to the addition of one vertebra per 1.5 million years. Under simple genetic control, such change might involve a selection coefficient of less than one in a thousand, if acting constantly throughout this period of time. This is far below the level that can be measured experimentally, or in natural populations (Endler, 1986). The average rate of change is probably less than might be attributed to random walk (Bookstein, 1988), but since it occurred in a more or less continuous manner over such a long period of time and led to animals of such different shape, it is logically attributed to selection.

The best evidence we have so far of the origin of a major group of Mesozoic aquatic reptiles is that of mosasaurs (deBraga and Carroll, 1993;Figure 5). The mid-Cretaceous aigialosaurs are almost ideal intermediates between primitive anguimorph lizards and the earliest mosasaurs (Caldwell et al., 1995). Unfortunately, the fossil record of the earlier stages of this transition are very poorly known. A gap of more than 60 million years separates the oldest known anguimorphs and the aigialosaurs. The known occurrence of aigialosaurs overlaps with that of the earliest mosasaurs, but this does not provide evidence of the actual duration of the transition between these two groups. It may have lasted tens of millions of years if the known aigialosaurs are relicts of an earlier established lineage.

Although it is impossible to establish actual rates of evolution when the fossil record is so incomplete, a pattern of evolutionary change was determined by adding up all the anatomical changes that occurred at different stages in the evolution from primitive terrestrial anguimorphs through the radiation of mosasaurs during the Late Cretaceous. The very complete specimens of aigialosaurs and many mosasaur groups allowed nearly all features of the skeleton to be evaluated. Forty-two character transformations occurred between primitive anguimorphs and mid-Cretaceous aigialosaurs. Approximately 23 transformations occurred between aigialosaurs and the most primitive adequately known mosasaurs. Sixty-three to 72 character changes occurred in the origin of four distinct mosasaur lineages. Continued change within each of the mosasaur subfamilies accounts for a total of 153 character transformations.

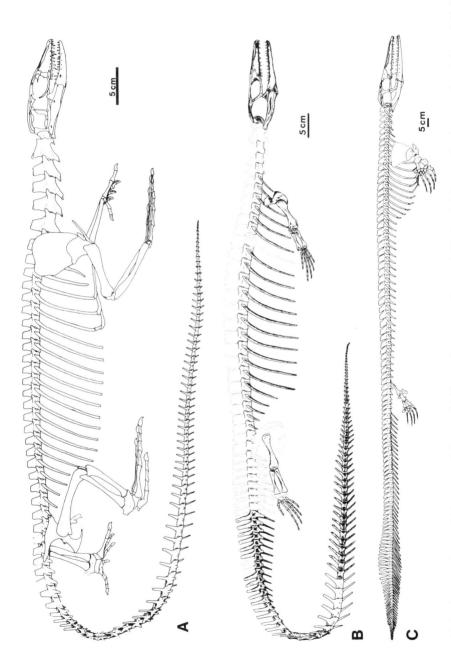

Figure 5. Skeletal comparisons of **A)** a terrestrial varanoid (the modern genus *Varanus*), **B)** the aigialosaur *Opetiosaurus buccichi*, and **C)** the mosasaur *Clidastes liodontus* (from deBraga and Carroll, 1993).

The origin of mosasaurs is logically taken as the time of the transformation of mosasaurs from aigialosaurs since this period encompasses major changes in limb structure from a terrestrial to an aquatic pattern. However, if one considers the entire skeleton, there were actually fewer changes during this time than there were within the lineage leading to mid-Cretaceous aigialosaurs, or within mosasaurs themselves. Although the limbs of aigialosaurs are indistinguishable from those of terrestrial varanoids, the skull is indistinguishable from that of early mosasaurs. The vertebral column and girdles display an intermediate morphology. Although the proportions and probably the function of the limbs changed at this time, much more extensive change occurred in the configuration of the individual bones and in the number of phalanges in later mosasaurs (Figure 6). Neither the number nor the nature of the changes that occurred during the time that may be identified as the origin of mosasaurs distinguishes this period from other times in the history of the mosasauroid clade. No measurable increase in evolutionary rate or dramatic changes in developmental pathways need be invoked to explain changes of this magnitude. Flattening of the tail and changes in the angle of articulation between the caudal vertebrae, as well as the nature of the deposits in which they were preserved, indicate that aigialosaurs had already adapted to an aquatic way of life. Reduction in limb size also may reflect commitment to aquatic locomotion. The earliest mosasaurs are distinguished by an approximate doubling of body length relative to that of described aigialosaurs, and significant changes in limb configuration that indicate the crossing of a threshold in locomotor capacity that enabled the lineage to extend its range from shallow, nearshore environments to the open ocean.

However, the most dramatic evolutionary event that occurred at this time was not a systematic reorganization of the skeleton, but the beginning of a major radiation that quickly resulted in worldwide domination. More anatomical changes occurred during this radiation and within the derived lineages than had during the origin of the group. If any special factor is necessary to explain the success of mosasaurs, it involves their capacity to radiate extensively, subsequent to achievement of the basic anatomical pattern that is common to the entire group. No special phenomena beyond natural selection of variables similar to those that occur in modern squamate groups are necessary to explain the anatomical changes observed.

DEVELOPMENT

Differences in developmental patterns and processes have been stressed by many authors (e.g., Gould, 1977; Thomson, 1988; Hall, 1992) as key to major changes in structure and way of life. Marine reptiles provide informative examples of several

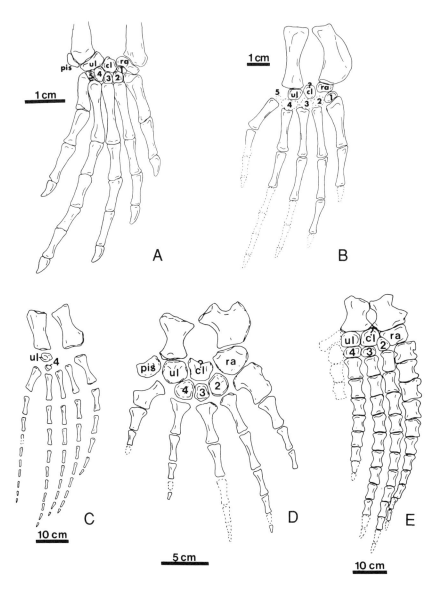

Figure 6. Lower forelimb of mosasauroids. **A)** The aigialosaur *Opetiosaurus buccichi*. **B)** The primitive mosasaur *Halisaurus sternbergi*. **C-E)** More advanced mosasaurs, showing divergent specializations of the carpus and phalanges. **C)** *Tylosaurus proriger*, in which the carpus is very incompletely ossified, but the number of phalanges is greatly increased. **D)** *Clidastes liodontus*, in which the carpus is fully ossified, and the manus is greatly expanded laterally, but with limited hyperphalangy. **E)** *Plotosaurus bennisoni*, exhibiting a well-ossified carpus and hyperphalangy. Abbreviations: **cl** = lateral centrale, **F** = femur, **f** = fibular, **H** = humerus, **pis** = pisiform, **r** = radius, **ra** = radiale, **t** = tibia, **u** = ulna, **ul** = ulnare, **1-5** = distal carpals and tarsals.

categories of developmental change. Heterochrony, the change in timing of developmental processes, has been emphasized by Gould (1977) as an extremely important mechanism for morphological change. Rieppel (1989) demonstrated the widespread occurrence of skeletal paedomorphosis among secondarily aquatic reptiles. The rate of ossification in the vertebrae, girdles, and limbs is slowed, so that much cartilage is retained in the adult. This has two adaptive advantages for aquatic animals. The body is lightened, since cartilage weighs about half of the equivalent volume of bone, and the limbs become much more flexible so that they can serve as effective paddles, rather than retaining the fixed articulating surfaces that are necessary for terrestrial support and locomotion.

Delay of ossification appears to be a general phenomenon throughout the skeleton rather than being limited to the limbs and girdles, which would be expected to be most strongly affected by selection. A striking feature of early pleurosaurs and sauropterygians is the retention of a line of cartilage between the neural arch and the centrum, although this area is strongly fused at an early developmental stage in their putative ancestors among Upper Triassic sphenodontids and Upper Permian diapsids. It is difficult to imagine a selective advantage for any degree of mobility in this area since the suture line runs through the transverse process within the area of articulation for the single rib head (Figure 7). This area could not possibly form a functional joint in living members of these groups. The retention of cartilage in the vertebrae must be attributed to selection for a general delay of ossification that is only incidently reflected in the neurocentral suture.

Gould and Lewontin (1979) argued that many anatomical changes, such as that affecting the neurocentral suture in these primitive aquatic reptiles, may be by-products of selection acting on functionally unrelated elements. The arch and centra are even more loosely attached in ichthyosaurs than in pleurosaurs and early

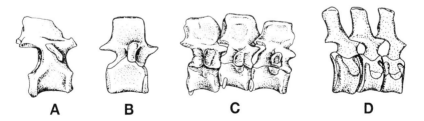

Figure 7. Trunk vertebrae, showing delay in ossification of the neurocentral suture. **A)** Isolated trunk vertebra of the Upper Triassic terrestrial sphenodontid *Planocephalosaurus* redrawn from Fraser and Walkden (1984). **B)** Reconstruction of a single trunk vertebra of the Lower Jurassic pleurosaur *Palaeopleurosaurus*. **C)** Posterior trunk vertebrae of the Upper Jurassic pleurosaur *Pleurosaurus*. **D)** Trunk vertebrae of the Middle Triassic ichthyosaur *Cymbospondylus buchseri* redrawn from Sander (1989).

sauropterygians as part of a more general restructuring of the vertebrae. In ichthyosaurs, the change in vertebral configuration is probably associated with significant changes in locomotor patterns. Delay in fusion of the neurocentral suture in early ichthyosaurs provides an example of what Gould and Vrba (1982) refer to as an "exaptation"---a character in an ancestral group that confers a new or different selective advantage in its descendants.

Reduction in ossification is a major feature in restructuring the limbs of aquatic reptiles. In some groups this has been shown to follow a specific pattern that can be considered as a developmental constraint. Carroll (1985b) showed that the sequence of reduction in ossification of the carpals and tarsals in nothosaurs reversed the sequence of their ossification in early terrestrial diapsids. The last element to be ossified during development in primitive diapsids was the first element to be lost during deossification in nothosaurs, and the first element to ossify is the last to be reduced and lost. Rieppel (1989) extended this comparison to other groups of primitive aquatic reptiles and found close correspondence.

Much more comprehensive study of developmental changes in the limbs of other groups of marine reptiles is now being carried out by Michael Caldwell (in progress). One particularly significant change is the timing of ossification of the carpals and tarsals relative to the rest of the limb. In all tetrapods, *chondrification* of the limb proceeds from proximal to distal (Shubin and Alberch, 1986). In contrast, the *ossification* of the carpals and tarsals in most reptiles and amphibians is delayed relative to the remainder of the limb. In immature individuals, both living and fossil, the areas of the carpus and tarsus remain cartilaginous after both more distal and more proximal elements are extensively ossified (Caldwell, 1994). The individual bones ossify in two stages. The perichondral bone that forms the surface of the shaft ossifies before the endochondral tissue that contributes to the inner portion of the shaft and the joint surfaces. The delay in ossification of the carpals and tarsals can be attributed to their extensive articulating surfaces and initial absence of perichondral bone. The delay in ossification of the carpals and tarsals may have evolved initially in the fish-tetrapod transition in conjunction with the formation of a complex joint between the proximal portion of the limb and the hand and foot, compared with the continuity of the fin in osteolepiform fish.

A reversal of this process can be documented among secondarily aquatic reptiles. The role of the carpus and tarsus as areas in which limb articulation is concentrated extends proximally and distally to encompass the entire flexible paddle. In ichthyosaurs and plesiosaurs, but not mosasaurs, bones adjacent to the carpals and tarsals reduce their perichondral ossification, and the epipodials, mesopodials, and metapodials become similar in overall appearance. Caldwell has shown that these changes occurred progressively over tens of millions of years in ichthyosaurs and plesiosaurs. Where successive stages of ossification can be followed during development, as in Jurassic ichthyosaurs, ossification of the limb appears to proceed from proximal to distal throughout, without a delay in the area of the carpus and tarsus (Figures 8 and 9).

These changes in the relative expression of perichondral and endochondral bone are specifically related to modification in the timing of developmental processes, but occur progressively over long periods of time within these groups. They do not occur rapidly, as has been assumed would be the case with changes in developmental processes, and they occur well after the initial shift in habitat and overall change in structure that typify each of these lineages.

TRENDS

Gould has repeatedly asserted that large-scale evolutionary changes, such as those involved in major transitions, result from sorting at the species level, rather than from the accumulation of phyletic changes within species. This hypothesis was discussed extensively in his 1990 paper. Many of his arguments can be countered by data from the fossil record of marine reptiles. One of Gould's major contentions is that selection pressures are rarely sufficiently constant for long enough periods of time to account for long-term directional change in morphological characters. This condition is met in all transitions between environments dominated by major differences in physical parameters. Among vertebrates, examples are provided by

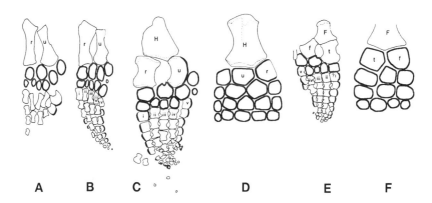

Figure 8. Adult condition in progressively more derived ichthyosaurs in which loss of perichondral ossification extends to elements proximal and distal to the carpals. **A-D)** Forelimbs. **E-F)** Hindlimbs. **A)** *Grippia.* **B)** *Parvinatator.* **C)** *Mixosaurus.* **D)** *Ophthalmosaurus.* **E)** *Mixosaurus.* **F)** *Ophthalmosaurus.* Thick lines around elements indicate absence of perichondral bone (from Caldwell, in preparation). Abbreviations: **i-v** = metacarpals and metatarsals. For other abbreviations, refer to Figure 6.

the transition from an aquatic to a terrestrial way of life seen in the origin of tetrapods, the numerous transitions from terrestrial to aquatic habitats as exemplified by aquatic reptiles, and the achievement of flight in pterosaurs, birds, and bats.

In all these cases, major changes in the skeleton can unquestionably be attributed to radically different requirements for support and locomotion. The characteristics of the different environments to which the ancestral and descendant lineages adapted are physical constraints. They act as agents of selection as long as the lineages survive. The rates of accommodation by the organisms are extremely irregular from group to group and from biological system to biological system. Within a particular system, evolution may appear to show a relatively constant rate of change, as in the very incomplete fossil record of pleurosaurs, in which there appears to be progressive increase in the number of presacral vertebrae and decrease in the relative length of the limbs. In other groups, change may be sporadic, with long periods of stasis interspersed with shorter periods of rapid change. In the case of the sauropterygians, nothosaurs, and plesiosaurs show relatively constant limb and trunk proportions and limb and girdle structure, between which significant changes occurred at a more rapid pace (Carroll and Gaskill, 1985). Ichthyosaurs also show two separate stages in their evolution, with markedly different evolutionary patterns and rates (McGowan, 1983; Massare and Callaway, 1990). During the Triassic, the limbs show progressive change from a pattern similar to that of early terrestrial diapsids, to that of a paddle. The general body form differs from family to family in overall size and proportions, suggestive of adaptive radiation into a variety of modes of swimming and feeding (Massare, 1988). Between the Late Triassic and Early Jurassic, a single lineage evolved toward the carangiform or thunniform pattern common to the fastest swimming sharks, teleosts, and whales. Once achieved in the Early Jurassic, the basic body form, distinguished by a spindle-shaped trunk and high aspect, lunate tail, remained essentially constant until the group's extinction in the Late Cretaceous. This is a clear example of stasis, resulting from the physical constraints of the environment within which this body shape is optimal for rapid swimming.

Gould (1990:7) argued that the few cases of long-term directional evolution that could be documented in the fossil record (he cited planktonic foraminifera as a possible example) actually contradicted the assumption of the significance of

---

Figure 9. Ontogenetic change in limb bones of ichthyosaurs. **A)** The smallest known ichthyosaur forelimb, *Ichthyosaurus communis*. **B-F)** Progressively larger specimens of the forelimb of *Stenopterygius quadriscissus*. **G-K)** Progressively larger specimens of the hindlimb of *Stenopterygius quadriscissus*. Ossification extends progressively from proximal to distal, without interruption in the area of the carpals (from Caldwell, in preparation). Arrows indicate areas of perichondral bone. For abbreviations, refer to Figure 6.

Mesozoic Marine Reptiles 483

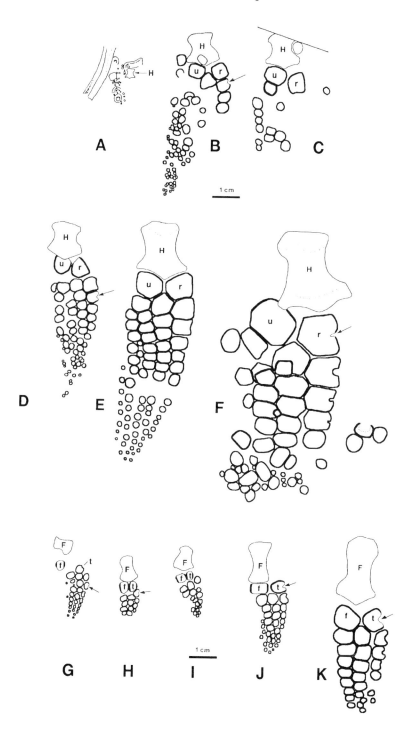

long-term evolutionary trends, since they took far too long to achieve minor changes. The history of marine reptiles shows many well-documented examples in which major changes were achieved over very long periods of time. Elongation of the pleurosaur body took place over a period of approximately 60 million years. It took 35 million years of evolution during the Triassic for ichthyosaurs to achieve a thunniform configuration. The fossil record is still too incompletely known to establish the detailed phylogenetic pattern of these changes, and it is probable that the rates of change varied considerably in all these groups. Nevertheless, there is no reason to deny the role of natural selection within individual lineages as a major force of evolutionary change. Numerous other examples are documented in fish (Long, 1990), terrestrial reptiles (Benton, 1990), and Cenozoic mammals (Janis and Damuth, 1990).

Gould (1990) also argued that biologists were wrong in ignoring speciation as a factor in phyletic evolution. If speciation occurred periodically within the period of an observed trend, how could the same selective factors be active throughout? He compared trends with a relay race in which the passing of the baton from runner to runner is a vital factor. He suggested that this break in continuity was comparable with change in genetic characteristics and factors of selection across speciation events. He admitted that a single selective regime might persist under conditions in which one descendant species retained most of the range and population size of the ancestral lineage, but argued that this would have been a rare occurrence. The problem raised by Gould does not apply in cases such as that exemplified by secondarily aquatic reptiles, in which the major selective features of the physical environment were common to all lineages. Trends to improve aquatic locomotion, buoyancy, and change in sense organs will be selected for no matter how many speciation events occurred. Different means of aquatic locomotion may lead to different anatomical solutions, as in the several different body forms of Triassic ichthyosaurs, and the emphasis on the limbs rather than the trunk for propulsion in sauropterygians, but in each lineage, similar trends in the structure of the limbs themselves have occurred, and were elaborated over tens of millions of years (Figure 8). None of the problems of accepting the phyletic nature of evolutionary trends raised by Gould apply to the well-documented patterns of evolution among marine reptiles.

Gould (following Stanley, 1975, 1979) has repeatedly emphasized that evolutionary trends are the result of sorting at the species level, not the result of adaptive changes within species. Differential survival of lineages certainly occurred among diverse groups such as nothosaurs, mosasaurs, and Triassic ichthyosaurs, but there is no evidence that this process contributed to overall directional changes such as the evolution of the thunniform body pattern that appeared in early Jurassic ichthyosaurs, or the major change in body and limb proportions that differentiate the

Jurassic plesiosaurs from their putative Triassic ancestors.

Unfortunately, knowledge of the fossil record is still too incomplete to provide any actual evidence of these transitions. In the better known transitions between terrestrial sphenodontids and pleurosaurs, and between aigialosaurs and mosasaurs, there is no evidence of a multiplicity of lineages showing a progressive accumulation of morphological changes as a result of differential survival. Knowledge of the fossil record is still far too incomplete to rule out such a possibility, but it seems unlikely that an extensive radiation at the species level would occur in the narrow transitional habitat occupied by a lineage undergoing transformation from one major adaptive zone to another. Each derived mosasaur lineage subsequently underwent extensive morphological change, but along adaptively divergent pathways. All of the major lineages persisted until near the end of the Cretaceous, without obvious evidence of any sorting processes leading to overall directional change in morphology. In contrast, homoplastic origin of similar derived characters was extensive throughout most of the lineages (deBraga and Carroll, 1993).

OBSERVATIONS

The following features of large-scale, long-term evolution are observed in the diversification of Mesozoic marine reptiles:

1. The period of transition between terrestrial and aquatic ways of life occurred over a much shorter period of time than the subsequent duration of the derived groups.

2. The amount of skeletal change that occurred during the most critical portion of the transformation may have been relatively small compared with subsequent change in the derived group, and overall change may not have occurred at a faster rate.

3. Changes in developmental processes and patterns are important to aquatic adaptation, but they may occur relatively slowly within the derived group, rather than quickly at its very base.

4. Changes in skeletal anatomy among aquatic reptiles may occur very slowly, over millions of years, although the fossil record is too incomplete to document the rate over short intervals of time.

The more we know of the transitions between primitive terrestrial diapsids and their aquatic descendants, the more gradual and continuous they appear. It is primarily the incompleteness of the fossil record that gives the impression of large gaps in morphology. Processes of anatomical and developmental change are not necessarily qualitatively or quantitatively different during transitions, compared with what occurs within established groups, but change is more clearly and sustainably directional, in response to fundamentally different selective pressures.

SUMMARY

The transition from a terrestrial to an aquatic habitat by many lineages of diapsid reptiles provides an informative model of long-term, directional change within a uniformly different physical environment. Significant gaps in time and morphology separate many aquatic lineages from their putative ancestors, but these gaps can be attributed to geological factors, and do not necessarily have any biological significance. Neither the rate nor the amount of change between major adaptive regimes is necessarily greater than within major groups. Significant change in limb structure can be associated with particular developmental processes that act to influence the direction of evolution. Changes in developmental processes occur within major lineages, and are not necessarily concentrated during transitions between adaptive types.

ACKNOWLEDGMENTS

I thank Michael deBraga and Rupert Wild for help in previous work on the evolution of aquatic diapsids, and Michael Caldwell for permission to quote his work now in progress. The diagram of the fossil record of nothosaurs was drafted by Lin Kebang. Other drawings were prepared by Pamela Gaskill and Elana Roman. This work was supported by grants from the Natural Sciences and Engineering Research Council of Canada.

REFERENCES

Bardet, N. 1994. Extinction events among Mesozoic marine reptiles. *Historical Biology* 7:313-324.

Benton, M. J. 1985. Classification and phylogeny of diapsid reptiles. *Zoological Journal of the Linnean Society* 84:97-164.

Benton, M. J. 1990. Reptiles. IN K. J. MacNamara (Ed.), *Evolutionary Trends*, pp. 279-300. Belhaven Press, London.

Bookstein, F. L. 1988. Random walk and the biometrics of morphological characters. *Evolutionary Biology* 23:369-398.

Caldwell, M. 1994. Developmental constraints and limb evolution in Permian and modern Lepidosauromorph diapsids. *Journal of Vertebrate Paleontology* 14:459-471.

Caldwell, M., R. Carroll, and H. Kaiser. 1995. The pectoral girdle and forelimb of *Carsosaurus marchesetti* (Aigialosauridae), with a preliminary phylogenetic analysis of mosasauroids and varanoids. *Journal of Vertebrate Paleontology* 15:516-531.

Carroll, R. L. 1981. Plesiosaur ancestor from the Upper Permian of Madagascar. *Philosophical Transactions of the Royal Society of London* B 293:315-383.

Carroll, R. L. 1985a. A pleurosaur from the Lower Jurassic and the taxonomic position of the Sphenodontia. *Palaeontographica,* Abteilung A 189:1-28.

Carroll, R. L. 1985b. Evolutionary constraints in aquatic diapsid reptiles. *Special Papers in Palaeontology* 33:145-155.

Carroll, R. L. 1987. *Vertebrate Paleontology and Evolution.* W. H. Freeman and Co., New York.

Carroll, R. L. and P. Currie. 1991. The early radiation of diapsid reptiles. IN H.-P. Schultze and L. Trueb (Eds.), *Origins of the Higher Groups of Tetrapods*, pp. 354-424. Comstock Publishing, Ithaca and London.

Carroll, R. L. and P. Gaskill. 1985. The nothosaur Pachypleurosaurus and the origin of plesiosaurs. *Philosophical Transactions of the Royal Society of London* B 309:343-393.

Carroll, R. L. and R. Wild. 1994. Marine members of the diapsid order Sphenodontida. IN N. Fraser and H.-D. Sues (Eds.), *In the Shadow of the Dinosaurs,* pp. 70-83. Cambridge University Press, Cambridge.

Currie, P. J. 1982. The osteology and relationships of *Tangasaurus mennelli* Haughton (Reptilia, Eosuchia). *Annals of the South African Museum* 86:247-265.

Darwin, C. 1859. *On the Origin of Species.* John Murray, London.

deBraga, M. and R. L. Carroll. 1993. The origin of mosasaurs as a model of macroevolutionary patterns and processes. *Evolutionary Biology* 27:245-322.

Eldredge, N. and S. J. Gould. 1972. Punctuated equilibria: an alternative to phyletic gradualism. IN T. J. M. Schopf (Ed.), *Models in Paleobiology*, pp. 82-115. Freeman, Cooper and Co., San Francisco.

Eldredge, N. and S. Stanley (Eds.). 1984. *Living Fossils.* Springer-Verlag, New York and Berlin.

Endler, J. A. 1986. *Natural Selection in the Wild.* Princeton University Press, Princeton, New Jersey.

Erwin, D. H. 1993. *The Great Paleozoic Crisis.* Columbia University Press, New York.

Evans, S. E. and M. Hecht. 1993. A history of an extinct reptilian clade, the Choristodera: longevity, Lazarus-taxa, and the fossil record. *Evolutionary Biology* 27:323-338.

Fraser, N. C. and Walkden, G. M. 1984. The postcranial skeleton of the Upper Triassic sphenodontid *Planocephalosaurus robinsonae. Palaeontology* 27:575-595.

Gingerich, P. D. 1983. Scaling of evolutionary rates. *Science* 222:159-161.

Gingerich, P. D., 1993. Rates of evolution in Plio-Pleistocene mammals: six case studies. IN R. A. Martin and A. D. Barnosky (Eds.), *Morphological Change in Quaternary Mammals of North America*, pp. 84-106. Cambridge University Press, Cambridge.

Gould, S. J. 1977. *Ontogeny and Phylogeny.* Belknap Press of Harvard University, Cambridge, Massachusetts..

Gould, S. J. 1990. Speciation and sorting as the source of evolutionary trends, or 'things are seldom what they seem.' IN K. J. McNamara (Ed.), *Evolutionary Trends*, pp. 3-27. Belhaven Press, London.

Gould, S. J. and N. Eldredge. 1977. Punctuated equilibria: the tempo and mode of evolution reconsidered. *Paleobiology* 3:115-151.

Gould, S. J. and N. Eldredge. 1993. Punctuated equilibrium comes of age. *Nature* 366:223-227.

Gould, S. J. and R. C. Lewontin. 1979. The spandrels of San Marco and the Panglossian paradigm: a critique of the adaptationist programme. *Proceedings of the Royal Society of London* B 205:581-598.

Gould, S. J. and E. Vrba. 1982. Exaptation---a missing term in the science of form. *Paleobiology* 8:4-15.

Greer, A. E. 1991. Limb reduction in squamates: identification of the lineages and discussion of the trends. *Journal of Herpetology* 25:166-173.

Hall, B. 1992. *Evolutionary Developmental Biology*. Chapman & Hall, London and New York.

Janis, C. M. and J. Damuth. 1990. Mammals. IN K. J. McNamara (Ed.), *Evolutionary Trends*, pp. 301-350. Belhaven Press, London.

Laurin, M. 1991. The osteology of a Lower Permian eosuchian from Texas and a review of diapsid phylogeny. *Journal of the Linnean Society of London* 101:59-95.

Long, J. A. 1990. Fishes. IN K. J. McNamara (Ed.), *Evolutionary Trends*, pp. 255-278. Belhaven Press, London.

Massare, J. A. 1988. Swimming capabilities of Mesozoic marine reptiles: implications for method of predation. *Paleobiology* 14:187-205.

Massare, J. A. and J. M. Callaway. 1990. The affinities and ecology of Triassic ichthyosaurs. *Geological Society of America Bulletin* 102:409-416.

McGowan, C. 1983. *The Successful Dragons*. Samuel Stevens, Toronto and Sarasota.

Reisz, R. R. 1981. A diapsid reptile from the Pennsylvanian of Kansas. *Occasional Papers, Museum of Natural History, University of Kansas* 7:1-74.

Rieppel, O. 1987. *Clarazia* and *Hescheleria*: a re-investigation of two problematical reptiles from the Middle Triassic of Monte San Giorgio (Switzerland). *Palaeontographica*, Abteilung A 195:101-129.

Rieppel, O. 1989. *Helveticosaurus zollingeri* Peyer (Reptilia, Diapsida) skeletal paedomorphosis, functional anatomy and systematic affinities. *Palaeontolographica*, Abteilung A 208:123-152.

Sander, P. M. 1989. The large ichthyosaur *Cymbospondylus buchseri*, sp. nov., from the Middle Triassic of Monte San Giorgio (Switzerland), with a survey of the genus in Europe. *Journal of Vertebrate Paleontology* 9:163-173.

Shubin, N. H. and P. Alberch. 1986. A morphogenetic approach to the origin and basic organization of the tetrapod limb. *Evolutionary Biology* 20:319-388.

Simpson, G. G. 1944. *Tempo and Mode in Evolution*. Columbia University Press, New York.

Simpson, G. G. 1953. *The Major Features of Evolution*. Columbia University Press, New

York.
Stanley, S. M. 1975. A theory of evolution above the species level. *Proceedings of the National Academy of Science* 72:646-650.
Stanley, S. M. 1979. *Macroevolution: Pattern and Process*. W. H. Freeman and Co., San Francisco.
Storrs, G. W. 1993. Function and phylogeny in sauropterygian (Diapsida) evolution. *American Journal of Science* 293A:63-90.
Thomson, K. S. 1988. *Morphogenesis and Evolution*. Oxford University Press, London.

Index

References to illustrations are in italics. Tables are indicated by "t," appendices by "app."

Africa, 259-260, 270, 272, 275, 367-369, 375, 393, 425, 446, 473
Agassiz, L., 108
Aigialosauridae, 324
Aigialosaurus dalmaticus, 285
Aldzadasaurus, 214
Aldzadasaurus columbiensis, 191
Alligator, 390
Alligator mississipiensis, 392
Allopleuron, 221, 234, 236, 238, 259, 260, 271, 274
Allopleuron hoffmanni, 227, *227*, *232-233*, 260, 262-263, 272, 275
Alps, 109-111, 122
  Alpine Triassic, 121-123, 125, 129, 132, *134*, 135-139
  "*Formazione a gracilis*," 123, 135
  Grenzbitumen horizon, 108, 135-136, 138, 140
  Partnach Plattenkalk, 135-136
  Perledo-Varenna Formation, 135-136, 138
  Prosanto Formation, 135
Amphichelydia, 220
Anarosaurus, 113, 133-135, 139
"*Anarosaurus multidentatus*," 122
Andrews, C. W., 18, 23, 191, 198
Angola, 286
  Cabinda, 367
Angolosaurus bocagei, 286
Anning, M., 146, 148, 162
Antarctica, 446
  Seymour Island, 287
Aphrosaurus, 214
Appleby, R. M., 6, 52, 440
Arabia, 368
Archaeonectrus, 179
Archelon, 221, 237-238, 270

Archelon ischyros, 229-230, 233
Argentina
  Neuquén Basin, 435-436, *437*, 438-440, 442-446
  distribution of marine reptiles, *438*
  Patagonia, 446
Argillochelys, 267-268
Argillochelys antiqua, 267-268
Argillochelys athersuchii, 267-268
Argillochelys crassicostata, 272
Argillochelys cuneiceps, 267-268
Asia, 40, 287
Askeptosauridae, 472
Atlantochelys, 238
Atlantochelys mortoni, 230, 237
Atlantosuchus, 379, 382
Attenborosaurus, 179
Austria, 431
  Carinthia, 37, 39

Baier, J. J., 3
Baird, D., 263
Bakker, R. T., 210
Bassani, F., 45
Baur, G. H., 4, 45, 285
Belgium, 238, 262, 265, 268-270, 272, 283, 285
Bell, G. L., 429
Biochronology, 423-432
  stratigraphic ranges and biozones, *425*
  temporal ranges, *424*
  Triassic ichthyosaurs, 429
  temporal ranges, *430*
Blainville, H. M. D. de, 4
Body forms, *402*
Bonaparte, J., 440
Bottosaurus, 393
Brachauchenius, 205, 208

491

Brancasaurus, 194-195, 202, 208, 214
Brancasaurus brancai, 206, *207*
Brazil, 368, 375
  Santana Formation, 236
Brinkman, D. B., 9-10, 47, 53
Broom, R., 5-6
Brown, D., 209
Buckland, W., 150
Buffetaut, E., 368, 376, 382, 391, 394
Burgundy Gate, 135-136, 138, 140

Calcarhichelys gemma, 229
Caldwell, M., 480
Californosaurus, 22, 33-34, 36
  rib articulation, 19-20
Callaway, J. M., 7-10, 19, 33, 35, 51, 53, 65, 92-93
Camp, C. L., 69, 92-93, 286
Canada, 272
  British Columbia, 10, 38, 78, 111, 429-430
    Vancouver Island, 245, 255
    Wapiti Lake, 39, 47
    Williston Lake, 10, 39, 61-63, *64*, 65, 68-69, 73, 77
  Pardonet Formation, 61
Cardiochelys eocaenus, 271
Cardiochelys rupeliensis, 269-270
Caretta, 219, 237, 269
Caretta caretta, 254
Carroll, R. L., 6, 47, 113, 208, 287, 321, 480
Carsosaurus marchesetti, 285
Casichelyidia, 271
"*Catapleura*" *arkansaw*, *227*, 228, 236
Catapleura repanda, 226
Ceresiosaurus, 111-112, 114, 136, 137, 140
Chaohusaurus, 52, 430
Chelone, 220
Chelone harvicensis, 265
Chelone mydas, 272
Chelonia, 440
Chelonia, 219, 237
Chelonia mydas, 219, *232*
Cheloniidae, 220-221, 225, 234, 245, 259-260, 263, 267-269, 442
  geographic distribution, 265
Chelonioidea, 220-221, 225, 230, 238-239, 260
  cladogram, *273*
  cladogram of interrelationships and distribution, *235*
  stratigraphic distribution, *261*
Chelosphargis advena, 229, *229*
Chensaurus, 52
Chensaurus chaoxianensis, 9
Chile, 435, 445-446
  Los Valdés Formation, 436
China, 46, 109-112, 132, 424-425, 429-430, 443
  Tibet, 431
Cimiochelys benstedi, 262
Claraziidae, 472
Claudiosaurus, 113, 473
Clidastes, 283, 297, 307, 314-315, 321, 324, 333, 340t-341t, 342, 347, 349-350
  bone microstructure, 337-338
Clidastes liodontus, 303-305, *319*, *476*, *478*
Clidastes propython, *301*, *339*
Colbert, E. H., 6, 113
Collins, J. I., 262
Colombia, 246, 254-255
Colymbosaurus trochanterius, 444
Compressidens, 285
Conybeare, W. D., 4, 146, 150-151, 282
Cope, E. D., 282-283, 375
Corosaurus, 110-112, 114
Corsochelys, 221, 239, 269
Corsochelys haliniches, 230, *231*, 237
Cosmochelys dolloi, 271
Crassachelys neurirregularis, 271
Cretaceous Chalk Basin, 259
Cretaceous-Tertiary boundary, 238
Croatia, 285, 294
Crocodylia, 357-370
Crocodylidae, 369
  Transoceanic Migration Hypothesis, 369-370
Crocodylomorpha, 439

Crocodylus, 369, 379, 382t, 391
Crocodylus acutus, 393
Crocodylus niloticus, 359, 384
Crocodylus porosus, 357-358, 366, 393, 401
Cryptoclidus, 172-173, 204, 445-446
Cryptodira, 220, 442
Ctenochelys, 234, 265
Ctenochelys stenoporus, 226, *226, 232*
Cuba, 445-446
Cuvier, M. G., 3, 219, 281-282
Cyamodontoidea, 108, 138
Cyamodus, 138-140
Cymatosaurus, 110-111, 114, 122, 127, 131-132, 135, 137, 139
Cymatosaurus erythreus, 132
Cymatosaurus latissimus, 110
Cymatosaurus multidentatus, 132
Cymbospondylinae, 95
Cymbospondylus, 10, 21-22, 29, 33-35, 37, 77, 430, 431
  dentition, 95, 98, 100
  rib articulation, 19-20
Cymbospondylus buchseri, 479
Cymbospondylus petrinus, 29, 29t, 50
  dentition, 88, 93, 95

Dactylosaurus, 109, 131, 133, 135, 139
Dakosaurus, 366, 439, *441*, 442, 444, 446
Dechaseaux, C., 47, 54
De la Beche, H. T., 4, 146, 150
Dermochelyidae, 220-221, 225, 230, 236-237, 239, 259-260, 268-270, 272
  geographic distribution, *264*
Dermochelys, 219, 237, 269-270
Dermochelys coriacea, 221, *231-233*, 275
Desmatochelyidae, 221, 245, 260, 263
  geographic distribution, *264*
Desmatochelys, 237, 260, 262, 267, 269
Desmatochelys lowi, 229, *229, 232-233*, 243-256, 262
  age distribution, *256*
  appendicular skeleton, 250, *251,* 253
  axial skeleton, 250, *251*
  carapace and plastron, 250, *251, 253*
  hyoid, 247
  mandible, 247, *248*
  skull, 246-247, *248-249*
Diapsida, 5-6, *7,* 22, 113-114, 145, 149
  phylogeny, *469*
Dolichorhynchops, 194-195, 197-198, 200-209, *209,* 210
  cranial measurements, 199t, *199*
Dolichorhynchops osborni, 192-193, *198-199, 202, 205,* 210, 214, 215app
Dollo, L., 283-284
Dollosaurus lutugini, 283
Dong, Z.-M., 86
Dyrosauridae, 357, 367-368
  stratigraphic range, *360*
Dyrosaurus, 361, 363, 367, 379, 382t, 385

East Carpathian Gate, 122, 139
Ectenosaurus, 296, 310, 312, 314, 320-321, 323-324
Ectenosaurus clidastoides, 303-305
Egypt, 260, 271
Elasmosauridae, 111, 193, 214
  phylogenetic position, *210*
Elasmosaurus, 214
"*Elasmosaurus*" *morgani*, 192-193, 210-211, 214
Elasmosaurus platyurus, 192, *192,* 193, 211, 214
England, 77, 107-108, 136, 145, 180, 260, 262-263, 265, 268, 270, 272, 282, 362
  Bath, 3
  Bristol, 146, 179
  Charmouth, Dorset, 179
  Fulbeck, Nottinghamshire, 179
  Gloucestershire, 179
  Keynsham, 179
  Lyme Regis, Dorset, 3, 146, 148-150, 162, 166, 179, 410t
  Oxford Clay, 412t
  Somerset, 179
Eochelone, 267-268
Eochelone brabantica, 267-268, 272

Eosauropterygia, 126-127, 131
Eosphargis, 231, 236, 238, 269-270
Eosphargis breineri, 269-270
Eosphargis gigas, 269-270, 275
Eretmochelys, 237
Erquelinnesia, 263, 265
Erquelinnesia molaria, 265
Erquelinnesia planimentum, 265
Eupleurodira, 271
Europe, 259-260, 263, 272, 275, 283, 368-369, 425-426, 439, 443-445
Euryapsida, 6, 113-114
Eurysaurus, 110
Eusarkia rotundiformis, 271
Eusauropterygia, 108, 110-114, *123, 126-127*, 138, 149
Evolution, 467-486
  heterochrony, 479
  quantum, 469
  stasis, 467-468
  trends, 481-482, 484, 485

Faunas through time, 405, 408-417
  changes in predator types, 416t
Feeding constraints, 451-464
Fernandez, M., 442
Fraas, E., 18, 45, 88, 99
France, 72, 99, 148, 179, 181, 260, 263, 270-272, 283, 430, 473
  Posidoniaschiefer, 146

Gaffney, E. S., 220, 254, 260, 262-263, 265, 267, 270-271
Gafsachelys phosphatica, 271
Gasparini, Z., 440, 442
Gaudry, A., 283
Gauthier, J. A., 7
Gavialus fraterculus, 375, 378
Geochelone radiata, 254
Geosaurus, 365-366, 390, 442, 444, 446
Geosaurus araucanensis, 438-439, *441*, 444
Germanic Basin, *134*
  Bundsandstein, 110, 122, 132
  fossil-bearing localities, *133*
  Gipskeuper, 111, 136-138, 140

Grundgipsschichten, 137
Lettenkeuper, 110-111, 135-139
Muschelkalk, 20-22, 33, 37, 39, 41, 46, 107-108, 111, 121-122, 125-127, 129, 132, 134-139, 425, 429, 431
Röt, 121-122, 132-133, 139
Schaumkalk, 124, 126-127, 132-133
Germanosaurus, 110
Germany, 72, 77, 137, 145, 179, 430
  Altorf, 3
  Bayreuth, 108
  Franconia, 20, 22-23
  Holzmaden, 181, 411t, 428, 473
  Posidoniaschiefer, 146
  Solenhofen, 473
Gervais, P., 282
Globidens, 286, 296, 310, 314, 323-324
Globidens alabamaensis, 285
Globidens dakotensis, 303-305
Globidens fraasi, 285
Glyptops, 253
Glytochelone, 236, 238
Glytochelone suyckerbuykii, 228, 262
Goldfuss, A., 282-283, 285
Goniopholis, 376, 379, 382t
Goronyosaurus, 286
Gould, S. J., 467, 473, 479-482, 484
Grippia, 9-10, 50, 430, *481*
  dentition, 98, 100
Grippia longirostris, 49, 52
  dentition, 81-82, 87, 99

Hagdorn, H., 429
Hainosaurus, 283, 321
Halisaurus, 287, 296, 310, 314, 321-322, 324
Halisaurus nov. sp., *301*
Halisaurus onchognathus, 283
Halisaurus platyspondylus, 283
Halisaurus sternbergi, 286, *303-305, 319, 478*
Harlan, R., 282
Hawker, J., 3
Himalayasaurus tibetensis
  dentition, 82, 86, 92

Index 495

Hirayama, R., 254, 262, 274
Hispanic Corridor, 445, *445*, 446
Home, E., 3-4
Hudsonelpidia, 68
Hudsonelpidia brevirostris, 10, 65, *67*
Huene, F. von, 5, 8, 18-19, 33, 37, 46, 48, 88-89, 376
Hupehsuchia, 6, 424
Hupehsuchus, 6, 453
Hydralmosaurus, 214
Hydrotherosaurus, 214
Hyposaurus, 221, 367, 375-381, 382t, 383-395
Hyposaurus bequaerti, 367
Hyposaurus derbianus, 375
Hyposaurus ferox, 378
Hyposaurus fraterculus, 378, 395
Hyposaurus natator, 367, 378, 382, 390
Hyposaurus rogersii, 367, 375-376, 378, 379, *380-381*, 383, *385-387, 389, 392-393*, 395

Iakovlev, N. N., 283
Ichthyopterygia, 4-6
Ichthyosauria, 3-11, 22, 438-439, 472
  biochronology, 426, 431
  dentition, 8-9, 29-30, 81-101
    ankylosed thecodonty, 84, 96, 98-100
    aulacodonty, 81-82, 84, 92, 96, 99-100
    ichthyosaurian thecodonty, 88, 93, 95-96, 98, 100
    pleurodonty, 88, 92
    stratigraphic distribution, *97*
    subthecodonty, 86-89, 91, 95-97, 99-100
    tooth implantation, 83t, *84, 96-97*
  hypothetical phylogeny, 7, *7*
  indeterminate, *70-72, 74-76*
  limbs, *481, 483*
  rib articulation, 18
  supratemporal bone, 9, 51
Ichthyosaurus, 4, 10, 28-29, 50, 65, 68, 76-78
  dentition, 85, 95
  rib articulation, 18
Ichthyosaurus acutirostris
  dentition, 94
Ichthyosaurus breviceps, 68, 76
Ichthyosaurus communis, 29t, 73, *483*
Ichthyosaurus quadriscissus
  dentition, 94
Ichthyosaurus tenuirostris
  dentition, 94
Igdamanosaurus, 287
Ireland, 148
Israel, 110
  Muschelkalk, 111, 136
Italy, 10, 45, 107, 136, 268-270, 430

Jaekel, O., 139
Japan, 237, 239, 287, 430

Keichousaurus, 109, 132
Kimmerosaurus, 453
Kolposaurus bennisoni, 286
Kronosaurus, 464
Kuhn, O., 47, 376
Kuhn-Schnyder, E., 6, 53, 113

Lamprosauroides goepperti, 131, *131*
Lanthanotus, 286
Lariosaurus, 111-112, 114, 136-137, 140
Lariosaurus buzzi, 136
Laurin, M., 7
Leidy, J., 283-284
Leiodon haumuriensis, 283, 286
Lembonax insularis, 238
Lepidochelys, 219, 237
Lepidosauromorpha, 7, *7*, 114
Leptopterygius, 69
Leptopterygius burgundiae, 72
Leptopterygius tenuirostris, 10, 73, 77
"*Leurospondylus*," 172
Lhwyd, E., 3
Libonectes, 193-195, 198, *199*, 200-209, *209*, 210
  cranial measurements, 199t
Libonectes morgani, *192*, 193, *196-197, 200-201*, 206, *207*, 214app
Liodon, 282

Liopleurodon, 195, 208, *209*, 438, *441*, 444, 446
Liopleurodon ferox, 203
Liopleurodon macromerus, 438-439
London-Paris-Belgium Basin, 259
Lophochelys, 234, 265
Lydekker, R., 220
Lytoloma cantabrigiense, 227, 236
Lytoloma crassa, 265
Lytoloma crassicostatum, 265
Lytoloma elegans, 265, 269
Lytoloma gosseleti, 265
Lytoloma planimentum, 265

Machimosaurus, 367
Machimosaurus hugii, 362
Machimosaurus mosae, 364
Macrobaenidae, 234
Macrosaurus, 282
Madagascar, 113, 473
Malayemys, 236
Mantell, G., 282
Marsh, O. C., 282-283
Massare, J. A., 7-8, 10, 33, 65, 92-93, 350
Mazin, J.-M., 6, 8-10, 46, 51-52, 82, 84, 87, 91, 94, 99
McGowan, C., 8-10, 18-19, 23, 38, 47, 50-51, 92, 440
McGregor, J. H., 5
Merriam, J. C., 8, 19-20, 23, 36, 47, 52, 65, 88, 91-92, 283
Merriamia, 22, 34, 36, 69
  dentition, 98
  rib articulation, 19
Merriamia zitteli
  dentition, 92, 98
Mesosaurus, 5, 10, 21, 453
Mesosuchia, 357, 368, 376, 391
Metriorhynchidae, 357, 364, 366, 439
  stratigraphic range, *360*
Metriorhynchus, 361, 363, 364, 366, 438-439, 442, 444-446
Metriorhynchus brachyrhynchus, 364
Metriorhynchus durobrivensis, 439
Metriorhynchus potens, 439

Metriorhynchus superciliosus, 364
Meyer, H. von, 107
Mexico, 10, 445-446
  Sonora, 431
Microcleidus, 179
Microleptosaurus schlosseri, 139
Micronothosaurus stensioeii, 110
Middle East, 425
Miochelys fermini, 270
Mixosauridae, 9, 35, 45, 47, 51, 53
  rib articulation, 19
Mixosauridae *incertae sedis*, 46
Mixosaurus, 38, 45-55, 68-69, 430-431
  appendicular skeleton, 52-53, *481*
  axial skeleton, 53-54
  cranium, 48-51
  dentition, 49, 95, 98, 100
  distribution, 46
  rib articulation, 19, 53
Mixosaurus atavus, 45-53, 55
  dentition, 88-89, 99
Mixosaurus cornalianus, 9, 45, 48, *48*, 49-55
  dentition, 81-82, 84, 89, 91, 99
Mixosaurus maotaiensis, 46
Mixosaurus natans, 46-47
Mixosaurus nordenskioeldii, 46-47
  dentition, 82
Moody, R. T. J., 254, 262-263, 265, 267-269, 271
Mook, C. C., 376
Morenosaurus, 214
Morocco, 260
Mosasauridae, 281-288, 296, 322, 336
  cladogram, *295*
  systematic relationships, *284-285*
Mosasauroidea, 293-351
  Adam's consensus trees from PAUP, *299*
  appendicular skeleton, 317-321, *478*
  biochronology, 426, 431
  bone microstructure, 333-351
  cladogram, *295*
  ingroup data matrix, 329app-332app
  phylogenetic analysis, 296
  postcranial axial skeleton, 315-317

Index

preferred hypothesis of relationships, *300*
quadrates, comparison of, *309*
revised phylogeny, 321-323
skull characters, 297-317, *301, 303*
strict consensus trees from PAUP, *298*
Mosasaurus, 283, 302, 313-315, 320-322, 324
Mosasaurus conodon, 429
Mosasaurus flemingi, 287
Mosasaurus hobetsuensis, 287
Mosasaurus hoffmani, 282
Mosasaurus mangahouangae, 286
Mosasaurus maximiliani, 282
Mosasaurus maximus, 429
Mosasaurus missouriensis, 303-305, 315, 429
Motani, R., 8-9, 49, 430
Münster, G., 108
Muraenosaurus, 203-204, 445, 446

Nanchangasaurus, 6
Natantia, 282, 322-324
Neodiapsida, 7, *7*
Netherlands, the, 148, 238, 259, 262, 263, 268, 272, 281
Neurochelys harvicensis, 265
Neusticemys, 443, 446
Neusticemys laticentralis, 441
Neusticemys neuquina, 438, *441*, 442
Neusticosaurus, 110, 137, 140
Neusticosaurus pusillus, 135, 137
New Caledonia, 431
New Zealand, 283, 287, 431
Nicholls, E. L., 9, 47, 111, 195, 245, 255
Niger, 271, 287
Nigeria, 271, 286
North America, 191, 237, 239, 259-260, 275, 294, 367-368, 393, 425
  Pacific Province, 112
  Western Interior, 191, 255, *256*, 426
Norway
  Svalbard, 429-431
    Spitsbergen, 10, 39-41, 47, 99
Nothosauria, 425
  stratigraphic record, *470*

Nothosauridae, 114
Nothosauriformes, 114
Nothosaurus, 110-111, 123, *124*, 125-128, 131-132, 135-136, 139-140
Notochelone, 237, 262
Notochelone costata, 229, *229*, 262
Notoemys, 443-444
Notoemys laticentralis, 438, 442

Omphalosauridae
  rib articulation, 18
Omphalosaurus, 51-52, 430
  rib articulation, 18
Opetiosaurus, 297, 317, 321
Opetiosaurus bucchichi, 285, *319, 476, 478*
Ophthalmosaurus, 5, 18, 28-29, 438-440, *441*, 442, 444, 446, *481*
  rib articulation, 18
Ophthalmosaurus icenicus, 18, 29, 29t
  dentition, 94
Ophthalmosaurus monocharactus, 438, 440
Osborn, H. F., 5
Osteopygidae, 221, 260, 263, 265
  geographic distribution, *265*
Osteopygis, 234, 236, 238, 263, 265
Osteopygis emarginatus, *226-227, 232-233*, 265
Owen, R., 4, 18, 108, 111, 113-114, 165, 282, 375

Pachypleurosauria, 132
Pachypleurosauridae, 109
Pachypleurosauroidea, 108-110, 114
Pakistan, 367
Palaeopleurosaurus, 473, *474, 479*
Paleotethys, 122, 139
Pangaea, 41, 46
"*Paranothosaurus*," 114, 136
Paraplacodus, 108, 138
Parapsida, 5-6
Paris Basin, 265
Parris, D. C., 376, 394
Partanosaurus zitteli, 139
Parvinatator, 50-51, *481*

Parvinatator wapitiensis, 9, 49
Peipehsuchus teleorhinus, 359
Pelagosaurus, 364
Pelomedusidae, 220, 259, 271
  stratigraphic distribution, *261*
  geographic distribution, *266*
Peloneustes, 195, 203, 208
Peritresius, 234, 238
Peritresius ornatus, 227
Pessosaurus, 10
  dentition, 98
Pessosaurus polaris
  dentition, 82
Phalarodon, 10, 49-53
  dentition, 98, 100
Phalarodon fraasi, 47
  dentition, 91
Phalarodon nordenskioeldii, 9, 47
  dentition, 91
Pholidosauridae, 357, 366
  stratigraphic range, *360*
Pholidosaurus, 366
Phosphatosaurus, 367, 379, 382t, 391
Phosphorosaurus, 283
Pistosaurus, 107, 111, 114, 138, 195, 203, 208-209, *209*
  phylogenetic position, *210*
Placodontia, 108, 114, 425
Placodontoidea, 108
Placodus, 108, 138-140
Placodus gigas, 108
Planocephalosaurus, *479*
Planocephalosaurus robinsonae, *475*
Platycarpus, 283, 296, 306, 310, 312, 321, 324, 333, 340t-341t, *343*, 349-350
  bone microstructure, 338, 342, 344
Platycarpus coryphaeus, 312
Platycarpus ictericus, 312
Platycarpus planifrons, 312, 314
Platycarpus somenensis, 283
Platycarpus tympaniticus, *301, 303-305* 306, *309*, 313, *319*
Platychelone emarginata, 262
Platychelys, 442
Platynota, 285

Platypterygius, 50, 440
Platypterygius compylodon
  dentition, 94
Plesiochelyidae, 220, 442
Plesiochelys, 220
Plesiochelys etalloni, 219
Plesiosauria, 111, 145, 210
  biochronology, 426, 431
Plesiosauridae, 111
  phylogenetic position, *210*
Plesiosauroidea, 111, 214
  phylogenetic position, *210*
Plesiosaurus, 111, 145-190
Plesiosaurus brachypterygius, 179
"*Plesiosaurus*" *cliduchus*, 180
"*Plesiosaurus*" *conybeari*, 179, 195
Plesiosaurus dolichodeirus, 146, *147*, 148-151, 178-181, 203-204, 208
  dentition, *152, 154, 163, 168*, 169-170
  dimensions, 186app-190app
  forelimb, 171-172, *172*, 173, *174*, 176
  hindlimb, *175*, 176, *177*, 178
  mandible, 165-167, *167-168*, 169
  skull, 151-165, *152, 154, 158-160, 163-164, 166-167*
  vertebral column, *147, 152*, 170-171
Plesiosaurus guilelmiiperatoris, 146, 150, 173, 179
Plesiosaurus hawkinsi, 153, 155-156, 169-170, 179, 208
"*Plesiosaurus*" *homalospondylus*, 179
"*Plesiosaurus*" *macrocephalus*, 204
Plesiosaurus macromus, 180
Plesiosaurus priscus, 150
"*Plesiosaurus*" *rostratus*, 179
Plesiosaurus tournemirensis, 146, 179
Plesiotylosaurus, 296, 307, 310, 318, 324
Pleurodira, 220, 442-443
Pleurosaurus, 473, *479*
Pleurosaurus ginburgi, *474*
Pleurosaurus goldfussi, *474*
Plioplatecarpus, 283, 296, 306, 310, 312-313, 321, 323-324
Pliosauridae, 193, 440, 444
  phylogenetic position, *210*
Pliosauroidea, 111

phylogenetic position, *210*
Pliosaurus, 111, 165, 208, 210, 438-439, 444, 446
Plotosaurus, 286, 302, 313, 320-321, 323, 324
Plotosaurus bennisoni, 303-305, 309, *319, 478*
Poland, 430
  Gogolin, Silesia, 123, 126-128, 130, 131, 133, 135, 138
  Karchowice Beds, 138
  Muschelkalk, 110-111, 122, 126, 128, 130, 133, 135, 138-139
Polycotylidae, 193, 215
  phylogenetic position, *210*
Polycotylus, 210, 215
Porthochelys, 234
Porthochelys laticeps, 226
Portugal, 260
Prionochelys, 234
Prionochelys nauta, 227
Procólpochelys melii, 268
Prognathodon, 283, 296, 310, 312, 318, 321, 324
Prognathodon overtoni, *303-305*, 310, 313-314
Prognathodon rapax, *309*, 310, 314
Prognathodon waiparaensis, 286
Proneusticosaurus, 110
"*Proneusticosaurus*," 127, 129, 131-132
"*Proneusticosaurus*" *madelungi*, 126
"*Proneusticosaurus*"*silesiacus*, 125, 127, *128-129*, 129, 132
Proteosaurus, 4
Protosphargis veronensis, 269-270
Protostega, 221, 270
"*Protostega*" *anglica*, 229, 237, 272
"*Protostega*" *copei, 230*
"*Protostega*" *eaglefordensis*, 229
Protostega gigas, 229, *230, 232*, 237
Protostegidae, 221, 228, 236, 238, 254, 262, 272
  geographic distribution, *264*
Psephophorus, 237, 270, 272, 275
Psephophorus polygonus, 269-270
Psephophorus pseudostracion, 270

Psephophorus scaldii, 270
Psilotrachelosaurus, 109
Puppigerus, 262, 267-268
Puppigerus camperi, 267, 272, 275
Puppigerus crassicostata, 267-268
Pythonomorpha, 283

Quenstedt, F. A., 45

Repossi, E., 45, 48, 54, 89
Rhabdognathus, 367, 379, 382, 391
Rhinochelys, 237, 253-254, 260, 262-263, 272, 275
Rhinochelys amaberti, 262-263
Rhinochelys cantabrigiensis, 262
Rhinochelys elegans, 262
Rhinochelys pulchriceps, 228, *232*, 262
Rhomaleosaurus, 157, 162, 165, 205-206, 208
Rieppel, O., 6, 114, 359, 479-480
Rikisaurus tehoensis, 287
Romer, A. S., 4-6, 9, 23, 47, 50-51, 220
Russell, D. A., 208, 286-288, 308, 312, 314-315, 320-321, 333-334, 428
Russellosaurus, 322
Russia, 283, 431

Sander, P. M., 10, 47
Santana protostegid, 228, *232*
Sauria, 7, *7*, 114
Sauropterygia, 107-115, 145, 149, 200, 214, 439, 472
  hypothetical relationships, *109*
  Triassic paleobiogeography, 121-140
Sauросphargis, 138
Scheuchzer, J. J., 3
Seeleyosaurus holzmadensis, 179
Selmasaurus russelli, 286
Serpianosaurus, 135, 137, 139-140
Shastasauridae, 8, 22, 35-36
  rib articulation, 19-20
Shastasauridae *incertae sedis*, 33
Shastasaurus, 8, 10, 17, 20-22, 34-39, 41, 64-65, 69, 77-78, 431
  paleobiogeography, 39, *40*
  rib articulation, 19-20

Shastasaurus altispinus
  dentition, 93
Shastasaurus careyi, 36, 38
"*Shastasaurus carinthiacus*," 37
Shastasaurus neoscapularis, 34, 36, 38-39, 65, *66*
  dentition, 82, 92
Shastasaurus neubigi, n. sp., 20, 22-23, *24-27*, 28-29, 29t, *31*, 33, *34-35*, 37-41
  appendicular skeleton, 33-34
  dentition, 29-30
  gastralia, 32
  holotype locality map, *21*
  lower jaw, 28
  skull, 23-27
  vertebrae and ribs, 30-32
Shastasaurus osmonti, 33, 36
Shastasaurus pacificus, 33, 38
Shingyisaurus, 110
Shonisaurus, 22, 34, 36, 61, 69, 71-72, 77, 457-458, 464
  dentition, 93, 95, 98, 100
  rib articulation, 19-20
Shonisaurus popularis
  dentition, 92, 95
Silesian-Moravian Gate, 135-137
Silvestrosaurus, 111, 114, 136-137
Simolestes, 208
Simosaurus, 108, 113-114, 138-140
Simosaurus gaillardoti, 110, 139
Slovakia, 270
Slovenia, 285, 293-294
Sokotosuchus, 379, 382t
Sollas, W. J., 23, 29, 95, 195, 198
South America, 259, 271, 367-369, 393, 435-447
Soviet Union, 443
Spain, 111, 260, 263, 271-272
  Muschelkalk, 136
Steneosaurus, 359, *361, 363*
Steneosaurus larteti, 362
Steneosaurus leedsi, 362
Steneosaurus priscus, 362
Stenopterygius, 50
  rib articulation, 18

Stenopterygius quadriscissus, 73, *483*
Storrs, G., 195, 200, 203, 208, 367
Stretosaurus, 439
Stretosaurus macromerus, 439
Stupendemys, 220
Styxosaurus, 198, 214
Sudan, 367
Svalbardosaurus, 430
Swimming behavior, 401-403
Switzerland, 425, 430-431
  Kalkschieferzone, 136
  Monte San Giorgio, 39, 45, 48, 61, 109, 111, 135-136, 138, 409t
Syllomus, 237
Synaptosauria, 113

Taniwhasaurus, 286
Taniwhasaurus oweni, 283
Taphrosphys, 259, 271, 275
Tarsitano, S., 6
Taylor, M. A., 334, 344, 349
Teleidosaurus, 364
Teleidosaurus bathonicus, 364
Teleidosaurus calvadosi, 364
Teleorhinus, *361*, 366
Teleosauridae, 357, 359, 364
  stratigraphic range, *360*
Teleosaurus, 359
Temnodontosaurus, 69, 72
Temnodontosaurus eurycephalus, 76
Testudines, 219-221
Tethys, 39, 46, 110-112, 122, 132, 135, 137-138, 140, 220, 368, 394, 425, 442, 444-446
Tethysuchia, 368
Tetrapod oropharyngeal cavities, *459*
Thailand, 430
Thaisaurus, 50-52, 430
Thaisaurus chonglakmanii, 49
  dentition, 94
Thalassemydidae, 442
Thalassemys, 220
Thalassochelys testei, 269
Thalassomedon, 198, 214
Thalassomedon hanningtoni, *192*, 214
Thalattosauria, 425

Thalattosauridae, 472
Thalattosuchia, 364, 366, 439
Thevinin, A., 283
Thinochelys, 234
Thinochelys lapisossea, 226
Tholodus, 139, 429
Thoracosaurus, 369, 379, 392, 395
Thoracosaurus neocesariensis, 378
Tilemsisuchus, 367
Timor, 430
Tooth form and prey preference, 403-405
  feeding guilds, 406t-407t, 452-454
Toretocnemus, 34-35
Toxochelyidae, 221, 234, 260, 263, 265, 268
Toxochelys, 234, 238, 265, 267-268, 271, 274
Toxochelys latiremis, 226, *226*, 232
Toxochelys moorevillensis, 226, *226*
Trans-Saharan Seaway, 259-260, 271
Tricleidus, 204, 210
Trinacromerum, 195, 215
Troxell, E. L., 376, 383, 390-391
Tuarangisaurus, 214
Tuarangisaurus keysi, 192
Tunisia, 260, 265, 268, 270-271
Turkey, 430
Tylosaurus, 283, 286-287, 296, 302, 306, 308, 310, 315-316, 321, 323-324, 333, 340t-341t, 342, 349-350
  bone microstructure, 344, 347
Tylosaurus nepaeolicus, 306, *309*
Tylosaurus nov. sp., 306
Tylosaurus proriger, 301, 303-305, 306, 314, 316, *319,* 320-321, *345-346, 478*

United States, 148, 271-272
  Alabama, 283, 286, 377
    Clayton Formation, 376, 394
  Alaska, 430
  Arizona, 255
    Mancos Shale, 243, *244*, 245-246, 256
  California, 36, 38, 270, 286, 431
    Hosselkus Limestone, 39

Idaho, 10
Kansas, 245, 254-255, 282, 285
Maryland, 270
Mississippi, 283
Nebraska
  Benton Formation, 245
Nevada, 39, 69, 111, 430, 431
  Luning Formation, 61
  Prida Formation, 47
New Jersey, 238, 282-283, 367-377,
  Hornerstown Formation, 375-376, 383, 394
Niobrara Chalk, 283, 333, 415t
Oregon, 431
Pierre Shale, 415t
Smoky Hill Chalk, 193, 215
South Carolina, 385, 390, 394
South Dakota, 255, 282, 286
Texas, 293, 322
  Eagle Ford Group, 215
Wyoming, 111
Utatsusaurus, 10, 52, 430
  dentition, 98-100
Utatsusaurus hataii
  dentition, 81-82, 88, *89*, 95
Utatsusaurus sp.
  dentition, 88

Varanus, 476
Venezuela, 220

Welles, S. P., 191-194, 198, 286
Williston, S. W., 4-6, 113, 191, 193, 209, 245, 283, 286
Wiman, C., 19, 47, 87, 91, 285
Woodward, A. S., 4

Young, C. C., 46
Young, G., 4
Younginiformes, 7, *7*

Zaire, 260, 267
Zangerl, R., 263, 265, 267-268, 272
Zittel, K. A., 4, 376